Ore Deposit Geology
AND ITS INFLUENCE ON MINERAL EXPLORATION

To our families

Veronica, Rachael, Faith, Anna
and Rebecca

Lesley, Caroline and Rosalind

*Who had to put up with preoccupied husbands
and fathers for three years*

Ore Deposit Geology

AND ITS INFLUENCE ON MINERAL EXPLORATION

Richard Edwards and Keith Atkinson
Camborne School of Mines, Redruth, Cornwall

LONDON   NEW YORK
CHAPMAN AND HALL

First published in 1986 by
Chapman and Hall Ltd
11 New Fetter Lane, London EC4P 4EE
Published in the USA by Chapman and Hall
29 West 35th Street, New York NY10001

© *1986 R. P. Edwards and K. Atkinson*

Printed in Great Britain at the University Press, Cambridge

ISBN 0 412 24690 2 (Hb) 0 412 24700 3 (Pb)

This title is available in both hardbound and paperback editions. The paperback edition is sold subject to the condition that it shall not, by way of trade or otherwise, be lent, resold, hired out, or otherwise circulated without the publisher's prior consent in any form of binding or cover other than that in which it is published and without a similar condition including this condition being imposed on the subsequent purchaser.
All rights reserved. No part of this book may be reprinted, or reproduced or utilized in any form or by any electronic, mechanical or other means, now known or hereafter invented, including photocopying and recording, or in any information storage and retrieval system, without permission in writing from the publisher.

British Library Cataloguing in Publication Data

Edwards, Richard
  Ore deposit geology and its influence on mineral exploration.
  1. Ore-deposits
  I. Title II. Atkinson, Keith
  553   TN263
  ISBN 0-412-24690-2
  ISBN 0-412-24700-3 Pbk

Library of Congress Cataloging in Publication Data

Edwards, Richard, 1940–
  Ore deposit geology and its influence on mineral exploration.
  Includes bibliographies and index.
  1. Ore-deposits. 2. Geology. 3. Prospecting.
I. Atkinson, Keith, 1942– . II. Title.
TN265.E33 1985   553'.1   85-11713
ISBN 0-412-24690-2
ISBN 0-412-24700-3 (pbk)

Contents

	ACKNOWLEDGEMENTS	page	ix
	PREFACE		xi
	GLOSSARY		xii
1	**INTRODUCTION**		1
	1.1 Objectives and reasons for the approach taken		1
	1.2 Mineral deposit or mine?		4
	1.3 A genetic model as the basis for exploration		12
	1.4 The scientific study of mineral deposits		15
	References		17
2	**MAGMATIC DEPOSITS**		18
	2.1 Introduction		18
	2.2 Chromite deposits		18
	2.3 Nickel sulphide deposits		37
	2.4 Kimberlites		54
	2.5 Concluding statement		63
	References		65
3	**MAGMATIC HYDROTHERMAL DEPOSITS**		69
	3.1 Introduction		69
	3.2 Porphyry copper deposits		69
	3.3 Exploration for porphyry copper deposits		87

Contents

3.4	Porphyry molybdenum deposits	92
3.5	Exploration for porphyry molybdenum deposits	104
3.6	Porphyry gold deposits	106
3.7	Porphyry tin deposits	106
3.8	Volcanic-associated massive sulphide deposits	107
3.9	Exploration for volcanogenic sulphide deposits	128
3.10	Concluding statement	137
	References	138

4 HYDROTHERMAL VEIN DEPOSITS — 143

4.1	Introduction	143
4.2	Classification of hydrothermal vein deposits	144
4.3	Classification of hydrothermal gold deposits	146
4.4	Hydrothermal gold deposits in Archaean terrain	146
4.5	Exploration for gold in Archaean terrain	164
4.6	Concluding statement	169
	References	170

5 PLACERS AND PALAEO-PLACERS — 175

5.1	Introduction	175
5.2	Placer deposits	175
5.3	Eluvial (residual), colluvial and fluvial (alluvial) deposits	181
5.4	Beach sand deposits	188
5.5	Marine placers	197
5.6	Palaeo-placer deposits	201
5.7	Concluding statement	211
	References	212

6 SEDIMENT-HOSTED COPPER–LEAD–ZINC DEPOSITS — 215

6.1	Introduction	215
6.2	Sediment-hosted copper deposits	216
6.3	Syngenetic and diagenetic lead-zinc deposits in shales and carbonates (sedimentary-exhalative deposits)	230
6.4	Epigenetic carbonate-hosted lead–zinc deposits (Mississippi Valley-type)	259
6.5	Exploration for Mississippi Valley-type deposits	266
6.6	Concluding statement	267
	References	269

7 ORE DEPOSITS FORMED BY WEATHERING — 274

7.1	Introduction	274
7.2	Bauxite deposits	276

	7.3	Lateritic nickel deposits	291
	7.4	Kaolin deposits	298
	7.5	Supergene manganese deposits	306
	7.6	Supergene sulphide enrichment	308
	7.7	Concluding statement	311
		References	311
8	IRON ORES OF SEDIMENTARY AFFILIATION	314
	8.1	Introduction	314
	8.2	Classification of iron ores	315
	8.3	General characteristics of iron-formation	317
	8.4	Genesis of iron-formation	324
	8.5	Enriched haematite ore deposits	325
	8.6	The Hamersley Basin – an example of banded iron-formation and associated enrichment ores	326
	8.7	Exploration	331
	8.8	Evaluation	332
	8.9	Concluding statement	333
		References	334
9	URANIUM ORES OF SEDIMENTARY AFFILIATION	337
	9.1	Introduction	337
	9.2	Geochemistry of uranium in the secondary environment	339
	9.3	Unconformity-type uranium deposits of the Northern Territory, Australia and Northern Saskatchewan, Canada	341
	9.4	Sandstone-hosted uranium deposits of the western USA	355
	9.5	Concluding statement	365
		References	366
10	ORES FORMED BY METAMORPHISM	370
	10.1	Introduction	370
	10.2	Skarns	371
	10.3	Skarn deposits	372
	10.4	Classification of skarn deposits	373
	10.5	Genesis of skarn deposits	392
	10.6	Exploration for skarns	394
	10.7	Concluding statement	396
		References	398
11	THE DESIGN AND IMPLEMENTATION OF EXPLORATION PROGRAMMES	400
	11.1	Introduction	400

11.2	Who undertakes exploration?	402
11.3	Factors affecting exploration programmes	403
11.4	The exploration programme	407
11.5	Concluding statement	422
	References	424

APPENDIX 427

    Mineral list 429

INDEX 445

Acknowledgements

The writing of this textbook would not have been possible without the help of individuals and organizations in both industry and education.

In particular we wish to thank the Governors of the Camborne School of Mines and the following organizations for financial assistance towards the cost of travel: the Institution of Mining and Metallurgy, the British Council, BP Minerals International, Consolidated Gold Fields PLC, Gencor (UK) Ltd and RTZ Services Ltd. We also wish to acknowledge a contribution by the Minerals Industry Research Organisation towards the cost of typing.

We are indebted to the following individuals for their valuable criticisms of individual chapters:

Dr J. D. Ridge
Dr C. A. Lee
Mr A. M. Killick
Dr P. H. Nixon
Dr P. W. Gregory
Dr D. I. Groves
Dr R. P. Foster
Professor D. A. Pretorius
Professor K. F. G. Hosking
Professor M. J. Russell
Mr C. A. J. Towsey
Dr S. Patterson
Mr R. H. Parker
Dr R. C. Morris
Mrs A. M. Giblin
Dr Graham Closs

Acknowledgements

We also wish to thank the many geologists who have given their time during mine visits and responded to our queries. We have benefited from the stimulus and advice of our colleagues.

We are grateful to the students who read the manuscript and suggested those words which appear in the glossary. We would like to acknowledge the extensive help we have received from the Library Staff at the Camborne School of Mines.

Finally we express our appreciation to Mrs Joan Edwards and Mrs Anne Taylor for their excellent typing.

Figures and tables from the journal *Economic Geology* have been reproduced with permission of the Economic Geology Publishing Company, USA.

Preface

Why another book about Ore Deposits? There are a number of factors which motivated us to write this text and which may provide an answer to this question. Firstly our colleagues are predominantly mining engineers and minerals processing technologists, which provides us with a different perspective of ore deposits from many academic geologists. Secondly we have found that most existing texts are either highly theoretical or merely descriptive: we have attempted to examine the practical implications of the geological setting and genetic models of particular ore deposit types. We have written the text primarily for undergraduates who are taking options in Economic Geology towards the end of a Degree Course in Geology. However, we hope that the text will also prove valuable to geologists working in the mining industry.

The text is to a large extent based on a review of the existing literature up to the end of 1984. However, we have visited most of the mining districts cited in the text and have also corresponded extensively with geologists to extend our knowledge beyond the published literature. Nonetheless writing a text-book on Ore Deposits is a demanding task and it is inevitable that sins of both omission and commission have been committed. We would therefore welcome comments from readers which can be incorporated in future editions.

RICHARD EDWARDS
KEITH ATKINSON

Camborne School of Mines
April 1985

Glossary

Adit	A horizontal, or near horizontal, passage from the surface into a mine.
Area of influence	The zone around a borehole to which values of grade, thickness, etc., recorded for that borehole, may be applied.
Astrobleme	An eroded remnant of a meteoritic or cometary impact crater.
Autolith	An inclusion in an igneous rock to which it is genetically related.
Bayer process	Chemical conversion process involving the digestion of bauxite for the production of purified anhydrous alumina *en route* to aluminium.
Blasthole (sampling)	Holes drilled during mining for blasting. Holes are usually shallow, closely spaced and uncased. Samples of the dust from these holes may be taken for evaluation and comparison with diamond drill hole exploration samples.
Block caving	Large sections (blocks) of the orebody are successively undercut and allowed to cave above the undercut portion. Drawing off the caved ore causes further caving, the ore caves and crushes by its own weight, and weight of overburden, into pieces of suitable size for handling. Drawing continues until hanging wall material appears in the drawpoints.
Boxwork (structure)	A network of intersecting blades or plates of limonite or other iron oxides, deposited in cavities and along fracture planes from which sulphides have been dissolved by processes associated with the oxidation and leaching of sulphide ores such as in porphyry copper deposits.

Glossary

By-product metal A subsidiary metal worked from ore deposits that are dominantly characterized by other metals. In some cases, such as porphyry copper deposits, the income from by-products (for example, gold) can equal or exceed that from the major metal type (copper).

Caliche Strongly indurated crust of soluble calcium salts plus gravels etc. commonly found in layers on or near the surface of stony soils of arid and semiarid regions. Term broadly applied in the SW USA and South America.

Carbonatization A type of hydrothermal alteration which results in the formation of secondary carbonates in the host rock, commonly developed in intermediate to basic rocks. The main process is the addition of CO_2.

Chalcophile Elements having a strong affinity for sulphur, characterized by the sulphide ore minerals.

China clay A commercial product predominantly sold to the paper industry. China clay is a mixture of minerals dominated by kaolinite with minor quantities of mica and feldspar. The particle size is mainly less than 10 microns.

Colloform (banding) Rounded, finely banded kidney-like mineral texture formed by ultra-fine-grained rhythmic precipitation.

Decline A downward inclined tunnel into a mine or potential ore body.

Doline A funnel-shaped depression found in association with karst surfaces in limestone terrain.

Duplex structure A structural complex consisting of a roof thrust at the top and a floor thrust at the base, within which a suite of more steeply dipping imbricate thrust faults thicken and shorten the intervening panel of rock.

Environmental Impact Statement (EIS) Statement prepared by mining companies which assesses the likely effects on the environment of a proposed mining venture. Such statements are legal requirements in certain countries such as the USA.

Epeiric sea A shallow sea located within a continental landmass.

Flotation (froth flotation) A selective process used to separate minerals from complex ores. It uses the differences in physico-chemical surface properties of minerals. After treatment with reagents fine particles of the desired mineral species attach to air bubbles and are separated as the froth, or float, fraction. (Froth flotation.)

Footwall The surface of rock beneath an orebody (compare hanging wall).

Framboidal A textural term applied to sub-rounded pyrite, literally raspberry shaped. The texture is normally developed in carbonaceous pelitic sediments.

Glossary

Fugacity A thermodynamic function used when dealing with gas pressure and defined by the equation

$$\bar{F}_1 - \bar{F}'_1 = RT \ln f_1/f'_1$$

where $\bar{F}_1$ and $\bar{F}'_1$ are partial molar energies and f_1 and f'_1 are the fugacities of the one substance at two different concentrations. Fugacity may be considered in qualitative terms as an idealized vapour pressure and is equal to the vapour pressure when the vapour behaves as a perfect gas.

Garnetite A metamorphic rock consisting chiefly of an aggregate of interlocking garnet grains.

Gondite A metamorphic rock consisting of spessartine and quartz, probably derived from manganese-bearing sediment.

Gossan A Cornish term which describes the iron-oxide-rich zone which lies above a sulphide-bearing ore deposit.

Green strength The mechanical strength of ceramic ware after it has been shaped but before it has been dried and fired.

Greisen A light-coloured, coarse-grained rock dominated by quartz and white mica with accessory fluorite and topaz. It is normally formed by the hydrothermal alteration of granite but pelitic sediments may also be altered to greisen.

Hanging wall The surface of rock above an ore body (compare footwall).

Hydrofracturing Fracturing of rock by water or other fluid under pressure. The effect is to increase permeability. Although this occurs naturally, as in the genesis of porphyry deposits, it is also undertaken artificially to increase permeability in oil, gas and geothermal reservoirs.

Iapetus An ocean which occupied the site of the present North Atlantic and which was at its widest extent in Lower Cambrian times.

Inductive-coupling effect The coupling of two bodies by a common magnetic field.

Klippen A nappe or pile of nappes detached by erosion or gravity gliding from the parental mass.

Kniest facies A facies developed in the footwall of the Rammelsberg ore deposit, FDR. The facies results from the localized silicification of shale horizons and is believed to be late syngenetic or early diagenetic in age.

Labile A term applied to minerals (such as feldspars) that are easily decomposed.

Ligand Anions or molecules which contain at least one unshared pair of electrons. Ligands may combine with a transition metal to form a complex.

Glossary

Lithophile — Elements having a strong affinity for oxygen, concentrated in the silicate minerals.

Photosynthesis — A process in which the energy of sunlight is used by green plants to build up complex substances from carbon dioxide and water.

Polygons method — A method used in reserve estimation where the area of influence around a borehole extends halfway to the next borehole. The value for the borehole is applied to the resulting polygon.

Proppants — Materials used in hydraulic fracturing of deep rocks to increase porosity and therefore improve recovery of petroleum and gas. Proppants made of fused abrasive-grade bauxite have been found to be more effective in this use than such materials as quartz sand, plastic balls, steel shot or ground walnut shells.

Protolith — The unmetamorphosed, or parent, rock from which a given metamorphic rock was formed.

Pseudosection — A display of resistivity or induced polarization data where the values are assigned to the intersection points of 45° lines drawn from the mid-points of the current and potential electrode pairs. The result is a vertical geophysical 'section' below a traverse. Depths in this section bear no simple relationship to the true geological section.

Pyritization — A process of hydrothermal alteration, normally developed in iron-rich host rocks such as basalts, which involves the crystallization of pyrite.

Raise — A shaft excavated upwards for connecting levels in a mine or ore body.

Ripping (rippability) — The removal of overburden, soft or fractured rock, using mechanical excavators without recourse to drilling and blasting. Rippability is an index which measures if this is possible.

Saprolite — That part of the weathering profile which remains *in situ* and retains original structures of the parent rock such as jointing.

Sericitization — A hydrothermal alteration process which results in the development of potash mica, generally sericite, as a result of the hydrolysis of feldspar. Chemically it involves the addition of K_2O and H_2O and usually some removal of SiO_2, FeO and CaO.

SIROTEM — A large loop, time-domain, fixed-transmitter, electromagnetic (EM) geophysical system. Designed in Australia by CSIRO (hence name).

Stockwork — Mineral deposit composed of a three-dimensional network of veinlets usually of large enough scale, and so closely spaced, so that the whole mass can be mined.

Glossary

Strategic mineral A mineral considered essential to the military and industrial strength of a nation but which is of limited availability.

Stylolite An irregular suture-like boundary developed in some limestones which is generally independent of the bedding planes. Stylolites are considered to be the result of pressure solution followed by immediate local deposition.

Talus An accumulation of rock fragments, usually coarse and angular, derived from the mechanical weathering of rocks and lying at the base of a cliff or steep rocky slope. (Synonym of 'scree'.)

UTEM A large loop, time-domain, fixed-transmitter electromagnetic (EM) geophysical system. The name UTEM is the registered trademark of Lamontagne Geophysics Ltd.

Wehrlite An ultramafic and ultrabasic igneous rock containing essential augite and olivine with accessory plagioclase and orthopyroxene.

Wire-line (drilling) A diamond drilling technique in which the inner tube of the core barrel is attached to a wire rope which passes within the drill rods. This enables the sections of core to be retrieved without withdrawing the entire string of drill rods each time.

1

Introduction

1.1 OBJECTIVES AND REASONS FOR THE APPROACH TAKEN

In their introduction to the *75th Anniversary Volume of Economic Geology*, Skinner and Simms (1981) recalled the philosophy of the founders of that prestigious journal. From the outset the objectives were defined as the scientific study of ore deposits, with particular emphasis on 'the chemical, physical and structural problems bearing on their genesis.' At the same time a deliberate decision was made to avoid the 'engineering and commercial aspects of mining.' The emphasis on the genesis of ore deposits and the avoidance of the practical aspects of mining geology have characterized not only the premier Western publication concerned with ore deposits, but also a tradition of teaching in the Western World. As a result newly graduating students who have taken courses in economic geology may be forgiven for believing that the study of ore genesis is the principal task of the industrial geologist. This attitude is very evident to us because we interview a large number of students who are on the point of graduating in geology. We continue to be surprised by the wide variations in the amount of time a new graduate may have devoted to the study of ore deposits during his or her university or college career.

Even students who have attended courses on ore deposit geology appear to have had their attention focused mainly on the genetic aspects of ore deposits with little concept of how this theoretical knowledge might be applied. Often there is a complete lack of awareness of the economic and technical factors which determine whether a concentration of minerals will remain an object of scientific curiosity or be developed as a mine.

One of the objectives of this textbook is to redress the balance between theory and practice. We do not wish to denigrate theoretical studies of ore deposits but rather to increase the emphasis on the practical implications which arise from changes in genetic theory. In particular we seek to examine the importance of models in determining the strategy for mineral exploration.

A subsidiary aim is to introduce the reader to some of the problems which face the mine geologist. The function of the mine geologist varies considerably with the style and complexity of an ore deposit. Most commonly the mine geologist is concerned with grade control, mine exploration and the calculation of ore reserves.

Table 1.1 Modified Lindgren classification of ore deposits (from Ridge, 1968)

Type	Conditions of formation		
	Temperature (°C)	Pressure (atm)	Depth (ft)
I. Deposits mechanically concentrated	Surface conditions		
II. Deposits chemically concentrated			
A. *In quiet waters*			
1. By interaction of solutions (sedimentation)			
a. Inorganic reactions	0–70	Low	0–600
b. Organic reactions			
2. By evaporation of solvents (evaporation)	0–70	Low	0–600
3. By introduction of gaseous igneous emanations and water-rich fluids	0–80	Low	0–600
B. *In rocks*			
1. By rock decay and weathering (residual deposits)	0–100	Low	Shallow
2. By ground water circulation (supergene processes) {with or without introduction of material foreign to rock affected}	0–100	Low–moderate	Shallow
C. *In rocks by dynamic & regional metamorphism*	up to 500	High–very high	Medium / Great
D. *In rocks by hydrothermal solutions*			
1. With slow decrease in heat and pressure			
a. Telethermal	50–150	Low–moderate (40–240)	Shallow (500–3000)
b. Leptothermal	125–250	Moderate (240–800)	Medium (3000–10 000)
c. Mesothermal	200–350	Moderate–high (400–1600)	Medium (5000–20 000)
d. Hypothermal			
(a) In non-calcareous rocks (Lindgren's hypothermal)	300–600	High–very high (800–4000)	Great (10 000–50 000)
(2) In calcareous rocks	300–600	High–very high (800–4000)	Great (10 000–50 000)
2. With rapid loss of heat and pressure			
a. Epithermal	50–200	Low–moderate (40–240)	Shallow–medium (500–3000)

However he may also be concerned with the geotechnical or hydrological aspects of a mine. In some cases, for example at Mount Isa mine in Queensland, Australia, the geologist is moved between different roles, from working in one of the underground sections of the mine to surface exploration. We attempt to provide the reader with some feel for the nature of the work involved and the type of problem that the geologist may be asked to solve.

Whilst our objectives have been clear from the outset we have found that the approach to be adopted was less evident. The format of a textbook concerned with ore deposits may be

Introduction

Table 1.1 *continued*

Type	Conditions of formation		
	Temperature (°C)	Pressure (atm)	Depth (ft)
b. Kryptothermal	150–350	Low–moderate (40–280)	Shallow–medium (500–3500)
c. Xenothermal	300–500	Low–moderate (80–700)	Shallow–medium (1000–4000)
E. In rocks by gaseous igneous emanations	100–600	Low	Shallow
F. In magmas by differentiation or in adjacent country rocks by injection			
1. Early separation–early solidification	500–1500	Very high (1200+)	Great (15 000+)
a. Disseminations			
b. Crystal segregations			
c. Crystal segregations, plus injection as crystal mush			
2. Early separation–late solidification	500–1500	Very high (1200+)	Great (15 000+)
a. Early immiscible sulphide melt accumulation			
b. Early immiscible sulphide melt accumulation, plus later fluid injection			
3. Late separation–late solidification, with or without fluid injection			
a. Silicate pegmatites			
1. Simple	575±	High–very high (800–4000+)	Great (10 000–50 000+)
2. Complex (Usually gradational)	200–550	High–very high (800–4000+)	Great (10 000–50 000+)
3. Barren Quartz	100–300	High–very high (800–4000+)	Great (10 000–50 000+)
b. Immiscible (metal-oxygen-rich) melts	500–1500	Very high (1200+)	Great (15 000+)
c. Immiscible (carbonate-rich)	500–1500	Low–very high (0–4000+)	Shallow–great (0–50 000+)
4. Late formation–deuteric alteration	Less than 575	Moderate–very high (400–4000+)	Medium–great (5000–50 000+)

designed in a number of ways: genetically based, commodity based or related to the tectonic environment in which the ore deposit has formed.

A textbook planned on a genetic basis requires more confidence in the shifting sands of genetic theory than we possess. In addition it requires an acceptable classification of mineral deposits which is sufficiently flexible to allow for changing ideas and the recognition of new styles of deposit. One of the most useful systems has been Ridge's (1968) modified version of the Lindgren classification of ore deposits (Table 1.1). Mechanical and chemical concentration and magmatic differentiation are still accepted as important ore-forming processes.

However, whilst hydrothermal solutions are seen increasingly to be of importance in the formation of ore deposits, few geologists would wish to use the subdivisions which are retained from a time when veins were of much more economic significance. In Chapter 4 we explain the reasons why few commodities are mined from veins at the present time.

The design of a textbook with an emphasis on commodities has some attractions. For example, most mining companies are orientated towards particular metals or minerals and as this textbook has an emphasis on mineral exploration there are good reasons for considering this approach. However, commodities such as copper may be concentrated in a wide range of geological environments and the description of all of these would have entailed considerable repetitiveness.

An approach based upon plate tectonics has the advantage of using fundamental divisions of the earth's lithosphere as a context in which to discuss mineral deposits. Table 1.2 shows an early attempt to classify ore deposits on the basis of plate tectonics. However it is evident that some ore deposits, such as podiform chromite, may be found in more than one plate environment. A further criticism is that many dissimilar styles of deposit such as evaporites, kimberlites and the Witwatersrand auriferous conglomerates are grouped together.

It therefore seems that no single grouping is wholly satisfactory. In the event we have selected a format which has a typological approach. Dixon (1979) suggested that a 'type' of mineral deposit may be taken as either a group of known deposits with significant numbers of common features or a group with sufficient features in common with a famous or well-understood 'type example'. We have utilized the former alternative and whilst it may seem inelegant it enables us to link exploration for a metal or group of metals to their geological setting and to genetic models. Table 1.3 tabulates the ore types described in each chapter and shows their principal commodities, main processes of formation and dominant tectonic settings.

1.2 MINERAL DEPOSIT OR MINE?

A mineral deposit is an abnormal concentration of minerals within the earth's crust. The geological processes which control the enrichment of metals are a main theme of this textbook. However, it is important that the geologist appreciates that the factors which determine whether a mineral deposit will be mined are mainly political and economic.

1.2.1 Political factors

One of the major changes which has shaped the present-day mining industry is the increasingly important part governments play in determining the direction of mining. For example, in 1983 about 51% of the Western World's copper production and 33% of the Western World's tin-mining production were government owned (Hargreaves and Williamson, 1984). In part this reflects a wish by developing countries to gain control over their mineral deposits, from what are sometimes seen as the symbols of an exploiting colonial regime. However, governmental control of mining is not confined to developing countries. For example, the entire mining industry in France has been nationalized.

There are a number of reasons why ore bodies are mined for political rather than economic gains. Firstly, many developing countries depend upon the export of mineral commodities as one of their principal sources of foreign exchange, and this aspect may take precedence over profitability. Secondly, some governments are prepared to subsidize unprofitable mining operations in order to alleviate the problems of unemployment. For example some underground coal mines in the UK and France continue to operate due to a substantial government subsidy. Similarly the Avoca copper mine was sustained for some years by the Irish government, although as a predominantly underground operation, with an average grade of 0.4% Cu, it was not economically viable.

Whilst the general trend has been for governments to press for increased mining they may also play an inhibiting role. For example,

Table 1.2 Proposed relationship of some ore-deposit types to lithospheric plates (from Guild, 1974)

Deposits formed	Types, possible examples
1. *At or near plate margins*	Orientation of deposits, districts and provinces tends to parallel margin
(a) Accreting (diverging)	Red Sea muds. Ancient analogues?
	Certain cupriferous-pyrite (massive sulphide) ores, Cyprus? Newfoundland?
	Podiform Cr (may be carried across ocean and incorporated in island arc or continental margin)
(b) Transform	Podiform Cr, Guatemala?
	Cu and Mn, Boleo, Baja California
(c) Consuming (converging)	Chiefly of continent/ocean or island arc/ocean type; deposits formed at varying distances on side opposite oceanic, descending plate
	Podiform Cr, Alaska
	FeS_2–Cu–Zn–Pb stratabound massive sulphide, New Brunswick, Japan (Kuroko ores), California, British Columbia
	Mn of volcanogenic type associated with marine sediments, Cuba, California, Japan
	Magnetite-chalcopyrite skarn ores, Puerto Rico, Hispaniola, Cuba, Mexico, California, British Columbia, Alaska
	Cu(Mo) porphyries, Puerto Rico, Panama, SW USA, British Columbia, Philippine Is, Bougainville
	Ag–Pb–Zn, Mexico, western USA, Canada
	Au, Mother Lode, California; Juneau Belt, Alaska
	Bonanza Au–Ag, W USA, W, Sn, Hg, Sb western and S America
2. *Within plates*	Deposits tend to be equidimensional, distribution of districts and provinces less oriented (may be along transverse lineaments)
(a) In oceanic parts	Mn–Fe (Cu, Ni, Co) nodules
	Mn–Fe sediments in small ocean basins with abundant volcanic contributions?
	Evaporites in newly opened or small ocean basins
(b) At continental margins of Atlantic (trailing) type	Black sands, Ti, Zr, magnetite, etc.
	Phosphorite on shelf
(c) In continental parts	Au (U) conglomerates, Witwatersrand
	Mesabi and Clinton types of iron formation
	Evaporites, Michigan Basin, Permian Basin; salt, potash, gypsum, sulphur
	Red-bed Cu; Kupferschiefer and Katangan Cu–Co
	U, U–V deposits, Colorado Plateau
	Fe-Ti-(V) in massif anorthosite, Canada, USA
	Stratiform Cr, Fe-Ti-V, Cu-Ni-Pt, Bushveld Complex
	Carbonatite-associated deposits of Nb, V, P, RE, Cu, F
	Kimberlite, diamonds
	Kiruna-type Fe(P), SE Missouri
	Mississippi Valley type deposits, Pb–Zn–Ba–F–(Cu, Ni, Co)

Table 1.3 Relationship between ore type, commodity, genetic process and tectonic setting

Chapter	Ore type	Commodity	Main process of formation	Principal tectonic settings
2	Magmatic	Cr–Pt	Magmatic segregation	Constructive margin transform fault, destructive margin, intracratonic
		Ni–Cu (Au, Pt)		
		Diamonds		Mainly intracratonic
3	Porphyry copper	Cu, Mo, Au, Ag	High-level calc-alkaline magmatic intrusion with subsequent alteration due to meteoric fluids	Destructive plate margin
	Volcanogenic	Cu, Pb, Zn	Submarine volcanic activity (dominantly mafic)	Spreading centre, back-arc basin
4	Hydrothermal vein	Au	Deposition from hot (>200°C) circulating hydrothermal fluids (epigenetic)	Destructive plate margin, Intracratonic
5	Alluvial/eluvial	Sn, Au, diamond		Back-arc basin
	Beach sand	Diamonds, Ti, Zr	Mechanical concentration	Passive continental margin
	Palaeo-placer	Au, U		Intracontinental
6	Sediment-hosted base metal	Cu	Deposition from hydrothermal fluids of highly variable temperature (50–260°C)	Intracontinental rift
		Pb, Zn, Ag		
		Pb, Zn		Passive continental margin
7	Ores formed by weathering	Al (Co, Ni)	Intense chemical weathering under tropical/subtropical climate	Continental
		Kaolinite		
8	Banded iron-formation	Fe	Protore formed as a chemical sediment. Supergene ore formed by prolonged leaching	Passive continental margin
9	Unconformity-type U	U		Intracontinental
	Sandstone-hosted U	U, V	Deposition from circulating hydrothermal fluid of variable temperature	Post-collision intramontane trough
				Back-arc basin
10	Ores formed by metamorphism	W, Cu, Zn–Pb, Mo, Sn	Metamorphic recrystallization or metasomatic reaction between dissimilar lithologies or infiltration metasomatism by hydrothermal fluids	Continental margin, orogenic belts
		Cu		Island arc
		Sn		Incipient rifted cratons

mining is normally prohibited in National Parks or Wilderness areas, and for this reason Riofinex were obliged to terminate their investigation of the Coed-y-Brenin copper deposit in North Wales, UK. The mining of uranium is a sensitive political issue and some governments and states have passed legislation which either prohibits or controls uranium production. For example the present (1984) policy of the Australian government is to allow the development of new uranium mines only when the uranium is a by-product metal. For this reason the Olympic Dam Cu–U deposit has been given permission to proceed. Existing uranium mines in the Northern Territory will be allowed to continue but no new projects will be allowed (Chapter 9.3).

Taxation is a further example of government control on mining which influences the style of deposit which may be mined. In Western Australia gold mining is exempt from taxation which explains the current frenzy of exploration in that state. More commonly taxation has a discouraging effect on exploration and mine development. Mackenzie (1984) has estimated the effect of taxation on 276 base metal deposits discovered in Canada before 1977, the majority of which were ultimately developed. He analysed the economic viability of each deposit under present-day conditions using an 8% return on capital cost (Fig. 1.1). Mackenzie concluded that 124 would be economic at the present time before taxation but only 98 would be viable under the existing rates of taxation (Fig. 1.1). In effect taxation can influence the type of deposit which may be exploited.

1.2.2 Economic factors

The first stage of a mining project is to define the size and grade of the deposit. This work is normally carried out by a mine geologist and differing approaches to evaluation are described in Chapters 5, 7 and 8. It is normal practice to quote reserves as either proved, probable and possible or alternatively as measured, indicated and inferred. Figure 1.2 shows how these terms are defined. If the measured reserves are sufficient to sustain a mining operation over a reasonable period of time (normally about 20 years), the next stage is to carry out a feasibility study to determine the economic viability of the deposit. Table 1.4 shows the main components of a feasibility study and indicates those areas where a geologist might contribute information. There is insufficient space to discuss all the factors which are considered during a

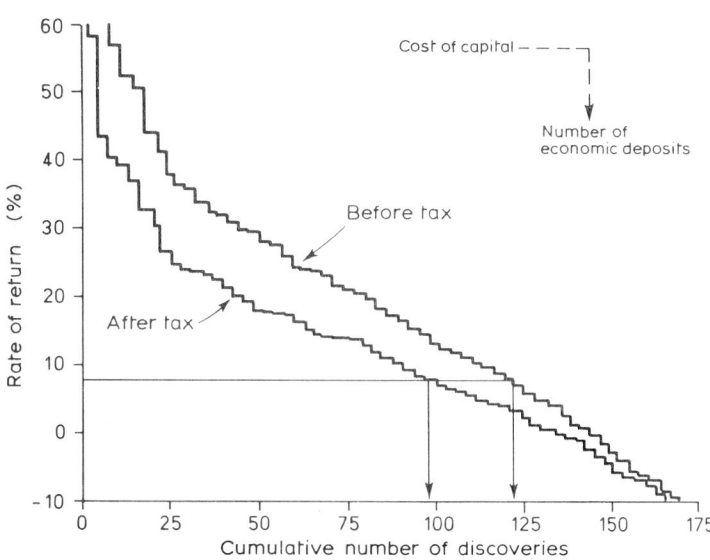

Fig. 1.1 Relationship of rate of return to cumulative number of discoveries (from Mackenzie 1984).

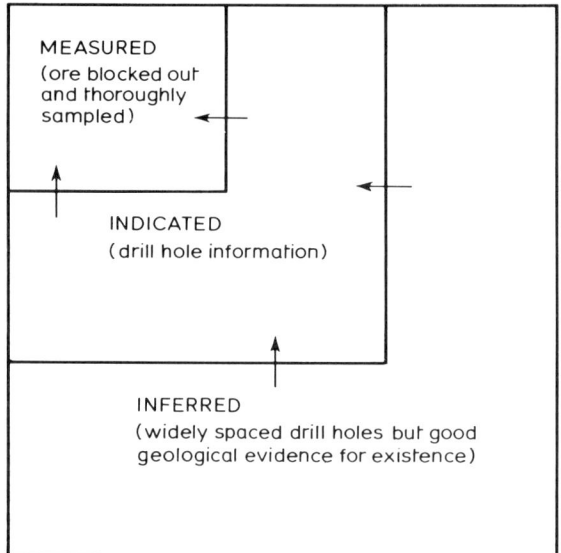

Fig. 1.2. Categories of ore reserves. Arrows indicate the upgrading of reserves with continued exploration.

(a) Metal price

Commodity prices are one of the key factors in determining the viability of a mineral deposit. The main problem is that the lead time between the feasibility study on an ore deposit and initial production is normally 4–7 years, and can be much longer. It is extremely difficult to predict how metal prices will vary during this time. In some cases the decision to develop a mine is based on a prediction of future metal prices which are not borne out in the ensuing years. The Beisa uranium mine in South Africa provides a good example to illustrate this point. The lead time between the feasibility study and full production at Beisa was about 6 years and the cost of bringing the mine into production was about £170 million. However, by the time the mine came into production in 1982 the uranium price had declined very dramatically and the mine was closed in 1984. In some cases companies who are bringing a new uranium mine into production may attempt to avoid this type of problem by negotiating long-term fixed-price contracts with particular consuming countries.

There are a number of approaches which mining analysts can take to forecast metal prices. A simple feasibility study but some of the factors of particular interest to the geologist are now considered.

Table 1.4 The main components of a mining feasibility study

Component	Level of geological input	Comments
Geology	High	Geophysical, geochemical, geological, geotechnical data
Ore reserves	High	Drilling and sampling data
Mining	Moderate	Orebody characteristics and structure
Production schedule	Low	
Mineral processing	Moderate	Mineralogical characteristics of ore
Metal extraction	Low	
Labour	Nil	
Infrastructure	Low	Geological advice on siting
Environmental considerations	Moderate	Pollution, dust, geological hazards
Marketing	Low	
Capital costs	Nil	
Operating costs	Low	Mine-based exploration
Fiscal factors	Nil	
Financial evaluation	Low	

Introduction

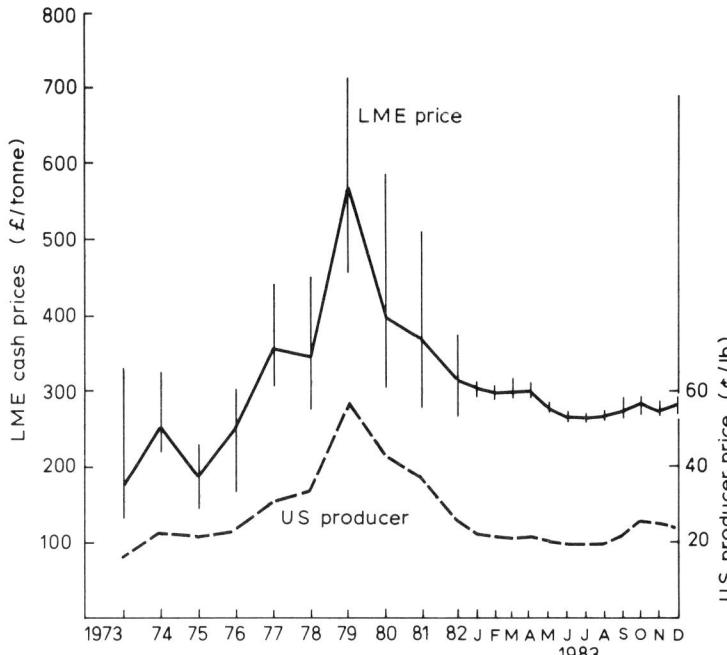

Fig. 1.3. High, low and average London Metal Exchange (LME) cash prices and US producer prices for lead 1973–1983 (from Hargreaves and Williamson 1984).

method is to extrapolate historical data. However Fig. 1.3 shows that, if this technique had been applied to lead in the mid-1970s, the forecast would have been wildly inaccurate. It is more common for analysts to use computer-based long-term economic models which include such factors as future trends of economic growth and the development of new mine capacity.

(b) Cut-off grade

Cut-off grade is the minimum metal content necessary to maintain production costs and sales income in balance and it therefore represents the break-even point for a mine. Once the cut-off grade has been determined the size and average grade of the ore deposit can be calculated. Changes in the cut-off grade can have a dramatic effect on the size of the ore reserve. For example Gaspe Copper Mines Ltd, a skarn porphyry copper deposit in eastern Canada, were reported to have a reserve of 51.7×10^6 tonnes at 1.08% Cu in 1970. The following year the cut-off grade was reduced and published reserves increased to 263×10^6 tonnes at 0.59% Cu (McAllister, 1976).

The determination of cut-off grade is a complex subject and only two of the factors involved will be mentioned here.

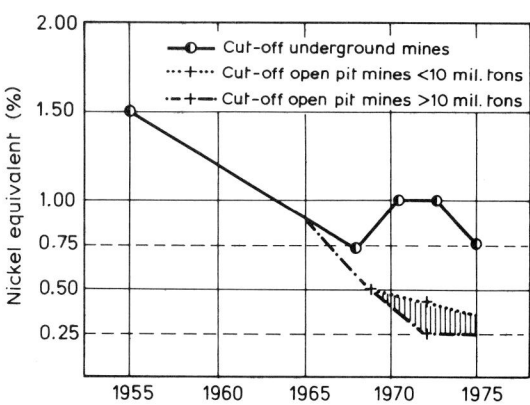

Fig. 1.4. Trend of cut-off grade in underground and open-pit nickel sulphide mines 1955–1975 (from Prokop 1975).

Table 1.5 Average grade of selected open pit copper deposits

Deposit	Average grade (%)	By-products
(1) Palabora, S. Africa	0.5 (0.2% cut-off)	U, ZrO, Fe, Au, Pt, Pd, P
(2) Panguna, Papua New Guinea	0.47	Mo, Au, Ag
(3) Bingham, USA	0.71	Mo
(4) El Abra, Chile	0.7	—

(i) Open pit or underground methods of extraction

In the first instance the selection of open pit or underground methods will determine the cut-off grade because open pit mining has lower production costs. Figure 1.4 shows the contrast in grades between open pit and underground nickel sulphide mines. If the decision is taken to mine underground the cut-off will often determine the size and geometry of the ore body.

(ii) By-product metals

The presence of by-product metals plays an important role in the economics of mining and may determine the cut-off grade. This point is demonstrated by Table 1.5 which shows the average grade of a number of open pit copper operations. It is evident that those operations without significant by-products must have a higher grade of copper in order to be economically viable.

(c) Mining method

During the feasibility study a decision must be made about the mining method. Whilst this stage lies mainly within the realm of the mining engineer, much of the information is provided by the geologist. It is therefore important that during the initial drilling programme thought is given to the geotechnical properties of the rock in addition to considerations of size and grade.

The following are the key factors which influence the choice of mining method:

(1) size and shape of orebody; (2) type and thickness of overburden or rock cover; (3) presence of aquifers overlying the ore body (see Chapter 6); (4) strength of the ore body and surrounding rocks; (5) tendency of the mineral to oxidize immediately after mining; (6) temperature gradient and the resulting demand on ventilation.

These aspects can normally be investigated within a period of about one year. The factors which may prolong the development period of a mine include delays for planning permission, the construction stages and unforeseen geological problems. In the case of Ok Tedi porphyry copper deposit, Papua New Guinea, production did not commence until 15 years after the discovery. This long lead time resulted from aborted negotiations, newly estab-

Table 1.6 Typical recoveries of some common minerals (source B. Wills personal communication)

Mineral	Ore type	Recovery (%)
Native gold	Quartz–pebble conglomerate	99
Chalcopyrite	Porphyry copper	90
Cassiterite	Quartz vein	50–80
Galena	Volcanogenic	50–60
Sphalerite	Volcanogenic	70–80

lished political independence, domestic political struggles and, finally, difficulties of planning restrictions for tailings disposal (Pintz, 1984).

The preferred method of mining is normally by open pit because the development of large-scale equipment, such as 200-tonne trucks, has dramatically decreased mining costs during the 1970s. Gentry (1976) estimated that 70% of World production of metallic and non-metallic ores and coal are mined by surface techniques. However, the economics of large open pit mines is very vulnerable to fluctuations in demand, because they require a constant high tonnage of production. For this reason many porphyry copper deposits in the USA were placed on a care-and-maintenance basis during the early 1980s. Open pit mines also have the disadvantage of being more conspicuous than underground operations so in countries with strong environmentalist pressure groups the choice of mining technique may be influenced by factors other than cost.

(d) Metallurgical factors

During the feasibility study it is necessary to investigate the proportion of valuable minerals which can be obtained from the ore. Initial studies are concerned with ore mineralogy which includes an assessment of the ore minerals which are present and how they are related to one another. Up to this point most of the information obtained from boreholes and trenches will be in the form of a chemical analysis. A chemical analysis indicates whether an ore is potentially valuable, but now it becomes necessary to determine how much of the mineral can be recovered before a decision can be made about economic viability. It is also important to identify any elements which may incur penalties from the smelter (see Section 8.8).

The processing of an ore involves two main stages: comminution and separation. Comminution involves breaking the ore down to a sufficiently small grain size for the ore minerals to be liberated. The problem is to define the optimum point when the maximum amount of ore minerals has been released with minimal

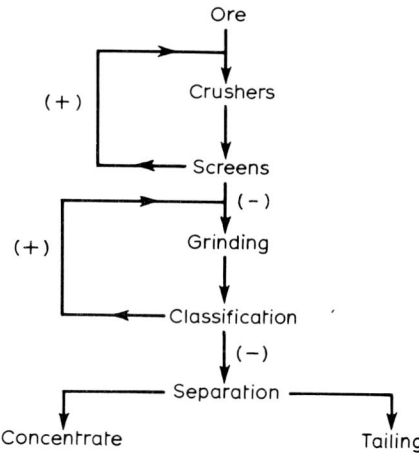

Fig. 1.5. Line flow sheet. (+) indicates oversized material returned for further treatment and (−) undersized material which is allowed to proceed to the next stage (from Wills 1981).

energy. Separation involves utilizing a physical property, such as density, or a chemical property, such as solubility, to divide the ore minerals from their associated gangue. The sequence of operations is illustrated in Fig. 1.5, which shows a simple line flow sheet. It is evident that two products result from the processing operation: concentrate and tailings. Ideally all the valuable material is contained within the concentrate. However in practice this is rarely achieved and the term 'recovery' is used to describe the efficiency of the process. The recovery is the percentage of the total metal contained in the ore that is recovered in the concentrate; a recovery of 90% means that 90% of the metal in the ore is recovered in the concentrate and 10% is lost in the tailings (Wills, 1981). Table 1.6 shows some typical recoveries for some common ore minerals.

A preliminary assessment of recovery can normally be obtained prior to processing from ore microscopy, combined with laboratory bench-scale tests. However it is common practice to mine a small portion of the ore body and feed it through a pilot plant. This permits a realistic estimate of recovery to be made and enables the mineral processing engineer to modify the original flow sheet.

1.3 A GENETIC MODEL AS THE BASIS FOR EXPLORATION

It is possible to recognize three main stages of mineral exploration which have evolved during this century. Each period has been characterized by a particular approach to exploration, which initially has been successful, but has gradually decreased in effectiveness. The dates which we have selected are not intended to serve as precise milestones, but as a general guide to changing attitudes and techniques.

1900–1940

During this period, exploration was based mainly on direct observation and prospectors played a dominant role in ore discovery. Examples of major discoveries include Mount Isa in Queensland (1933), Roan Antelope, Zambia (1925) and the Merensky Reef, S. Africa (1924). Most deposits discovered during this era either cropped out at surface or had a clear surface expression and for this reason geological theory did not play an important part in ore discovery. Commonly the role of the geologist during this period was to carry out the more detailed stages of evaluation after the initial discovery and then to search for extensions of the deposit. Geophysical methods of exploration were introduced in the later part of this period. For example gravity surveys were utilized to define extensions of the Witwatersrand Basin in the late 1930s (Weiss, 1983).

1940–1965

This period was characterized by regional surveys using indirect methods with the main emphasis on geophysics. This was partly as a result of the development of geophysical techniques for military purposes during World War II. The approach proved to be of particular success in glaciated terrain, such as the Canadian Shield. It involved the selection of an area of general favourability for base metals, such as a greenstone belt, which was then investigated using airborne and ground-based geophysics. Magnetic and electromagnetic methods were normally the preferred techniques. This approach is sometimes disparagingly referred to as 'anomaly hunting', partly because there was often only a minimal geological input into some exploration programmes, and partly because the airborne survey produced many spurious anomalies which had to be laboriously checked out by ground geophysics and drilling. During the 1950s and 1960s regional geophysical surveys of this type were responsible for the discovery of many new deposits in Canada. Examples include the Highland Valley area, Valley-Copper, and Lornex. A similar approach was adopted in Africa and Australia but without the same degree of success (see Section 2.4).

During the 1950s a number of American-based companies pioneered a new approach to exploration which foreshadowed the concept of the genetic model. This involved the detailed study of many examples of a particular style of ore deposit, in order to determine a set of characteristics which could be used to focus attention on a limited number of target areas. The new approach led to the discovery of the spectacular Henderson and Kidd Creek deposits (Mannard, 1983), and prepared the path for the use of genetic models in the ensuing decades.

1965 to the present day

During this period new ore discoveries have continued to be made by prospectors (Kambalda, 1966), and indirect methods of exploration have remained an important part of the geologist's armoury. Developments of particular significance include the application of geochemical techniques to semi-arid and glaciated terrain, and the increased sophistication in geophysical equipment. However, there has been a growing awareness of the need to apply geological principles to the problems of exploration. In part this has reflected the increasing difficulty of discovery for the prospector and partly arises from the increasing cost of the regional geophysical approach. Therefore, during this period, the exploration geologist has attempted to restrict the

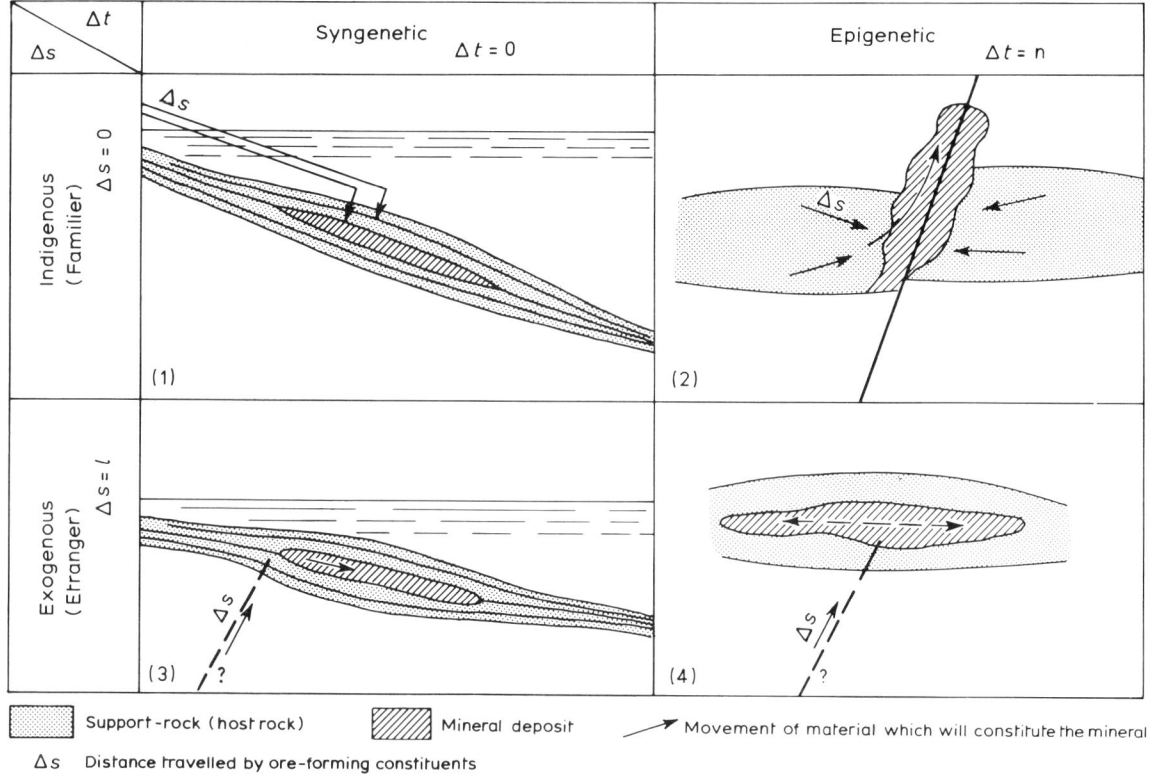

Fig. 1.6. Four fundamental genetic models (from Routhier 1967).

area for geophysical and geochemical work by highlighting areas with good ore potential. The main theoretical basis for this approach has been the genetic model. As we have seen the strategy was first developed by a number of American companies. However, to our knowledge Routhier (1967) was amongst the first geologists to express ideas of ore formation in terms of genetic models in the sense that they are understood at the present time.

For Routhier the genesis of an ore deposit is best analysed in terms of source, transport and environment of deposition. Single or multiple sources may provide the valuable components. Transport may involve the mechanical movement of fragments, or migration of elements in solution. In either case 'filters' may cause a separation of elements. For example in the formation of a lateritic nickel deposit the differing solubilities of iron, magnesium and nickel determine that iron will be enriched as a residual material whilst nickel and magnesium will migrate. Routhier distinguished the general environment of ore deposition (milieu de dépôt) from the more restricted trap (piège) within which the mineral is concentrated. For example lead and zinc deposits of Mississippi Valley type occupy restricted depositional sites within a relatively large expanse of carbonate (Section 6.4).

Routhier (1967) recognized four main genetic models (Fig. 1.6). These are based in part on the concepts of syngenesis and epigenesis which will be familiar to the reader. However he also utilized the concept of 'familier-etranger' which is little used in the English-speaking world. The term 'familier' may be translated as indigenous but the sense of 'belonging to the family' is better expressed in the original French term. Etranger may be translated as foreign or exogenous.

Routhier concluded the discussion of his four genetic models as follows: 'the adoption of a particular genetic model, or combination of models, has a great influence on the strategy and tactics of mineral exploration. This is why it is very important from a practical viewpoint not to adopt a rigid attitude.'

How are genetic models applied in practice? One of the aims of this book is to provide an answer to this question. We approached the problem with an open mind expecting that the development of a genetic model would be crucial in some cases and less important in others. We will develop this argument in later chapters but at this stage we propose to describe the differing strategies of exploration teams exploring for base metals in Ireland, using differing genetic models. Much of the information is based on a personal communication from C. Bowler.

Favourable taxation legislation contained in the Finance Act of 1956 led to an exploration boom in Ireland during the 1960s. In the first instance exploration was confined to areas with a previous history of mining, or where there was a record of a mineral occurrence. These investigations led to the discovery of the Tynagh deposit in 1961. Tynagh marked an important landmark because it indicated that Ireland had not only favourable taxation, but also contained worthwhile exploration targets. However, the paucity of outcrops and scarcity of information about Tynagh meant that in order to take out prospecting licences in an appropriate area a genetic model was required – the alternative would have been the awesome task of investigating the entire Central Plain of Ireland.

In the event four differing approaches were adopted:

(1) Company A utilized an 'exogenous-epigenetic' model. That is they identified the fault structure at Tynagh as the principal control of deposition for an ascending ore fluid. Exploration was therefore concentrated along the strike of the Tynagh fault.

(2) Company B utilized the 'indigenous-syngenetic' model. Palaeoenvironment was considered to be the critical factor controlling the concentration of ore at Tynagh and therefore prospecting was concentrated around the inliers of Lower Palaeozoic within the Irish Plain (Fig. 6.14).

(3) Company C also utilized a variation of the 'indigenous-syngenetic' model but considered that the presence of tuffaceous material within the Carboniferous sequence indicated a possible volcanogenic source for the metal.

(4) Company D used an 'indigenous-syngenetic' model integrated with geophysical and geochemical surveys in areas where there had been some indications of previous mining activity.

In the event Company D discovered the Mogul deposit at Silvermines and Company B discovered subeconomic mineralization at Keel. A fuller discussion of genetic models in Irish exploration is contained in Chapter 6.

Genetic models of the type defined by Routhier have two important parameters – the timing of mineralization relative to the enclosing rock, and the relative distance of transport of the valuable constituents. Whilst the syngenetic or epigenetic characteristics of an ore body can usually be determined from field relationships and studies of mineral paragenesis the distance of migration is normally a matter for speculation. For this reason the model is partly theoretical, and in part based on observations of geological characteristics. Ridge (1983), in a highly challenging paper, has questioned the value of the theoretical component in a genetic model. He argues that in all styles of mineralization it is the geological setting, i.e. stratigraphy and structure of the ore deposit, which provides most information of value for the geologist. This provides a different type of framework for exploration, and we propose to use the term 'empirical model' to describe this approach.

Introduction

Ultimately, the genetic or empirical model must be integrated with the physiographic and climatic factors of the specific search area to produce an exploration model.

The distinction between genetic models and empirical models is important and we will return to this subject in ensuing chapters. Each style of deposit is discussed in terms of geological setting and the recent ideas concerning ore genesis. The relevance of this information for the exploration geologist is then debated at the end of each chapter. The implications of our conclusion are important for both the mining industry and for universities. Is the current bias towards an understanding of genetic theory within many university departments justified? Will genetic models provide an adequate basis for mineral discovery during the closing years of this century? We hope the answers to these questions are contained within this textbook.

1.4 THE SCIENTIFIC STUDY OF MINERAL DEPOSITS

Mineral deposits, like all other geological phenomena, are best studied by careful observations made in the field. In some cases observations will be restricted to surface mapping, and in other cases this information will be supplemented by the results of drilling. If the deposit becomes exploited, a three-dimensional model can be constructed from underground geological mapping.

In many cases the geologist's perception of the ore deposit changes during exploitation. For example, the initial description of the Wheal Jane mine in Cornwall was based mainly on drilling and a limited amount of underground development (Rayment et al., 1971). At this early stage geologists observed that the stanniferous ore bodies were restricted to the footwall of gently dipping feldspar porphyry dykes and the dykes were therefore regarded as having an important structural control on the deposition of cassiterite and associated sulphides. The mine has been in intermittent production for 13 years (1984) and development has been extended to a depth of 400 m. It is now apparent that the ore bodies and dykes have a more casual relationship and were probably emplaced along planes of weakness of similar orientation.

Information about the geological history of a mineral deposit may also be obtained from petrographical and analytical studies utilizing standard techniques. In addition, a range of laboratory techniques has become available in recent years which may be less familiar to some readers. The type of information provided by these techniques is summarized in Table 1.7 which also cites key references explaining the general principles on which the method is based.

Table 1.7 Some experimental techniques applicable to ore deposits

Technique	Type of information provided	References on general principles	Application (examples cited in text)
(I) Stable isotopes (1) Sulphur	(a) Limited range of $\delta^{34}S$ about 0‰ indicates a magmatic source of sulphur		(a) Dominantly magmatic source for the Dianne and O.K. ore bodies. Northern Queensland, Australia (3.8.4)
	(b) Wide range of $\delta^{34}S$ values and enrichment in $\delta^{32}S$ indicates a biogenic source of sulphur	Hoefs (1980)	(b) Pyrite in Mount Isa lead/zinc ore body, Queensland, Australia (6.3.3)
	(c) The pattern of variation in $\delta^{34}S$ may indicate localized migration of sulphur		(c) Derivation of sulphur from fault at Tynagh mine, Ireland (6.3.6)

Table 1.7 continued

Technique	Type of information provided	References on general principles	Application (examples cited in text)
	(d) Enrichment in $\delta^{34}S$ indicates that sulphur was derived from sea-water/evaporites		(d) Source of sulphur at Pine Point, Canada (6.4.3)
	(e) Differing $\delta^{34}S‰$ values in adjacent ore bodies suggests differing modes of formation		(e) Genesis of Pb–Zn and Cu ore bodies at Mount Isa, Queensland, Australia (6.3.3)
	(f) Isotopic fractionation between sphalerite and galena provides information about the temperature of deposition of ore minerals	Ohmoto and Rye (1979)	(f) Temperature of ore fluid at McArthur River, Northern Territory, Australia (6.3.3)
	(g) Isotope systematics provide data on the possible sources of metals	Doe and Zartman (1979)	(g) Source of metals in the Central Plain of Ireland (6.3.6)
(2) D/H and $^{18}O/^{16}O$ ratios	(a) Relative proportion of $\delta D‰$ and $\delta^{18}O‰$ in inclusion-fluids or minerals may identify the principal source of water in hydrothermal fluids	Sheppard (1977) Taylor (1979)	Use of oxygen isotopes in mine-based exploration at the Amulet 'A' orebody, Amulet Mine, Noranda, Canada (3.9)
			Characterization of ore fluid which has deposited quartz in auriferous deposits (4.4.3)
	(b) Differing $\delta D‰$ values obtained on water removed from minerals may suggest that adjacent minerals have disparate ages	Taylor (1974)	Evidence concerning the position of chlorite in the mineral paragenesis of the Saskatchewan unconformity-type deposits (9.3.3)
	(c) Partitioning of $^{18}O-^{16}O‰$ and D-H‰ between minerals deposited in equilibrium with each other provides unambiguous temperature of deposition (i.e. no pressure corrections)	Taylor (1979)	Temperature of formation of auriferous quartz veins in Canada (4.4.3)
(II) Investigations of fluid inclusions	(a) Composition of ore fluid	Roedder (1979)	(a) Composition of the fluid from which ore minerals deposited in Irish Pb–Zn deposits (6.3.6)
	(b) Temperature and pressure conditions of ore deposition		(b) Temperature of deposition of gold in hydrothermal vein deposits (4.4.3) Temperature of fluids in kaolinization in south west England (7.4.6)
(III) pH/Eh diagrams	Stability of mineral and ionic species under differing acidity and oxidation potential conditions in the surface or near-surface environments	Krauskopf (1979)	Studies of the mobility of uranium in the secondary environment (9.2) Controls in weathering of sulphides and supergene enrichment (7.6)
(IV) fO_2/pH diagrams	Characterization of ore-forming environments and of metal-transport mechanisms	Barnes (1979)	—

Introduction

REFERENCES

Barnes, H. L. (1979) Solubilities of ore minerals. In *Geochemistry of Hydrothermal Ore Deposits*, 2nd edn (ed. H. L. Barnes), John Wiley, New York, pp. 404–460.

Dixon, C. J. (1979) *Atlas of Economic Mineral Deposits*, Chapman and Hall, 143 pp.

Doe, B. R. and Zartman, R. E. (1979) Plumbotectonics, The Phanerozoic. In *Geochemistry of Hydrothermal Ore Deposits*, 2nd edn (ed. H. L. Barnes), John Wiley, London, pp. 22–70.

Gentry, D. W. (1976) Development of deep mining techniques. In *World Mineral Supplies Assessment and Perspective* (eds G. J. S. Govett and M. H. Govett), Elsevier, Amsterdam, pp. 397–416.

Guild, P. W. (1974) Distribution of metallogenic provinces in relation to major earth structures. In *Metallogenetische und geochemische Provinzen* (ed. W. E. Petrascheck), Springer-Verlag, Vienna, pp. 10–24.

Hargreaves, D. and Williamson, D. (1984) *The Annual Review of Metal Markets 1983-4*, Shearson/American Express Ltd, 262 pp.

Hoefs, J. (1980) *Stable Isotope Geochemistry*, Springer-Verlag, New York, Heidelberg and Berlin, 208 pp.

Krauskopf, K. B. (1979) *Introduction to Geochemistry*, 2nd edn, McGraw-Hill, New York, 617 pp.

Mackenzie, B. W. (1984) *Economic Mineral Exploration Targets*. Centre for Resource Studies. Queens University, Kingston, Ontario. Working Paper No. 28, 63 pp.

Mannard, G. W. (1983) Predictive metallogeny: a two-edged sword. *Geosci. Can.* **10(2)**, 97–99.

McAllister, A. L. (1976) Price, technology and ore reserves. In *World Mineral Supplies Assessment and Perspective* (eds G. J. S. Govett and M. H. Govett), Elsevier, Amsterdam, pp. 37–62.

Ohmoto, H. and Rye, R. O. (1979) Isotopes of sulphur and carbon. In *Geochemistry of Hydrothermal Ore Deposits*, 2nd edn, (ed. H. L. Barnes), John Wiley, London, pp. 509–561.

Open University Press (1984) S238, Block 1, *The Earth's Physical Resources*, The Open University, Milton Keynes.

Pintz, W. S. (1984) *Ok Tedi – Evolution of a Third-World Mining Project*, Mining Journal Books Ltd, London, 206 pp.

Prokop, F. W. (1975) *The Future Economic Significance of Large Low-Grade Copper and Nickel Deposits*, Monograph Series on Mineral Deposits No. 13, Gebruder Borntraeger, Berlin and Stuttgart, 67 pp.

Rayment, B. D., Davis, G. R. and Wilson, J. D. (1971) Controls to mineralisation at Wheal Jane, Cornwall. *Trans. Instn Min. Metall. (Sect. B. Appl. Earth. Sci.)*, **80**, 224–237.

Ridge, J. D. (1968) Changes and developments in concepts of ore genesis 1933–1967. In *Ore Deposits of the United States 1933–1967, The Graton-Sales Volume* (ed. J. D. Ridge), American Institute of Mining, Metallurgical and Petroleum Engineers, New York, pp. 1713–1834.

Ridge, J. D. (1983) Genetic concepts versus observation data in governing ore exploration. *CIM Bull.*, **76 (852)**, 47–54.

Roedder, E. (1979) Fluid inclusions as samples of ore fluids. In *Geochemistry of Hydrothermal Ore Deposits*, 2nd edn (ed. H. L. Barnes), John Wiley, London, pp. 684–737.

Routhier, P. (1967) Le modele de la genèse. *Chron. Mines Rech. Min. Paris*, **363**, 177–190.

Sheppard, S. M. F. (1977) Identification of the origin of ore-forming solutions by the use of stable isotopes. In *Volcanic Processes in Ore Genesis*, Geol. Soc. London Spec. Publ. No. 7, pp. 25–41.

Taylor, H. P. Jr (1974) The application of oxygen and hydrogen isotope studies to problems of hydrothermal alteration and ore deposition. *Econ. Geol.*, **69**, 843–883.

Taylor, H. P. Jr (1979) Oxygen and hydrogen isotope relationships in hydrothermal mineral deposits. In *Geochemistry of Hydrothermal Ore Deposits*, 2nd edn (ed. H. L. Barnes), John Wiley, London, pp. 236–277.

Weiss, O. (1983) The discovery of two goldfields: a tribute to an inspired mining geologist – Alfred Frost. *IMM Bull.*, **924**, 4–5.

Wills, B. A. (1981) *Mineral Processing Technology: An Introduction to the Practical Aspects of Ore Treatment and Mineral Recovery*, 2nd rev. edn, Pergamon Press, Oxford, 525 pp.

2

Magmatic deposits

2.1 INTRODUCTION

Some ore deposits have such an intimate association with igneous rocks that a common heritage can be inferred. In many cases it can be confirmed by a study of field relationships, from ore textures and from the results of experimental petrology that the ores have segregated during the crystallization of a magma.

The principal styles of mineral deposit which have formed by magmatic segregation are listed in Table 2.1. It is evident that nickel, copper, chromium, vanadium, titanium and the platinum group elements (PGEs) are the major elements produced from this class of deposit. In addition cobalt and gold may be significant by-product elements.

Cu-, Ni-, PGE-, Cr- and V-rich ores are associated with mafic and ultramafic lithologies. The principal ore minerals in this association are pentlandite, pyrrhotite, pyrite, chalcopyrite, chromite and vanadiferous magnetite. The gangue minerals are those which constitute the enclosing host rock and include olivine, serpentine, plagioclase, ortho- and clino-pyroxenes.

The bulk of World ilmenite production is from beach sand deposits (Chapter 5). However, in North America more than half the ilmenite is mined from anorthosites and gabbros (Lynd and Lefond, 1975). The mineral assemblages are of three types: ilmenite–magnetite, ilmenite–hematite and ilmenite–rutile. The gangue minerals are predominantly plagioclase and pyroxenes.

In this chapter we have selected nickel sulphides, chromite and diamondiferous kimberlites for particular consideration. These deposits are of importance on a world-wide scale, and they serve to illustrate contrasting approaches to mineral exploration.

2.2 CHROMITE DEPOSITS

2.2.1 Introduction

Chromite is the only ore mineral of chromium. The bulk of chromite is converted to ferrochrome, of which about 70% is used in the manufacture of stainless steel. Other uses of chromite include refractories and chemicals. There is no substitute for chromite in ferrochrome manufacture and this factor, combined with the very limited geographical distribution of chromite resources, makes it an important strategic mineral.

In 1982 World production of chromite was 7.6×10^6 tonnes (*Mining Annual Review*, 1983). In that year South Africa was pre-eminent with about

31% of production with the USSR, Albania, Turkey and the Philippines also having a substantial production (*Mining Annual Review*, 1983). According to von Gruenewaldt (1980) 97% of the World's chromite reserves are present within the stratiform deposits of South Africa and Zimbabwe. Whilst the discovery of new reserves in other producing countries will almost certainly modify von Gruenewaldt's estimate, it is probable that an increasing proportion of chromite will be mined from stratiform deposits.

Until the 1970s a high chromium content (Cr/Fe 2.8:1) was essential in chromite designated for conversion to ferrochrome. This precluded the use of South African ore for metallurgical purposes. However, changes in smelting technology have resulted in this ratio becoming less critical (Streicher, 1980).

2.2.2 Classification of deposits

The bulk of the World's chromite is mined from rocks of mafic and ultramafic composition. Traditionally the ores are subdivided into two types. Stratiform ores are associated with mafic/ultramafic layered intrusions, are Precambrian in age and have an intracratonic setting. Podiform ores are mainly Phanerozoic in age, have a highly irregular geometry and are associated with ophiolites.

Most of the chromite associated with layered intrusions is mined from the Bushveld Complex in South Africa and the Great 'Dyke' in Zimbabwe (Fig. 2.1). Chromite from podiform deposits accounted for about 55% of World production in 1982 but this proportion is likely to decline during the next decade (Duke, 1983). The reasons for this are partly the greater technical problems of evaluating and mining podiform chromite compared with stratiform deposits. In addition the high Cr/Fe ratio of podiform chromite is becoming less critical in the manufacture of ferrochrome. The principal countries producing chromite from this style of deposit are the USSR, Albania, Philippines, Turkey and India.

2.2.3 General characteristics of stratiform chromite deposits

(a) Distribution in space and time

Stratiform chromite deposits are known in the Americas, Europe, India and southern Africa. They are associated with ultramafic/mafic intrusions which are characteristically emplaced into stable cratonic environments. In many cases the intrusives are emplaced into gneissic basement rocks but the Stillwater Complex and Bushveld Complex intrude supracrustal rocks. The evidence of the tectonic setting of the layered intrusions is not always clear. However, the Bushveld Complex and Great 'Dyke' are believed to occur within the Proterozoic Megafracture system of southern Africa which is possibly the root zone of an aborted intracontinental rift (Mitchell and Garson, 1981).

Layered intrusions which host economic concentrations of chromite are of Archaean and Proterozoic age, but the main inrusives of economic significance range from about 2000 to 2700 million years. Significantly the Muskox intrusion (North West Territory, Canada) has an age of 1150–1250 million years but contains no chromite of economic significance (Irvine and Smith, 1969). The reasons for this are not clear but are likely to be changing tectonic processes and/or changing chemistry within the mantle during the evolution of the earth.

(b) Form and petrology of intrusions

The igneous intrusions which host stratiform chromite deposits may be divided into two morphological types. The Bushveld Complex (South Africa) and Great 'Dyke' (Zimbabwe) which are of principal economic importance consist of adjacent funnel-shaped intrusions. However, in most other cases the intrusive is a sill, which may have retained its original attitude, or have been tilted or deformed by subsequent tectonism.

Most of the layered intrusives relevant to this discussion can be subdivided into lower ultramafic and upper mafic zones. Chromite normally occurs

Table 2.1. Occurrence and importance of magmatic ores. Modified from Naldrett and Watkinson (1981). Information on diamonds from Garlick (1982) and Allen (1982a)

Type	Rock and/or tectonic association	Examples	Typical grade to be expected											Relative importance within overall type						
			wt %			ppm					ppb									Pt, Pd,	Ru, Ir,
			Ni	Cu	Co	Pt	Pd	Rh	Ru	Ir	Os	Au	Ni	Cu	Co	Au	Rh	Os
Sulphide ores																		
Ni- and Cu-rich	Noritic rocks associated with astrobleme	Sudbury	1.5	1.2	0.08	0.6	0.7	50	30	20	10	200	Maj	Maj	Min	Min	Min	Tr
	Mafic intrusions associated with continental rifting	Duluth complex Noril'sk	0.3–1.0	0.4–2.0		1–2 5–10	5–10 500	500				300	(But probably unique) Maj	Maj	Min	Min	Maj	?
	Gabbro and gabbro periodotite intrusions in orogenic belts					(Duluth unknown. Figures are for Noril'sk only.)												
	Precambrian greenstone belts	Lynn Lake, Manitoba Carr Boyd, Western Australia Montcalm, Ontario Pechenga, USSR	1.5	1.2	0.08	0.3	0.2					100	Min to Maj	Min	Min	Min	Tr	?
	Phanerozoic orogenic belts	Rånå, Norway Crillon–La Perouse, Alaska	0.5 0.5	0.2 0.3		0.1	0.1						Min Min	Min Min	Tr Tr	Tr Tr	Tr Tr	Tr Tr
	Komatiite association in Precambrian greenstone belts	Western Australia Manitoba Rhodesia Ungava	2–3	0.1–0.2	0.04	0.15 (Can be very variable.)	0.3	30	100	50	50	60	Maj	Min	Min	Tr	Tr to Min	Min
PGE-rich	Large intracontinental stratiform intrusions	Bushveld Stillwater	0.2	0.1		4	1.5–10	200	400	50	50	300	Min	Min	Min	Min	Maj	Maj

			wt.% Cr₂O₃	Total Fe FeO	Cr/Fe	Cr₂O₃	
Oxide ores Cr-rich	Stratiformly layered intrusions						
	Intracontinental	Bushveld	35–45	25	1–2	Maj	
		Stillwater					
	Precambrian greenstone belts	Selukwe, Rhodesia	45–55	10–15	3–4	Maj but apparently unique	
	Alpine-type	Greece, Turkey, Philippines, S. Urals, USSR	35–65	10–20	2–5	Maj	

			wt.% V₂O₅	TiO₂	P	V	TiO₂
V- and Ti-rich	Stratiformly layered intrusions	Bushveld Complex	1.4–1.7	12–14	<0.05	Maj	Min
		Duluth Complex	0.1	4–5			
	Anorthosite complexes	Allard Lake, Quebec	0.2	24	<0.01		Maj
		Magpie Mt., Quebec	0.17	11	0.078	Tr	Maj
		St. Charles, Quebec					
		P-rich		15	3–5		
		P-poor		19	<0.13		
		Sanford Hill, New York		18	Tr	Tr	Maj
		Iron Mt., Wyoming		22	Tr	Tr	Maj

			Diamonds
Diamonds	Kimberlites in Archaean cratons	AKI, Australia	5.0 carats/tonne
		Letlhakane } Botswana	0.42–0.66 carats/m³
		Orapa	1.48–4.36 carats/m³

Fig. 2.1. Location of magmatic ore deposits.

in chromitite layers of variable thickness within the lower ultramafic zone. However, in the case of the Bushveld Complex the chromitite layers are largely confined to the Critical Zone of the ultramafic component. Further details of the petrology of the Bushveld Complex are provided in Section 2.2.4. The ultramafic zones of the Great 'Dyke' and Stillwater Complex (Montana, USA) comprise a series of cyclic units (Wilson, 1982; Page, 1977). For example, the ultramafic component of the Hartley Complex in the Great 'Dyke' is divisible into fourteen units most of which contain chromitite layers in association with dunite or harzburgite (Wilson, 1982). However, cyclic units are not evident in the Campo Formoso (Brazil) and Kemi (Finland) intrusives where the chromitite layers are associated with peridotites (Duke, 1983).

(c) Mineralogy

Within layered igneous complexes chromite occurs mainly as layers of chromitite which range from 1 cm to more than 1 m in thickness. Typically chromitite is composed of 50–90% fine-grained

Magmatic deposits

(0.2 mm) cumulus chromite with interstitial olivine, orthopyroxene, plagioclase and clinopyroxene (Duke, 1983). The average grade of ore mined from layered intrusions is summarized in Table 2.2.

Chromite belongs to the spinel group of minerals and has the general formula $(MgFe^{2+})(CrAlFe^{3+})_2O_4$. There is considerable compositional variation both between deposits and within particular intrusions. Figure 2.2 shows the compositional variation in chromite from a number of layered intrusions. It is evident that chromite from the Great 'Dyke' has a consistently high chromium content whilst chromite from the Stillwater Complex and the Bushveld Complex has more variable compositions. Wilson (1982) reports that chemical variation in the Great 'Dyke' is related to textural and mineralogical environments. In the Bushveld Complex there is a cryptic variation of chromite composition, with Cr/Fe ratios decreasing upwards through the Critical Zone (Cameron, 1977).

The Bushveld Complex and the Stillwater Complex are also important sources of the

Ore deposit geology and its influence on mineral exploration

Table 2.2 Some characteristics of stratiform chromite deposits

Location		Size of intrusive	Grade	Reserves (tonnes $\times 10^6$)
Bushveld Complex S. Africa	Area Thickness	480×380 km★ 9000 m	46–47.6% Cr_2O_3★ Cr/Fe 1.6:1	7260 to† depth of 1200 m
Great 'Dyke' Zimbabwe	Length Width	480 km‡ 8 km	43.4–51.4% Cr_2O_3 ‖ Seams 4–11 Cr/Fe 2.8:1	560 to§ inclined depth of 200 m
Stillwater Complex Montana, USA	Strike length Width up to	48 km★ 5.5 km	Mouat Mine★ 22.5% Cr_2O_3 Cr/Fe 1.6:1	Mouat Mine★ 4
Kemi Finland	Length Width up to	15 km★ 2 km	26% Cr_2O_3 Cr/Fe 1.55:1	50★
Campo Formoso Brazil	Length Width	40 km 1.1 km	33–42% Cr_2O_3★ Cr/Fe 1.3:1 to 2.4	3★

★Duke (1983).
† von Gruenewaldt (1980).
‡Wilson (1982).
§Hopkins (1980).
‖ *Metal Bulletin Monthly* (May 1981).

platinum group elements. Platinum in the Bushveld Complex is mainly exploited from the Merensky Reef pyroxenite but the Platreef and UG2 chromitite also contain large reserves (Section 2.2.4). Within the Stillwater Complex economic concentrations of platinum are confined to the J-M reef (Todd *et al.*, 1982). The J-M reef varies from 1 to 3 m in thickness and contains minor concentrations of chalcopyrite, pyrrhotite and pentlandite which enclose small grains of the platinum group minerals.

2.2.4 The Bushveld Complex

(a) Geological setting

The Bushveld Complex extends over some 182 000 sq.km. within the Kapvaal craton, S. Africa. The mafic rocks which host chromite and platinum occur as four arcuate zones, the most southerly of which is obscured by younger rocks (Fig. 2.3). The mafic rocks have been dated at 2095 ±24 million years using the Rb/Sr method (Hamilton, 1977).

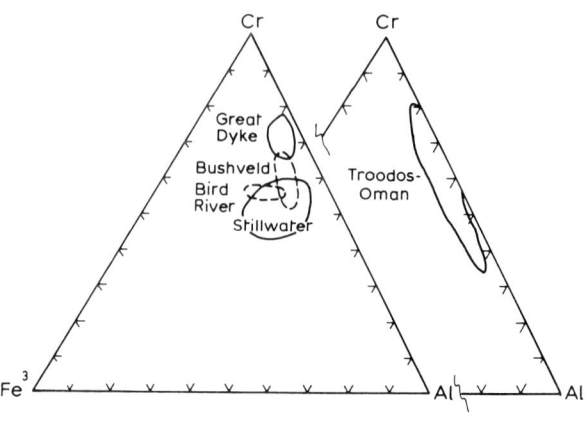

Fig. 2.2. Proportions of trivalent cations in chromite from certain stratiform (left) and podiform deposits. (From Duke 1983, who quotes the data sources).

Magmatic deposits

The original suggestion that the mafic rocks formed part of a single lopolith was subsequently modified by mapping, combined with gravity and airborne magnetic surveys. The initial interpretation of these data indicated that the mafic rocks formed a series of overlapping funnel-shaped intrusions. More recent gravity surveys show that the mafic rocks do not extend beneath the centre of the Complex but form separate and only partly overlapping structures (von Gruenewaldt, 1979).

The evolution of the Kapvaal craton has been characterized by the persistence of NNE, ENE and NNW structural trends. These trends also appear to have played a role in the emplacement of the mafic rocks and have influenced the configuration of the Bushveld Complex (von Gruenewaldt, 1979). The structural controls are evident in the ENE and WNW trending axes of the Complex, in the alignment of gravity highs which probably reflect feeder zones for the mafic rocks and in the orientation of a basement high which controls the western boundary of the eastern Bushveld (Fig. 2.4).

The Bushveld Complex is the product of a prolonged sequence of magmatic events. Igneous activity was heralded by sills of amphibolitic composition (Sharpe, 1981). The first stage of

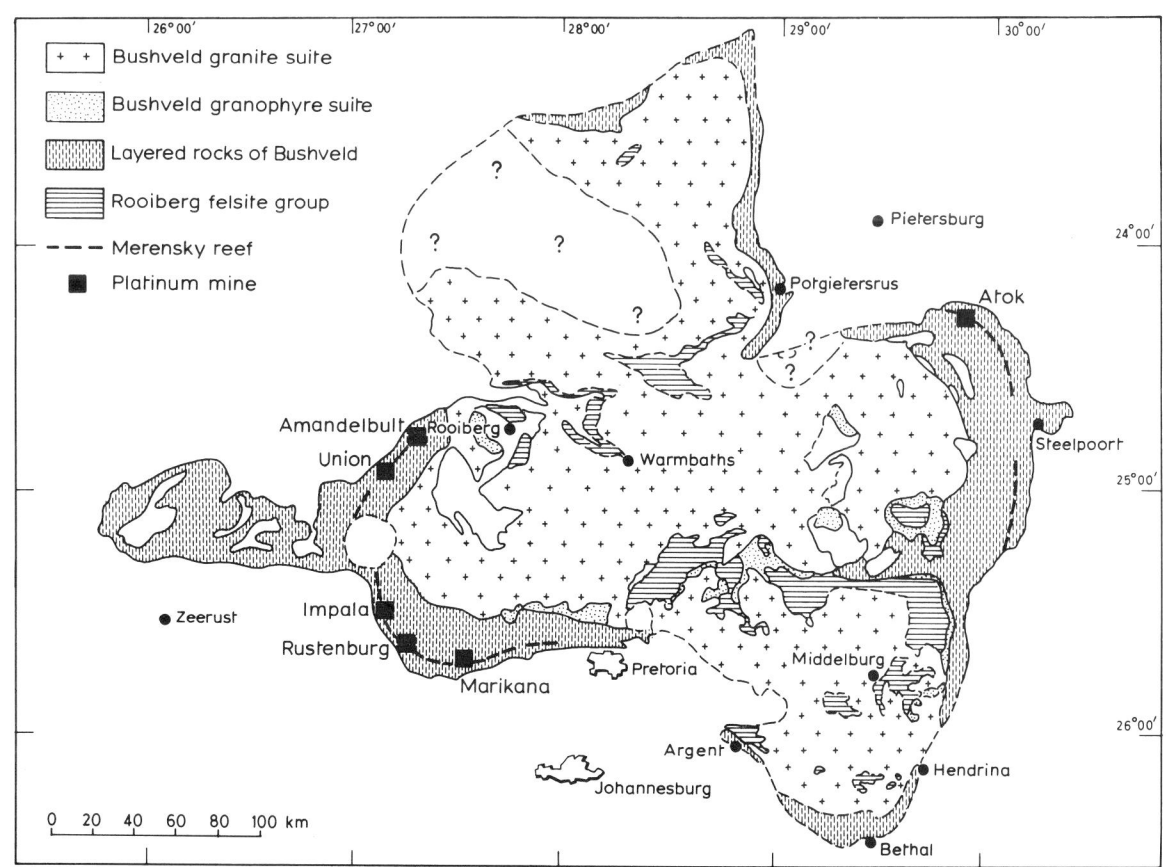

Fig. 2.3. Geological map of the Bushveld Complex, (from Campbell *et al.* 1983, after Willemse 1969).

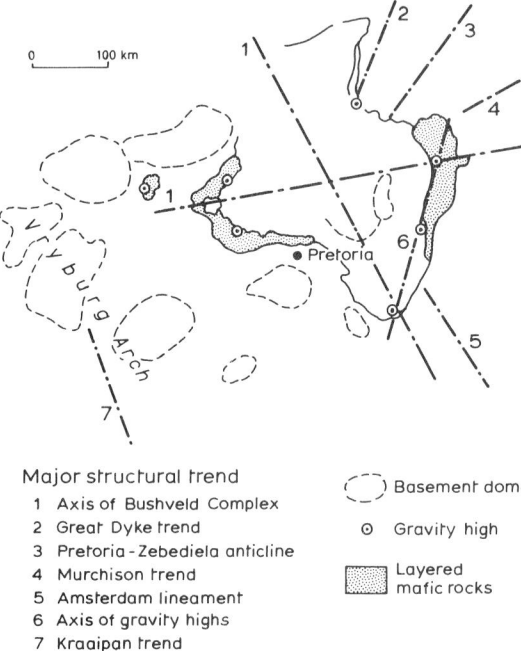

Fig. 2.4. Structural setting of the Bushveld Complex (modified from von Gruenwaldt 1979).

Bushveld activity was marked by the emplacement of the Laminated Marginal Zone, which comprises mainly alternating sheets of norite and pyroxenite (Sharpe, 1981). The second, and major, stage of magmatism resulted in the intrusion of a series of mafic and ultramafic lithologies which are referred to as the Layered Series. The final episode of magmatism is represented by the Bushveld Granite. The Layered Series locally exceeds 9000 metres in thickness and is divisible into five zones (Fig. 2.5). The Lower and lower Critical Zones are characterized by ultramafic to mafic lithologies which are commonly monomineralic. Anorthosites, norites and gabbros are the dominant rock types within the upper Critical, Main and Upper Zones. The Critical Zone is of major economic importance and hosts both the chromite and platinum group elements.

Geophysical studies and detailed mapping have revealed that the five zones of the Layered Series are not uniformly represented throughout the Bushveld Complex. It is evident that the ultramafic rocks commonly have a more restricted distribution than the mafic lithologies of the Main and Upper Zones (von Gruenewaldt, 1979). For example, the full stratigraphic succession in the northern lobe is restricted to an area near Potgietersrus, and the gabbros and norites of the Upper and Main Zones transgress the Lower zones and occupy a larger area (von Gruenewaldt, 1979).

The parent magma of the Layered Series has been the subject of much speculation. Many workers (Willemse, 1969; Gibjels et al., 1974; Hamilton, 1977) have supported the two-magma theory: one magma is considered to have been responsible for the rocks of the Lower and Critical Zones and the second magma was the parent of rocks in the Main and Upper Zones. Sharpe (1981) cites evidence from field work and geochemistry to support the concept of multiple pulses of magma of contrasting chemical composition.

(b) Chromium

All the chromite layers and the principal platiniferous layers are confined to the Critical Zone, which is therefore the unit of overwhelming economic importance in the Bushveld Complex. The Critical Zone is distinguished by a prominent regular layering and may be divided into lower pyroxenitic and upper anorthositic subzones (Cousins and Feringa, 1964; Cameron and Desborough, 1969). Chromite is disseminated within silicates throughout most of the Critical Zone but is only of commercial value when concentrated into layers of chromitite (Fig. 2.6). Within the chromitite layers chromite is the cumulus phase with bronzite and plagioclase being the main interstitial silicates. Chromitite layers are distributed in both subzones of the Critical Zone, but are mainly exploited from the pyroxenite subzone because of their higher Cr/Fe ratios.

In the western sector of the Bushveld Complex the Critical Zone ranges from about 915 to 1750 m

Magmatic deposits

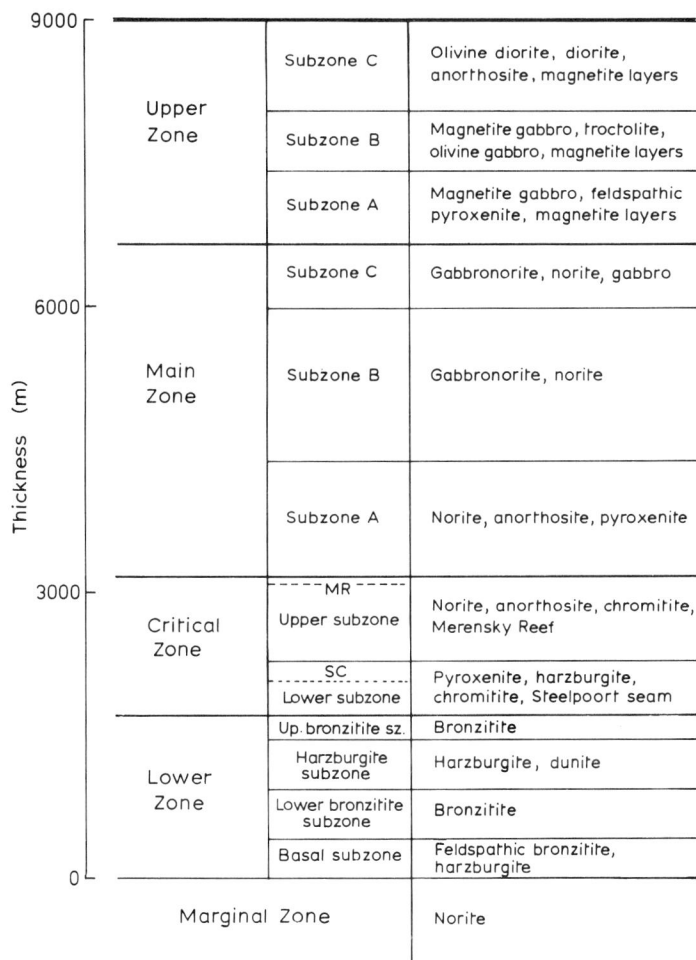

Fig. 2.5. Subdivision of the Layered Series of the Bushveld Complex. MR, Merensky Reef; SC, Steelpoort chromite (from Duke 1983, after Vermaak, C. F. and von Gruenewaldt, G. (1981) *The Bushveld Complex Excursion Guide*, Geocongress 1981, 3rd International Platinum Symposium, 62 pp.

in thickness and contains 13 chromitite layers of potential economic value (Cousins and Feringa, 1964). The layers have a remarkable lateral continuity, for example the LG3 and LG4 layers can be traced along strike for 63 km (von Gruenewaldt, 1977). The layer of prime importance in the Western Bushveld is the LG6 which is associated with pyroxenites (Table 2.3).

In the Eastern Bushveld, the Critical Zone ranges from about 1220 to 1370 m in thickness. The number of chromitite layers ranges from five to 28. Cameron and Desborough (1969) stated that chromitite layers in the pyroxenite subzone are regular and persistent, but this is not the case in the anorthosite subzone. In the Eastern Bushveld, the layers which are currently exploited are the Steelpoort and F layers, and their characteristics are summarized in Table 2.3.

The genesis of chromite in the Layered Series of the Bushveld Complex is a controversial subject. Genetic models must explain the virtually monomineralic nature of the chromitite layers, their repetition throughout the Critical Zone, and virtual absence from the Lower and Main Zones.

Fig. 2.6. Chromitite seams within anorthosite, Bushveld Complex, South Africa.

Table 2.3 Chemical characteristics of pure chromite from Bushveld chromitite layers associated with mining activity (from Buchanan, 1979)

Area and chrome mine	Layer	Cr_2O_3 content (%)	FeO content as total iron (%)	Cr/Fe ratio	Layer thickness (m)
North of Pilansberg					
Zwartkop	LG6 (magazine)	46.36	26.23	1.56	0.86
West of Pilansberg					
Ruighoek	LG6 (main)	47.50	26.00	1.65	0.76
Rustenburg-Brits					
Kroondal	LG6 (main)	46.00	26.00	1.58	0.94–1.27
Eastern Belt (north of Steelpoort)					
UCAR	Steelpoort	47.12	25.15	1.66	0.60
Winterveld	Steelpoort	47.65	25.12	1.67	0.90–1.20
Eastern Belt (south of Steelpoort)					
Grootboom	F Bed	46.98	26.12	1.58	1.30

The evolution of ideas on these problems has drawn both from detailed petrographical and geochemical studies of the Critical Zone (Cameron and Desborough, 1969; Cameron, 1980) and from comparisons with other layered intrusions (Jackson, 1961; Irvine and Smith, 1969).

The genesis of the chromitite layers has been debated in the context of two separate processes: fractional crystallization from one or two parent magmas, and the repeated introduction of successive heaves of magmas. Until recently most writers considered that gravitational settling of chromite, to form cumulus layers, was an essential part of the ore-forming process (Cameron and Emerson, 1959). However, a simple gravitational model is inconsistent with the distribution of chromitite layers within the Critical Zone. It also conflicts with the result of experimental petrology (Duke, 1983) and the reinterpretation (Campbell, 1978) of cumulus textures.

A number of alternative mechanisms have been proposed. Cameron and Desborough (1969) suggested that chromitite seams might be precipitated by changes in oxygen fugacity. This mechanism has been investigated by Snethlage and von Gruenewaldt (1977), who measured pO_2 in natural chromitites from the western part of the Bushveld Complex, and compared their results with those obtained from synthetic systems. Snethlage and von Gruenewaldt (1977) concluded that there is a strong case for oxygen fugacity as a major factor for chromitite layer formation. They reviewed the mechanisms which might be responsible for changes in oxygen fugacity, but did not support a particular model.

Cameron (1980) has carried out extremely detailed petrographic and geochemical studies of the Critical Zone. In discussing the genesis of the chromitite bands, he rejects the role of changing oxygen fugacity, on the grounds that it is unlikely to have produced the uniformity of composition which is such a characteristic feature of the Critical Zone. Cameron (1980) favoured variation in pressure as the main control on the formation of chromitite layers. He suggested that pressure changes could result either from tectonic activity or by the addition or removal of batches of magma from the magma chamber. However, Lee (personal communication) considers that the range of pressure change required exceeds the total pressure at the base of the magma chamber (∓ 6.5 kbars).

(c) Platinum group elements (PGE)

Platinum group elements in the Bushveld Complex are contained within three layers, the Platreef, Merensky Reef and UG2 chromitite. Table 2.4 summarizes the main characteristics of the three layers. It is evident that the UG2 chromitite has the highest grade and largest reserves, but production has only recently commenced due to a combination of metallurgical and mining problems. There is no production at the present time from the Platreef, although this may be developed in the future under favourable economic circumstances.

In this section the main emphasis is placed on the Merensky Reef, which is the main producing horizon for the PGEs, and the subject of most published data on platinum in the Bushveld Complex (Fig. 2.7).

The Merensky Reef occurs in both western and eastern sectors of the Bushveld Complex. Stratigraphically it is located near the top of the Critical Zone. The Merensky Reef, together with overlying spotted and mottled anorthosites, constitutes the Merensky Unit which is overlain by a petrographically similar sequence called the Bastard Unit (Cousins, 1969). Von Gruenewaldt (1979) defined the Merensky Reef as the basal pyroxenitic portion of the Merensky cyclic unit which includes porphyritic pyroxenite, pegmatitic pyroxenite and chromitite stringers. The dominant PGEs in both the Merensky Reef and the other platiniferous horizons are platinum and palladium (von Gruenewaldt, 1977).

In general the Merensky Reef dips at a shallow angle. However, locally it is disturbed by depressions called 'potholes'. Potholes are irregular in shape, have diameters estimated to range from 1 to 300 m and vary in depth up to 60 m (Cousins, 1969). The mechanisms responsible for the

Table 2.4 Summary of characteristics of Pt-bearing layers in the Bushveld Complex

Layer and location	Thickness and strike length	Recovery grade	Reserves	Mineralogy	Lithology
Platreef Potgietersrus area	Estimated payable over 30 km Thickness 25 m‡	3.0 g/t₃‡ Grades are higher in A reef	12.2 × 10⁶ kg ‡	Cooperite and sperrylite	Pyroxenite†
UG2 chromitite	Total strike‡ length 250 m Thickness 0.9 m	6.0 g/t‡ Considerable lateral variation	32.5 × 10⁶ kg ‡	Laurite,★ cooperite, braggite. In part of E. Bushveld Complex alloys formed with a range of metals	Chromitite★
Merensky Reef	Total strike‡ length 230 km Thickness 0.8 m	5.5 g/t‡ Highest values with Cr stringers	18.1 × 10⁶ kg‡	Mineral species vary with locality. Pt sulphides dominant at Rustenburg. Arsenides and tellurides at Western Platinum Mine	Pyroxenite★

★von Gruenewaldt (1979).
†Buchanan and Rouse (1984).
‡von Gruenewaldt (1977).

potholes are not fully understood but have been attributed to rotating eddies in the magma (Ferguson and Botha, 1963), slump structures (Cousins, 1964) and resorption of cumulate layers by an unsaturated hybrid magma (Irvine *et al.*, 1983). The implications of potholing in the mining of the Merensky Reef will be discussed in a later section.

Two hypotheses have been proposed for the genesis of the Merensky Reef: differentiation by gravity settling in the mafic portion of the Bushveld Complex and the introduction of a restricted pulse of magma.

Cousins (1969) drew attention to the petrological discontinuity above and below the Merensky Unit and noted the mineralogical evidence for gravitational segregation within the Merensky Reef. He concluded that the Merensky represented a separate pulse of magma.

Campbell *et al.* (1983) consider that both the high PGE content and petrological characteristics of the Merensky Reef can be explained by the dynamic behaviour of the new magma pulse. They estimate that the temperature difference between the new pulse of magma and the fractionated magma in the chamber was about $100 \pm 50°C$. Under these conditions the new magma would have risen as a turbulent plume for a short distance above the floor of the magma chamber. It would then have spread out at its own density level to form a hybrid layer. Cooling of the hybrid layer would have resulted in liquation of a sulphide liquid, nucleation of silicates and, eventually, convective overturn which would

Fig. 2.7. Merensky Reef with anorthositic footwall rocks, Rustenburg, South Africa. The scale and blackboard indicating date and location are used routinely on this mine. The geologist can assess the nature of each stope and decide which parts of the mine require his immediate attention.

have brought the hybrid melt to the floor of the magma chamber.

How does this help to explain the high PGE content and pegmatitic texture which are important characteristics of the Merensky Reef? Campbell *et al.* (1983) state that if sulphides have a high PGE content they must have achieved equilibrium with a large column of silicate melt. They suggest that this is achieved during the turbulent entry of the plume into the magma chamber. Campbell *et al.* (1983) explain the pegmatitic texture of the Merensky Reef in terms of gravitational settling. They consider that the pyroxenes in the Merensky Reef are true cumulates and would therefore have had a high initial porosity. As a consequence the cumulate layers would have acted as traps for volatiles expelled during the solidification of cumulates with lower initial porosities.

(d) Mining geology

The chromitite layers within the Critical Zone have a gentle dip, are normally persistent along strike for tens of kilometres, and are relatively undisturbed. As a result there are few problems of exploration or mine development. In evaluation, drilling is normally carried out at 0.5 to 1.0 km intervals with one to three deflections per intersection (C. A. Lee, personal communication).

The Merensky Reef normally has a gentle, uniform dip. Persistent marker horizons in both the hangingwall and footwall zones are extremely valuable in mine development because they indicate very accurately the position of the ore body (Fig. 2.8). The marker horizons are particularly valuable when the Merensky Reef is displaced by faulting or disturbed by potholes.

The main geological problems associated with the Merensky Reef are caused by localized structural disturbances, called potholes, which are described in an earlier section. In the western sector of the Bushveld Complex the potholes are classified as shallow and Brakspruit types. In the shallow potholes the reef is rarely displaced by more than 2 m and stoping may be either maintained at the same elevation with some loss of reef or modified slightly to include Merensky Reef. The Brakspruit-type potholes pose a more serious mining problem and in some districts up to 50% of the ground may be unmineable (Fig. 2.9). In the Brakspruit-type potholes, the Merensky Reef may be displaced as much as 30 m and is commonly associated with bad hangingwall conditions. It is therefore important for the mine geologist to be able to assess the potential loss of ground due to potholes before mine development.

2.2.5 General characteristics of podiform chromite deposits

(a) Distribution in space and time

Podiform chromite deposits have a consistent association with ophiolites. However, the distribution of economic concentrations of chromite within ophiolite complexes is highly variable. For example, the Alpine belt which extends in an arc through Yugoslavia and Albania to Greece is notably rich in chromite, whereas the New Zealand ophiolites have no important concentrations of chromite.

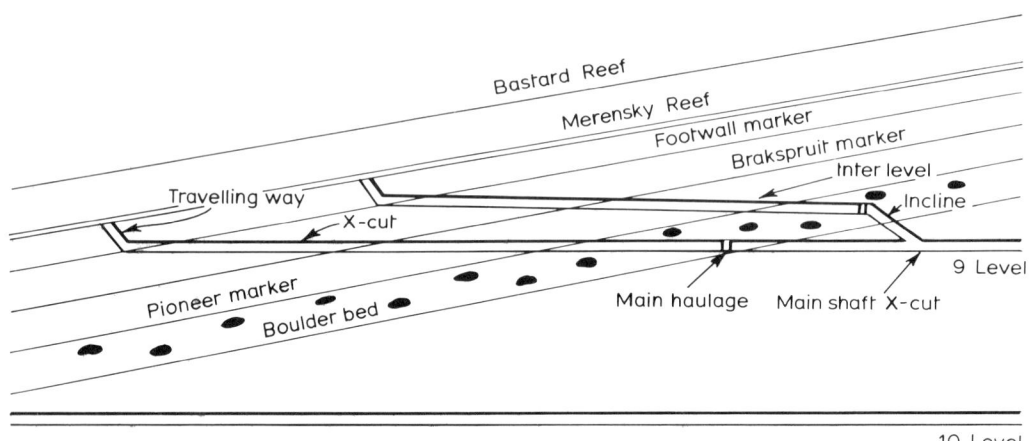

Fig. 2.8. Section showing the relationship of mine development to geology at Rustenburg platinum mine (from Wetherelt 1982).

Magmatic deposits

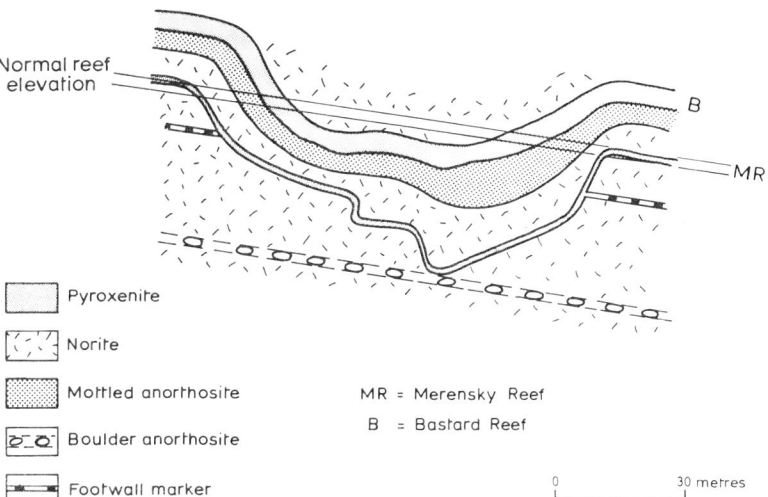

Fig. 2.9. Cross-section through a Brakspruit-type pothole Rustenburg, S. Africa (modified from Cousins 1964).

The USSR is the major producer of chromite from this style of deposit. The bulk of production is concentrated in the Perm and Kempirsai mining districts in the Urals (De Young et al., 1984). Other countries with a substantial production from podiform chromite deposits include Albania, Turkey, the Philippines, New Caledonia and India (Thomson, 1983). Less important deposits of podiform chromite occur in Cuba, Yugoslavia, Iran, Greece, Pakistan and Brazil. Hodge and Oldham (1983) report that the Ramu River deposit in Papua New Guinea has recently been evaluated and contains a significant proportion of the World's chromite resources.

Podiform chromite deposits are formed initially within the oceanic lithosphere, and are sometimes transported tectonically to shallower structural levels during orogenesis. The age of the chromite deposits is usually assigned to the period of deformation when the ophiolite was emplaced into its present tectonic setting. Apart from the Indian deposits most podiform chromite is Phanerozoic in age. Deposits are associated with Caledonian (Urals), Hercynian (Iran, Turkey) and Alpine (Yugoslavia, Albania) earth movements (Pavlov and Grigor'eva, 1977).

(b) Size and grade

The size of podiform chromite deposits is variable but most production comes from bodies containing ≥100 000 tonnes of ore (Evans, 1980). The chromite may occur as either high-grade massive or low-grade disseminated ore, but the proportion of massive ore which is mined has tended to decrease during this century. At the present time published grades vary from 28–30% Cr_2O_3 in Turkey (Erganulp, 1980) to 43% Cr_2O_3 in Albania (Thomson, 1983). The Cr/Fe ratio is higher in chromite from podiform deposits compared with most layered intrusions (Fig. 2.2). For example Albanian ore has a Cr/Fe ratio of 3:1 (Thomson, 1983) compared with 1.6:1 for South African ore. As a consequence chromite from podiform deposits has been traditionally preferred for metallurgical uses and remains the sole ore type for refractories.

The geometry of podiform chromite deposits can most commonly be described as tabular lenses or irregular 'pencils' (Thayer, 1969). In Turkey many of the smaller ore bodies are lenticular but in the Kavak mine the larger ore bodies are pipe shaped (Erganulp, 1980) (Fig. 2.10).

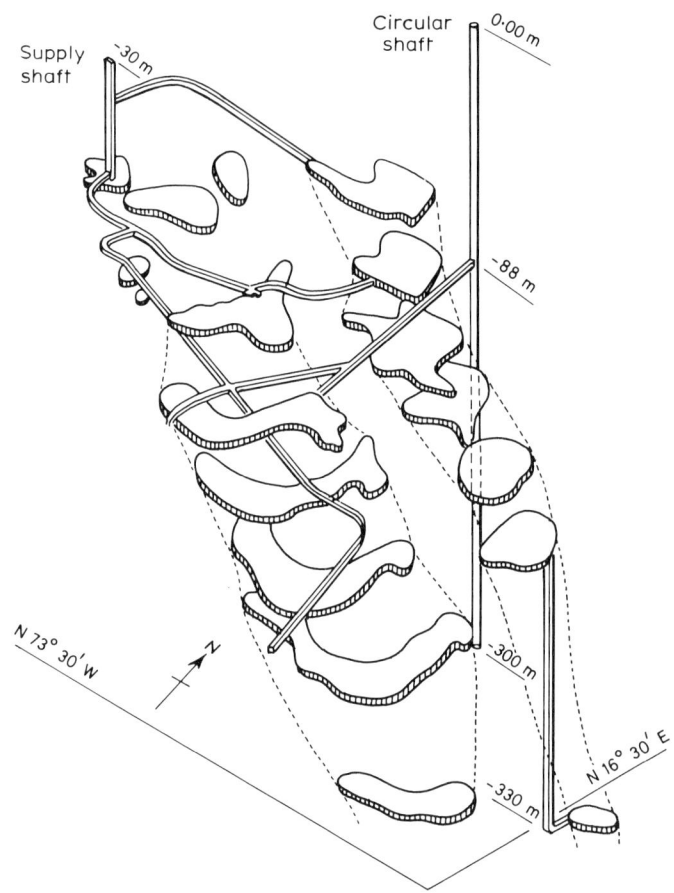

Fig. 2.10. Isometric diagram of main ore body at Kavak chrome mine (from Erganulp 1980).

(c) Mineralogy

Chromite is the only chromium-bearing ore mineral, and in podiform deposits is normally associated with olivine. By-product minerals are not characteristic of this style of mineralization. Platinum group elements are detectable in podiform chromite but not in sufficient concentration to merit their extraction (Page et al., 1984). The ophiolite complexes which host podiform chromite may, however, also be a source of copper (Section 3.8.3) and are sometimes mined for magnesite, chrysotile asbestos and talc.

The textures and structures of podiform chromite are varied. According to Thayer (1969) nodular textures are particularly characteristic of podiform ores, a point which must be borne in mind in discussing genetic models. Greenbaum (1977) and Dickey (1975) considered that some of the textures described from podiform chromite are of cumulate origin. However, in many cases the original textures are destroyed during tectonism.

(d) Host lithology and tectonic setting

Ophiolites are considered to be portions of oceanic lithosphere which have been emplaced into a continental setting during orogenesis. The zoning of oceanic lithosphere is well established from the study of mid-ocean ridges, and an identical stratigraphy can be recognized in some ophiolite complexes. In essence oceanic crust comprises a veneer of sediments, underlain by distinctive layers which are dominated successively by pillow lavas, sheeted dykes and gabbros. The base of the oceanic crust is defined by the Moho – a distinctive seismic discontinuity. Below the Moho occurs a zone of ultramafic cumulates underlain by tectonized mantle material. The ultramafic cumulates consist of dunites, wehrlites and clinopyroxenites that accumulated on the floor of a shallow magma chamber. The lithology of the tectonites comprises variably serpentinized harzburgite (essentially olivine + orthopyroxene + chrome spinel) which shows considerable flow folding. The harzburgites are considered to represent mantle material depleted by partial melting processes (Hughes, 1982).

Chromite deposits occur mainly within the tectonized harzburgite but may also occur in the lower part of the ultramafic cumulates (Duke, 1983). According to Dickey (1975) the podiform chromite tends to be most abundant near the top of the harzburgite zone. There is always an association between chromite and olivine which may be reflected in the dunitic composition of the host rock or in an envelope of olivine surrounding the chromite ore within harzburgite. The ophiolites which host podiform chromite deposits commonly form discontinuous curvilinear features which commonly extend for hundreds of kilometres. Most ophiolites are considered to be allochthonous and occur in klippen with low-angle thrust contacts at their base (Hughes, 1982).

2.2.6 Genesis of podiform chromite deposits

Prior to the development of plate tectonic theory, Alpine-type peridotites were interpreted as intrusions of either liquid or crystalline ultrabasic magma into the continental crust. Both the ultrabasic composition of the magma and the occasional presence of podiform chromite, were interpreted as resulting from the gravitational settling of crystals, which produced relatively homogeneous layers within the mantle. For example, Guild (1947) described the chromite deposits from the Moa district in Cuba, and suggested that the podiform nature of the ore could be attributed to the fracturing of originally homogeneous chromite masses during the emplacement of partly crystalline ultrabasic magma into the continental crust.

The development of plate tectonic theory has changed both the interpretation of Alpine-type intrusions and our understanding of podiform chromite deposits. Most geologists who have discussed the origin of podiform chromite within the last decade accept that the chromite was originally concentrated within the oceanic lithosphere at constructive plate margins, and subsequently emplaced on to or within continental crust during orogenesis. The main divergence of opinion lies in whether the chromite originated within the tectonized peridotite or was introduced by gravitational settling from an overlying magma chamber.

Dickey (1975) considered that podiform chromite originated within the magma chamber, at the base of the oceanic crust below mid-ocean ridges. The main basis of Dickey's hypothesis is the distinctive chemistry of podiform chromite and the results of experimental petrology. He considered that the chromite forms by incongruent melting and early precipitation from silicate liquids, and suggested that rapid changes of temperature and pressure may have been important controls. Dickey (1975) suggested that the nodular texture of the ore may be attributed to the snowballing of chromite crystals within zones of turbulence in the magma chamber. He explained the concentration of chromite within the tectonized harzburgite as being due to the sinking of chromite autoliths through the enclosing olivine crystals. Panayiotou (1978) considered that this mechanism explained the distribution of chromite ore within tectonized harzburgite from the Limassol Forest Complex in Cyprus.

On the other hand Lago and co-workers (1982) have proposed another hypothesis for chromite formation within the zone of tectonized harzburgite. Lago's genetic model is based on the field relationships of podiform chromite with the

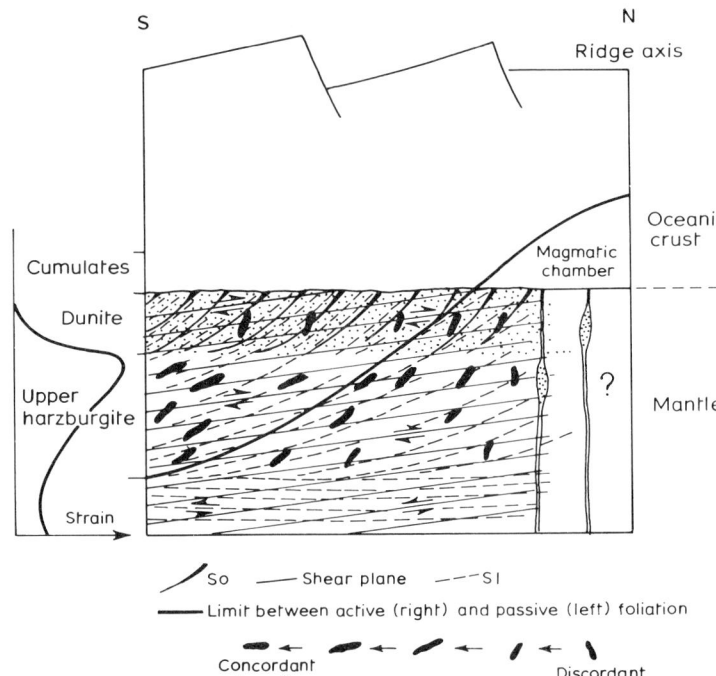

Fig. 2.11. Sketch of genesis and evolution of the chromite pods in uppermost oceanic mantle beneath an active spreading ridge (from Lago *et al.* 1982).

enclosing harzburgites and dunites in southern New Caledonia. Detailed structural mapping, by Cassard and his colleagues (1981), has demonstrated that there are three structural styles of podiform chromite in New Caledonia, which are defined by the relationship of the chromite with the penetrative fabric of the enclosing peridotite. Lago and his colleagues (1982) consider that those chromite-rich zones which have a discordant relationship with the foliation of the harzburgite represent the sites of dykes which have acted as channelways to the overlying magma chamber. Two mechanisms are important: the upward pressure of magma within the dyke, and convective circulation within bulges in the dykes which encourage the mixing of chromite grains and the growth of nodules. Eventually the chromite-enriched dykes are caught up by plastic deformation, and become orientated parallel to the metamorphic fabric of the enclosing harzburgite (Fig. 2.11).

2.2.7 Exploration for podiform chromite deposits

Chromium plays a significant role in modern industrial society, with both civilian and military applications. However, the main reserves of chromite are confined to Zimbabwe and South Africa. As a consequence of its industrial importance and limited geographical distribution chromite is classed as a strategic mineral. It is therefore necessary to improve the geological concepts and exploration techniques which will lead to the discovery of future reserves.

The most favourable environment for new chromite reserves outside southern Africa is within ophiolite complexes. These are widely distributed and many are known to be repositories of chromite. However, the search for new podiform chromite deposits in this environment is hindered by a paucity of consistent exploration guides, and a lack of appropriate indirect methods.

Magmatic deposits

The most valuable guide to exploration, which seems to have global applicability, is the association of podiform chromite with the ultramafic zone of ophiolite complexes. In particular, dunite zones underlying gabbros are recognized as favourable lithologies for chromite (Cassard *et al.*, 1981; Flint *et al.*, 1948; Peters and Kramers, 1974; Greenbaum, 1977). Other field guides have been described from New Caledonia (Cassard *et al.*, 1981) which may have wider relevance. These include the restriction of some chromite deposits to particular stratigraphic levels, and the frequent alignment of chromite zones with the lineation of the enclosing peridotite. Cassard and his colleagues (1981) also describe field guides which are applicable to mine exploration. For example, in New Caledonia sharp changes in the dip and strike of foliation are observed in host rocks near to chromite ore bodies. There is a great need for similar descriptions of field associations and structural characteristics from other districts where chromite is mined.

If favourable lithologies are located during field mapping it is necessary to ascertain whether blind deposits of chromite occur. Are there any appropriate indirect methods of exploration? Wynn (1983) reports the results of geophysical orientation studies over known chromite ore bodies in California, USA. The techniques used include magnetics, VLF–EM, complex resistivity and seismic refraction. The most promising approach indicated by this study is a combination of complex resistivity and seismic refraction, although neither technique produced anomalies of dramatic dimensions.

At the present time it appears that there is little alternative to detailed geological mapping within ophiolite complexes, with closely spaced percussive drilling in areas with favourable geological characteristics.

2.3 NICKEL SULPHIDE DEPOSITS

2.3.1 Introduction

Nickel is mined from two fundamentally different ore types: nickel sulphides and nickel silicates or lateritic nickel. Lateritic nickel is formed by weathering processes, and therefore this ore type is discussed in Chapter 7. Figure 2.12 shows the growth of World nickel production from 1900 to 1980. It is evident that sulphide ores have been dominant throughout this period, although

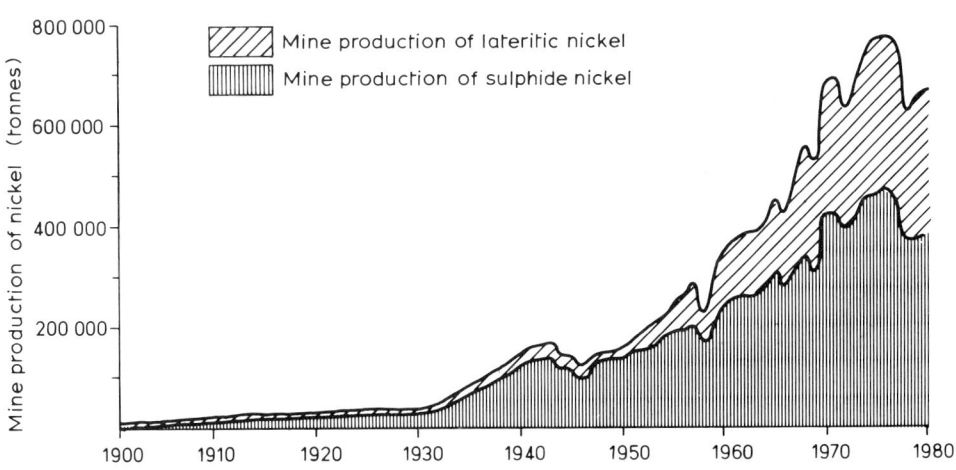

Fig. 2.12. World mine production of sulphide and laterite nickel 1900–1980 (simplified from Ross and Travis 1981).

lateritic ores have grown increasingly important since the late 1950s.

In 1980 the total non-Communist production of nickel was 516 000 tonnes. The mining of nickel is also important in some Communist countries, particularly in the USSR where annual production is in the order of 160 000 tonnes of nickel (Buchanan, 1982). Sulphide ores accounted for 58.7% of non-Communist production in 1980 (Buchanan, 1982). Canada is the dominant nickel producer with about 37% of Western World production in 1980. Other countries with a significant production from sulphide ores include Australia, South Africa, Botswana, Zimbabwe and Finland.

The bulk of nickel is used in stainless-steel production, due to the increased strength and resistance to corrosion imparted by the nickel. Other minor uses include batteries, dyes, pigments, catalysts and insecticides.

2.3.2 Classification of nickel sulphide ores

A comprehensive classification of mafic and ultramafic bodies, with examples of their associated ores, is presented in Table 2.5, and shows that nickel sulphides are associated with a range of mafic and ultramafic lithologies in a diversity of tectonic settings. The relative economic importance of nickel sulphides within mafic/ultramafic complexes is documented by Ross and Travis (1981) who present the following classification of nickel sulphide ores:

(1) The dunite–peridotite class
    (a) Intrusive dunite association
    (b) Volcanic–peridotite association
(2) The gabbroid class
    (a) Intrusive mafic/ultramafic complexes
    (b) Large layered intrusions
    (c) Sudbury
    (d) Other categories

Table 2.5 Classification of mafic and ultramafic bodies (modified from Naldrett, 1981)

Class of body	Examples of associated ore deposits
A. Synvolcanic bodies	
1. Komatiitic suites	
(a) Lava flows	Kambalda, W. Australia
(b) Layered sills	Madziwa, Zimbabwe
(c) Dunite–peridotite lenses	Agnew, W. Australia
(d) Uncertain types	Thompson belt, Manitoba
2. Tholeiitic suites	
(a) Synvolcanic layered intrusions	Pechenga, USSR
(b) Anorthositic bodies	No known deposits
3. Undocumented parentage	
(a) Stratiform intrusions	Montcalm, Ontario
(b) Tectonically reworked terrain	Selebi-Phikwe, Botswana
B. Intrusions in cratonic areas	
1. Intrusions related to flood basalts	Duluth Complex, Canada
	Noril'sk, USSR
2. Large layered complexes with no documented relation to flood basalts	
(a) Sheet-like	
(i) With repetitive layering	Bushveld Complex, S. Africa
(ii) Without repetitive layering	Sudbury, Ontario
(b) Dyke-like	Great 'Dyke', Zimbabwe
3. Alkalic ultramafic intrusions	Råna, Norway

Magmatic deposits

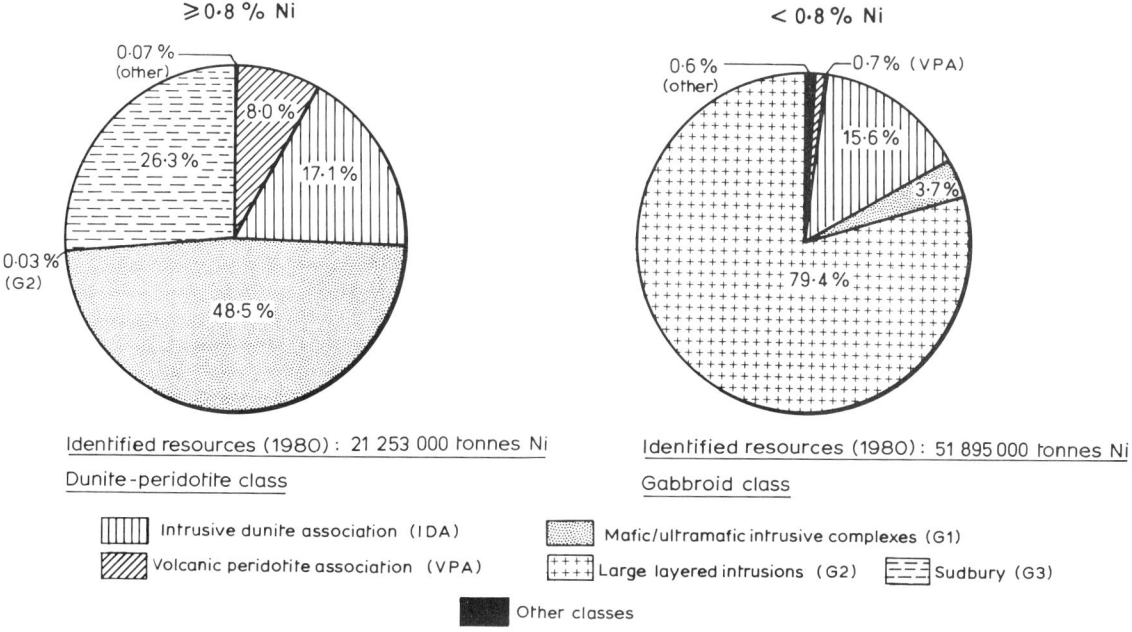

Fig. 2.13. World total of identified resources of sulphide Ni by deposit type in ores with ≥0.8% Ni and <0.8% Ni (from Ross and Travis 1981).

Figure 2.13 shows the distribution of nickel sulphide resources within these classes. It is clear that in the case of resources with > 0.8% Ni, the mafic and ultramafic intrusive complexes are of dominant importance. This category includes 54 deposits in Archaean greenstone belts in Western Australia, Canada and Zimbabwe, post-Archaean mobile belts in Scandinavia, the USSR, Botswana and Canada and Phanerozoic fold belts in the USSR and North America (Ross and Travis, 1981). The second most important deposit type in this resource category is the Sudbury Complex in Ontario, Canada. The Sudbury deposits formed nearly 40% of the original resource of nickel, and have dominated Western nickel sulphide production for most of this century. However, during the early 1980s the recession in the industrialized World has resulted in a much sharper decline in production from Sudbury compared with other nickel-mining districts (*Mining Annual Review*, 1983). We have omitted further discussion of Sudbury – partly because its uniqueness as an astrobleme (Peredery and Morrison, 1984) precludes the search for similar styles of deposit.

The dunite–peridotite class comprises about one-quarter of nickel resources in both the grade categories selected for illustration. This is particularly interesting when one considers that the full economic potential of this style of mineralization was not appreciated until the discovery of the Kambalda deposit in Western Australia in 1966.

Figure 2.13 shows that if nickel resources containing <0.8% Ni are expressed in terms of deposit type, the large layered intrusions, such as the Bushveld Complex, become highly significant. However, nickel and copper are produced only as by-products of platinum production from the Merensky Reef (Section 2.2.4) and it is unlikely that they would be viable as base metal deposits under foreseeable economic conditions.

It is clear that nickel sulphide deposits have a wide range of geological settings and there is insufficient space in this chapter adequately to describe each class of ore. We have therefore selected the dunite–peridotite class for particular consideration. The main reasons are that it is well documented and has been the object of considerable exploration in recent years.

2.3.3 General characteristics of nickel sulphides of the dunite–peridotite association

(a) Distribution in space and time

Nickel sulphide deposits which are grouped within the dunite–peridotite class are irregularly distributed in Archaean greenstone belts. They are best developed in the eastern sector of the Yilgarn block, Western Australia. They have a more limited distribution in the Superior and Churchill provinces of the Canadian Shield and in Zimbabwe. Their virtual absence from Archaean greenstone belts in India, South America, South Africa and the Pilbara block of Western Australia suggests that particular conditions are required for the concentration of nickel sulphide ores (see also Section 2.3.5).

(b) Size and grade of deposits

Ore deposits of the dunite–peridotite class are subdivided into volcanic–peridotite and intrusive dunite associations. Deposits belonging to the volcanic–peridotite association are usually small (1×10^6–5×10^6 tonnes) and high grade (1.5–3.5% Ni). The sulphides normally contain 10–15% Ni and 0.5–1.5% Cu (Naldrett, 1981). In Western Australia the sulphide ores contain significant concentrations of gold and platinoids (Groves and Hudson, 1981).

Mineral deposits belonging to the intrusive dunite association are subdivided into those of medium size, which contain both high- and low-grade ores, and those of very large size which contain only low-grade disseminated ores. An example of the former type is the Agnew (formerly Perseverance) deposit in Western Australia. This ore body had an *in situ* reserve of 45×10^6 tonnes at 2.05% Ni and 0.10% Cu (Billington, 1984). The Mount Keith deposit in Western Australia is an example of a very large, low-grade type. Published reserves are 290×10^6 tonnes at a grade of 0.60% Ni (Butt and Nickel, 1981).

(c) Form and petrology of host rocks

Nickel sulphide deposits of the dunite–peridotite class occur within Archaean greenstone belts. Greenstone belts are linear, irregularly shaped synformal supracrustal successions, which range in width from 5 to 250 km and are normally several hundred kilometres in length. Greenstone belt lithologies are dominated by pillowed mafic volcanics, and in some cases they contain abundant ultramafic (komatiitic) lavas in the lower part of their stratigraphy. Greenstone belts always occur as enclaves within a wider expanse of granitic terrain.

(i) The volcanic–peridotite association

Ores which belong to the volcanic–peridotite association normally occur within flows, which may range up to 100 m in thickness. The ultramafic lavas are classified as komatiites. The term komatiite was originally defined by Viljoen and Viljoen (1969) to describe a suite of Mg-rich lavas from the vicinity of the Komati River in the Barberton district of South Africa. However, petrologists have had difficulty in applying the original definition to similar rocks within other greenstone belts. The criteria which have been suggested by different petrologists are summarized by Condie (1981). It is sufficient for our purposes to consider komatiites as a distinctive series of volcanic rocks which range in composition from dunite through peridotite and pyroxenite to basalt. They usually contain a high MgO content and may have a characteristic quench texture called a spinifex texture in the upper part of the flow (Fig. 2.14).

The nickel sulphide ores tends to be concentrated near to the basal part of the first flow

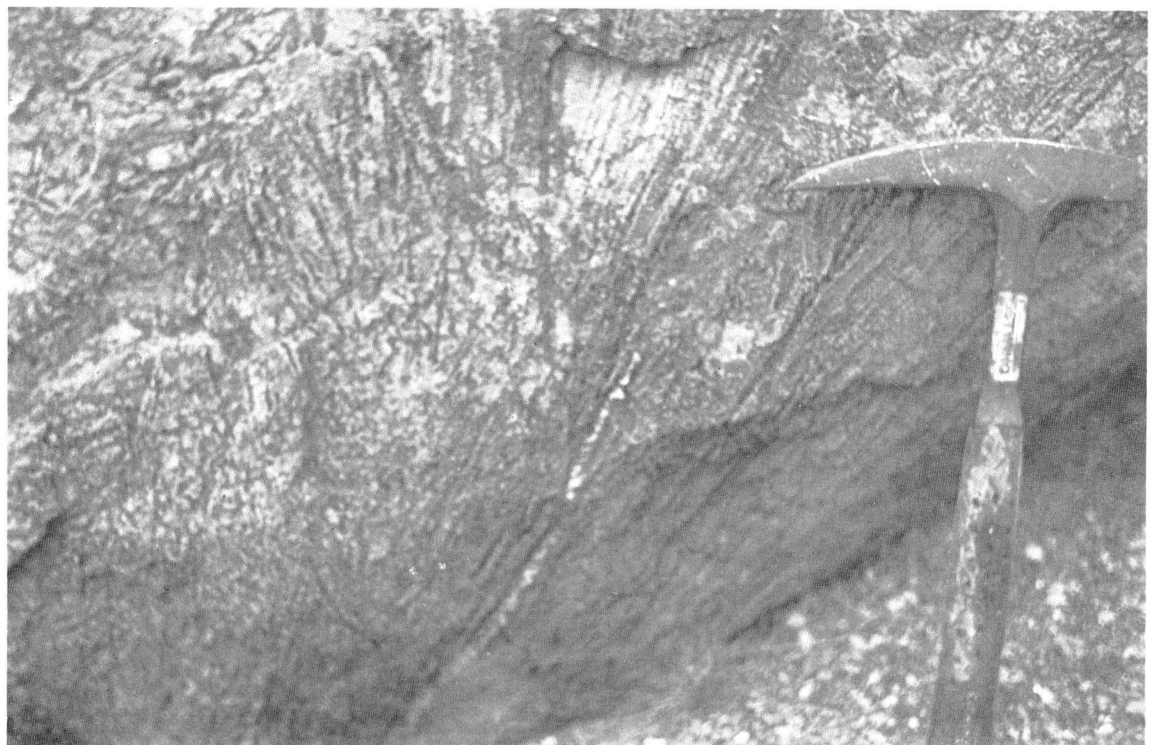

Fig. 2.14. Spinifex texture in komatiite, Kalgoorlie, Western Australia.

within an ultramafic sequence, although the second and third ultramafic units may also be mineralized (Groves and Hudson, 1981). In addition to occurring within flows, ores belonging to this class of deposit may also be associated with intrusives. For example, the Shangani deposit, in Zimbabwe, is associated with a mushroom-shaped peridotite which is intrusive into felsic lavas (Viljoen et al., 1976). Williams (1979) has suggested that the ore-bearing body may be a fissure-like extrusive centre.

(ii) The intrusive dunite association

Sulphide ore bodies which are grouped within the intrusive dunite association occur within lens-shaped bodies of variable size. In both the Manitoba nickel belt in Canada and in the Norseman–Wiluna belt in Western Australia the dunite lenses are confined to a linear tectonic zone (Naldrett, 1981; Marston et al., 1981). The composition of the lenses is dominated by olivine or its alteration products. The dunite lenses are commonly intrusive into felsic volcanics or metasediments, and are thought to have acted as feeder zones for overlying komatiitic volcanism (Naldrett, 1981).

(d) Mineralogy

(i) The volcanic–peridotite association

Nickel sulphide ores associated with volcanic–peridotites are characterized by a limited number of common opaque minerals, and in some cases by a larger suite of minerals which are normally rare but may become locally abundant. Pentlandite is always the most significant economic ore mineral.

In Western Australia the most common opaque minerals include pyrrhotite, pentlandite, pyrite, chalcopyrite, magnetite and ferrochromite. Pentlandite is subordinate in quantity to pyrrhotite and may contain up to 0.8% Co. The more obscure ore minerals are large in number and are documented by Groves and Hudson (1981). In Canada the nickel sulphide ore associated with this style of deposit exhibits some regional variation. For example, whilst pentlandite and pyrrhotite are normally major sulphides, pyrrhotite may be absent, and at several mines millerite occurs as a major ore mineral. Chalcopyrite is a minor but consistent ore mineral (Naldrett and Gasparini, 1971). In Zimbabwe, the Damba nickel deposit is characterized by pentlandite, millerite and pyrite (Williams, 1979).

In Western Australia and Canada, the nickel sulphide ores are subdivided using the relative proportion of sulphide and silicate minerals. Normally the lowermost part of an ore lens comprises >90% sulphides and is designated as massive ore (see Fig. 2.16). The massive ore is overlain by a zone in which sulphide is interstitial to silicates and this is termed either matrix or net-textured ore. The uppermost zone consists of weakly disseminated sulphides in an ultramafic host.

(ii) The intrusive dunite association

The intrusive dunite association is characterized by disseminated sulphide mineralization. According to Marston *et al*. (1981), the higher grade sulphide ores in Australia are dominated by pyrrhotite and pentlandite with subsidiary magnetite, chromite, pyrite and chalcopyrite. Pyrite and pyrrhotite also characterize the low-grade disseminated ores, but a wide range of nickel and cobalt sulphides may also be present (Marston *et al*., 1981).

The gangue minerals in both associations are dominated by olivine and its alteration products. These minerals often contain up to 0.3% Ni. The nickel content is reflected in the assay of ore samples, but nickel contained in silicates is not recovered in the treatment of sulphide ores.

(e) Tectonic setting

The restriction of nickel sulphide ores of the dunite–peridotite class to Archaean greenstone belts has already been mentioned. The factors which controlled tectonism in Archaean times are uncertain, and geologists disagree about the role which plate tectonics might have played at this early stage in earth history. Some of the arguments which have been presented are summarized by Condie (1981).

The tectonic setting of the Western Australian nickel sulphides within the Yilgarn block is described by Groves *et al*. (1984). These authors draw attention to the fact that the nickel sulphide ores are concentrated in the Norseman–Wiluna belt, which is younger than the other greenstone belts in the Yilgarn block. Groves *et al*. (1984) also state that ores belonging to the volcanic–peridotite and intrusive dunite associations are spatially separated and occupy distinctive structural settings. They conclude that ores of both associations occupy different sites within a major fault-controlled rift system.

2.3.4 The Kambalda mining district – an example of the volcanic–peridotite association

The intersection of nickel mineralization at Kambalda in 1966 marked the first major discovery of nickel sulphides in Archaean greenstones in Australia, and triggered an important exploration boom. The Kambalda field represents the largest concentration of nickel within the volcanic–peridotite association in Western Australia. Current annual production varies from 1.3 to 1.4 million tonnes of ore with an average grade of 3.15% Ni. The Kambalda district has been the subject of numerous reports and publications, but this summary is mainly drawn from Gresham and Loftus-Hills (1981).

The Kambalda mining district occurs within the southern part of the Eastern Goldfields province of

Magmatic deposits

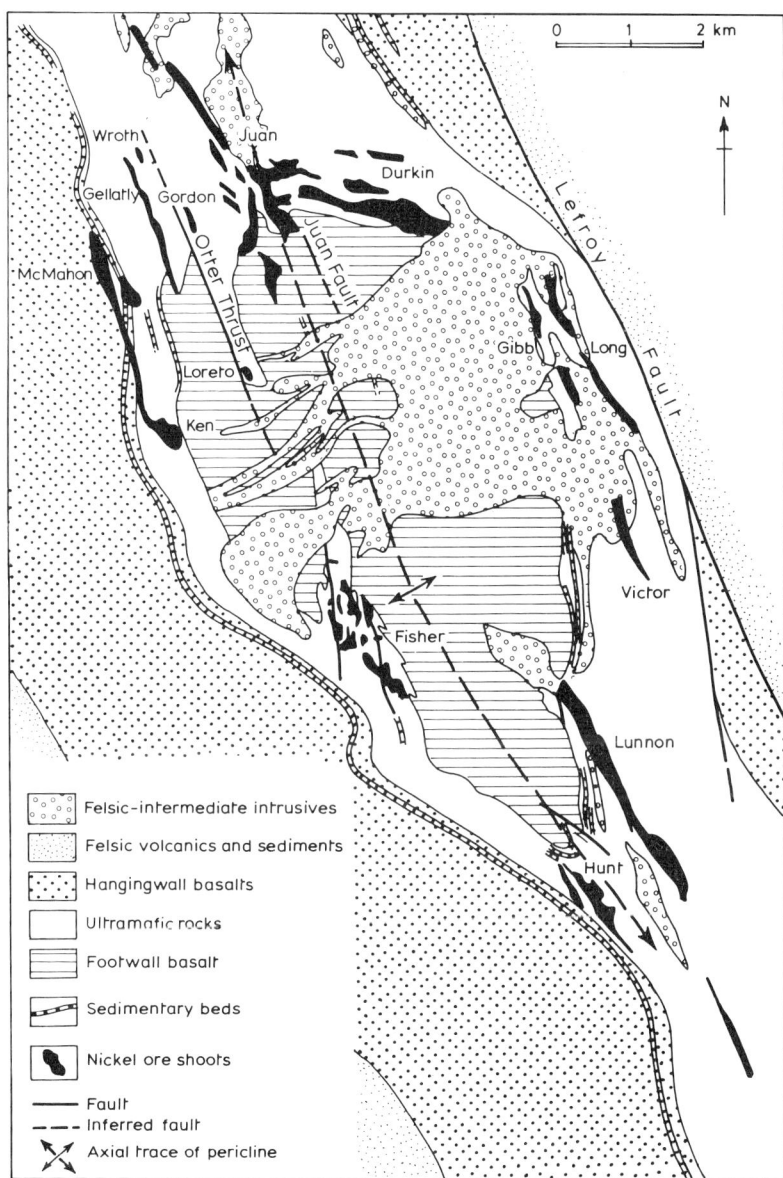

Fig. 2.15. Geological map of the Kambalda dome showing the nickel shoots in plan projection (modified from Gresham and Loftus-Hills 1981).

the Yilgarn block, Western Australia. The stratigraphy comprises two separate sequences (the Kambalda and Bluebush sequences) both of which are dominated by ultramafic and mafic volcanics with minor sediments. Four phases of folding are recognized, and interference folding is responsible for the domed structure at Kambalda (Fig. 2.15). NNW trending faults have been active throughout the Archaean; they control basin margins and lines of vulcanicity and sometimes define the limits of sulphide mineralization.

Virtually all the Kambalda nickel ores are

Fig. 2.16. Massive ore with minor component of matrix ore, Hunt Mine, Kambalda, Western Australia.

Fig. 2.17. Veinlets of sulphide ore penetrating footwall basalt, Hunt Mine, Kambalda, Western Australia.

Magmatic deposits

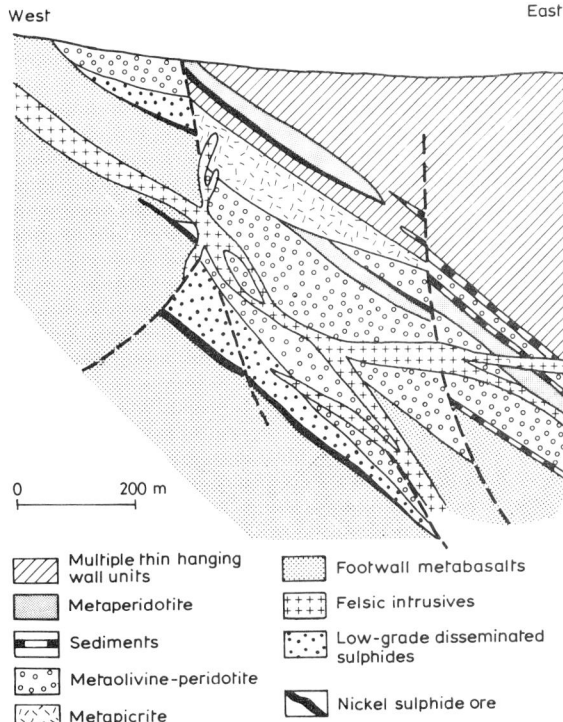

Fig. 2.18. Section through the Lunnon and neighbouring ore shoots, Kambalda, Western Australia (from Evans 1980, after Ross and Hopkins 1975).

associated with the Kambalda sequence which comprises footwall basalts, ultramafic rocks, hangingwall basalts and associated sedimentary and intrusive rocks. The ultramafic rocks are the principal hosts for nickel sulphide mineralization (Figs 2.16 and 2.17). They include picrites (15–28% MgO), peridotites (28–36% MgO) and olivine peridotites (>36% MgO) and range in thickness from 0 to 1000 m. Individual units may exceed 100 m and have thin but well-developed spinifex textured tops.

Most of the sulphide ore occupies depressions or trough structures in the top of the footwall basalt. Major troughs vary from 1000 to 2300 m in length and 150 to 250 m in width and control the size and geometry of the ore shoots. The contacts of the troughs are commonly defined by either normal or reverse faults (Fig. 2.18).

2.3.5 Agnew deposit – an example of the intrusive dunite association

The Agnew nickel deposit is located 330 km north of Kalgoorlie in Western Australia. The *in situ* ore reserve is 45×10^6 tonnes with a grade of 2.05% Ni and 0.10% Cu. This summary of the geology of the Agnew deposit is from Billington (1984).

The Agnew mine is situated within the Norseman–Wiluna greenstone belt in the Eastern Goldfields province of the Yilgarn block (Fig. 2.19). In the vicinity of Agnew the greenstones are dominated by mafic igneous rocks, and deformed by large-scale NNW trending faults and folds. Regional metamorphism is believed to have attained mid-amphibolite facies. The Perseverance ultramafic intrusion, which is the main host for nickel mineralization, occurs near the base of the greenstone sequence and is overlain by felsic volcanics, psammitic and pelitic metasediments and metagabbros and metabasalts. The ultramafic unit varies from a few metres to 800 m in width and extends for 10 km in length. Figure 2.20 shows that the Perseverance ultramafic intrusion consists in plan of a dunite core, surrounded by a series of concentric zones in which the mineralogy is dominated by the alteration products of olivine. The nickel sulphide ore is interpreted as occurring with the footwall of the ultramafic unit. Three types of mineralization are recognized:

(a) Massive sulphides with >4% Ni (Fig. 2.22);
(b) Strongly disseminated sulphides 1–3% Ni;
(c) Weakly disseminated sulphides <1% Ni;

Contacts between the deposit types are normally distinct. Strongly disseminated sulphide is the main ore type which is located within three large elliptical zones (Fig. 2.21).

2.3.6 Genesis of nickel sulphides of the dunite–peridotite class

(a) The volcanic–peridotite association

Most geologists who have studied this class of ore deposit accept that the nickel sulphide ores have formed by the segregation of an immiscible

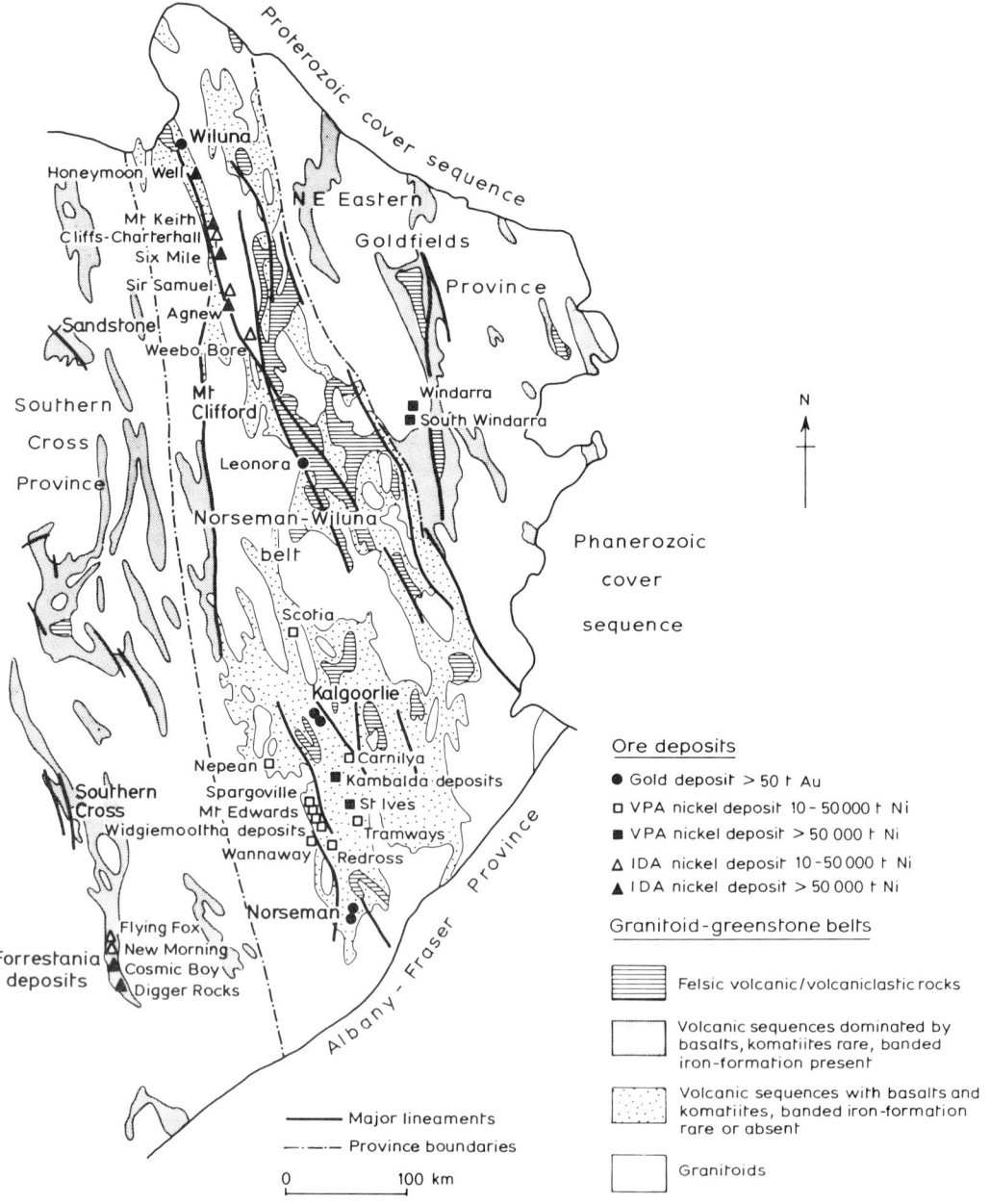

Fig. 2.19. Solid geology of eastern part of Yilgarn Block illustrating setting of major nickel deposits in terms of lithofacies, structure and major tectonic subdivisions of granitoid–greenstone terrains (from Groves *et al.* 1984).

Magmatic deposits

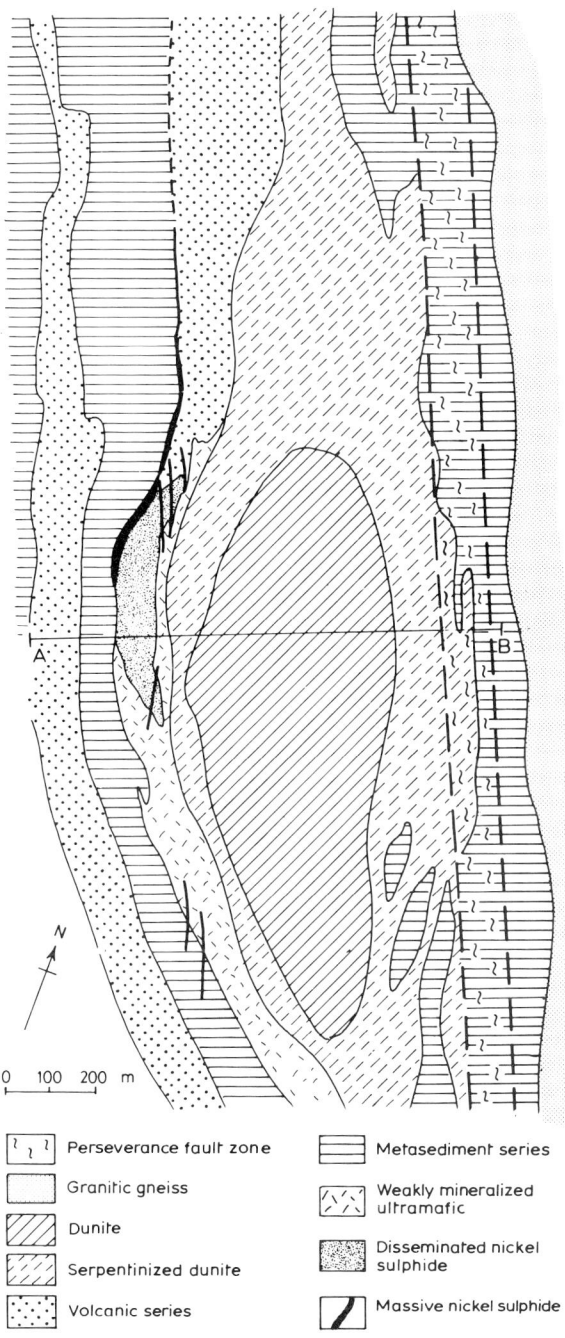

Fig. 2.20. Geological map of Agnew embayment area (from Billington 1984).

sulphide liquid from an ultrabasic magma. However, although there is agreement about the general nature of the ore-forming process, there remains some controversy about the precise nature of the mechanisms involved. The main thrust of the debate has centred on whether a liquid massive sulphide layer can exist beneath a thick ultramafic flow.

Naldrett (1973) sought to explain the close spatial association of massive, net-textured and disseminated ores within ultramafic flows using a 'billiard-ball' model. In essence this is a static model, involving the gravitational settling of sulphides within a cooling ultramafic flow. Critics of this model argue, from theoretical considerations, that under these circumstances the sulphides would be displaced upwards into the overlying cumulate zone by the weight of the overlying column of olivine crystals (Barrett et al., 1977). The assumptions on which these criticisms are based have in turn been challenged by Naldrett (1981).

Australian geologists have preferred to invoke a dynamic model of sulphide separation. Ross and Hopkins (1975) have suggested a model involving the segregation of sulphide and silicate components during vertical flow due to viscosity contrasts. Later modifications of this model have emphasized the separation of sulphide liquid, silicate liquid and olivine crystals during horizontal flow. The lower-viscosity sulphide phase would tend to precede the silicate magma, with partial crystallization having already occurred before being overwhelmed by the slower, more viscous silicate magma. Whilst these models avoid Barrett's criticism of the 'billiard-ball' model, they imply the presence of separate lava flows for the massive and net-textured ores.

The source of sulphur for the sulphide ores is uncertain. Both mantle-derived sulphur and sulphur enrichment from the assimilation of sediments have arguments in their favour, and it is probable that both have contributed to the sulphide component of the magma.

Not all geologists have interpreted nickel sulphide associated with volcanic–peridotites as

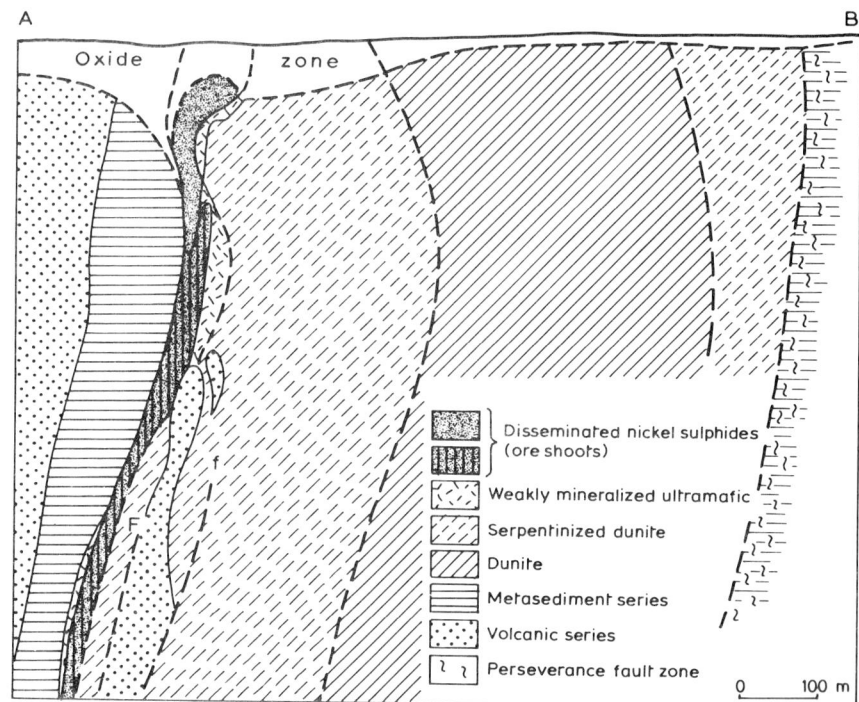

Fig. 2.21. Cross-section through Agnew mine (simplified from Billington 1984).

resulting from magmatic segregation. A volcanic-exhalative model has been proposed by Lusk (1976), but there are many arguments against this proposal which are summarized by Groves and Hudson (1981). Barrett *et al.* (1977) have proposed that the massive nickel sulphide ores can be formed by the upgrading of net-textured ore during regional metamorphism. All geologists who have studied the Western Australian ores accept that their textures and mineralogy have been modified by metamorphism. However, Groves and Hudson (1981) consider that the transformation of matrix to massive ores occurs on only a small scale. In Zimbabwe low-grade ores may be enriched as a result of serpentinization due to metamorphism. At the Epoch and Damba deposits nickel released during serpentinization can change pyrrhotite to millerite, thus making a high-grade concentrate from disseminated ore (A. M. Killick, personal communication).

(b) The intrusive dunite association

The consensus amongst geologists who have discussed the genesis of the intrusive dunite association is that the sulphides have segregated from an ultrabasic magma. The main evidence comes from the intimate association of sulphides with the enclosing dunite and serpentinite. In addition the close spatial and compositional relationship between intrusive dunite and overlying komatiites in the Forrestania district (Porter and McKay, 1981) suggests that both are part of the same magmatic system. Therefore acceptance of magmatic segregation for ores of the volcanic–peridotite association implies a similar mode of origin for the ores of the intrusive dunite association.

The contrast in ore types and grades within the intrusive dunite class reflects a difference in timing in the segregation of the sulphide ores. The dominance of disseminated sulphides indicates

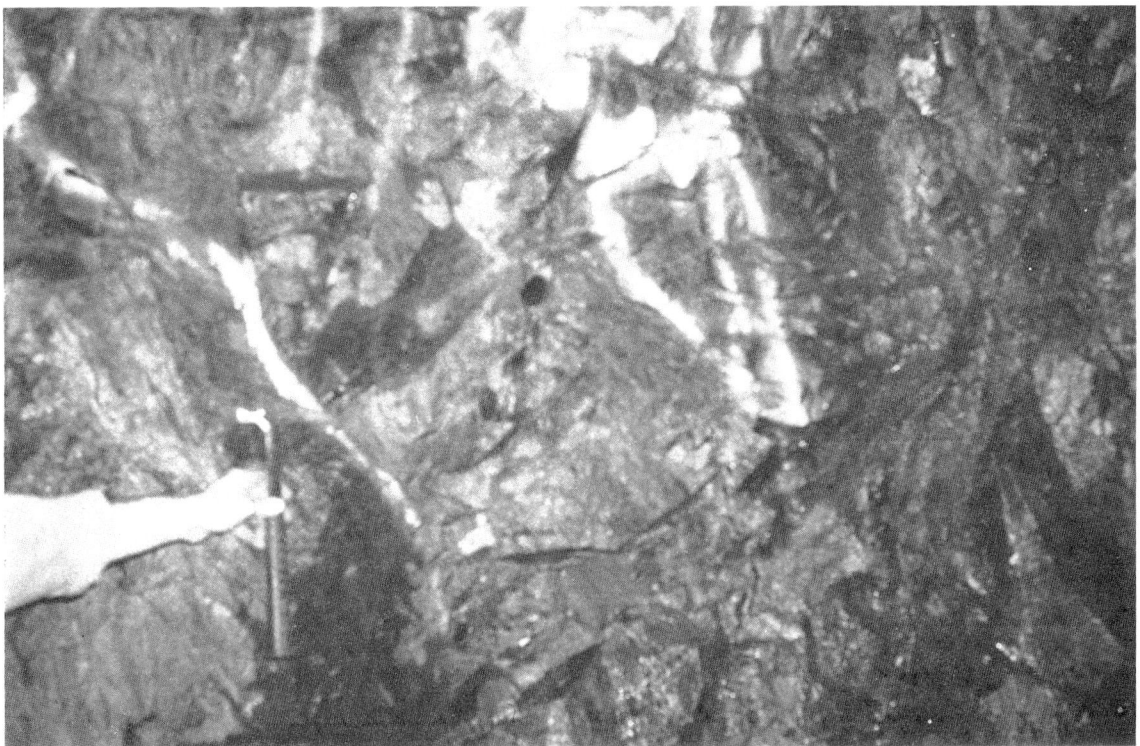

Fig. 2.22. Massive nickel ore (1A ore body) remobilized into country rock. Agnew mine, Western Australia.

that they have segregated at a relatively late stage, to be trapped within an olivine framework. However the occurrence of massive ore near the footwall contact of some dunites indicates that gravitational settling has played some part in the ore-forming process. The remobilization of some ores into the adjacent country rock (Agnew), combined with textural evidence, suggests that in many cases regional metamorphism has modified the original magmatic features of many deposits (Fig. 2.22).

Why are some Archaean greenstone belts devoid of nickel sulphide mineralization? This question is of crucial importance for exploration as well as being highly relevant to ore genesis. The subject has been discussed by Groves and Hudson (1981) and Groves et al. (1984). They consider that the tectonic setting of the greenstone belt is of fundamental importance. From a study of the greenstone belts within the Yilgarn block, Western Australia, they conclude that nickel sulphide ores are concentrated within younger greenstone belts, in more active fracture-controlled basins, in which interflow sulphidic sediments formed an integral part of volcanic activity.

2.3.7 Exploration for nickel sulphide deposits

The first task of the geologist is the selection of terrain which is underlain by rock types known to be potential hosts for mineralization. The development of soundly based classifications of mafic/

ultramafic bodies and of nickel sulphide deposits (Naldrett, 1981; Ross and Travis, 1981) has aided the identification of tectonic environments and host rocks which have good potential for economic nickel deposits. For example, Archaean greenstone belts are recognized as the geological environment in which ores of the dunite–peridotite class are found. In contrast ophiolite complexes are usually devoid of nickel sulphide mineralization.

The extent to which direct geological techniques can be used in exploration for nickel sulphides varies with the class of deposit and the amount of outcrop. Ores of the dunite–peridotite class are always associated with a komatiitic sequence in the lower part of a greenstone belt succession. Potter (1984) has carried out an analysis of success rates of nickel exploration in Zimbabwe. One hundred and fifty three exploration projects led eventually to the discovery of six new mines. This represents a success rate of 4% which is significantly higher than average in other countries. The success probably stems in part from the ability to concentrate the exploration effort within particular sectors of the Archaean greenstone belts and in part stems from the success of exploration geochemistry in this environment.

Intrusive mafic/ultramafic complexes are known to be favourable repositories for nickel sulphides, but their geological setting and age are quite variable, and their potential for economic nickel deposits is more difficult to assess. Normally gravitational processes have played an important role during the segregation of nickel sulphide ores which therefore tend to be concentrated near the base of intrusions.

(a) Remote sensing

Mafic and ultramafic rocks are known to be the most appropriate lithology for nickel sulphide mineralization. In areas where remoteness, dense vegetation or poor outcrop reduce the effectiveness of conventional geological mapping remote sensing may be valuable at the reconnaissance stage of mineral exploration. Normally rocks which contain abnormally high contents of copper and nickel have a distinctive flora or exhibit particular geobotanical characteristics such as dwarfism. Where this is the case Landsat data may be used as a mapping tool. Raines and Wynn (1982) investigated the value of Landsat data for constructing a geological map in heavily forested terrain, underlain by an ultrabasic complex in northern California, USA. Ground checking of the geological map constructed from Landsat data showed few major discrepancies.

In Western Australia the interpretation of aerial photographs plays an important role at the reconnaissance stage of exploration. Normally the first stage of photointerpretation is carried out at a scale of 1:25 000 using true colour photography and this information is combined with the results of airborne magnetometry to produce a preliminary geological map. A second stage of photointerpretation then focuses on areas of particular interest. The second stage utilizes both false colour and true colour photography. The advantage of false colour photography is that it helps the geologist to identify areas which are underlain by ultrabasic rocks. Areas of lateritic cover and felsic lithologies can also be distinguished (D. Harley, personal communication). In Zimbabwe black-and-white photography is found to be satisfactory for detecting changes in vegetation (A. M. Killick, personal communication).

(b) Geochemistry

Geochemistry has been particularly valuable in the search for nickel sulphide ores both in helping to identify the host rocks with good potential for mineralization and in the definition of secondary dispersion patterns.

Nickel sulphides of the dunite–peridotite class are associated with komatiites. However, both komatiitic and tholeittic rocks may be represented within a greenstone sequence and if the characteristic spinifex texture of komatiites is not evident it may be difficult to distinguish the appropriate lithology. Naldrett and Arndt (1976) pointed out that the two series may be readily identified from their major element chemistry.

Magmatic deposits

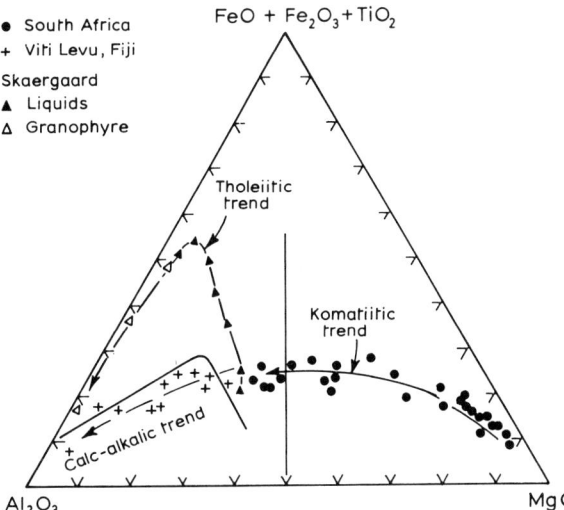

Fig. 2.23. Jensen Cation Plot comparing the patterns of variation of komatiitic, tholeiitic and calc-alkalic rock suites. Arrowed lines serve to indicate recognized patterns; solid lines serve to separate komatiitic, tholeiitic and calc-alkalic rock suites. Oxides are calculated in cation percent. (Modified from Jensen 1976).

However the most effective discriminator between komatiitic, tholeiitic and calc-alkaline volcanics is the Jensen cation plot (Fig. 2.23).

The concentration of TiO_2 is also a good discriminator: komatiites contain less than 1% TiO_2 whereas tholeiites contain more than 1% TiO_2 (Naldrett and Arndt, 1976). When a komatiite sequence has been recognized within a greenstone belt the task remains of selecting areas for more detailed investigation. Studies of the geochemistry of komatiites which host nickel sulphide deposits indicate that rocks with >40% MgO have the highest probability of containing nickel sulphides (Naldrett and Arndt, 1976).

Nickel sulphide deposits form as the result of a sequence of events which includes the formation of an immiscible sulphide liquid. A number of geologists have attempted to characterize those ultramafic and mafic bodies from which an immiscible sulphide phase has separated. Cameron et al. (1971) analysed 1079 samples of ultramafic rock from mineralized and barren intrusives in the Canadian Shield for Cu, Ni, Co and S. Whilst some distinction could be drawn between barren and mineralized rocks using discriminant analysis the contrast was insufficient for the technique to be used in routine exploration.

Naldrett et al. (1984) have attempted to characterize ore-bearing mafic and ultramafic rocks by their depletion in nickel. Their approach is plainly stated, 'because chalcophile elements partition strongly into sulphides in preference to silicate melts and minerals, magmas from which sulphides have segregated will be depleted in these elements.' The approach adopted by Naldrett et al. (1984) has been to model the compositional trend of silicate magmas derived from the fractionation of a komatiitic liquid. Both sulphide-unsaturated and a series of sulphide-saturated conditions were selected. Comparisons of the results of theoretical modelling, with analyses from mineralized and barren ultramafics, suggest that the depletion of nickel in the silicates of mineralized ultramafics may be a valuable exploration guide (Fig. 2.24).

Additional studies by Naldrett et al. (1984) on mineralized gabbroic intrusions in the USA and Southern Africa suggest that the Ni/forsterite ratio of olivine can indicate the potential of a mafic or ultramafic intrusion for nickel sulphides.

Geochemical exploration for nickel sulphides using secondary dispersion patterns in soils, and to a lesser extent in stream sediments, has been highly successful in a wide range of climatic regimes. Instructive case histories are described by Thomson (ref. in Bradshaw and Thomson, 1979), Shilts (ref. in Bolviken and Gleeson, 1979), Rugman (1982) and Leggo and McKay (1980).

In most of the surveys described copper and nickel have been the elements of principal importance in defining areas for detailed exploration. Thomson (ref. in Bradshaw and Thomson, 1979) described the results of exploration for nickel sulphides in gabbroic rocks in Brazil. The main problem was distinguishing base metals derived from sulphides in gabbroic rocks, from nickel and copper associated with silicates in the associated barren ultrabasics. Thomson found that the use of copper/nickel ratios acted as a valuable discriminator of sulphide mineralization.

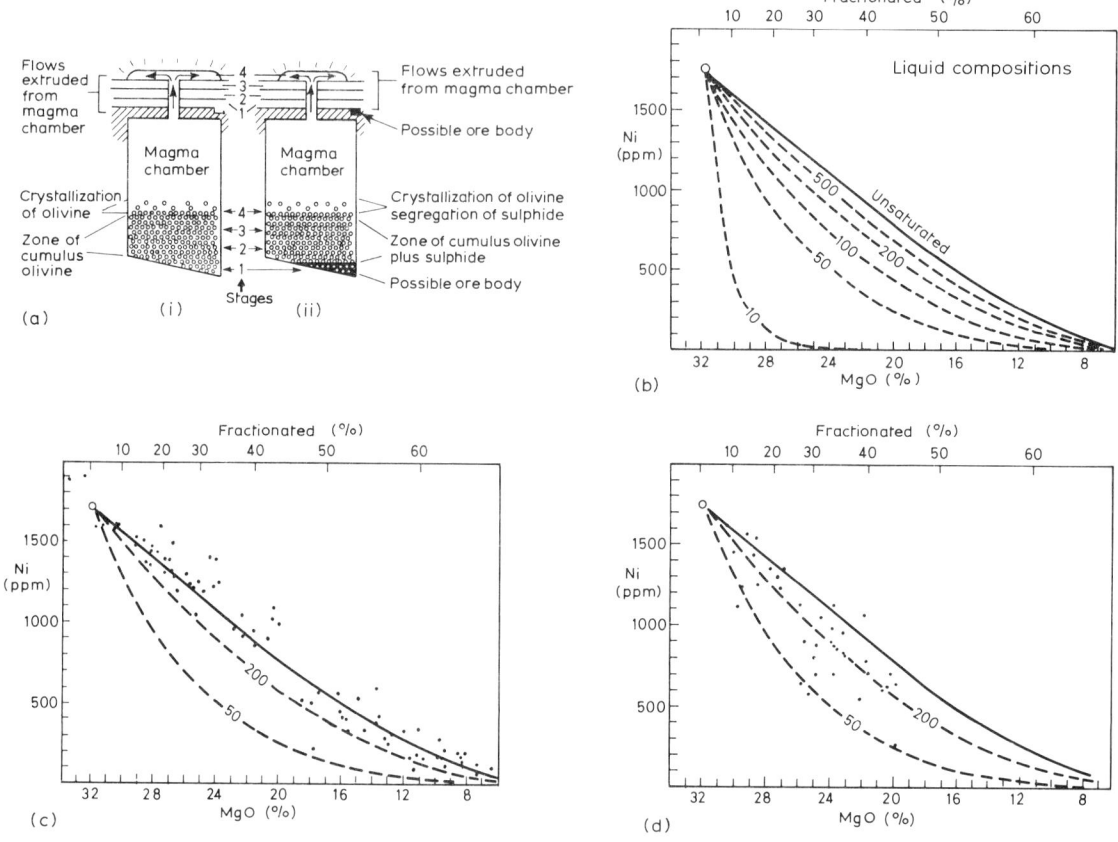

Fig. 2.24. Comparison of the results of computer modelling of the composition of komatiitic magmas compared with analyses from barren and mineralized ultramafics. (From Naldrett *et al.* 1984, who quote data sources).

(a) Fractional crystallization of olivine and fractional segregation of sulphide from a cooling komatiitic magma. (i) Sulphide-unsaturated case; (ii) sulphide-saturated case. Note that the magma is assumed to erupt periodically to give rise to a series of komatiitic flows of successively lower MgO content.

(b) Computer simulation of the fractionation of a komatiitic magma containing 32% by wt MgO as a consequence of the separation of olivine (unsaturated) or a mixture of different proportions of olivine plus sulphide. The numbers indicate the weight ratio of olivine/sulphide.

(c) Composition of 'barren' spinifex-textured peridotites (STP) and komatiitic basalts (B) from Yakabindie, Western Australia, Belingwe, Zimbabwe, the Abitibi belt, Canada and Barberton, S. Africa. The data are in reasonable agreement with the sulphide-unsaturated trend from (b).

(d) Compositions of spinifex-textured peridotites (STP) from Kambalda, Western Australia are plotted together with the model trends from (b). The rocks are significantly depleted in Ni in comparison with the sulphide-unsaturated trend, and fractionation under sulphide-saturated conditions with an average olivine/sulphide ratio of 200:1 is suggested.

In Western Australia soil geochemistry has been successfully used in the discovery of nickel sulphide deposits. Initial geochemical soil surveys entailed the analysis of samples for Ni, Cu, Co, Cr and Zn. However, high Zn contents usually indicate a metasedimentary influence, high Co can reflect Mn scavenging, and high Ni and Cr can be secondarily enriched by lateritization of barren ultramafics. As a consequence copper is the key element used in the search for nickel sulphides in soil surveys in Western Australia (Leggo and McKay, 1980).

The discovery of many of the nickel sulphide deposits in Western Australia can be attributed to the investigation of gossans. Initially many of the gossans could be interpreted as being of economic significance from their high nickel content. For example, the gossan overlying the Kambalda deposit contained up to 1% Ni. However, in many cases extensive weathering has depleted the base metal content of the gossans, so that they are easily confused with gossans developed on barren sulphides, or even the ferruginous products of weathering. Travis *et al.* (1976) have demonstrated that in Western Australia the palladium and iridium content of gossans is an excellent guide to their ore-bearing potential. Figure 2.25 shows that the palladium and iridium contents can even provide a semiquantitative guide to the original nickel sulphide content of the weathered rock material. An additional approach to the identification of gossans developed above nickel sulphides is the use of multi-element geochemistry. Moeskops (1977) reported that Cu, Ni, Cr, Mn, Zn, Pb, Pt, Pd and Se are useful elements in the characterization of gossans developed on nickel sulphide ores.

(c) Geophysics

Nickel sulphide ores are dense, conductive and commonly magnetic and therefore in principle a wide range of geophysical techniques is

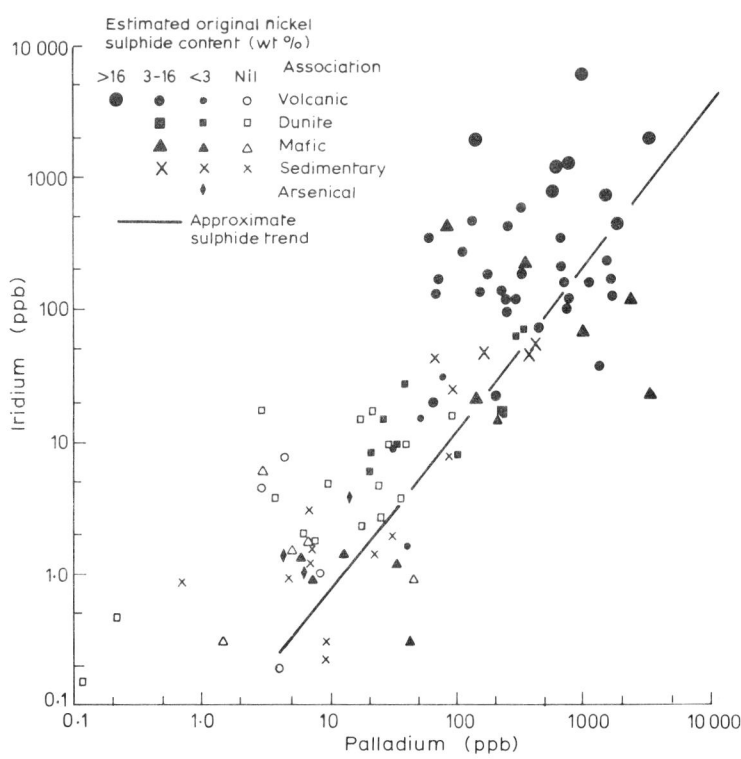

Fig. 2.25. Variation of Pd–Ir in the oxidized products of rocks containing nil to massive accumulations of nickel sulphide (from Travis *et al.* 1976).

appropriate for their exploration. However, in practice, there is considerable regional variation in the techniques which are preferred.

In Canada the extensive cover of glacial drift precludes direct geological observation, and geophysical techniques have served as the main approach to reconnaissance exploration. Canadian geophysicists prefer magnetic and electromagnetic methods because they are cheap, provide good physical contrast and satisfactory depth penetration and may be used in both airborne and rapid ground-based surveys (Dowsett, 1967). Becker (1979) has reviewed developments in the use of airborne electromagnetic methods in Canada, whilst ground-based electromagnetic methods are described by Ward (1979).

In Western Australia, geophysics is used both as an aid in reconnaissance mapping and, to a lesser extent, in the evaluation of gossan zones and geochemical anomalies. Detailed airborne magnetic surveys are used for reconnaissance mapping and may be used in conjunction with aerial photography to define the ultramafic zones which may host nickel sulphide deposits. The effectiveness of electromagnetic methods is severely limited by the deep zone of oxidation, the presence of a conductive overburden and the association of sulphide ores with graphitic metasediments. However, the use of a single-loop transient electromagnetic system has reduced the problem posed by conductive overburden (Pridmore *et al.*, 1984).

2.4 KIMBERLITES

2.4.1 Introduction

Kimberlites are the ultimate source of natural diamonds, although not all kimberlites are diamondiferous. About 50% of the World's natural diamonds are mined from kimberlites (Garlick, 1982). The remainder are obtained from placer deposits which have been derived from the erosion of kimberlites and these are described in Chapter 5.

The principal countries from which diamonds are produced from kimberlite are the USSR, South Africa, Botswana and Tanzania. The exciting discovery of the AK 1 kimberlite in the northern part of Western Australia has increased the World's reserves of natural diamonds by 40% and it is anticipated that production will commence in 1985 from this deposit (Garlick, 1982).

Natural diamonds are classified as being of gemstone quality, near-gem, industrial and boart. The classification is based upon the assessment of an expert but includes such factors as crystal perfection, colour, size and shape. Both the value and production of diamonds are reported in carats (5 carats = 1 g). The value varies dramatically from $US2 – 3 per carat for boart to $US65 000 per carat for a high-quality gemstone (Jones, 1983). The average value of diamonds as mined varies from about $US6.50 per carat for the Argyle AK 1 stones to $US150 per carat for South African alluvial diamonds (Jones, 1983).

Normally placer deposits contain a high proportion of gemstone-quality diamonds because the imperfect crystals are destroyed during transport. In contrast kimberlite, which normally occurs as pipe-shaped intrusions, contains a high proportion of boart and industrial stones.

Some 10–15% of the diamonds obtained from the Orapa pipe, Botswana, are of gemstone quality which is a favourably high proportion (Allen, 1982b), whereas the current assessment of the AK 1 pipe in Australia indicates that only 5% of the diamonds are of gemstone quality (Garlick, 1982).

2.4.2 General characteristics

(a) Distribution in space and time

Kimberlites have a very specific regional distribution and are confined to highly stable Precambrian cratons which are older than 2000 million years. The association between kimberlites and ancient cratons is an important guide to area selection in the early stages of exploration planning (Section 2.4.4). At the present time diamondiferous kimberlites are recognized in Africa (S. Africa, Botswana, Angola, Lesotho, Zaire, Swaziland, Tanzania, Sierra Leone), the USSR (Yakutia) and Australia (Western Australia). Kimberlite is also recorded from the USA and Canada, and the presence of alluvial diamonds in Guiana and Venezuela indicates kimberlitic sources in South America.

The age of kimberlites varies from Proterozoic to Tertiary. In southern Africa the most prolific period of kimberlite activity was during the Cretaceous period. The kimberlites are believed to have been intruded during a period of strong uplift of the continent, with attendant downwarping around the periphery (Dawson, 1970).

(b) Geological setting

The restriction of kimberlites to Precambrian Shield areas has already been noted. More specifically they are located on the margins of cratons or on the transition zone between large tectonic domes and basins within a large craton. Commonly the kimberlites occur in linear zones and may be associated with a dyke swarm of similar trend.

In this section we restrict our attention to the distribution of kimberlites within the African continent. Fig. 2.26 shows the distribution of major occurrences of kimberlite in Africa; they are confined to cratons which are older than 2500 million years. Figure 2.27 shows the distribution of diamondiferous and barren kimberlites in southern Africa. It can be seen that the diamondiferous kimberlites are clustered in the Kimberley district, and that this is the centre of three fundamental fracture trends: NW–SE; NE–SW and E–W (Dawson, 1970). Some of the pipes have been excavated sufficiently to show that they contract downwards into dykes or groups of dykes, the intersection of which is the cause of the location of the pipe (Hawthorne, 1975).

In Lesotho and adjoining districts of South Africa the kimberlites strike WNW and are parallel to the dominant direction of dolerites of Karoo (Mesozoic) age.

In Tanzania there is no alignment of kimberlites,

Fig. 2.26. Map of Africa showing the major cratons and distribution of kimberlites within the cratons (from Dawson 1980). 1, Liberia; 2, Sierra Leone; 3, Ivory Coast; 4, Mali; 5, Gabon; 6, Namibia; 7, South Africa; 8, Swaziland; 9, Botswana; 10, Zambia; 11, Angola; 12, Bakwanga area, Zaire; 13, Kundelungu Plateau, Zaire; 14, Southern Tanzania; 15, Tanzania Main Province. *Abbreviations*: AKC, Angola-Kasai Craton; KB, Kibaride Belt; TC, Tanzania Craton; IB, Irumide Belt; RTC, Rhodesia-Transvaal Craton; ORB, Orange River Belt.

but the kimberlite intrusions lie within a broad NNW–SSE trending zone which may reflect a fundamental fracture at depth. Individual kimberlites are related to lines of weakness, such as the contact between granite and roof pendants of metasediment (Dawson, 1970).

In the eastern part of Sierra Leone three kimberlite pipes are known, which are situated on and post-date dyke swarms which trend ENE (Dunbar, 1975).

(c) Petrography of kimberlite

Kimberlite is a complex and variable rock and there are at least five published definitions which

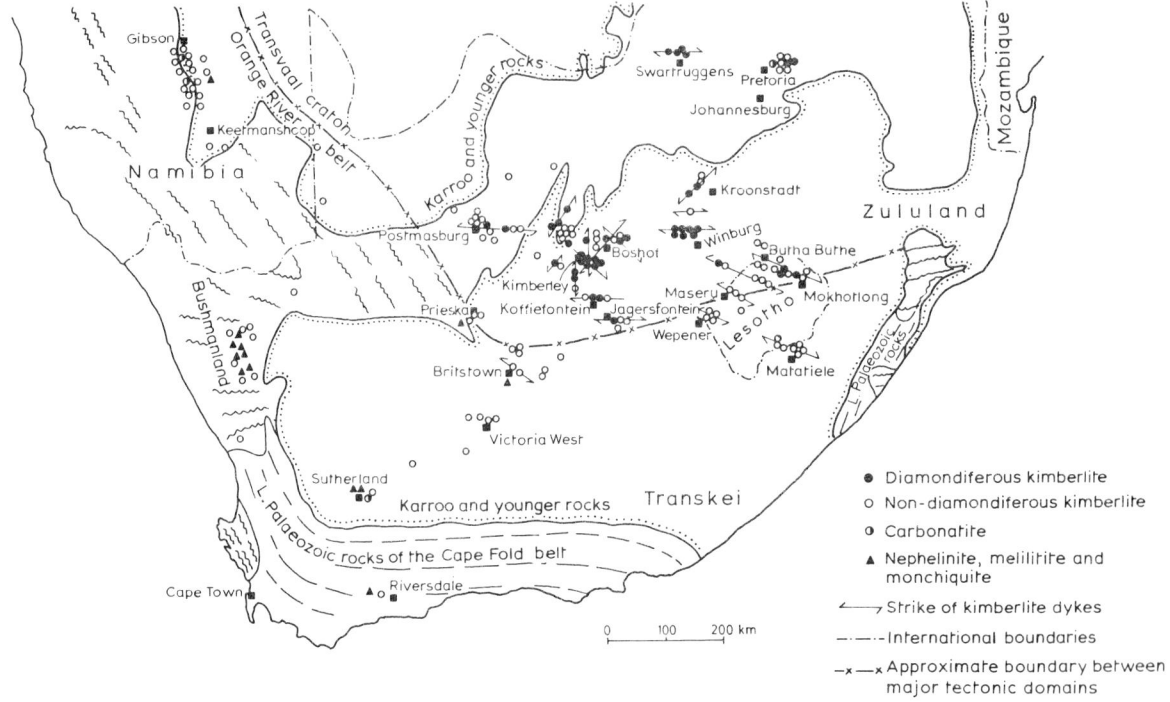

Fig. 2.27. Distribution of kimberlites and associated alkalic rocks in Namibia, South Africa and Lesotho (from Dawson 1970 and 1980).

attempt to describe its variable mineralogy. Dawson (1967) defined kimberlite as 'a serpentinised and carbonated mica peridotite of porphyritic texture, containing nodules of ultrabasic rock types characterised by such high pressure minerals as pyrope and jadeitic diopside; it may or may not contain diamond.' The mineralogy of kimberlite is explained more fully in Mitchell's (1970) definition: a porphyritic, alkalic peridotite containing rounded and corroded phenocrysts of olivine, phlogopite, magnesian ilmenite, pyrope and Cr-rich pyrope set in a fine-grained groundmass composed of second-generation olivine and phlogopite together with calcite, serpentine, magnetite, perovskite and apatite. Diamond and garnet–peridotite xenoliths may or may not occur.

Kimberlites may be classified on the basis of texture which in turn reflects the conditions of intrusion (Rathkin et al., 1962).

(i) Massive kimberlite

A massive porphyritic rock containing few inclusions of country rock – the result of comparatively gentle intrusion of kimberlite magma into dykes or cavities previously cleared of country rock by explosive surges of earlier magma. The kimberlite being exploited at present from the Kimberley district, South Africa, is dominantly of this type.

(ii) Intrusive kimberlite breccia.

This type consists of 20–60% fragments of various rock types set in a kimberlite matrix. The kimberlite exposed in the upper part of the Koidu pipe, Sierra Leone, is of this type.

(iii) Kimberlitic tuff

This type consists of 60–90% fragments of kimberlite and country rock cemented by hydrothermal minerals. Many of the pipes in the Daldyn–Alakit district of Yakutia are of this type (Davidson, 1967).

(d) Form, size and grade

Kimberlites occur as volcanic pipes, dykes, sills and 'blows' or enlargements along dykes. Kimberlite pipes are of prime economic importance although dyke-like bodies may be mined in the vicinity of pipes.

Hawthorne (1975) has described the anatomy of a kimberlite pipe based on detailed observations in the Kimberley area and other parts of Africa (Fig. 2.28). At the surface a kimberlite pipe which has been little eroded is often expressed as a large volcanic crater or mar. In some cases the crater is filled with lacustrine sediment which attains a thickness of 300 m at Mwadui, Tanzania (Nixon, 1980). Immediately below the crater the walls of the kimberlite pipe are flared outwards, but they quickly steepen to an angle of about 80°. In depth the pipe often divides into separate 'root-like' forms. The surface expression of kimberlites is therefore variable and is partly dependent upon erosional level. Figure 2.29 shows the variable size and shape of kimberlite pipes in plan. Pipes vary in their surface area from 0.4 hectare (Roberts Victor, S. Africa) to 146 hectares (Mwadui, Tanzania) (Bruton, 1978).

The grade of kimberlites is highly variable, and on a world-wide basis only about one in 100 pipes is economic. It is important to appreciate that a kimberlite is evaluated on the basis of both grade and the quality of its diamonds, and it is therefore possible for a low-grade kimberlite to be more valuable than a high-grade one. For example the Koidu pipe in Sierra Leone has a relatively low grade of 0.6 carat/tonne but has a high proportion of gemstone-quality diamonds (R. Jones, personal

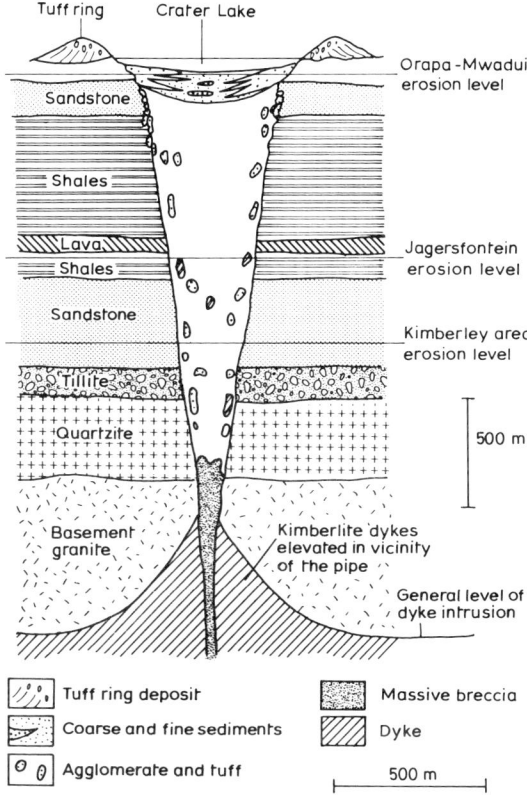

Fig. 2.28. Generalized model of a kimberlite diatreme and its subdiatreme dykes. (From Dawson 1980, based mainly on Hawthorne, 1975).

communication), whereas the AK 1 pipe in Western Australia has a high grade of 5 carats/tonne but a low proportion of gemstones (Garlick, 1982). The grade and quality of diamonds may also vary in depth within a kimberlite pipe, and both improvement (Koidu) and deterioration (Jagersfontein) with depth are recorded.

2.4.3 Genesis of kimberlites

An understanding of kimberlite genesis is important for geologists because it is the host rock for diamonds and also because it contains

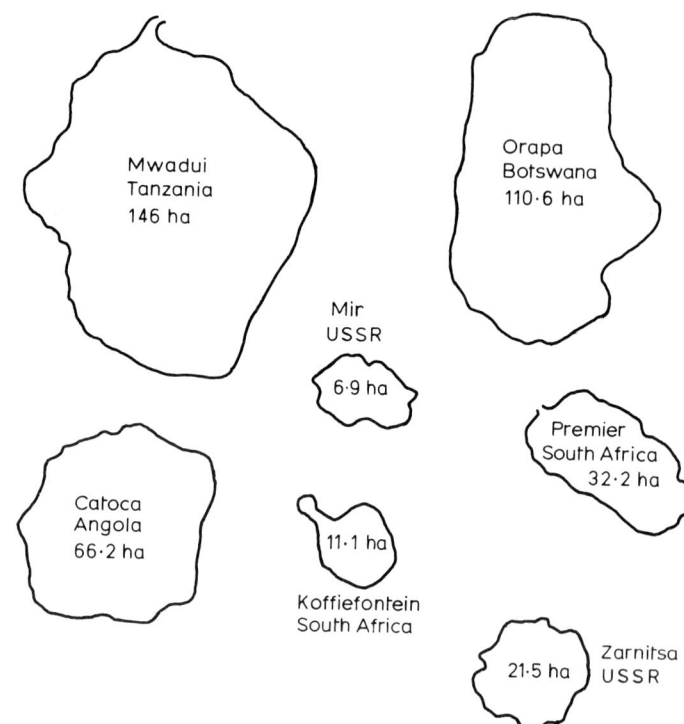

Fig. 2.29. Size and shape of some kimberlitic pipes (from Bruton 1978).

ultra-basic xenoliths, which provide valuable evidence about the nature of the upper mantle. As a result there is a voluminous literature on kimberlite petrogenesis but most of the key problems, including that of diamond distribution, remain unresolved.

Kimberlites combine an ultrabasic mineralogy, similar in many respects to that of peridotite, with a high concentration of incompatible elements which are usually found in association with highly differentiated magmas. Harris (1957) has proposed that the geochemical characteristics of kimberlite can be explained by a process of zone refining, in which a liquid generated at a depth of 600 km within the mantle moves upwards by a process of solution stoping. During ascent the major components remain in equilibrium but the incompatible elements undergo continuous enrichment in the liquid phase. The process is analogous to that used in metallurgical refining. Dawson (1980) argues that the kimberlite mineralogy and geochemistry can be better explained by the partial melting of a phlogopite-bearing garnet lherzolite magma, with primary titaniferous phlogopite providing the source for many of the incompatible elements.

What is the source of diamonds within kimberlite? Opinion is divided between those who consider that the diamond is derived by the fragmentation of diamondiferous ecologite and peridotite (xenocrystal) and those who think that it has precipitated from a kimberlite magma (phenocrystal). There are good arguments for both alternatives. Evidence in support of a xenocrystal theory is provided by Cretaceous kimberlites which contain diamonds of Precambrian age (Kramers, 1979). A phenocrystal origin is suggested by delicate growth structures in diamond and the presence of liquid inclusions of carbonated picritic composition (Dawson, 1980).

Magmatic deposits

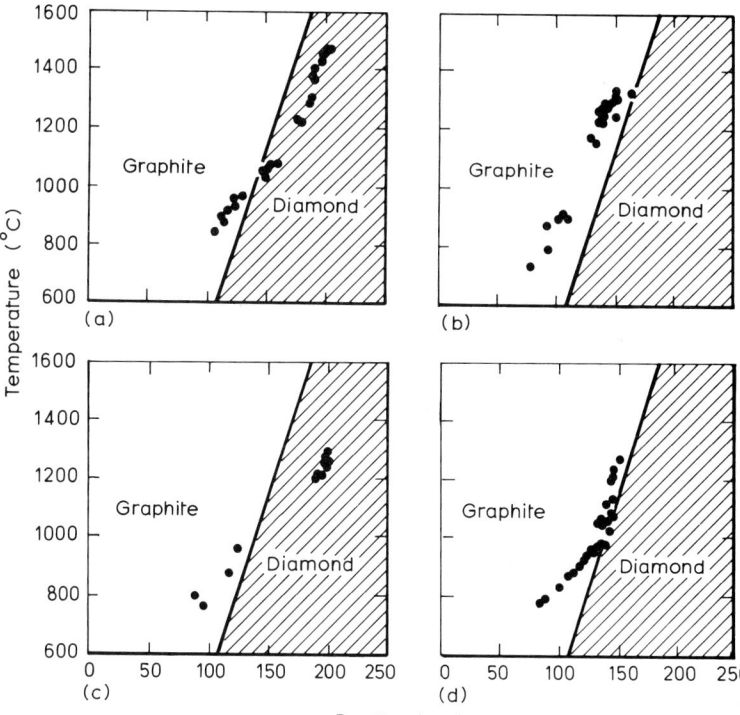

Fig. 2.30. Estimates of the temperatures and depths of origin of peridotite nodules included in kimberlites from four areas. Each point represents one nodule. (a) Northern Lesotho (diamondiferous); (b) East Griqualand, S. Africa (barren); (c) Udachnaya, Yakutia, USSR (diamondiferous); (d) Gibeon, S.W. Africa (barren). (From Boyd 1982).

The obvious conclusion to be drawn from this evidence is that diamonds may be of more than one origin.

The confinement of kimberlites to cratons is an aspect of both genetic and exploration significance. The key factor here is the low geothermal gradient which is associated with cratons. The result is that melting takes place at depth in these areas.

The factors which determine whether kimberlite is barren or diamondiferous are of genetic interest and are also very important in the evaluation of kimberlites. Kimberlites may originate at differing depths, and the distinction between diamond-bearing and barren kimberlite depends upon whether kimberlite has originated within the stability field of diamond (>150 km). Research by Boyd (1982) has shown that the degree of solution between particular silicates in ultrabasic nodules can provide valuable evidence about the depth of origin of kimberlites. The degree of solution between enstatite and diopside is temperature-dependent whereas the extent of solution of garnet in pyroxene depends upon pressure. Microprobe studies on the degree of mutual solution between these mineral pairs in ultrabasic nodules can provide pressure/temperature constraints within which the nodule (and by implication, the kimberlite magma) has originated. Figure 2.30 shows that kimberlites from diamond-producing areas contain nodules which have originated within the diamond-stability field. In contrast barren kimberlites have originated at a shallower depth, outside the diamond-stability field. This research has considerable potential in the evaluation of kimberlite pipes as the nodules can be obtained at surface or from shallow trenches.

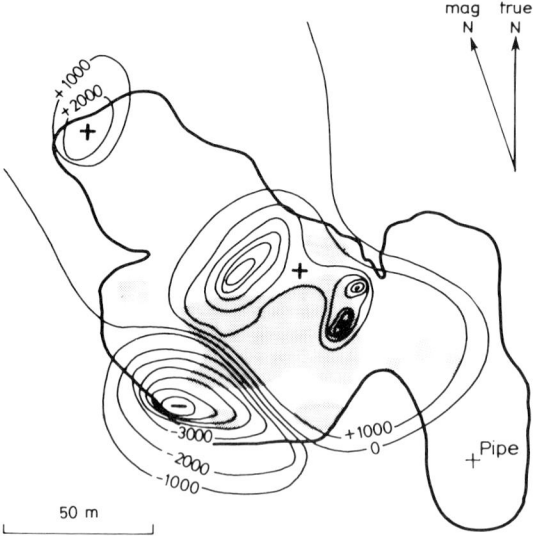

Fig. 2.31. Vertical magnetic field over Kolo kimberlite pipe intruded into sandstone using a Jalander fluxgate magnetometer (values in gammas). Stippled areas within the pipe are resistant hardebank kimberlite. (From Nixon 1980, after Burley and Greenwood 1972).

2.4.4 Exploration for diamondiferous kimberlites

The search for diamondiferous kimberlites involves the initial selection of areas appropriate for exploration (area selection) in which the main emphasis is placed on the appropriate geological setting for kimberlite intrusion. This is then followed by a programme of detailed exploration based upon the characteristics of the kimberlite itself.

(a) Area selection
Most mining companies utilize a combination of the following criteria in their choice of suitable areas for exploration.

(i) Age of craton
Cratons (in which the basement rocks exceed 1500 million years or preferably 2000 million years) are considered to be the most suitable areas for exploration.

(ii) Regional structure
The most favourable structures are the intersections of large regional faults or lineaments; interruptions in the regularity of the regional structure or evidence of vertical tectonism are also important.

(iii) Local structure
Kimberlite intrusions often occur on the downwarped crests of pitching anticlines, along large strike-slip faults, and on the intersections of fault zones.

(iv) Favourable petrological indications
The known occurrence of kimberlitic rocks is an obvious key to target selection, and was the basis of a United Nations Development Programme exploration programme in Lesotho in the early 1970s (Nixon, 1980). Other favourable petrological considerations include the presence of numerous dykes and sills of dolerite which provide evidence of tensional stress in the earth's crust. The occurrence of plateau/flood basalt with the later development of alkali-rich undersaturated extrusives and intrusives may also prove a useful guide (Garlick, 1982).

(b) Detailed exploration
Kimberlites are relatively small exploration targets, and their lack of strike continuity requires that exploration should be carried out on a more detailed scale than would be required for many types of base metal deposit. Most of the exploration techniques for kimberlites were pioneered in southern Africa and applied with spectacular success to the Yakutia district of Siberia and the Kimberley district of Western Australia. Table 2.6 lists the principal exploration techniques and summarizes their importance in the discovery of kimberlites.

Table 2.6 Summary of comments on methods used in kimberlite exploration (from Nixon, 1980)

Method	General applicability (xxxx, indispensable x, of limited use)	Remarks
Heavy mineral	xxxx	Wide variety of 'indicator' minerals, some of which are found in other rocks. Both stream sediment and interfluvial soils applicable. Wide dispersion by water and wind. Low values near some in-filled craters. Fairly labour-intensive method.
Geochemistry	x	Samples easily collected during heavy mineral survey. Nb characterizes kimberlite and a few other alkaline volcanics. Other elements, e.g. Cr, Ni, Sr, Ba, may be used if background is appropriate. Occasional observed change in proportion of some plant species on pipes may reflect the rock composition, e.g. high P.
Scintillometer	x	Some kimberlites slightly radioactive, some dykes especially rich in phlogopite (K). Overburden masks radiation.
Satellite imagery	xxx	Good for structural analysis; many kimberlites too small to be observed directly. Easy and cheap to obtain but sophisticated electronic processing can increase costs. Spectral bands unlikely to be precisely diagnostic but one particular climatic condition may 'fit' better than others.
Air photography, etc.	xxxx	Scale of 1:30 000 recommended. Basic tool for studying structural environment of kimberlite and identification of intrusions, particularly dykes. Fracture analysis may aid location of kimberlites. Colour photography may be of little additional assistance in poorly exposed areas.
Airborne multispectral scanning	xxx?	Requires rigorous preliminary field reflectance studies over known intrusions under specified climatic conditions to obtain a significant spectral signature.
Magnetic, ground and airborne methods	xxxx	Applicable in areas of low 'magnetic' background, e.g. limestones and other sedimentary rocks, granites, etc., preferably in areas with uniform overburden. Reduced anomalies over sediment-infilled kimberlite craters and weathered kimberlites. Useful for both dykes and pipes on the ground. Used successfully in the air, but topographic irregularities may give false anomalies.
Resistivity	xx	Kimberlites generally conductive compared with host country rocks and large bodies can be delimited on the ground. Airborne radiophase methods will theoretically detect such bodies.
Electromagnetic	xx	Ground measurements useful in defining extent of kimberlite and shape of dykes. Airborne input EM feasible.
Seismic	x	Of little use except in oceanographic surveys where reflection studies are capable of locating pipes.
Gravity	xx	Bouguer profile may be a characteristic negative superimposed on a large positive anomaly over weathered kimberlite pipes. Useful for detecting 'blind' kimberlites or extensions to existing pipes.

(c) Mineralogical exploration

Kimberlites are characterized by a distinctive suite of minerals of which the most significant are magnesian ilmenite (geikielite) and chrome diopside. Chromite, enstatite, olivine, Cr spinel and zircon may also be significant. During the weathering of kimberlite these minerals become concentrated in residual soil and may become dispersed in the drainage system. The identification of kimberlite indicator minerals was utilized by prospectors in the early development of the Kimberley district and has been successfully applied to many other areas.

The size of the dispersion pattern formed by kimberlite indicator minerals depends mainly upon climate and relief. According to Mannard (1968) in areas with a tropical or semi-arid climate the less-resistant minerals such as diopside are rarely found more than 1 mile (>1.6 km) from source. However, garnet and ilmenite may persist for a few miles. In those climatic regions where mechanical weathering is dominant the diopside may travel 30 miles (48 km), garnet 95–125 miles (144–200 km) and ilmenite much further.

In much of southern Africa exploration is based on systematic soil sampling on a grid system. In more mountainous terrain stream sediment sampling may be preferred. Nixon (1980) described a programme of stream sediment sampling in Lesotho which was based on a sample density of 1.1 km^{-2}. According to Jones (1983), exploration in the Kimberley district of Western Australia in the late 1970s involved collecting 30 kg samples every 40–50 km^2. Later follow-up work involved sample collection at a density of 1–5 samples km^{-2}. The discovery of the AK 1 pipe in the Kimberley district was based upon the identification of diamonds in the sediment rather than the conventional kimberlite indicator minerals.

(d) Geophysical exploration

Geophysics is commonly used as a follow-up technique where favourable mineralogical sites have been identified. In Australia airborne geophysics is used as the main reconnaissance method where the drainage network is inadequate for stream sediment sampling.

Airborne and ground magnetics are the principal methods used. Normally kimberlite contains from 5 to 10% iron oxides which are present as ilmenite and lesser concentrations of magnetite. Where the kimberlite is emplaced into rocks such as limestones or granites with a low magnetite content a distinctive magnetic anomaly may be detected (Fig. 2.31). Patterson and MacFayden (1977) found magnetic anomalies over all known kimberlites in Lesotho. They attributed the large variation in magnetic response to variations in host rock magnetization, depth of weathering and kimberlite mineralogy. In other parts of Africa there are kimberlites which have no magnetic response.

Kimberlites characteristically weather to a yellow montmorillonitic clay near the surface. This zone may be highly conductive in which case electromagnetic or resistivity surveys may be appropriate. Macnae (1979) described the results of an airborne electromagnetic (EM) survey in southern Africa in which all of the eight kimberlite pipes discovered had distinctive EM signatures.

The application of other geophysical techniques to kimberlite exploration is described by Macnae (1979) and Nixon (1980).

(e) Aerial photography

The size and surface expression of kimberlite pipes are partly related to erosional level. In most cases kimberlites are characterized by a shallow depression, although sometimes they may form a low mound. In either situation the drainage system is likely to have a curvilinear form compared with the rectilinear drainage pattern that is commonly associated with Precambrian basement terrain. The yellow ground, which results from the chemical weathering of kimberlite, consists mainly of a montmorillonitic clay which will tend to support a poorly drained soil with a distinctive trace element content. A characteristic vegetation is sometimes developed on kimberlite due to these specific soil conditions.

Contrasts in drainage pattern and vegetation are

commonly utilized by geologists as the basis for photointerpretation, and photogeology often plays a significant role in the search for kimberlites. According to Barygin (1961) the most effective approach is to use black-and-white photography at scales of 1:10 000 to 1:50 000 supplemented by colour photography at a larger scale. Nixon (1980) utilized photointerpretation in the search for kimberlites in Lesotho where both black-and-white (scale 1:40 000) and colour (1:30 000) photography was used. Nixon reported that the larger scale was more useful than colour.

2.5 CONCLUDING STATEMENT

There are few other classes of ore deposit which manifest such a wide range of form and grade as those of magmatic affiliation. In many cases the only common factor is their intimate association with mafic and ultramafic lithologies. For example, the massive habit, high tenor of metal and lateral continuity of stratiform chromite deposits contrast very markedly with the irregular distribution and very low concentration of diamonds which are dispersed in kimberlites. However, both were originally derived by the partial melting of mantle material and both share a similar tectonic setting.

Now we come to the key issue. How does our understanding of ores formed by magmatic processes aid in the search for new deposits? In particular, we wish to review the relative importance of geological setting, genetic models and indirect techniques, such as geophysics, in the exploration for this class of deposit. Our discussions will focus on podiform chromite, nickel sulphide deposits of the dunite–peridotite type and kimberlites.

2.5.1 Podiform chromite

Chromite was originally mined from high-grade podiform deposits in Norway in the early part of the 19th century and there is therefore a long tradition of mining and prospecting for this style of deposit in Europe. However, our understanding of the ultrabasic lithologies which host podiform chromite deposits has undergone fundamental changes in recent years with the recognition of oceanic crust as the environment in which podiform chromite has originally concentrated. We therefore have an excellent opportunity to assess the impact of new geological ideas on ore discovery.

Let us briefly recall the most important advances in our understanding of Alpine-type peridotites. The most significant step forward has stemmed from the realization that the ultrabasic sequence with which podiform chromite is associated is comparable with present-day oceanic crust. The recognition of ophiolites as preserved fragments of oceanic crust has resulted in a greatly increased intensity of field work and petrological study, which has aided our understanding of those ultramafics which contain concentrations of chromite. Of particular importance is the recognition that most of the chromite is contained within harzburgites (Jackson and Thayer, 1972). This has allowed exploration geologists to focus their attention on particular parts of an ophiolite complex. Further important guides to exploration have stemmed from detailed structural mapping programmes such as those reported by Cassard et al. (1981).

Modern genetic models of podiform chromite must be closely related to our knowledge of processes which take place at constructive plate margins. The mafic dykes and pillow lavas which characterize layer 2 of the oceanic crust are considered to be derived from a magma chamber at the base of layer 3. Two main genetic models have been proposed. One envisages the chromite forming as a cumulate phase from the magma chamber which continues to sink down into the harzburgite unit (Dickey, 1975). The alternative model identifies dykes which transgress the harzburgite, and within which magma movement is dominantly upwards, as the environment in which chromite is concentrated by elutriation (Lago et al., 1982). It appears that neither genetic model fundamentally affects the approach a geologist would make in selecting areas for future exploration.

There is little published information on the use of indirect geophysical and geochemical methods in the search for podiform chromite. This is an area where more research and the publication of case histories would be of benefit.

2.5.2 Nickel sulphide ores of the dunite–peridotite class

The economic potential of this class of deposit was not fully appreciated until the discovery of the Kambalda deposit in Western Australia in 1966. Both Kambalda and several subsequent discoveries were found by prospectors who utilized nickel-bearing gossan as their guide to ore. Detailed drilling followed by careful geological studies have permitted geologists to characterize the geological setting of the deposits and several genetic models have been proposed.

Nickel sulphide deposits of the dunite–peridotite class occur in Archaean greenstone belts associated with ultramafic volcanics of komatiitic type and intrusive dunites which are probably their intrusive equivalents. Ores belonging to the volcanic–peridotite class tend to be concentrated near to the base of the ultramafic sequence. Exploration for this class of nickel sulphide is concerned in the first instance with obtaining a geological base map within which the ultramafic lithologies are well defined. The geological mapping may be aided by photogeology, airborne magnetics and geochemistry and so here we see conventional geology closely related to indirect methods of exploration. The knowledge that the massive sulphide zones occur near to the base of ultramafic units enables geologists to concentrate the more detailed aspects of follow-up work, such as soil sampling, in more restricted parts of the greenstone sequence.

Most genetic models for ores of this class involve the segregation of an immiscible sulphide phase from the crystallizing silicate magma. Dissension has mainly centred on the timing and mechanism of the separation of the sulphides. This aspect of genetic modelling has played little part in the choice of new exploration targets. However, the ability to characterize a mafic or ultramafic rock which has crystallized from a sulphur-saturated magma is an important exploration tool because nickel may have partitioned into the sulphide phase. Naldrett *et al.* (1984) have modelled the partitioning of nickel between silicate and sulphide phases and comparison of their results with the composition of barren and mineralized ultramafics confirms that this approach can be used in a predictive sense.

2.5.3 Kimberlites

The association between kimberlites and diamonds was first established by prospectors in South Africa. By examining the panned concentrate from stream sediments they were able to trace alluvial diamonds to their source. Surprisingly little has changed. In southern Africa the emphasis is now placed on the search for an assemblage of indicator minerals and the success of this technique has been repeated in other African countries and in the USSR. Ironically the discovery of the AK 1 pipe in Australia came from the discovery of small alluvial diamonds rather than from the more usual indicator minerals. The point which we wish to emphasize is that the mineralogical expression of kimberlites has always been the most effective exploration technique. Additional indirect exploration techniques have been successfully developed for kimberlites and are described in Section 2.4.4.

Ultrabasic and basic xenoliths in kimberlites provide a unique source of information about the composition of the mantle and lower crust, and scientific interest in this subject has resulted in a massive bibliography. As a result genetic models for kimberlites are based on a very considerable accumulation of geochemical and petrological data. Until recently this research appeared to be of only academic interest. However, it is now becoming evident that genetic models based upon detailed studies of kimberlite petrology are able to assist the geologist in his evaluation of kimberlite pipes (Boyd, 1982).

REFERENCES

Allen, H. E. K. (1982a) Development of Orapa and Letlhakane diamond mines, Botswana. *Trans. Instn Min. Metall. (Section A: Min. Indust.)*, **91**, A208–210.

Allen, H. E. K. (1982b) Some diamond mining and recovery methods. *Trans. Instn Min. Metall. (Section A: Min. Indust.)* (discussion p. 140).

Barrett, F. M., Binns, R. A., Groves, D. I., Marston, R. J. and McQueen, K. G. (1977) Structural history and metamorphic modification of Archaean volcanic-type nickel deposits, Yilgarn block, Western Australia. *Econ. Geol.*, **72**, 1195–1223.

Barygin, V. M. (1961) Prospecting for kimberlites from the air. *Min. Mag.*, **107 (2)**, 73–78 (transl. from the Russian by N. W. Wilson).

Becker, A. (1979) Airborne electromagnetic methods. In *Geophysics and Geochemistry in the Search for Metallic Ores, Geological Survey of Canada Economic Geology Report 31* (ed. P. J. Hood), pp. 33–43.

Billington, L. G. (1984) Geological review of the Agnew nickel deposit, Western Australia. In *Sulphide Deposits in Mafic and Ultramafic Rocks*, Institution of Mining and Metallurgy, London (eds D. L. Buchanan and M. J. Jones), pp. 43–54.

Bolviken, B. and Gleeson, C. F. (1979) Focus on the use of soils for geochemical exploration in glaciated terrain. In *Geophysics and Geochemistry in the Search for Metallic Ores, Geological Survey of Canada Economic Geology Report 31* (ed. P. J. Hood), pp. 295–326.

Boyd, F. R. (1982) Predicting the occurrence of diamondiferous kimberlites. *Ind. Diamond Q.*, **33**, 31–34.

Bradshaw, P. M. D. and Thomson, I. (1979) The application of soil sampling to geochemical exploration in non-glaciated regions of the World. In *Geophysics and Geochemistry in the Search for Metallic Ores, Geological Survey of Canada Economic Geology Report 31* (ed. P. J. Hood), pp. 327–338.

Bruton, E. (1978) *Diamonds*. N.A.G. Press, London, 532 pp.

Buchanan, D. L. (1979) Chromite production from the Bushveld Complex. *World Min.*, **32(10)**, 97–101.

Buchanan, D. L. (1982) *Nickel: a Commodity Review*. Occasional papers of the Institution of Mining and Metallurgy Paper 1. 28 pp.

Buchanan, D. L. and Rouse, J. E. (1984) Role of contamination in the precipitation of sulphides in the Platreef of the Bushveld Complex. In *Sulphide Deposits in Mafic and Ultramafic Rocks* (eds D. L. Buchanan and M. J. Jones), Institution of Mining and Metallurgy, London, pp. 141–146.

Burley, A. J. and Greenwood, P. G. (1972) *Geophysical Surveys over Kimberlite Pipes in Lesotho*. Geophys. Div. Rep. Inst. Geol. Sci., London.

Butt, C. R. M. and Nickel, E. H. (1981) Mineralogy and geochemistry of the weathering of the disseminated nickel sulphide deposit at Mt. Keith, Western Australia. *Econ. Geol.*, **76**, 1736–1751.

Cameron, E. N. (1977) Chromite in the central sector of the Eastern Bushveld Complex, S. Africa. *Am. Mineral.*, **62**, 1082–1096.

Cameron, E. N. (1980) Evolution of the Lower Critical Zone, Central Sector, Eastern Bushveld Complex and its chromite deposits. *Econ. Geol.*, **75**, 845–871.

Cameron, E. N. and Desborough, G. A. (1969) Occurrence and characteristics of chromite deposits – Eastern Bushveld Complex. In *Magmatic Ore Deposits, a Symposium* (ed. H. D. B. Wilson), *Econ. Geol. Monogr.*, **4**, 23–40.

Cameron, E. N. and Emerson, M. E. (1959) The origin of certain chromite deposits in the eastern part of the Bushveld Complex. *Econ. Geol.*, **54**, 1151–1213.

Cameron, E. N., Siddeley, G. and Durham, C. C. (1971) Distribution of ore elements in rocks for evaluating ore potential: nickel, copper, cobalt and sulphur in ultramafic rocks of the Canadian Shield. *CIM Special Volume* No. **11**, 298–313.

Campbell, I. H. (1978) Some problems with cumulus theory. *Lithos*, **11**, 311–323.

Campbell, I. H., Naldrett, A. J. and Barnes, S. J. (1983) A model for the origin of the platinum rich sulfide horizons in the Bushveld and Stillwater Complexes. *J. Petrol.*, **24(2)**, 133–165.

Cassard, D., Rabinovitch, M., Nicholas, A., Moutte, J., Leblanc, M. and Prinzhofer, A. (1981) Structural classification of chromite pods in southern New Caledonia. *Econ. Geol.*, **76(4)**, 805–831.

Condie, K. C. (1981) Archaean greenstone belts. In *Developments in Precambrian Geology*, Vol. 3, Elsevier, Amsterdam, 434 pp.

Cousins, C. A. (1964) The platinum deposits of the Merensky Reef. In *The Geology of some Ore Deposits in Southern Africa*, (ed. S. H. Haughton), Geological Society of South Africa, Johannesburg, pp. 225–238.

Cousins, C. A. (1969) The Merensky Reef of the Bushveld Igneous Complex. In *Magmatic Ore Deposits, a Symposium* (ed. H. D. B. Wilson). *Econ. Geol. Monogr.*, **4**, 239–251.

Cousins, C. A. and Feringa, G. (1964) The chromite

deposits of the western belt of the Bushveld Complex. In *The Geology of some Ore Deposits in Southern Africa* (ed. S. H. Haughton), Geological Society of South Africa, pp. 183–202.

Davidson, C. F. (1967) The kimberlites of the U.S.S.R. In *Ultramafic and Related Rocks* (ed. P. J. Wyllie), John Wiley & Sons, London, pp. 251–261.

Dawson, J. B. (1967) A review of the geology of kimberlite. In *Ultramafic and Related Rocks* (ed. P. J. Wyllie), John Wiley & Sons, London, pp. 241–251.

Dawson, J. B. (1970) The structural setting of African kimberlite magmatism. In *African Magmatism and Tectonics* (eds T. N. Clifford and I. G. Gass), Oliver & Boyd, Edinburgh, pp. 321–336.

Dawson, J. B. (1980) Kimberlites and their Xenoliths. Springer-Verlag, Berlin, 250 pp.

De Young, J. H., Lee, M. P. and Lipin, B. R. (1984) *International Strategic Minerals Inventory Summary Report*, Chromium U.S. Geological Survey Circular 930–B, 41 pp.

Dickey, J. S. (1975) A hypothesis of the origin of podiform chromite deposits. *Geochim. Cosmochim. Acta*, **39**, 1061–1074.

Dowsett, J. S. (1967) Geophysical exploration methods for nickel. Mining and groundwater geophysics. *Canadian Geological Survey Economic Geology Report* No. 26, part 2, pp. 310–321.

Duke, J. M. (1983) Ore deposit models 7, Magmatic segregation deposits of chromite. *Geosci. Can.*, **10(1)**, 15–24.

Dunbar, G. L. (1975) *Economic Geology and Exploration Potential of the Sierra Leone Diamond Fields*, Unpublished M.Sc. dissertation, Mining Geology Division, Imperial College, London.

Erganulp, F. (1980) Chromite mining and processing at Kavak mine, Turkey. *Trans. Instn Min. Metall. (Sect. A: Min. Indust.)*, **89**, A179–184.

Evans, A. M. (1980) *An Introduction to Ore Geology*, Blackwell, Oxford, 231 pp.

Ferguson, J. and Botha, E. (1963) Some aspects of igneous layering in the basic zone of the Bushveld Complex. *Trans. Geol. Soc. S. Africa*, **66**, 259–278.

Flint, D. E., De Albear, J. F. and Guild, P. W. (1948) Geology and chromite deposits of the Camaguey district, Camaguey Province, Cuba. *U.S. Geol. Surv. Bull.*, **954–B**, 39–63.

Garlick, H. J. (1982) *Current Status of the Diamond Exploration Effort in Australia*, Unpublished report prepared for Mackay & Schnellmann Pty Ltd.

Gibjels, R. H., Millard, H. T., Jr, Desborough, G. A. and Bartel, A. J. (1974) Osmium, ruthenium, iridium and uranium in silicates and chromite from the eastern Bushveld Complex, South Africa. *Geochim. Cosmochim. Acta*, **38**, 319–337.

Greenbaum, D. (1977) The chromitiferous rocks of the Troodos ophiolite complex, Cyprus. *Econ. Geol.*, **72**, 1175–1194.

Gresham, J. J. and Loftus-Hills, G. D. (1981) The geology of the Kambalda nickel field, Western Australia. *Econ. Geol.*, **76**, 1373–1416.

Groves, D. I. and Hudson, D. R. (1981) The nature and origin of Archaean strata-bound volcanic associated nickel–iron–copper sulphide deposits. In *Handbook of Strata-Bound and Stratiform Ore Deposits* (ed. K. H. Wolf), Elsevier, Amsterdam, Vol. 9, pp. 305–403.

Groves, D. I., Lesher, C. M. and Gee, R. D. (1984) Tectonic setting of the sulphide nickel deposits of the Western Australian Shield. In *Sulphide Deposits in Mafic and Ultramafic Rocks* (eds D. L. Buchanan and M. J. Jones), Institution of Mining and Metallurgy, London, pp. 1–13.

Guild, P. W. (1947) Petrology and structure of the Moa district, Oriente province, Cuba. *Am. Geophys. Union Trans.*, **28**, 218–246.

Hamilton, J. (1977) Sr isotope and trace element studies of the Great Dyke and Bushveld mafic phase and their relation to early Proterozoic magma genesis in Southern Africa. *J. Petrol.*, **18**, 24–52.

Harris, P. G. (1957) Zone refining and the origin of potassic basalts. *Geochem. Cosmochim. Acta*, **12**, 195–208.

Hawthorne, J. B. (1975) Model of a kimberlite pipe. *Phys. Chem. Earth*, **9**, 51–59.

Hodge, B. L. and Oldham, L. (1983) Mineral exploration. *Min. Annu. Rev.*, 125–149.

Hopkins, D. A. S. (1980) Chromium; Discussion and contributions to extended general meeting of the Institution of Mining and Metallurgy. *Trans. Instn Min. Metall. (Sect. A. Min. Indust.)*, **89**, A99.

Hughes, C. J. (1982) Igneous petrology. *Developments in Petrology*, Elsevier, Amsterdam, Vol. 7, 551 pp.

Irvine, T. N., Keith, D. W. and Todd, S. G. (1983) The J-M Platinum-Palladium Reef of the Stillwater Complex, Montana: II Origin by double-diffusive convective magma mixing and implications for the Bushveld Complex. *Econ. Geol.*, **78**, 1287–1334.

Irvine, T. N. and Smith, C. H. (1969) Primary oxide minerals in the Layered Series of the Muskox intrusion. In *Magmatic Ore Deposits, a Symposium* (ed. H. D. B. Wilson), *Econ. Geol. Monogr.*, **4**, 76–94.

Jackson, E. D. (1961) Primary textures and mineral associations in the ultramafic zone of the Stillwater Complex, Montana. *U.S. Geol. Survey Prof. Paper*, **358**, 106 pp.

Jackson, E. D. and Thayer, T. P. (1972) Some criteria for distinguishing between stratiform, concentric and Alpine Peridotite-Gabbro Complexes. *Int. Geol. Congr. 24th Session, Sect. 2*, 289–296.

Jensen, L. S. (1976) A new cation plot for classifying subalkalic volcanic rocks. *Ont. Div. Mines MP*, **66**, 22 pp.

Jones, D. (1983) *A Review of Diamond Exploration in the Kimberley Region of Western Australia*. Unpublished M.Sc. dissertation, James Cook University, North Queensland, 40 pp.

Kramers, J. D. (1979) Lead, uranium, strontium, potassium, and rubidium in inclusion-bearing diamonds and mantle-derived xenoliths from southern Africa. *Earth Planet. Sci. Lett.*, **42**, 58–70.

Lago, B. L., Rabinowicz, M. and Nicolas, A. (1982) Podiform chromite ore bodies: a genetic model. *J. Petrol.*, **23(1)**, 103–125.

Leggo, M. D. and McKay, K. G. (1980) Forrestania Ni deposits, Yilgarn block, Western Australia. In *Conceptual models in exploration geochemistry Australia* (eds C. R. M. Butt and R. E. Smith), *J. Geochem. Explor.*, **12(2/3)**, 178–183.

Lusk, J. (1976) A possible volcanic-exhalative origin for lenticular nickel sulphide deposits of volcanic association with special reference to those in Western Australia. *Can. J. Earth Sci.*, **13**, 451–469.

Lynd, L. E. and Lefond, S. J. (1975) Titanium minerals. In *Industrial Minerals and Rocks* (ed. S. J. Lefond), American Institute of Mining, Metallurgical and Petroleum Engineers, New York, pp. 1149–1208.

Macnae, J. C. (1979) Kimberlites and exploration geophysics. *Geophysics*, **44(8)**, 1395–1416.

Mannard, G. W. (1968) The surface expression of kimberlite pipes. *Geol. Assoc. Can. Proc*, **19**, 15–21.

Marston, R. D., Groves, D. I., Hudson, D. R. and Ross, J. R. (1981) Nickel sulphide deposits in Western Australia: a review. *Econ. Geol.*, **76**, 1330–1363.

Mitchell, A. H. G. and Garson, M. S. (1981) *Mineral Deposits and Global Tectonic Settings*, Academic Press, London, 405 pp.

Mitchell, R. H. (1970) Kimberlite and related rocks – a critical re-appraisal. *J. Geol.*, **78**, 686–704.

Moeskops, P. G. (1977) Yilgarn nickel gossan geochemistry – a review with new data. *J. Geochem. Explor.*, **8**, 247–258.

Naldrett, A. J. (1973) Nickel sulphide deposits – their classification and genesis, with special emphasis on deposits of volcanic association. *CIM Bull.*, **66(739)**, 45–63.

Naldrett, A. J. (1981) Nickel sulfide deposits: classification, composition and genesis. *Econ. Geol.*, *75th Anniv. Vol.*, 628–685.

Naldrett, A. J. and Arndt, N. T. (1976) Volcanogenic nickel deposits with some guides for exploration. *Trans. Soc. Min. Eng. AIME*, **260(1)**, 13–15.

Naldrett, A. J., Duke, J. M., Lightfoot, P. C. and Thompson, J. F. H. (1984) Quantitative modelling of the segregation of magmatic sulphides: an exploration guide. *CIM Bull.*, **77(864)**, 46–56.

Naldrett, A. J. and Gasparini, E. L. (1971) Archaean nickel sulphide deposits in Canada: their classification, geological setting and genesis with some suggestions as to exploration. In *Symposium on Archaean rocks* (ed. J. E. Glover), *Geol. Soc. Austr. Spec. Publ.*, **3**, 201–226.

Naldrett, A. J. and Watkinson, D. H. (1981) Ore formation within magmas. In *Mineral Resources: Genetic Understanding for Practical Applications*, National Academy Press, Washington D.C., pp. 47–61.

Nixon, P. H. (1980) Regional diamond exploration – theory and practice. In *Kimberlites and Diamonds* (eds J. E. Glover and D. I. Groves), Publication No. 5, Geology Dept and Extension Service, University of Western Australia, 80 pp.

Page, N. J. (1977) Stillwater Complex, Montana: rock succession, metamorphism, and structure of the Complex and adjacent rocks. *U.S. Geol. Surv. Prof. Paper*, **999**, 79 pp.

Page, N. J., Engin, T., Singer, D. A. and Haffty, J. (1984) Distribution of platinum-group elements in the Bati Reef Chromite Deposit, Guleman-Elazig area, Eastern Turkey. *Econ. Geol.*, **79**, 177–184.

Panayiotou, A. (1978) The mineralogy and chemistry of the podiform chromite deposits in the serpentinites of the Limassol Forest, Cyprus. *Mineral. Deposita*, **13**, 259–274.

Patterson, K. N. R. and MacFayden, D. A. (1977) *Geophysical Exploration for Kimberlites with Special Reference to Lesotho*. Paper presented at the Canadian Exploration Geophysical Society meeting, Toronto.

Pavlov, N. V. and Grigor'eva, I. I. (1977) Deposits of chromium. In *Ore Deposits of the U.S.S.R.* (ed. V. I. Smirnov), Pitman, London, pp. 179–236.

Peredery, W. V. and Morrison, G. G. (1984) Discussion of the origin of the Sudbury structure. In *Geology and Ore Deposits of the Sudbury Structure*, Ontario

Geological Survey Special Publication.
Peters, T. and Kramers, J. D. (1974) Chromite deposits in the ophiolite complex of northern Oman. *Mineral. Deposita*, **9**, 253–259.
Porter, D. J. and McKay, K. G. (1981) The nickel sulphide mineralisation and metamorphic setting of the Forrestania Area, Western Australia. *Econ. Geol.*, **76**, 1524–1549.
Potter, M. (1984) Success rates for nickel exploration in Zimbabwe, *Trans. Instn Min. Metall. (Sect. B: Appl. Earth Sci.)*, **93**, B31–34.
Pridmore, D. F., Coggon, J. H., Esdale, D. J. and Lindeman, F. W. (1984) Geophysical exploration for nickel sulphide deposits in the Yilgarn Block, Western Australia. In *Sulphide Deposits in Mafic and Ultramafic Rocks* (eds D. L. Buchanan and M. J. Jones), Institution of Mining and Metallurgy, London, pp. 22–34.
Raines, G. L. and Wynn, J. C. (1982) Mapping of ultramafic rocks in a heavily vegetated terrain using Landsat data. *Econ. Geol.*, **77**, 1755–1760.
Rathkin, M. I., Krutoyarskii, M. A. and Milashev, V. A. (1962) Classification and nomenclature of Yakutian kimberlites. *Tr. Issled. Inst. Geol. Arktiki*, **121**, 154–164 (Russian).
Ross, J. R. and Hopkins, G. M. (1975) Kambalda nickel sulphide deposits. In *Economic Geology of Australia and Papua New Guinea. I. Metals* (ed. C. L. Knight). Australasian Institute of Mining Metallurgy, Melbourne, 100–121.
Ross, J. R. and Travis, G. A. (1981) The nickel sulfide deposits of Western Australia in global perspective. *Econ. Geol.*, **76**, 1291–1329.
Rugman, G. M. (1982) Perseverance Mine – a prospecting case history. *Min. Mag.*, **147**, 381–391.
Sharpe, M. R. (1981) The chronology of magma influxes to the eastern compartment of the Bushveld Complex as exemplified by its marginal border groups. *J. Geol. Soc. London*, **138**, 307–326.
Snethlage, R. and von Gruenewaldt, G. (1977) Oxygen fugacity and its bearing on the origin of chromitite layers in the Bushveld Complex. In *Time and Strata-Bound Ore Deposits* (eds D. D. Klemm and H. J. Schneider), Springer Verlag, Berlin, pp. 352–372.
Streicher, P. E. (1980) Chromium, discussions and contributions to extended general meeting of the Institution of Mining and Metallurgy. *Trans. Instn Min. Metall. (Sect. A: Min. Indust.)*, **89**, A102.
Thayer, T. P. (1969) Gravity differentiation and magmatic re-emplacement of podiform chromite deposits. In *Magmatic Ore Deposits* (ed. H. D. B. Wilson). *Econ. Geol. Monogr.*, **4**, 132–146.
Thomson, A. G. (1983) Chromite. *Min. Annu. Rev.*, 69–70.
Todd, S. G., Keith, D. W., LeRoy, L. W., Schissel, D. J., Mann, E. L. and Irvine, T. N. (1982) The J–M Platinum-Palladium reef of the Stillwater Complex, Montana, 1, Stratigraphy and Petrology. *Econ. Geol.*, **77**, 1454–1480.
Travis, G. A., Keays, R. R. and Davison, R. M. (1976) Palladium and iridium in the evaluation of nickel gossans in Western Australia. *Econ. Geol.*, **71**, 1229–1243.
Viljoen, M. J., Bernasconi, A., Van Coller, W., Kinloch, E. and Viljoen, R. P. (1976) The geology of the Shangani nickel deposit, Rhodesia. *Econ. Geol.*, **71**, 76–95.
Viljoen, M. J. and Viljoen, R. P. (1969) Evidence of the existence of a mobile extrusive peridotitic lava from the Komati formation of the Onverwacht Group. *Geol. Soc. South Africa Spec. Publ.*, **2**, 87–113.
von Gruenewaldt, G. (1977) The mineral resources of the Bushveld Complex. *Miner. Sci. Eng.*, **9(2)**, 83–95.
von Gruenewaldt, G. (1979) A review of some recent concepts of the Bushveld Complex with particular reference to the sulfide mineralisation. *Can. Mineral.*, **17**, 233–256.
von Gruenewaldt, G. (1980) Chromite resources and production in South Africa. *Trans. Instn Min. Metall. (Sect. A: Min. Indust.)*, **89**, A99.
Ward, S. H. (1979) Ground electromagnetic methods and base metals. In *Geophysics and Geochemistry in the Search for Metallic Ores* (ed. P. J. Hood), Geological Survey of Canada, Economic Geology Report 31, pp. 45–62.
Wetherelt, A. (1982) *Rustenburg Platinum Mine*. Unpublished B.Sc. report, Camborne School of Mines, 113 pp.
Willemse, J. (1969) The geology of the Bushveld Igneous Complex, the largest repository of magmatic ore deposits in the World. In *Magmatic Ore Deposits, a Symposium* (ed. H. D. B. Wilson). *Econ. Geol. Monogr.*, **4**, 1–22.
Williams, D. A. C. (1979) The association of some nickel sulphide deposits with komatiitic volcanism in Rhodesia. *Can. Miner.* **17(2)**, 337–350.
Wilson, A. H. (1982) The geology of the Great 'Dyke', Zimbabwe: the ultramafic rocks. *J. Petrol.*, **23(2)**, 240–292.
Wynn, J. C. (1983) Strategic minerals geophysical research: the chromite example. *Min. Eng.*, **35(3)**, 246–251.

3

Magmatic hydrothermal deposits

3.1 INTRODUCTION

In the previous chapter we have described some of the mineral deposits that have formed by magmatic activity, often by segregation within the magma chamber itself. In this chapter we turn our attention to other magmatic activity which results in important mineral deposits. These are not ores that have segregated during crystallization of the magma but are those that arise by hydrothermal activity associated with igneous bodies emplaced at high levels in the earth's crust. Fluid inclusion and isotope studies indicate that magmatically derived fluids and circulating meteoric water have played varying roles in the genesis of these hydrothermal ore bodies. The tectonic setting, primary mineralogy and alteration characteristics are very important in guiding exploration for these deposits.

The principal types of deposit that are formed by magmatic hydrothermal processes are porphyry deposits and volcanic-associated sulphide deposits. These are the major suppliers of the World's copper and molybdenum. Porphyry copper deposits are becoming increasingly important sources of gold. The volcanic-associated ore bodies supply large quantities of zinc and lead.

Because of their enormous economic significance these deposits have received considerable geological attention both experimentally and in the field. The wealth of geological literature is such that we have had to restrict our attention to those features which we consider important in characterizing these deposits and useful in constructing models to assist in exploration for them.

3.2 PORPHYRY COPPER DEPOSITS

3.2.1 Introduction

It is estimated that up to 2000 BC the grade of copper ore being worked was as high as 15% but by 1500 AD this grade had reduced to 9%. The grade by 1800 was about 6 or 7% and this was maintained until the beginning of the current century. There was a marked decrease in the average grade mined from about 1904 onwards (Fig. 3.1). This very sudden change coincided with the development of Bingham Mine, Utah, from a small mine producing vein copper, with a grade around 6%, to a large-scale operation with reserves in excess of 290 million tonnes with

Ore deposit geology and its influence on mineral exploration

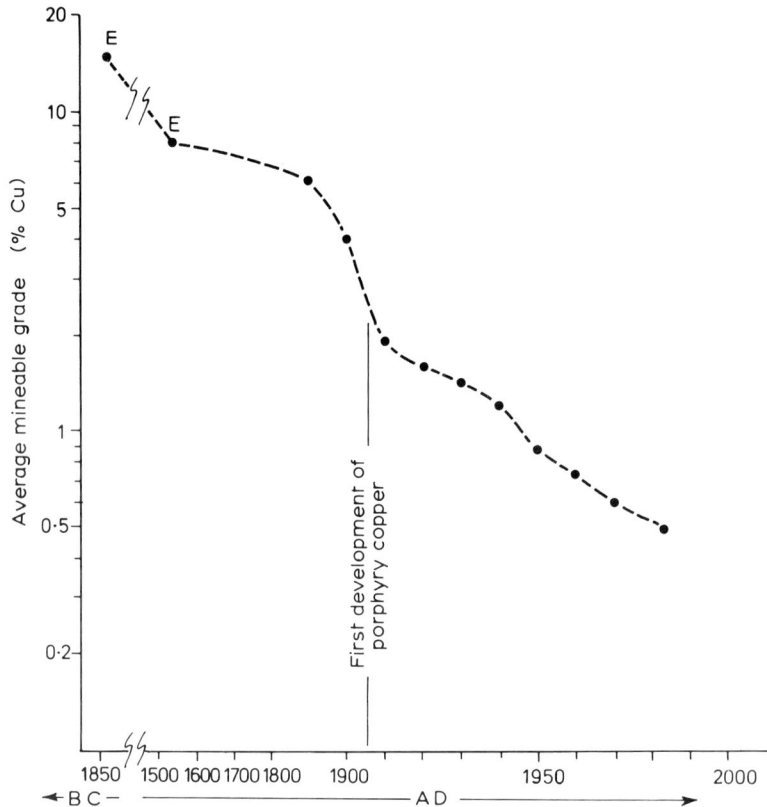

Fig. 3.1 Decline in grade of copper mined. E, estimated.

grades from 0.75 to 2.5% Cu. This large-tonnage, low-grade operation was the first of the many deposits subsequently known as the 'porphyry coppers'. Bingham Canyon is still a very large productive mine (Fig. 3.2) and has reserves quoted around 1700 million tonnes at 0.71% copper.

The success of Bingham was quickly followed by the development of other similar deposits at Globe-Miami, Arizona, Eli, Nevada, Santa Rita, New Mexico and Chuquicamata in Chile. The success of these operations was made possible by the almost contemporaneous development of froth flotation in mineral processing for the selective separation of copper sulphides. Porphyry copper deposits currently supply over half the World's copper.

3.2.2 General characteristics of porphyry copper deposits

(a) Distribution of deposits in time and space

The ages of formation of porphyry copper deposits have been considered by Meyer (1972), Hunt (1977) and Gustafson (1979). The majority of porphyry copper deposits were emplaced in the last 75 million years (Fig. 3.3). Figure 3.3 may be incomplete as some deposits lack absolute dating, others must have disappeared through erosion and the figure will also be biased by exploration philosophy. However, we believe the overall trend is reliable. Very few porphyry copper type occurrences have been identified older than 450 million years.

Fig. 3.2 Bingham Mine, Utah was the first porphyry copper mine to be developed. With reserves quoted at 1700 million tonnes, this open pit removes about 400 000 tonnes of ore and waste per day. The trains on the lower benches provide a scale for this massive operation.

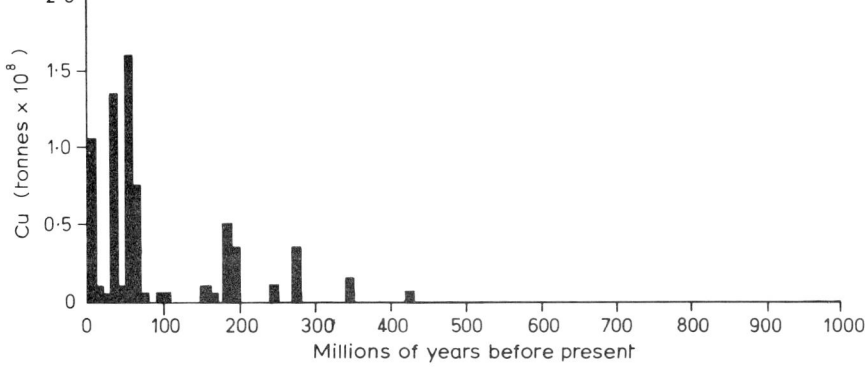

Fig. 3.3 Age distribution of porphyry copper deposits (after Hunt 1977, quoted in Gustafson 1979).

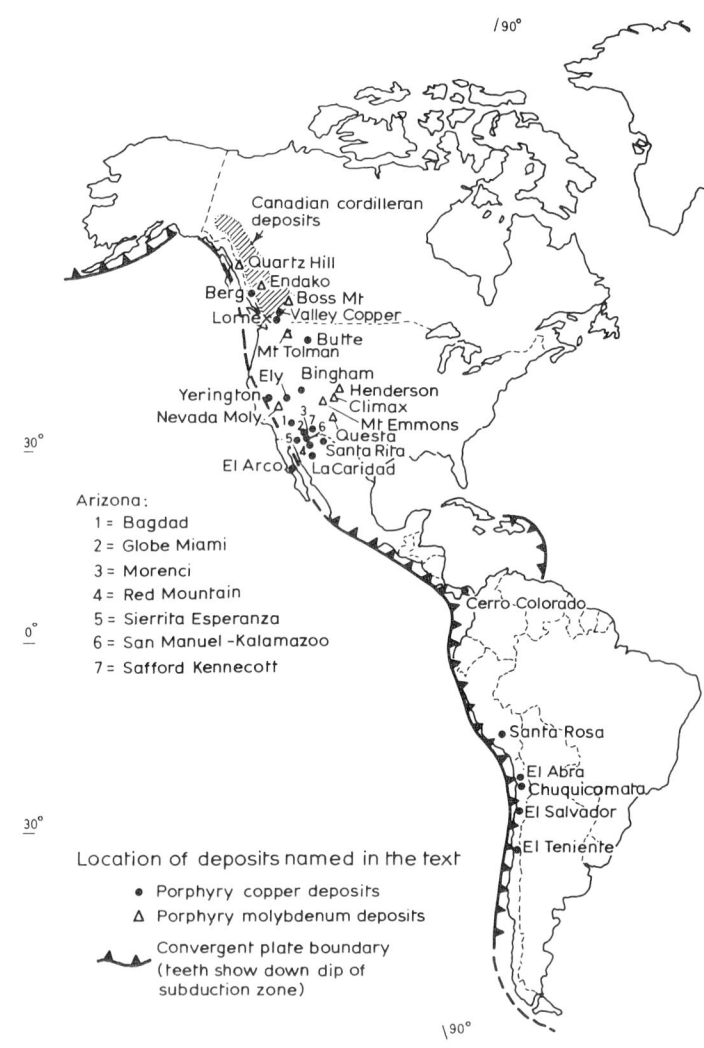

Fig. 3.4 World map of porphyry deposits (Cu and Mo) referred to in text.

Probably the copper–molybdenum porphyry at Haib, Namibia, is the best known Proterozoic example. Deposits with porphyry copper affinities have been described from the Archaean of Canada and Western Australia (Gustafson, 1979) but are of very low copper grade even by porphyry standards. We are unable to cite any deposit older than Devonian being mined, although some of the Chinese deposits may be Proterozoic. In Queensland, Australia, more than 40 porphyry copper prospects have been found associated with intrusives of Palaeozoic and Mesozoic age but none of commercial grade and tonnage (Horton, 1978). Also from Australia, Ambler and Facer (1975) reported a Silurian porphyry copper prospect in New South Wales. Porphyry deposits are fairly numerous in late Palaeozoic mountain belts, such as in the northern Appalachians from Maine to Quebec (Kirkham and Soregaroli, 1975). Large deposits of Palaeozoic age are known from the USSR, from the Kazakhstan and the Uzbek region.

The circum-Pacific belt is the predominant site of porphyry copper deposits (Fig. 3.4). A few of

Magmatic hydrothermal deposits

these deposits are Triassic; Yerington in western Nevada is Jurassic in age. There are other Jurassic and Cretaceous porphyries in Canada and a few in the United States and northern Argentina (Sillitoe, 1977). The vast majority of the deposits on the rim of the Pacific Basin are Tertiary in age. Those of the south west USA are mostly Palaeocene with some from the late Eocene. Some of the deposits in South America, mainly Chile, Peru and Ecuador, are also Paleocene but many are later with a large number being late Miocene and Pliocene. The deposits of the Cascades, western Canada and Alaska span a similar period of time to those in South America. The Philippines and the south-western Pacific Islands have deposits that are generally Miocene and Pliocene. Among the youngest of porphyry copper deposits are those in Papua New Guinea where, for example, Ok Tedi is some 1.2 million years old. It is possible that porphyry copper deposits are forming today beneath active volcanic areas.

(b) Size and grade of deposits

Porphyry copper deposits are often defined as very large-tonnage, low-grade copper deposits amenable to bulk mining methods. This description ignores the geological connotations of the word 'porphyry' but draws attention to the characteristic tonnages, measured in hundreds of millions of tonnes, and grades of often less than 1% Cu. Data from 103 deposits worldwide give the arithmetic average as about 550 million tonnes of ore grading 0.6% copper (Singer et al., 1975).

Table 3.1 lists some of the largest porphyry copper deposits in the World. We have only included those with quoted reserves in excess of 500 million tonnes. Some of the largest deposits lie in southwest USA and South America with the largest of all, Chuquicamata, in Chile, containing in excess of 10 000 million tonnes at 0.56% copper. The largest group of 'giant' porphyry copper deposits lies in the southwestern USA which contains six in addition to Bingham. Because of their enormous size and very low grade most of the porphyry copper deposits of the World are worked by open pit methods (Fig. 3.2). In the Bingham open pit more than 400 000 tonnes of material (ore plus overburden) are removed daily; furthermore approximately 10% of that mine's copper production is produced by leaching the overburden dumps. San Manuel, Arizona, is one of a small number of porphyry copper deposits worked by underground mining, block caving techniques (Fig. 3.5). In this the porphyry copper deposits differ from the porphyry molybdenum deposits where the major producers – Climax and Henderson in Colorado – are both underground mines and the most recent major prospect, Mt Emmons, will also be worked underground.

The average Canadian porphyry copper deposit carries about 150 million tonnes of ore grading about 0.45% copper. Here the calc-alkaline-associated deposits have higher tonnages (arithmetic mean of twenty deposits = 206 million tonnes) and lower grades (arithmetic mean = 0.39% Cu) than the alkaline-associated ones (49 million tonnes at 0.76%) (Drummond and Godwin, 1976).

Table 3.1 provides information of the grades of other metals present in deposits. The more common by-products are molybdenum, gold and silver although rhenium, tin and tungsten may be present in workable quantities. These by-product metals are very important to the commercial exploitation of porphyry copper deposits. Ok Tedi, Papua New Guinea, which commenced production early in 1984 (reserves about 350 million tonnes at 0.7% copper and 0.59% g/t gold) has a leached capping ore of around 34 million tonnes grading 2.86 g/t gold. This leached ore is to provide the sole production for the first 2 years of operation, amounting to 12 000 tonnes per day gold ore at 4.21 g/t with an anticipated recovery of 92% (Pintz, 1984). Without this 'front end capitalization' from gold, Ok Tedi would not have come into production as a porphyry copper deposit because of the prevailing financial climate. Similarly Panguna, Bougainville Copper Ltd, Papua New Guinea, treats about 130 000 tonnes per day and yielded 170 004 tonnes of copper, 17 528 kg of gold and 43 153 kg of silver in 1982. This produced a sales value for the year of US$276 million of which gold contributed 47% (Sassos, 1983).

(c) Mineralization and zoning

Hypogene mineralization consists of disseminations, fracture fillings and quartz veinlets carrying varying amounts of pyrite, chalcopyrite, bornite and molybdenite (Figs 3.6 and 3.7). Porphyry copper mineralization is often referred to as 'disseminated' and we have maintained this usage in Fig. 3.6. Although on a large scale, immense volumes of ore may contain disseminated values, on a small scale the occurrence of ore and gangue sulphides is controlled by fractures. Beane and Titley (1981) estimate that in excess of 90% of the sulphides are in, or adjacent to, fractures. Even apparently disseminated sulphide minerals are often aligned with microveinlets or lie in a chain-like fashion. The chains mark early fractures which have been sealed, or annealed, and camouflaged by quartz and potassium feldspar (Fig. 3.8). Such tiny fractures

Table 3.1. Reserves of porphyry copper deposits with tonnages greater than 500 million tonnes (source of data Gilmour, 1982).

Country	Deposit	Tonnage (million tonnes)	Grade Cu (%)	Mo (%)	Au (g/t)	Ag (g/t)
USA	Bagdad (Arizona)	800	0.50	0.03	—	0.6
	Morenci (Arizona)	500	0.9	0.007	—	—
	Safford-Kennecott (Arizona)	2000	0.5	—	—	—
	San Manuel-Kalamazoo (Arizona)	1000	0.74	0.015	—	—
	Butte (Montana)	500	0.8	—	Yes	Yes
	Santa Rita (New Mexico)	500	0.95	—	—	—
	Bingham (Utah)	1700	0.71	0.053	—	—
Mexico	El Arco (Baja California)	600	0.6	—	—	—
	La Caridad (Sonora)	750	0.67	0.02	—	—
Canada	Lornex	500	0.41	0.015	—	—
	Valley Copper	900	0.48	—	—	—
Papua New Guinea	Frieda River	800	0.46	0.005	0.2	—
	Panguna	1000	0.47	0.005	0.48	1.6
Cebu Island	Atlas	1100	0.55	—	Yes	Yes
Panama	Cerro Colorado	2000	0.6	0.015	0.06	4.6
Chile	Chuquicamata	10000	0.56	0.06	—	Yes
	El Abra	1200	0.7	—	—	—
	El Teniente	8000	0.68	0.04	—	—
Peru	Santa Rosa	1000	0.55	—	—	—

Fig. 3.5 San Manuel Mine, Arizona, is one of the few underground mines producing from porphyry copper deposits. The surface subsidence in the foreground results from the caving method used to mine the San Manuel ore body. Beyond the headframes lies the Kalamazoo ore body.

probably occurred in immense numbers and were very important in increasing the permeability of the host rocks during mineralization. A typical pattern of the zonal arrangement of sulphide ore minerals and gangue is shown in Fig. 3.7. Centred on the intrusion is a barren or weakly mineralized zone with minor chalcopyrite, molybdenite and rare bornite; pyrite is generally less than 2%. Outwards there is firstly an enrichment in molybdenite and then chalcopyrite. Pyrite also increases outwards in the ore shells. The ore shells are enclosed by a pyrite-rich halo with 10–15% pyrite with only minor amounts of chalcopyrite and molybdenite. Surrounding this halo is a peripheral zone of lower pyrite content which may contain radial fractures with base metal mineralization and gold and silver in the veins (McMillan and Panteleyev, 1980). Generally pyrite is the most abundant and widespread

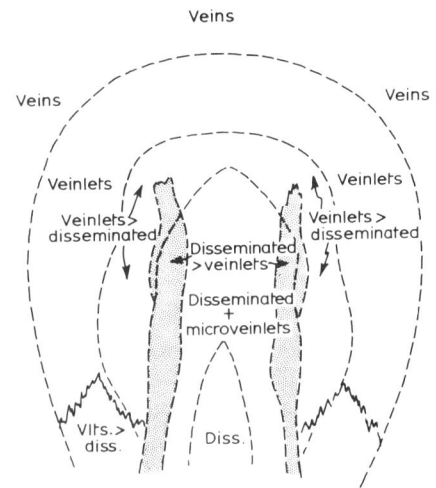

Fig. 3.6 Schematic drawing of sulphide occurrence zoning in a typical porphyry ore deposit (after Guilbert and Lowell 1974).

Magmatic hydrothermal deposits

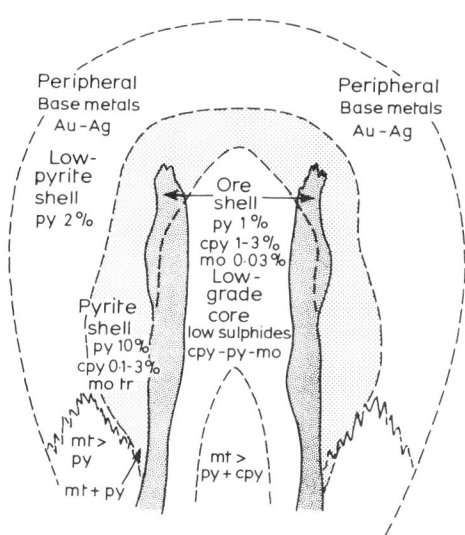

Fig. 3.7 Schematic drawing of mineralization zoning in a typical porphyry ore deposit (after Guilbert and Lowell 1974). Ag, silver; Au, gold; cpy, chalcopyrite; mo, molybdenite; mt, magnetite; py, pyrite; tr, trace.

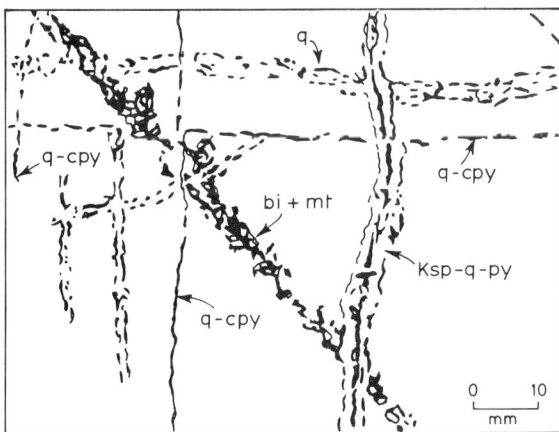

Fig. 3.8 Sketch of veins in a sample of quartz monzonite from the Sierrita deposit in southern Arizona. Note the cross-cutting relations which imply a number of generations of fracturing and fluid flow. Ksp, potassium feldspar; q, quartz; py, pyrite; bi, biotite; mt, magnetite; cpy, chalcopyrite. (Modified from Titley 1982, and Anthony 1983.)

sulphide mineral in porphyry copper deposits. Magnetite is a minor constituent of the deep and low-grade core zone but is abundant in, and diagnostic of, the deeper parts of the deposit surrounding the ore shells (Fig. 3.7). Chalcocite can occur from the ore shell to the outer peripheral zones in the higher levels of the deposit. Pyrrhotite and native copper are sometimes present in the peripheral shell and the immediate vicinity but only in minor or trace amounts.

Secondary or supergene enrichment is important in most porphyry copper deposits. The supergene blanket contains minerals such as chalcocite, djurleite and digenite. Covellite, cuprite and native copper may be present in small amounts. The development of an extensive supergene blanket is favoured by a hot arid climate, copious pyrite and a relatively inert host rock. The shape of the supergene zone depends on surface topography, intrusion geometry and structure and hydrology. The zone is usually quite variable in thickness (Fig. 3.9). The supergene zone is overlain by a limonite, jarosite or haematite capping, the characteristics of which may be utilized in exploration (Anderson, 1982).

(d) Host lithology

The intrusion upon which mineralization is centred is usually quite small, 1 or 2 km in diameter, and is porphyritic in texture. Because of the extent of the alteration phenomena it may be difficult to determine the primary rock compositions of the intrusives associated with porphyry mineralization. Generally they are of a calc-alkaline trend. The most common hosts are acid intrusives of the granite suite and these range from adamellite through granodiorite to tonalite. Diorites, quartz monzonites, syenites may also be important host rock types. The compositions of the intrusives are a function of the environment in which the deposit formed, granodiorite to quartz diorite in an island arc environment and quartz monozite through granodiorite in a continental arc.

In all porphyry copper provinces some intrusions occur which are barren but are of

Ore deposit geology and its influence on mineral exploration

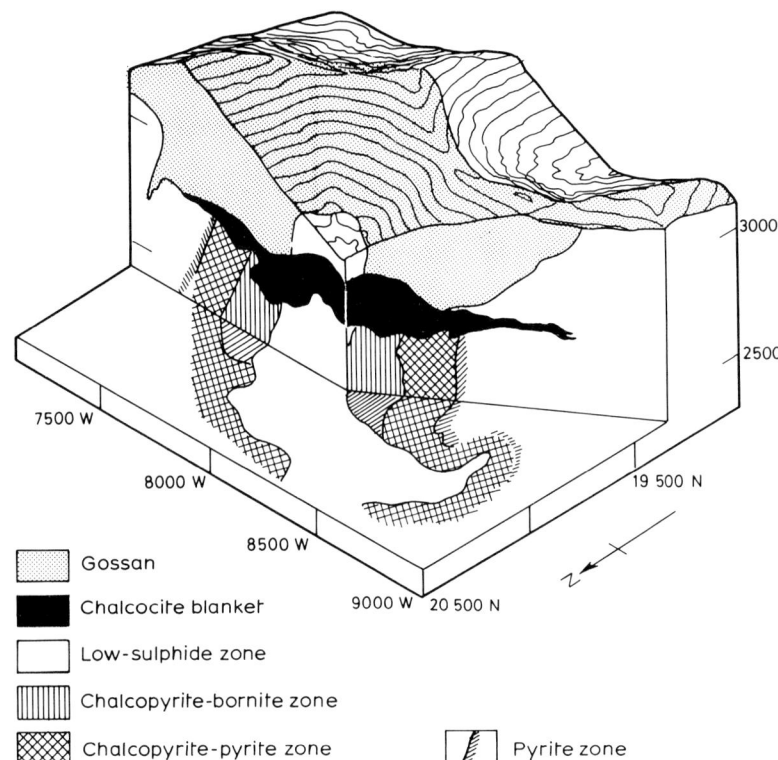

Fig. 3.9 Isometric diagram of El Salvador ore body showing mineral zones chalcocite blanket and gossan (after Gustafson and Hunt 1975).

apparently the same age as, and comparatively similar composition to, others which have associated economic mineralization. Discrimination of those which will host economic concentrations of minerals from those which are likely to be barren is of primary importance in exploration geology. Unfortunately, most attempts to separate the plutons into productive and barren ones have been unsuccessful but certain characteristics have been recognized which may be useful in this matter (Baldwin and Pearce, 1982). Firstly, pyroxene is rare or virtually absent from copper-related porphyries; hornblende and biotite are abundant. Secondly, the degree of fracturing may be important since mineralized intrusives are always extensively fractured.

Brecciated zones are common and may occur within the intrusion or in the wall rocks. Breccias which appear to have resulted from hydrothermal activity are frequently referred to as pebble dykes. These have rounded clasts but others associated with porphyry deposits are angular collapse breccias.

(e) Alteration

Although the porphyry copper deposits are genetically related to epizonal intrusions the hydrothermal effects usually extend far beyond the ore zone itself. The alteration effects are seen in the country rocks of all types and in early differentiates from the magma which subsequently gave rise to the porphyry-related intrusive. The alteration phenomena are selective, pervasive and vein–veinlet in occurrence. The extent of this alteration suggests that the porphyry copper mineralizing environment is one that involves considerable permeability in the host rocks. Facilitated by the high permeability, both selective and pervasive alteration begin at various times and places within the deposits, presumably guided by, and progressing from, fractures. Selective alteration is the conversion of one or two mineral species to other mineral species. The most common example of this in the porphyry systems is the selective conversion of amphibole or hornblende into secondary biotite in large volumes of rock. This biotite may, in turn, be converted to chlorite. Pervasive alteration on the other hand is the widespread and wholesale conversion of one rock type to another.

Lowell and Guilbert (1970) developed the concept of a 'typical' porphyry deposit as a result of studying 27 of the best-known deposits in Canada, South America and the southwest USA and comparing them to the model they had erected for the San Manuel-Kalamazoo deposit in Arizona. This model is referred to as the 'Lowell-Guilbert model' for porphyry coppers and shows that porphyry coppers have clearly recognizable lateral and vertical zoning. In 1974 the same workers published details of the variations in the zoning pattern which may arise from the size of the deposit, the composition of the pre-ore wall rock and pre-ore structures as well as those variations due to the chemistry of the igneous host rock (Guilbert and Lowell, 1974). The zoned arrangement of the Lowell and Guilbert model, suitably modified in the light of these other factors, is widely used by geologists to interpret geochemical anomalies and to guide drilling for porphyry copper deposits. The model typically recognizes four zones of hydrothermal wall rock alteration centred on the core of the causative intrusive (Fig. 3.10). These are as follows:

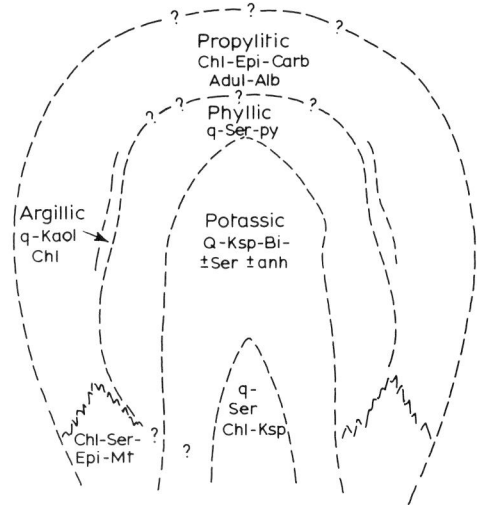

Fig. 3.10 Schematic drawing of alteration zoning in a typical porphyry ore deposit (after Guilbert and Lowell 1974). Adul, adularia; Alb, albite; Anh, anhydrite; bi, biotite; Chl, chlorite; Epi, epidote; Kaol, kaolinite; Ksp, potassium feldspar; Mo, molybdenite; py, pyrite; q, quartz; ser, sericite.

(i) The potassic zone

The potassic alteration results from potassium metasomatism which may be accompanied by leaching of calcium and sodium from rocks containing original aluminosilicate minerals. When this zone is present the characteristic minerals are biotite, orthoclase and quartz. These may have been introduced or are the stable remnants of original rock-forming phases. Accompanying these may be accessory albite, sericite, anhydrite and apatite. Magnetite, chalcopyrite, bornite and pyrite are commonly present. The pyrite is always minor and is antithetic to bornite and magnetite (Titley and Beane, 1981). Although the potassic alteration commonly occurs in or near the intrusive centre a broad zone of biotization may pervade igneous wall rocks.

(ii) The phyllic zone

This is sometimes referred to as the zone of sericitic alteration and results from the leaching of magnesium, sodium and calcium from aluminosilicate rocks. Potassium may be introduced or derived from original rock-forming feldspar. The phyllic assemblage is characterized by quartz, sericite and pyrite, the first two often totally replacing the original rock-forming silicates and resulting in the complete destruction of the original rock structure.

Pyrite may grow from the iron of pre-existing

mafic minerals and introduced iron and sulphur. The pyrite content may rise as high as 10% by volume. Chalcopyrite is generally present but rarely exceeds 0.5 vol. percent.

(iii) The argillic zone

Argillic alteration is characterized by the formation of new clay minerals. Intermediate argillic alteration is evidenced by montmorillonite, illite, chlorite and perhaps kaolinite. Advanced argillic alteration is marked by kaolinite with diaspore, quartz or amorphous silica, andalusite and, occasionally, corundum. Pyrite is the main sulphide present although chalcopyrite and bornite may occur. Pyrite is less abundant than in the phyllic zone and tends to occur in veinlets.

(iv) The propylitic zone

This, often quite extensive, zone is always present and is characterized by chlorite, epidote and calcite. Potassium released by the chloritization of biotite occurs as accessory sericite. Other accessory minerals include apatite, haematite, anhydrite and ankerite. Plagioclase may be unaltered. Sulphide occurrence within the propylitic zone varies from nil to minor amounts of pyrite and occasional economic concentrations of chalcopyrite.

(v) Vertical extent of alteration

In their original model Lowell and Guilbert tentatively close the alteration zones over the top of the causative intrusion suggesting a vertical zoning of the silicate minerals comparable to that demonstrated laterally (Fig. 3.10). The spatial distribution of these alteration zones is much used by exploration geologists. If buried porphyry systems are sought the vertical variation must also be investigated.

Firstly the overlying, pre-ore, rocks may be as varied in lithology as the surrounding wall rocks into which the intrusion has been emplaced. There may be an argument that in some cases these so-called 'lithocap' rocks will be dominated by comagmatic, but pre-ore, intrusives. This is certainly the case in some of the major porphyry molybdenum systems such as Henderson, Colorado. It also seems probable that the pre-ore lithocap will have experienced hypogene mineralization. There is some evidence for this in that the tonnages due to supergene enrichment in some deposits cannot be accounted for by the leaching of reasonable thicknesses of cap rocks having the same low grade of mineralization as the underlying intrusion.

Because of the present state of exploration and development of porphyry copper deposits most of the data available apply to the upper parts of the porphyry intrusion rather than the lithocap. We feel current economics dictate that porphyry systems buried beneath a reasonable covering of lithocap will not be exploited in the foreseeable future and therefore details of vertical variations from a range of deposits will be lacking for some time yet.

Beane and Titley (1981) quote the Red Mountain prospect in Arizona as one of the better documented cases of the alteration in the lithocap, although the causative intrusive was not revealed by the extent of drilling undertaken. At Red Mountain increasing depth indicates intermediate argillic alteration (kaolinite, montmorillonite, alunite, amorphous silica, tennantite, enargite and pyrite) underlain by phyllic and potassic alteration. As the intrusive porphyry body, which is assumed to lie at depth, is not intersected in the drillholes it is clear that the potassic alteration extends beyond it into pre-ore volcanic rocks. The downward succession in alteration zones is accompanied by decreasing sulphide contents and pyrite/chalcopyrite ratios. This evidence from Red Mountain would appear to justify the closure of the Lowell and Guilbert zones above the core-intrusive porphyry body.

(vi) Variations from the Lowell and Guilbert model

(a) Due to different intrusion types

Extensive studies of the alteration of porphyry

copper deposits worldwide at various levels within the deposits have led to descriptions which do not fit exactly the alteration zone model outlined above. A major departure occurs when the causative bodies are quartz dioritic rather than quartz monzonitic. Atypical alteration assemblages and zoning also appear in porphyries that are quartz-deficient, commonly dioritic but ranging to syenite.

In quartz diorite intrusions the phyllic alteration may be absent or is later than the potassic and propylitic alteration and sometimes only poorly developed. The more mafic character of the rock type is reflected by the dominance of biotite rather than orthoclase in the potassic zone. This zone also contains chlorite rather than sericite as an accessory mineral. The absence of the phyllic zone means that the propylitic zone lies peripheral to the potassic zone. Propylitization occurs principally by the alteration of biotite to chlorite ± epidote. Sulphide mineralization may occur in both the potassic and the propylitic zones and hypogene copper values may be sufficiently high (>0.4%) to constitute ore grade without supergene enrichment. Pyrrhotite may accompany weak development of the phyllic zone as in some of the intrusions in the Cascade district of Washington (Patton et al., 1973).

Hollister (1978) developed a dioritic model for alteration in porphyry copper environments related to low SiO_2 intrusions, this model having been developed primarily from observations in the Canadian Cordillera. One of the main reasons for the difference between the dioritic model and the Lowell and Guilbert model appears to be the relatively low sulphur concentrations in the mineralizing fluids. As a result much iron remains in the chlorites and biotites, not all the iron oxides in the host rocks being altered to pyrite. The alteration zones in the dioritic model appear restricted to the potassic and propylitic varieties. In the potassic zone biotite may be the most common potassium mineral and when orthoclase is not well developed, plagioclase may be the principal feldspar.

(b) Due to different wall-rock characteristics
The alteration zones are not restricted to the intrusions associated with the mineralization but extend into the surrounding wall rocks. In the 'type' example, at San Manuel-Kalamazoo, Arizona, where the wall rocks and the intrusion have similar compositions, alteration assemblages develop in response solely to changes in temperature and fluid chemistry. In this case the alteration will be uncomplicated by differences in rock compositions. Alteration minerals in silicate wall rocks may closely resemble those in the intrusions. Arkosic sediments undergo potassic, phyllic and propylitic alteration similar to granite wall rocks, as at Morenci, Arizona. Dioritic intrusions cross-cutting andesites or basalts develop high Ca–Mg alteration assemblages as at Panguna, Papua New Guinea.

The most marked difference between the alteration assemblages in the intrusion and the wall rocks occurs where the wall rocks are carbonates –limestones or dolomites. In this case the typical potassic or phyllic alteration of the intrusion abuts against skarn. We discuss the development of skarns in this and other environments in Chapter 10.

(c) Due to the location of the sulphide mineralization
The Lowell and Guilbert model showing alteration and mineralization (Figs 3.6, 3.7 and 3.10) sites the hypogene copper mineralization in the outer parts of the potassic zone. Outwards from this, sulphides become more abundant but are represented almost exclusively by pyrite. Detailed studies worldwide have shown that copper mineralization may be associated with any of the alteration zones. Care must be exercised to ensure that hypogene mineralization is recognized and distinguished from secondary supergene enrichment.

In most porphyry copper deposits the core is barren or very low grade but some, such as Bingham, contain bornite–molybdenum mineralization (Beane and Titley, 1981). Others have high-grade sulphides in quartz pods and veins, as at Yandera, Papua New Guinea. Where the phyllic alteration is missing, as is the case with dioritic intrusions, significant mineralization may occur associated with the propylitic alteration, for example at Sierrita-Esperanza, Arizona.

(f) Tectonic setting

Figure 3.4 shows that most porphyry copper deposits are associated with convergent plate boundaries and areas of andesitic volcanism. Most formed along destructive plate margins above subduction zones of oceanic crust. Some arose from continent-to-continent collision, those of the western Pacific formed in island-arc regimes.

To explain the occurrence of the largest concentration of porphyry copper deposits so far discovered, those in the southwestern states of the USA, we shall consider the processes of subduction in that region in more detail. These deposits are dominantly Laramide in age. This was the time when the oblique convergence of the continental and Farallon plates changed to near-normal convergence. The net convergence rate increased dramatically accompanied by a flattening of the subduction zone. Heidrick and Titley (1976) suggested that this convergent style caused epicrustal extension which in turn facilitated access for the porphyry magmas into the upper crust.

Nearly all the porphyry copper deposits of the American southwest had formed by 50 million years ago. The slow convergence rate of the Pacific and North American plates since that time corresponds to a period of reduced igneous activity in the area. The deposits of the Canadian Cordillera, the Philippines and many of the Andean deposits had also formed by this time. The porphyry copper deposit at Bingham Canyon, Utah, is younger and related to a pulse of igneous activity between 37 and 41 million years which is about the age of the El Salvador deposit in Chile (Fig. 3.9). The Andean province, mainly Chile and Peru, contains deposits with an age range from 59 to 4.3 million years. The deposits tend to occur in a linear belt approximately 100 km from the present-day coast and parallel to the plate boundary.

3.2.3 Genesis of porphyry copper deposits

It is generally accepted that these deposits are the result of the emplacement of hydrous magmas into highly permeable cover rocks at relatively shallow depths in the crust. These intrusions and their fluid phases produce the heat and energy necessary to create the fractures and convective fluid flow for the ground preparation for ore deposition. There is less certainty whether magmatic fluid provides the chemical components for alteration and mineralization (Anthony, 1983). There is much evidence showing that the fluids in these porphyry copper systems are derived from two different sources. Initially there is outward flow of exsolved solutions from the crystallizing magma and subsequently meteoric and connate waters are involved in the system. McMillan and Panteleyev (1980) describe the two ends of such a hydrothermal system – one dominated by fluids of magmatic origin and the other dominated by fluids of meteoric origin (Fig. 3.11).

Although the actual source of the magma has yet to be decided (Anthony, 1983), at a convergent margin a magmatic plume would rise from the more ductile lower crust to the more brittle higher levels in the crust (0.5 to 2 km from the contemporary surface). Conductive heat flow through the cover rocks would result in thermal metamorphism centred on the intrusion. These early metamorphic effects may be difficult to recognize if the country rocks were granitic, otherwise biotite hornfelses would be formed. The brittle nature of these hornfelses would facilitate the development of the fracturing which is very important in the formation of productive intrusions.

Early crystallization gave rise to an equigranular carapace to the intrusion. Continued crystallization would lead to increasing quantities of exsolved fluids and possibly scavenged elements being concentrated in the apical region of the intrusion (Burnham, 1979). The internal vapour pressure would increase until it exceeded the confining pressure and the tensile stress of the carapace. Retrograde boiling occurred (Phillips,

Magmatic hydrothermal deposits

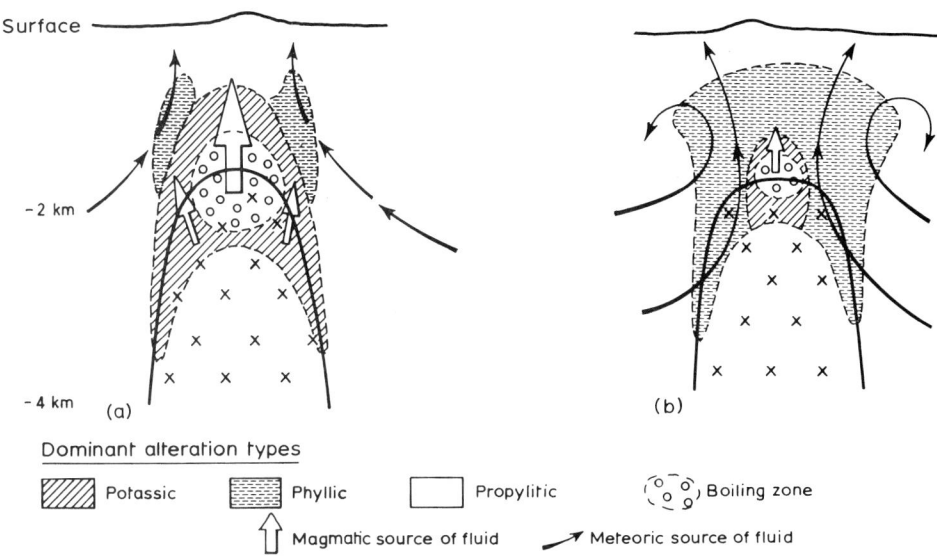

Fig. 3.11 Models of hydrothermal systems for porphyry copper deposit genesis (after McMillan and Panteleyev 1980). Two end members shown: (a) orthomagmatic; (b) convective.

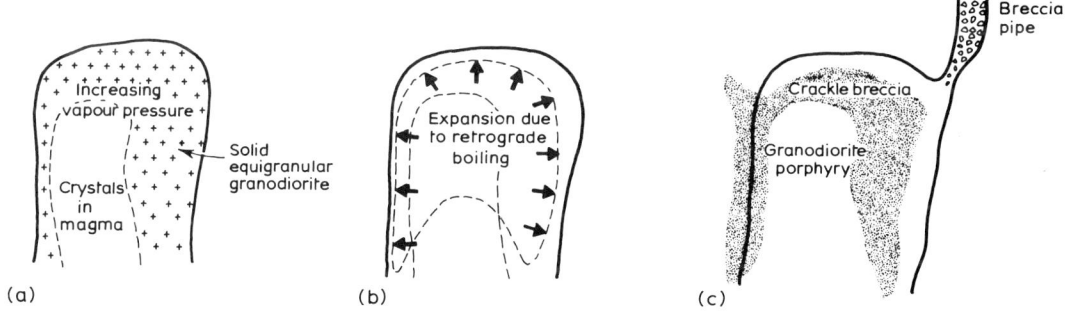

Fig. 3.12 Development of crackle brecciation. (a) Vapour pressure build-up; (b) retrograde boiling; (c) brecciation. (After Phillips 1973.)

1973). Hydrofracturing of the intrusion and host rocks followed as the hydrothermal fluids broke through the crystalline carapace. Probably these early formed fractures were sealed by mineral precipitation from the fluids and this 'throttling' allowed the internal fluid pressures of the intrusion to increase again until the retrograde boiling phase repeated. This may have repeated many times, the small chain-like occurrences of sulphides, to which we have referred already, possibly indicate the positions of these earlier fractures. When the fracture system coupled to the surface the internal pressure was suddenly decreased from lithostatic to hydrostatic and massive boiling took place. This event was accompanied by the generation of the breccias and pebble dykes (Fig. 3.12).

Fluid inclusion studies provide evidence that

boiling occurred since highly saline fluid inclusions, with co-existing gas and fluid-rich inclusions, are preserved. Release of the hydrothermal fluids caused an increase in the rate of cooling which would result in more rapid crystallization, creating the fine-grained groundmass and therefore the porphyritic texture of the intrusive.

The upward and outward flow of hydrothermal fluids led to the initial mineralization and caused the potassic core and the surrounding propylitic alteration. The upward moving, magmatically derived, hydrothermal fluids entrained some meteoric water which would also be involved in the alteration. The presence of the pluton would have created thermal convection currents in meteoric fluids in the neighbourhood of the intrusive (Fig. 3.11). As the cooling period of the intrusive extended, the role of the convecting groundwater became more important. The groundwater flowed towards and through the crackled zones resulting in widespread phyllic overprinting of the previous alteration zones. Often extensive remobilization and enrichment of early formed copper sulphides by hypogene leaching took place at this stage (Gustafson and Hunt, 1975).

Further cooling of the intrusion led to the collapse of the thermally driven convection cell (Fig. 3.11). This resulted in a dilute acid hot-spring environment with accompanying argillic overprinting. At this stage any further, post-ore, intrusives would react with the cooler groundwater and give rise to explosive activity such as pebble breccia pipes and diatremes. Subsequently porphyry copper deposits would be subjected to weathering and the downward percolation of meteoric water. The high pyrite content of the alteration zones resulted in an efficient leaching agent, and secondary or supergene enrichment of the ores took place. The enrichment is caused by leaching of the copper at high levels, in an oxidizing environment above the water table, and the redeposition of this copper at lower levels principally as chalcocite or djurleite. This zone of secondary enrichment is sometimes referred to as the 'chalcocite blanket' (Fig. 3.13).

3.2.4 Panguna porphyry copper deposit, Bougainville, Papua New Guinea

(a) Geology

Panguna, on the island of Bougainville at the western end of the Solomon Islands, is the largest producing porphyry copper deposit in the western Pacific island arc setting. The geology has been described by Baumer and Fraser (1975) and Baldwin et al. (1978). The genesis of the ore body has been discussed by Eastoe (1982). The oldest rocks in the region of the deposit are basaltic and andesitic lavas and pyroclastics of Lower Oligocene or Upper Miocene age. The copper deposit was formed 5 to 3.4 million years ago during, or following, the emplacement of quartz diorite to granodiorite intrusives. The deposit is located at the southern margin of the Kaverong Quartz Diorite stock (Fig. 3.14). The ore body consists of veins and disseminations in this rock and in the Panguna Andesite. Marginal intrusive phases of the Kaverong stock, the biotite granodiorite, the Biuro granodiorite, and the leucocratic quartz diorite, are all to some extent mineralized.

(b) Mineralization

The Panguna deposit is similar to other western Pacific island arc deposits, such as Ok Tedi, in having relatively low molybdenum (average 33 ppm) and high gold (average 0.55 g/t). The dominant copper sulphide is chalcopyrite with bornite relatively significant in a few places. Minor free gold is found associated with bornite. Gold is sporadically developed in the chalcopyrite, it is anomalously high in the bornite and negligible in the pyrite. The average magnetite content of the ore body is 2.7 wt% but rises to 10% in some high-grade areas close to the leucocratic quartz diorite and biotite granodiorite. Rare sphalerite and galena associated with late stage faults occur in all rock types over most of the area cutting earlier copper mineralization. The copper sulphides are mainly located on fractures, irrespective of lithology, and only 5% occurs as diseminations in the intrusives. Quartz veining and sulphide veins

Magmatic hydrothermal deposits

Fig. 3.13 Morenci Mine, Arizona. At the end of the 19th century a small underground operation here was producing copper at grades approaching 20%. Part of the supergene-enriched blanket is still visible overlain by the extensive gossan.

are necessary for high-grade ore, except in some breccias which have massive sulphides in the matrix.

(c) Genesis of the deposit

The country rocks of the Panguna Andesite formed part of a stratovolcano. Into these was intruded the Kaverong Quartz Diorite. Three mineralizing events account for the sulphide mineralization:

(1) the Kaverong Quartz Diorite/biotite diorite intruded into the Panguna Andesite. Assimilation and remobilization of existing copper mineralization may have resulted from the intrusion of the biotite granodiorite along the biotite diorite–Panguna Andesite contact then the dumping of this copper into the surrounding breccias and biotite diorite remnants.

(2) a mineralization phase associated with the emplacement of the leucocratic quartz diorite.

(3) a phase centred on an area close to the western margin of the biotite granodiorite and postdating the intrusion of the Biuro granodiorite.

Pebble dyke formation is post mineral and these are structurally controlled. According to Eastoe (1982) the copper was deposited mainly by salt-rich liquid expelled directly from the magma.

(d) Exploration, evaluation and mining

The Panguna open pit began stripping overburden in 1970 and came into production in 1972. This

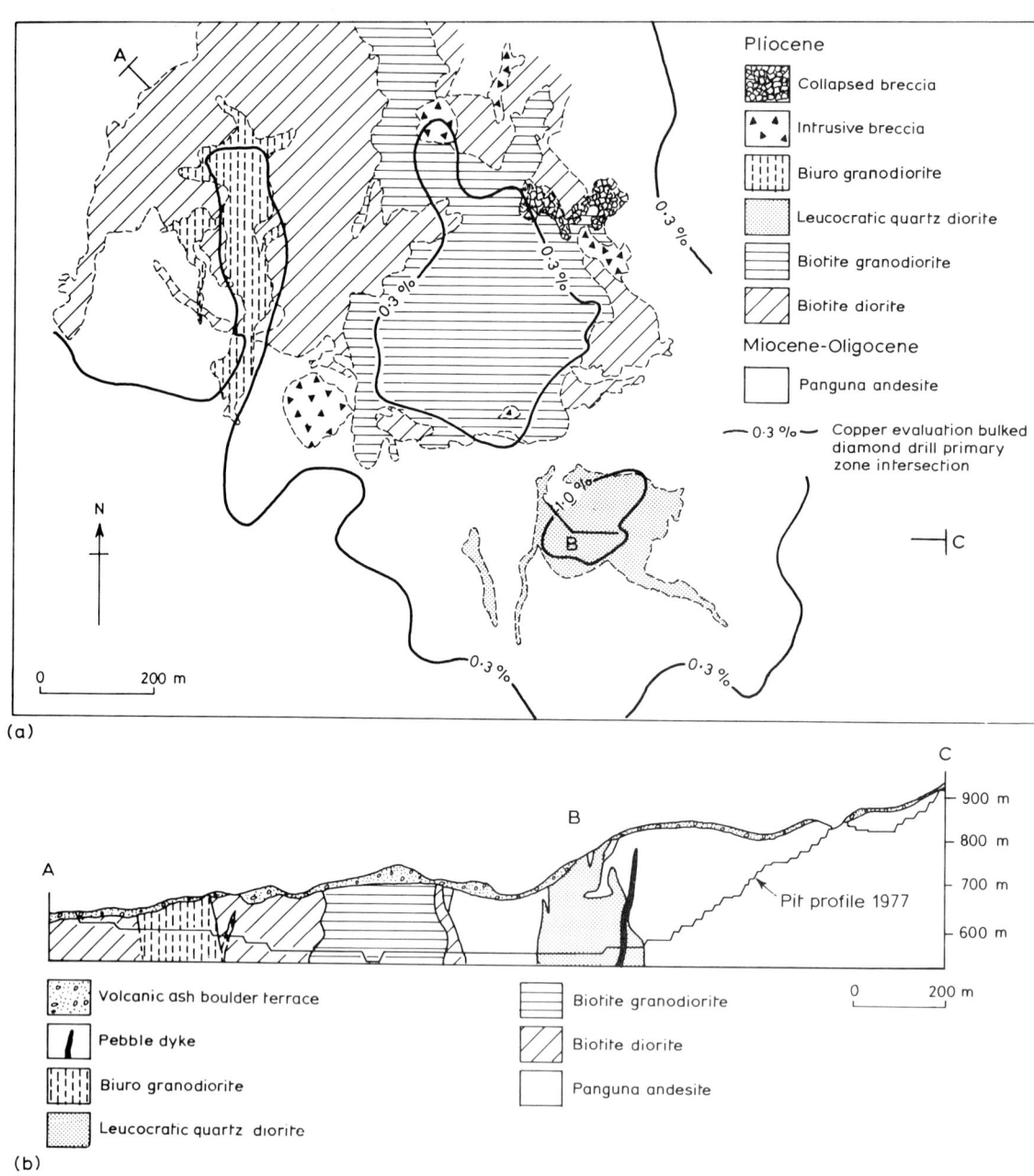

Fig. 3.14 (a) Geological map, Panguna Mine, Bougainville, Papua New Guinea, showing copper-grade evaluation. (b) Geological section along line A–B–C in (a). (After Baldwin et al. 1978.)

was the culmination of an extremely successful exploration programme; the discovery of the extensive mineralization followed a reconnaissance stream sediment survey (MacNamara, 1968). A soil geochemistry follow-up was combined with detailed geological mapping. A distinctive vegetation assemblage occurred over the copper deposit here (Cole, 1971).

The deposit was evaluated on a 122 m grid and the evaluation programme involved 238 vertical diamond drill holes. Diamond drill hole material was assayed and this was used for the ore-zoning programme (Baldwin et al., 1978). The grades were calculated at node points on a regular three-dimensional grid. There was 50% overlap on all axes of the grid and this eventually produced a $75 \times 75 \times 15$ m matrix for open pit bench planning. Baldwin et al. state that this was a computer-generated study without the input of geological constraints. The node values were contoured, irrespective of geological boundaries, to delineate ore-grade zones. Each node assay value was assigned to its area of influence ($75 \times 75 \times 15$ m), and the grade and tonnage within viable pit limits was calculated. Reserve estimates were also calculated by the polygons method using the evaluation diamond drill holes as the centres of polygons.

Once mining commenced these reserve calculations needed modification. Firstly the lack of consideration of geological controls on the mineralization became evident within the biotite granodiorite. Here the higher-grade copper values outside the intrusion had been averaged with the lower-grade values within the intrusive. This resulted in more ore being indicated than was actually present. Consequently the contact of the intrusion was inserted upon the bench plans to outline an area of waste. Secondly it was shown that the diamond drill holes had undervalued the deposit. This was first shown by bulk samples from the preproduction adit and raise exploration. Some of these were taken in adits and raises which followed the line of evaluation diamond drill holes. The samples assayed up to 29% higher than the diamond drill holes. This was explained by washing out of friable chalcopyrite from veins and fractures during diamond drilling. The overall core recovery was 93.4% in primary ore.

Subsequent sampling of blast holes has consistently shown a difference from the diamond drill hole data for the lower-grade mineralization (0.2 to 0.3% Cu for diamond drill hole as against 0.31–0.35% Cu for blasthole sampling). There is no significant change for the higher grade (>1% Cu) mineralization. Since the commencement of blasthole sampling this is used for pit planning, but only for two bench heights at a time. A drainage tunnel driven below the planned pit bottom also supplied bulk samples to indicate the continuity of the deposit in depth.

Reserves at Panguna are estimated at 1000 million tonnes (Table 3.1). This is over twice the estimated reserves when Conzinc Riotinto negotiated the concession with the Australian administration in Papua New Guinea in 1967 (Davies, 1978). The operation produces a copper–gold–silver concentrate with a mine output of about 130 000 tonnes per day. In 1982 the yield was 598 634 tonnes of 28.4% copper concentrate containing 170 004 tonnes of copper, 17 528 kg of gold and 43 153 kg of silver (Sassos, 1983). The income from this porphyry copper mine in 1982 came almost equally from its copper and gold sales. Gross sales value for the year was about US$ 276 million with 51% from copper and 47% from gold. This emphasizes the value of by-products to the commercial viability of porphyry copper deposits during the late 1970s and early 1980s.

3.3 EXPLORATION FOR PORPHYRY COPPER DEPOSITS

Today exploration for porphyry copper deposits is guided by the knowledge of the geotectonic setting in which they occur and the detailed knowledge of their structure and alteration zones. However, it is worth noting that many of the major porphyry copper deposits were discovered before the theory of plate tectonics had received universal acclaim and before the Lowell and Guilbert model was expounded in 1970. Therefore

Ore deposit geology and its influence on mineral exploration

Fig. 3.15 Fracture analysis is important in porphyry deposits, not only during exploration but also in stability analysis of operating open pits such as here at Morenci Mine, Arizona.

major deposits were discovered by careful interpretation of detailed geological mapping and sampling. Geochemistry and geophysics have been widely used and of course no economic deposits have been outlined without extensive drilling programmes.

The present understanding of porphyry copper deposits means that although broad regions may be identified as favourable prospective areas it is not possible to predict exactly where within the region productive intrusions will lie unless there is some surface showing.

(a) Remote sensing

With the advent of satellite imagery the detection of suitable areas for porphyry copper deposits should be easier. Features which may be visible on such images and aerial photographs include gossans, vegetation changes and fracture intensity or density. We have already described the extensive fracturing that is common in these deposits (Fig. 3.15), and Haynes and Titley (1980) describe the change in fracture density with distance from the progenitor intrusion for the Sierrita deposit in Arizona. Careful structural interpretation of aerial photographs should reduce considerably the area of subsequent ground search. O'Driscoll (1981) has described the recognition of structural corridors in Landsat lineament interpretation and shown their relationship to mineral deposits such as Bingham Canyon, Utah.

Experimental work by Abrams *et al.* (1983) using the deposits of the southwest USA has shown how remote sensing may be used for that area. With advances such as thematic mappers (TM) the recognition of the extensive alteration phenomena associated with porphyry copper deposits may be possible during a desk study.

(b) Geophysics

As part of reconnaissance, an aero-magnetic survey should be undertaken. Gravity surveys may be used to locate the most likely site of the progenitor intrusion and accurate modelling may indicate depth. During detailed ground reconnaissance, induced polarization (IP) is the most applicable method because of the disseminated nature of these deposits. Pseudosections constructed from IP surveys over chosen target areas may be used to guide subsequent drilling.

(c) Geochemistry

Undoubtedly geochemistry has been one of the more successful prospecting techniques in the search for, and extension of, porphyry copper deposits. The reasons for this may be summarized (Coope, 1973):

(1) The target is large, the alteration zone extending well beyond the ore shells.

(2) The host rocks are fractured and therefore there is easy access for water and oxygen. There is relatively easy exit for weathered products.

(3) The abundance of pyrite in the deposits gives rise to acid weathering conditions favouring the mobilization of copper and other metals in solution.

A geochemical exploration programme would follow traditional lines – a stream sediment or water survey followed by soil geochemistry. Additionally rock sampling, vegetation sampling or soil gas analysis may be applied.

(i) Stream sediments

Geochemical surveys using stream sediments have proved most useful in detecting copper mineralization, and dispersion trains up to 20 km have been recorded downstream from a porphyry copper showing. This is true in areas of humid climate and particularly in areas of high relief. Coope (1973) and other authors have warned about the use of cold-extractable copper (cxCu). The pyrite content in the alteration halo of large deposits results in acid ground- and stream-water which in turn can give misleading cxCu data without significantly affecting patterns in the total metal data.

In the exploration for the Yandera deposit in Papua New Guinea one sample of stream sediment per 1–2 km^2 effectively delimited the porphyry system. Samples from the $-185\mu m$ were analysed for copper, molybdenum and gold but the molybdenum was not so effective in this case.

Stream-sediment surveys are not as widely used, or applicable, in arid areas. Possible causes of contamination during the dry period in these areas are windblown dust and material from bank collapse. However, the La Caridad deposit, in northern Mexico, was located by a stream-sediment survey, probably because the deposit was high grade and there was a good drainage system. At La Caridad the molybdenum anomaly extended more than 32 km downstream from the deposit (Chaffee, 1982). In panned concentrates tungsten was detectable 11 km downstream from the La Caridad deposit. Copper, while obviously anomalous, may not be sufficiently diagnostic and should be correlated with anomalous values of other elements such as molybdenum, tungsten, gold, or even tellurium, selenium or rubidium.

(ii) Soil sampling

Once the area for more intensive exploration has been delineated then soil sampling is very important in the search for porphyry systems. Here copper may be less useful as a geochemical guide than either gold or molybdenum. There may be a zoned appearance to the anomaly with copper and molybdenum in the centre and anomalous zinc, manganese, silver, lead, cadmium and bismuth in the surrounding alteration aureole (Chaffee, 1982). Molybdenum does not appear to be such a useful tool in areas of deeper weathering and oxidation.

Gold has been shown to be a very good pathfinder element for porphyry systems, for example in Puerto Rico (Learned and Boissen, 1973). Gold was also useful in Papua New Guinea and seems most effective in areas of steep mountainous terrain. Here it may be the most definitive guide despite low concentrations in the primary mineralization (average 0.1 g/t in Yandera deposit).

Soil sampling may be undertaken on a regular grid system or, if the terrain is mountainous and thickly forested, ridge and spur sampling can be undertaken and combined with contour 'trail' sampling. At the Yandera deposit, Fleming and Neale (1979) recommended a trail spacing of 60 m vertically with channel samples cut across the trail at 10 to 15 m intervals.

In semi-arid areas the caliche layer may be sampled but since this may not be sufficiently widespread for an even distribution of samples it can result in sampling bias. Similarly the sampling of the desert varnish on pebbles may be useful in semi-arid areas (Chaffee, 1982) but this also suffers from the limited distribution of potential sampling sites within the chosen area.

(iii) Rock sampling

Where possible, rock sampling is particularly beneficial and may permit the recognition of element zoning around a system. This zoning will not be coincident with the alteration zones (Fig. 3.16). Detailed examination of the physical characteristics and trace element content of accessory minerals such as apatite and rutile can help in target identification.

(iv) Vegetation sampling

This can be divided into geobotanical sampling, by which we mean the recognition of the distribution of distinct plant species, and biogeochemical sampling where the trace element content of plants is analysed. Geobotanical sampling is very effective if diagnostic, or indicator, species can be recognized. It may be expensive if specialized botanical assistance is required but it can be very rapid. In arid areas it may have limited application since there may be no germination, flowering, or a stunting of growth.

Biogeochemical sampling may be useful because of the depth sampled by plant roots. At Mineral Butte the copper content in selected plants (mesquite and acacia) gave anomalous values 2 km from the nearest mineralized outcrop (Chaffee, 1982). Biogeochemical sampling suffers from the drawback we mentioned above when discussing caliche and desert varnish sampling, namely the problem of uniformity in sample density. Since there is unlikely to be equal plant distribution, some plants having very limited distribution, at least two species must be sampled. Other problems with sampling plants are the variation in element content of the ashed material with the time of sampling and the effects of minor climatic variations upon the uptake of elements by the plants.

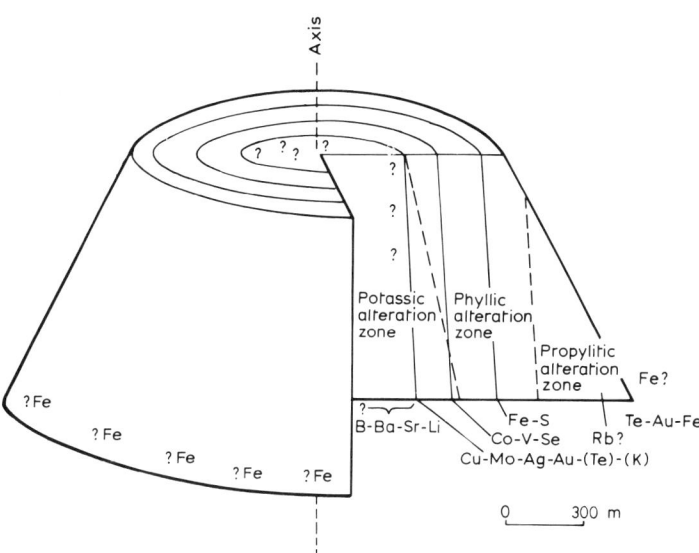

Fig. 3.16 Geochemical model of the Kalamazoo–San Manuel porphyry copper deposits (after Chaffee 1982).

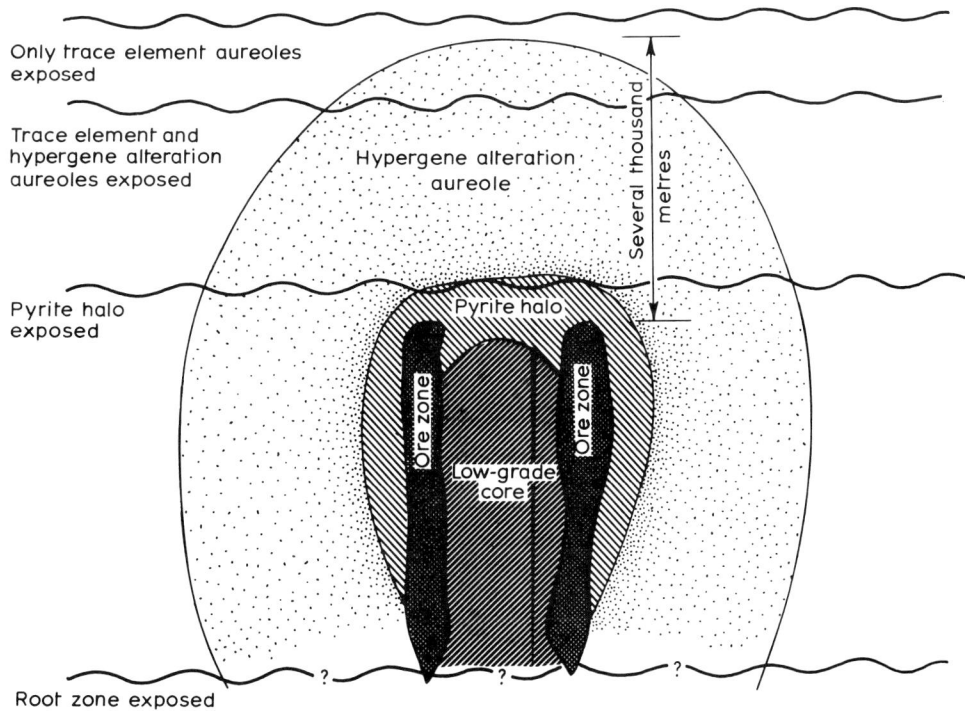

Fig. 3.17 Schematic diagram of porphyry copper system, showing hypothetical levels of erosion (after Lowell and Guilbert 1970, and Chaffee 1982).

(v) Groundwater and soil gas sampling

Both these techniques are more useful in arid areas than in humid areas. In arid areas wells may be sampled and the pumping effect of the well may mean that the groundwater is representative of wide areas. Soil gas sampling is probably to become very important in the search for 'blind' porphyry systems, that is those with no surface expressions. SO_2, H_2S and Hg emanations can all be sought because of the size of oxidizing sulphide bodies.

(d) Detailed geological mapping

Although the indirect methods, geochemistry and geophysics, provide many data to indicate suitable targets for further investigation, the role of careful geological mapping must not be underestimated. Detailed geological, structural and alteration mapping has to be undertaken at all available outcrops. The exploration geologist must consider the various features of the porphyry system which may be exposed by erosion (Fig. 3.17). The structural and alteration characteristics of porphyry copper systems are such that this geological approach is amongst the most significant ones adopted. All float material must be examined for alteration characteristics. Detailed laboratory petrological and mineralogical trace element studies on collected specimens will be undertaken to identify the mineralized zone. As we have seen, this mineralized area may be overlain by a leached capping. Certain elements may be enriched in these leached cappings. In a highly leached deposit, the Berg porphyry copper in British Columbia, Heberlein *et al.* (1983) showed a surface enrichment of Mo–Pb–Ag–F over the mineralized zone.

(e) Characteristics of the leached capping or gossan

Lithogeochemistry is used during the detailed reconnaissance stage to delimit the alteration zones as we have outlined above. It is also used in more detail, once the target has been selected, to characterize the leached capping (Anderson, 1982). Field estimates are made of the limonite and copper mineralogy of the capping. From these, estimates are made of the maximum and minimum copper grades of the former sulphide zone. Detailed mapping of the limonite and copper mineralogy and geochemical analyses for copper in the rock lead to semi-quantitative predictions of the primary grade and grade of the chalcocite enrichment zone. These predictions can be proven only by extensive diamond drilling. Anderson (1982) quotes figures for comparisons at 18 porphyry copper properties in southwestern USA. Here drilling results showed that over 75% of the primary-grade predictions agreed within 30% of the actual (drilling-determined) values and over 80% of the chalcocite-enrichment predictions agreed within 20% of the values obtained by drilling.

Although capping interpretation is very useful it is not a universally applicable definitive method of porphyry copper evaluation. The petrology of the cap may be affected by carbonate host rocks below the surface which inhibit copper migration or faults may be present which offset the sulphide mineralization. Consequently the cap evaluation must form only part of an integrated survey involving geology, geophysics, geochemistry and, subsequently, drilling.

3.4 PORPHYRY MOLYBDENUM DEPOSITS

3.4.1 Introduction

Although commercial quantities of molybdenum occur as by-products or co-products in porphyry copper deposits this is not the major source of molybdenum. A variety of other deposits contain economic concentrations of molybdenite including skarns, quartz veins and greisens. The most important source of molybdenum is the porphyry molybdenum deposits. White *et al.* (1981) provide an outline of molybdenum deposits and divide the porphyry molybdenum deposits into the Climax type and the quartz monzonite type based upon the petrology of the causative

Magmatic hydrothermal deposits

Table 3.2 Comparison of types of porphyry molybdenum deposits (from White et al., 1981).

Characteristics	Climax type	Quartz monzonite type
Co-genetic rock type	Granite porphyry	Quartz monzonite porphyry
Intrusive phases	Multiple intrusions of granite	Composite intrusions of diorite to quartz monzonite
Intrusive type	Stock	Stock or batholith
Ore body type	Stockwork	Stockwork
Ore body shape	Inverted cup	Inverted cup, tabular
Ore grade, average % MoS_2	0.30 to 0.45	0.10 to 0.20
Ore body tonnage	50 to 1000 million tonnes	50 to 1000 million tonnes
Disseminated MoS_2	Rare	Rare
Age	Middle to late Tertiary	Mesozoic and Tertiary
Fluorine minerals	Fluorite, topaz	Fluorite
Bismuth minerals	Sulphosalts	Sulphosalts
Tungsten minerals	Wolframite (huebnerite)	Scheelite
Tin minerals	Cassiterite, stannite	Rare
Copper minerals	Rare chalcopyrite	Minor chalcopyrite
Silicification	High silica core	No high silica
Greisenization	Greisen common	No greisen
Ore zone Cu: Mo ratio	1:100 to 1:50	1:30 to 1:1

igneous body. Table 3.2 shows some of the differences between the two types as recognized by White and co-authors. Sillitoe (1980) distinguished two broad categories of porphyry molybdenum deposits based on their tectonic setting. These were subduction related and rift related, a distinction which he stated accounted for some of the geological differences in known porphyry molybdenum deposits. Westra and Keith (1981) proposed a broad twofold division into (a) calc-alkaline stockworks related to granodioritic to granitic differentiates of calc-alkalic to high-potassium calc-alkalic magma series and (b) alkali-calcic and alkalic stockworks related to granitic differentiates of alkali-calcic and alkalic series.

3.4.2 General characteristics of porphyry molybdenum deposits

(a) Distribution of deposits in time and space
Porphyry molybdenum deposits have been recognized in western Canada, western USA, Mexico, Peru, Greenland, Yugoslavia, China and the USSR. Comparison with the global distribution of porphyry copper deposits (Fig. 3.4)

shows a marked similarity but not exact coincidence. Far fewer deposits of porphyry molybdenum have been worked commercially than porphyry coppers. Their more limited occurrence suggests somewhat narrower geological constraints on their formation than those applicable to porphyry copper. To illustrate the difference in localized geographical distribution we may quote examples from two states in the USA. Arizona, where so far no economic porphyry molybdenum deposits have been developed, contains in excess of 30 porphyry copper deposits; Colorado, on the other hand, which has a paucity of porphyry copper deposits, contains three of the largest porphyry molybdenum deposits in the World – Climax, Henderson and Mt Emmons (Fig. 3.18).

Most porphyry molybdenum occurrences are Mesozoic and Tertiary in age. Ages of the deposits in the western USA, the Colorado–New Mexico molybdenum province, range from about 17 to 140 million years. The oldest in British Columbia is the Endako deposit, dated at about 140 million years. The Malmbjerg deposit in Greenland is Middle Oligocene, and Jurassic porphyry molybdenum deposits have been recorded from the USSR.

(b) Size and grade of deposits

The molybdenite content of the Climax type of porphyry molybdenum deposits ranges from less than 0.1% to more than 1%, the average grade being between 0.3 and 0.45% MoS_2. The quartz–monzonite type has average grades nearer 0.15% MoS_2. The ore body tonnages for both types are similar, ranging from approximately 50 million tonnes to over 1000 million tonnes (Fig. 3.19). The largest concentration of Climax-type deposits, and of porphyry molybdenum deposits worldwide, occurs in Colorado; the sizes of these deposits are given in Table 3.3. Two of the largest quartz monzonite-type deposits are Mt Tolman, Washington, and Quartz Hill in Alaska.

(c) Mineralogy

In Climax-type deposits most, perhaps 90%, of

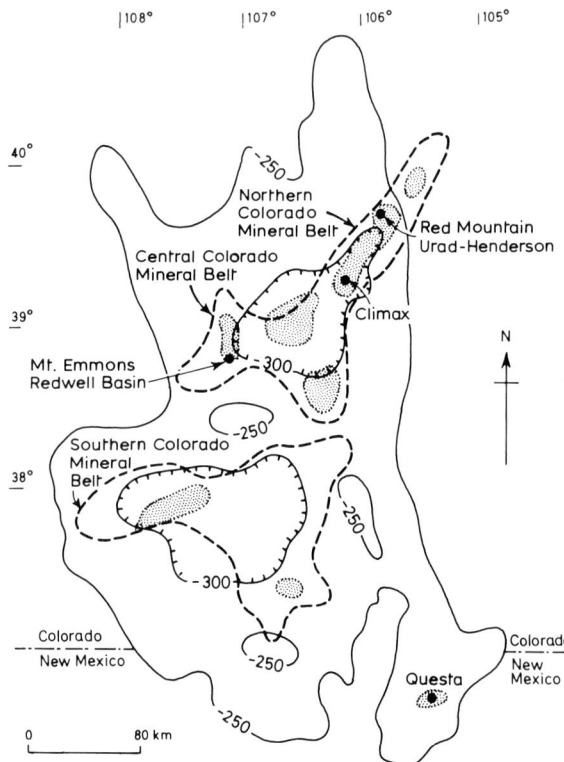

Fig. 3.18 Areas of known molybdenum occurrences (stippled) and selected major molybdenum ore deposits in Colorado and Northern New Mexico. Bouger gravity contours in milligals. (After White *et al.* 1981.)

the molybdenum is in quartz–molybdenite veinlets forming a stockwork. The molybdenite commonly occurs along the veinlet walls but may appear as discontinuous layers and disseminations within the quartz veins. Minor molybdenite may form coatings on fractures and can occur as disseminated crystals in porphyries, breccias and other associated igneous rocks.

In the majority of these systems the molybdenite is associated with pyrite and with fluorine- and tungsten-bearing species. Scheelite is a common accessory mineral in the quartz monzonite type and may be a by-product. These quartz monzonite systems may contain copper but chalcopyrite is rare in the Climax type.

Magmatic hydrothermal deposits

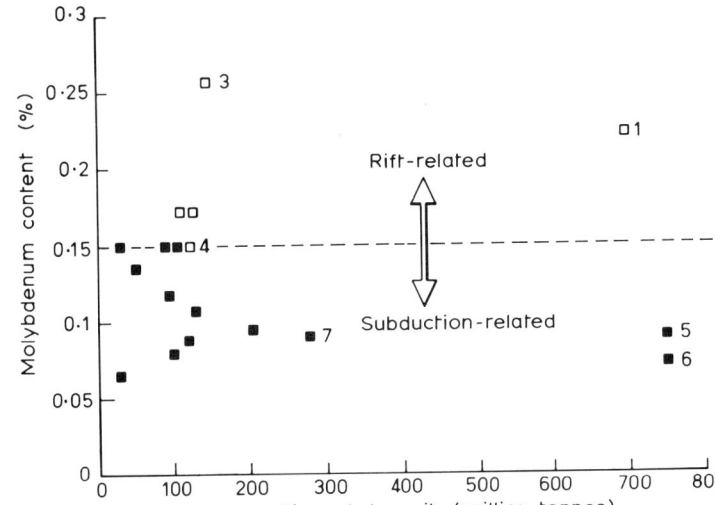

Fig. 3.19 Tonnages and grades of selected subduction-related (■) and rift-related (□) porphyry molybdenum deposits (after Sillitoe 1980). Deposits mentioned in the text: *Rift related:* (1) Climax, Colorado; (2) Henderson, Colorado; (3) Mt Emmons, Colorado; (4) Malmbjerg, Greenland. *Subduction-related:* (5) Quartz Hill, Alaska; (6) Mt Tolman, Washington; (7) Endako, B.C.

Table 3.3. Climax-type molybdenum deposits of Colorado. (Sources: White *et al.* 1981; Thomas and Galey 1982.)

Deposit	Size (0.2% MoS_2 cut off)	Thickness (m)	Reserves (million tonnes)	Grade (%) MoS_2	Status
Climax	1300 × 900 m	250	400	0.45	Underground mine
Henderson	1000 × 750 m	<300	300	0.49	Underground mine
Mt Emmons	700 m diam.	100	150	0.44	Mine design

It appears that some of the quartz monzonite systems are intermediate between porphyry molybdenum deposits and porphyry copper deposits. In such deposits the copper-to-molybdenum ratio ranges from 1:1 to 1:10. An example of such an intermediate deposit type is Nevada Molybdenum, USA. Tin may occur as cassiterite or stannite in Climax-type deposits but is rare in the quartz monzonite type.

(d) Host lithology

Porphyry molybdenum deposits are associated with porphyritic intrusions ranging from quartz monzonite to high-silica alkali-rich granites. It is on the basis of the composition of the co-genetic intrusion that White *et al.* (1981) have subdivided porphyry molybdenum deposits into a quartz monzonite subclass and a Climax- (granite-) type subclass. The latter subclass is named after the Climax mine, Colorado, which has been the World's largest single producer of molybdenum for many years.

In Climax-type deposits the dominant intrusives are granite porphyries with rhyolite porphyry caps if the system vented. The intrusives are highly differentiated. Ore deposits belonging to the quartz monzonite type are associated with small, usually composite, stocks and late stages of batholiths. Differentiation trends are frequently observed in the systems, beginning with diorite or quartz diorite and passing through granodiorite to quartz monzonites. The quartz monzonites are often well differentiated and in places approach granite in composition (White *et al.*, 1981). The geological characteristics of some of the more important quartz monzonite types, those occurring in British Columbia, are described by Soregaroli and Sutherland Brown (1976).

Important in the genesis of Climax-type ore bodies is the cyclical nature of the intrusion of the host plutons and the associated mineralization. The recognition of this cyclicity and the possibility of the occurrence of multiple phases of mineralization within a relatively short span of geological time led to the discovery of the Henderson ore body and may be crucial to the discovery of further Climax-type bodies. Where multiple intrusion and mineralization have occurred the later intrusions do not rise as high in the vertical sequence as the earlier intrusions and their associated ore bodies. This may be seen at Red Mountain, where the early Urad ore body occurs some 800 m higher than the later Henderson ore body (Fig. 3.20), and is also shown at Climax and Mt Emmons (Figs 3.21 and 3.23).

(e) Alteration

As with porphyry copper deposits the hydrothermal alteration of the rocks surrounding the ore bodies is characteristic of porphyry molybdenum deposits. The alteration halos greatly enlarge the target for exploration and may be the only surface expression of underlying molybdenum mineralization (Fig. 3.23). The major zones are comparable to those of the porphyry coppers – potassic zone, phyllic zone, an argillic zone which may be divisible into upper and lower zones, and the prophylitic zone. In addition to these there are minor zones defined by specific alteration mineral

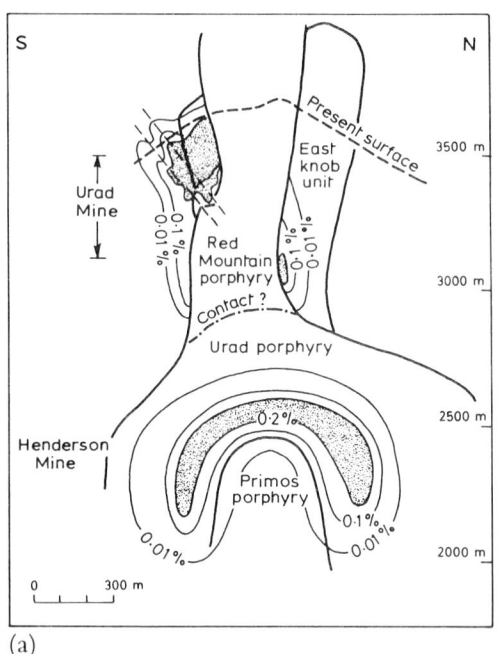
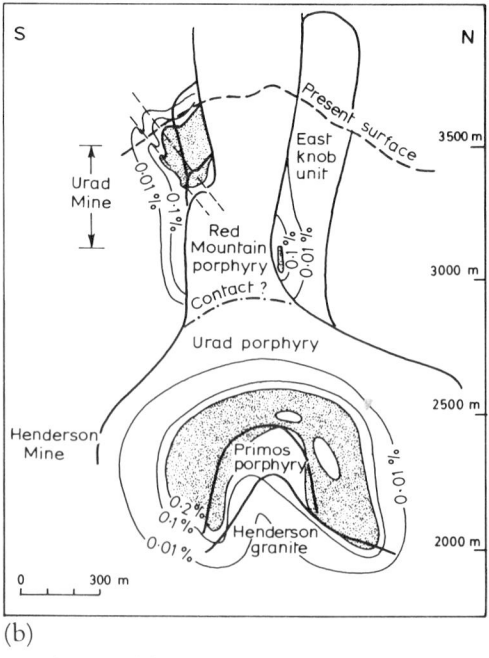

Fig. 3.20 (a) Diagrammatic section showing intrusion of Primos porphyry and formation of upper lobe of Henderson ore body. (b) Diagrammatic section showing intrusion of Henderson granite and renewed molybdenite mineralization. Stippled areas, $>0.2\%$ MoS_2. (After Wallace *et al.* 1978.)

Magmatic hydrothermal deposits

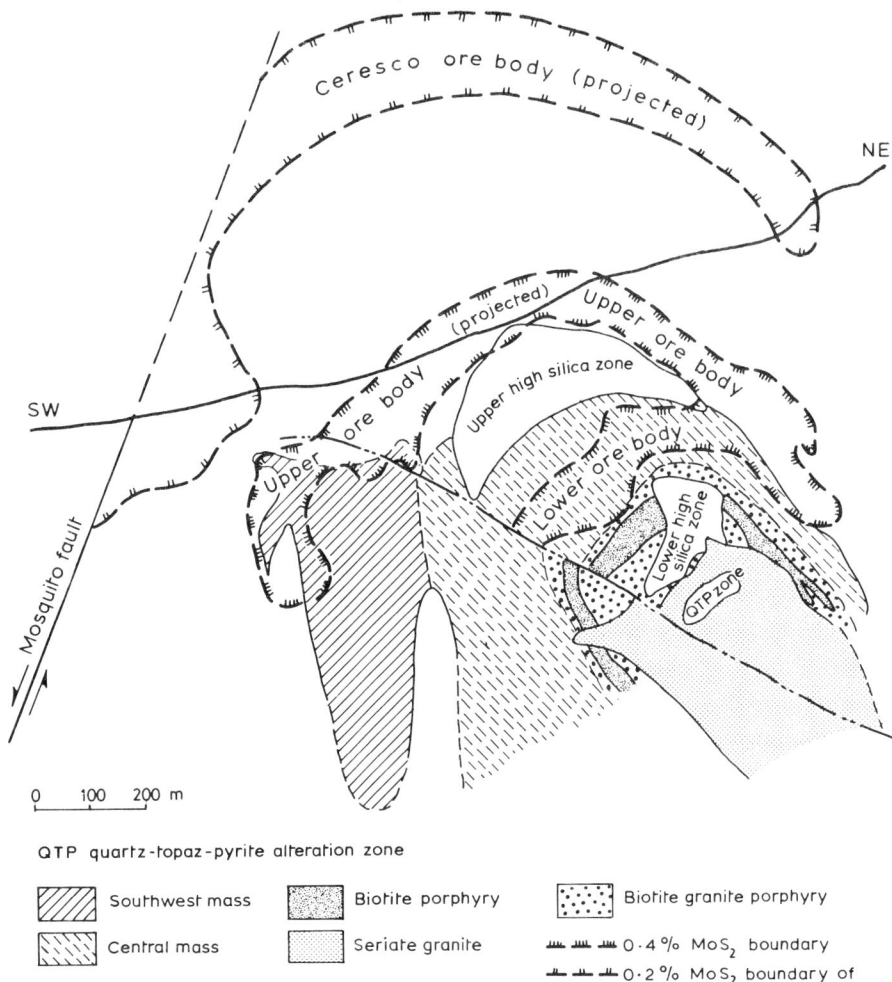

Fig. 3.21 Climax ore bodies, Colorado (from White *et al.* 1981).

species, which are distinct from the major alteration assemblages in which they occur. These minor zones are a vein silica zone, pervasive silica zone, magnetite and topaz zones, a greisen zone and a garnet zone.

(i) Potassic zone

Here the diagnostic feature is the total replacement of plagioclase by potassium feldspar and the zone is often bordered by a rock which shows pink potassium feldspar, relatively fresh primary biotite and plagioclase feldspars only partially replaced by potassium feldspars. The potassic zone almost exactly coincides with the ore body in the Henderson deposit (Fig. 3.20). Potassic alteration takes place before, during and after molybdenite mineralization in most deposits and the close spatial relationship noted for Henderson occurs in many other deposits (for example, Climax).

At Mt Emmons the potassium feldspar zone is veined, and pervasively flooded, by quartz, magnetite and biotite; the grade of molybdenite averages less than 0.1% MoS_2 and this represents a low-grade footwall zone to the molybdenite ore body. Potassic alteration is prominent in all deposits in the Canadian Cordillera and most deposits show a close spatial relationship between this alteration and molybdenum mineralization.

Field evidence supports experimental data in the hypothesis that late magmatic hydrous potassium-rich silicate melts play a significant role in concentrating molybdenum (Westra and Keith, 1981).

(ii) Phyllic zone

This alteration zone is characterized by the sericitization of potassium feldspar and plagioclase, and by abundant pyrite. It may show a change in dominant mineral component and pervasive alteration downwards, towards the intrusion, from sericite to pyrite to quartz. In the Henderson deposit this zone partially overlaps the ore body and extends almost 450 m upwards from it. At Mt Emmons it extends about 400 m above the upper limit of potassic alteration. The pyrite content of the phyllic zone at Mt Emmons averages 6%. At Climax the pyrite content of the pyritic zones averages between 3 and 6% whereas the pyrite zone at Henderson contains between 6 and 10% pyrite generally carrying little or no molybdenite. A phyllic zone occurs in most porphyry molybdenum deposits in the Canadian Cordillera, and at Endako and Boss Mountain sericite is directly related in variable degrees to molybdenum mineralization.

(iii) Argillic zones

The upper portion of the argillic zone is marked by the argillization of plagioclase feldspar, the feldspar being replaced by montmorillonite, kaolinite and sericite. According to White *et al.* (1981) montmorillonite is more common on the margins and kaolinite is more abundant in the interior. Potassium feldspar is relatively unaltered and quartz remains fresh. Biotite is altered, being replaced by muscovite, sericite, small quantities of rutile, leucoxene, pyrite, carbonate and fluorite. With depth the lower argillic zone is reached which is characterized by argillization of primary and secondary feldspars. Kaolinite is the most common clay mineral. In porphyry molybdenum deposits of the Canadian Cordillera argillic alteration is reported from only three deposits and is genetically important at one, Endako, where it accompanies molybdenite introduction (Soregaroli and Sutherland Brown, 1976).

(iv) Propylitic zone

The typical propylitic assemblage is chlorite–epidote–pyrite but will also contain calcite, clay and sericite which probably derive from the plagioclase. The chlorite of this zone is an alteration product of biotite and the appearance of unaltered biotite may mark the outward boundary of this zone. Propylitic alteration is common in the outer portions of many deposits in the Canadian Cordillera, for example Boss Mountain where early propylitized fragments are found in breccias which have subsequently been biotitized.

(v) Vein silica zone

This occurs in the potassium-feldspathized rock and is marked by much quartz veining. The average quartz content may be increased from 40 to 70%.

(vi) Pervasive silica zone

In the hydrothermally altered rocks, areas may occur which contain more than 90% quartz. These are within the K-feldspathized zone and may have destroyed most of the characteristic features of that zone. The borders of this zone may contain concentrations of green biotite, topaz and chlorite which White *et al.* (1981) suggest may have been purged from the areas of intense silicification.

(vii) Magnetite and topaz zones

Although these two minerals occur together they do not appear to be genetically related. The magnetite content of the potassic alteration zone may reach 30% (Thomas and Galey, 1982), the magnetite occurring in discontinuous monomineralic veinlets less than 6 mm wide. The topaz occurs in veins up to this width, or adjacent to such veins which also carry quartz and pyrite. Where the topaz-bearing veinlets cut quartz–sericite–pyrite rock or feldspar rock the topaz replaces feldspar or its pseudomorphs.

(viii) Greisen zone

A typical greisen vein is largely quartz and molybdenite with some topaz, pyrite, magnetite and muscovite. Greisen veins may cut through the ore body as at Henderson and they cut the K-feldspathized rock.

(ix) Garnet zone

The garnet zone is typified by the occurrence of spessartine garnet and at Henderson the zone overlaps the argillic and phyllic zones. The garnet here is intimately associated with a suite of galena–sphalerite–rhodochrosite and is part of a late-stage hydrothermal event.

(f) Tectonic setting

Sillitoe (1980) proposed the division of porphyry molybdenum deposits into two main types, subduction related and rift related. He assigns Climax, Henderson, Mt Emmons and Questa to a subdivision of the rift-related class, characterized by back-arc rifting induced by slowing, steepening and eventual cessation of subduction. He noted that during rifting, the back-arc region of the western USA changed from calc-alkaline to basalt-rhyolite magmatism and the metallogeny of the back-arc region changed from chalcophile, dominantly copper, to lithophile. The Climax-type magmas and ore bodies apparently formed during a relatively quiescent, atectonic interval that followed subduction-related compressional tectonism and calc-alkaline magmatism but preceded strong rift-related extensional faulting and occurred without, or before, known local rift-related basaltic volcanism. An illustration of the relationship between the stockwork molybdenum deposits, the porphyry copper deposits and a convergent plate margin is shown in Fig. 3.22.

The porphyry molybdenum deposits of the Canadian Cordillera also contain examples of ore bodies formed after a major phase of compressional stress had waned. This may be illustrated by the Endako deposit, at 140 million years the oldest economic porphyry molybdenum occurrence in the Cordillera. Here internal graben and antithetic vein structures within the deposit suggest distension but the age dating places this deposit within the Columbian compressional event. This suggests that the deposit formed late in the event when the compressional stresses had waned.

3.4.3 Genesis of porphyry molybdenum deposits

(a) Genesis of Climax type

The lower crustal partial melting which formed the calc-alkaline batholiths of the Colorado mineral belt was followed by the Climax-related episode of partial melting. This latter event probably occurred in the preheated upper portion of the lower crust; however, Westra and Keith (1981) propose that the enrichment process started in the mantle. They then invoke deep crustal material to result in the formation of a fluorine-rich, high silica and K-rich granitic melt. This part of the crust had not been depleted in granitic constituents and incompatible trace elements. The molybdenum present is likely to be concentrated in the K-rich residual melt. As the final fluid crystallizes to quartz and K-feldspar, excess potassium is introduced into the wall rock. A reduction of confining pressure allows the release of a large volume of less saline magmatic hydrothermal fluid which is responsible for the main mineralization. Westra and Keith suggest magma convection as the most likely process to supply the volume of hydrothermal fluids, the molybdenum and the thermal energy.

Extensive mixing of magmatic and meteoric

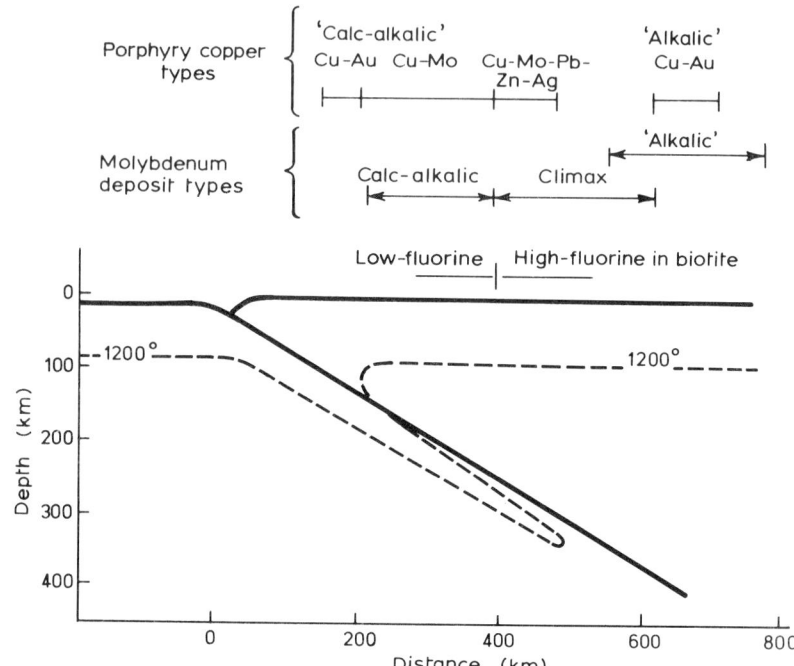

Fig. 3.22 Schematic relationship between stockwork molybdenum deposits and porphyry copper deposits in a convergent plate margin setting (from Westra and Keith 1981).

hydrothermal fluids occurred in the quartz–sericite–pyrite and argillic zones but within the ore zone magmatic fluid pressures prevented the influx of meteoric fluids beyond the edge of the zone of stockwork fracturing. The importance of meteoric waters for the formation of the molybdenite ore shell appears minor (Westra and Keith, 1981).

The intrusions associated with Climax-type ore bodies may have risen diapirically and the multiple intrusions typical of these deposits suggests repeated activity, possibly with successive intrusions rising along zones of weakness created by previous diapirs. The successive intrusions indicate increasing differentiation of a parent magma. As the diapirs rose higher they encountered zones of stronger thermal gradients and changes in stress patterns. The end result was the extreme upward concentration of volatiles and molybdenum. At Climax the intrusive cycle repeated at least four times, at Mt Emmons–Redwell at least three times and at Henderson at least five times.

The depth of formation for the Urad ore body was about 600 to 800 m below surface; for the Henderson ore body it was about 1600 to 2000 m below surface. The Ceresco ore body at Climax appears to have formed between 3300 and 4000 m below surface. Therefore in the Urad ore body the pressure was almost entirely hydrostatic and the flow of heat and volatiles would have been primarily upward from the magma through fractured permeable rocks, towards the surface. Such a stress regime would result in an almost cylindrical deposit with the Urad ore body little more than an aberrant bulge on the cylinder. Climax and Henderson, on the other hand, were well insulated and the hydrothermal solutions moved outwards as well as upwards because of the more homogeneous conditions. The more general 'umbrella'-shaped deposits were formed here. The greater depth of formation of the Climax ore bodies and the more gradual thermal gradient may be one of the reasons that the ore bodies formed further from the causative plutons than either those at Urad or Henderson. Another reflection of the more gradual thermal gradient may be the extent of low-grade molybdenite mineralization on the hangingwalls of the deposits. At Climax the 0.1% MoS_2 zones may be up to 170 m thick whereas at Henderson the equivalent zone is only 65 m thick.

At Climax and Red Mountain it is possible to correlate a specific ore body with a discrete intrusive phase of a composite intrusion. At Mt Emmons only one episode of economic molybdenum mineralization has been recognized to date, that associated with the Red Lady stock. At Climax, following the intrusion of the first phase of the Climax stock, all subsequent phases produced ore and were emplaced close to, but generally below, their predecessors, and each ore body partially overlaps the next older. At Red Mountain the cycle of intrusion-productive mineralization was interrupted by the intrusion of the Red Mountain porphyry with which there is no associated economic mineralization. This porphyry separates the Urad and Henderson ore bodies in time; these ore bodies are also separate in space by about 800 m (Fig. 3.20). Wallace *et al.* (1978) proposed that the Red Mountain porphyry vented and the accompanying mineralizing fluids were dissipated at surface.

The genesis of the Henderson ore body is viewed by White *et al.* (1981) as related to the intrusive suite of the Primos intrusive. This intrusive injected and fractured the host Urad porphyry stock. Inward cooling and fractional crystallization of the Primos magma produced an aqueous phase which collected in areas now marked by micrographic textures and pegmatites. Continued accumulation of fluids and build-up of hydraulic overpressures caused hydrofracturing of the chilled margin of the intrusive and its surrounding host. The ore fluids were released to form part of the Henderson ore body. Quenching, along with the self-sealing nature of the quartz-molybdenite mineralization, formed a new relatively impermeable carapace that allowed the process to repeat again and again until exhausted.

(b) Genesis of calc-alkaline molybdenum deposits

These are associated with calc-alkaline and high-K calc-alkalic arc magmas. The parent magmas are low in fluorine and molybdenum. They may occur as stock-type or as plutonic-type stockworks. The lack of co-magmatic extrusive rocks suggests the magma had a low water content or was emplaced at rather deep levels in the crust. Formation of the zone of stockwork fracturing started a complex magmatic hydrothermal convective system. A central zone dominated by magmatic hydrothermal fluids at lithostatic pressure was surrounded by a large meteoric convection system under hydrostatic pressure (Westra and Keith, 1981). In the plutonic stockworks high confining pressures prevent boiling of the magmatic hydrothermal fluid. Lack of boiling stops effective H_2S fractionation into the vapour phase and the pyrite halo is weakly developed or lacking in these plutonic stockworks.

Owing to the low initial molybdenum content of the parent magma and the relatively inefficient concentration mechanisms in the magma, the final grade in the hydrothermal system will rarely exceed 0.25% MoS_2.

3.4.4 The Mount Emmons porphyry molybdenum deposit

The Mt Emmons deposit is one of three molybdenite deposits which underlie this mountain, 30 km north of Gunnison in west central Colorado. The other two are the Upper and Lower Redwell deposits which are smaller and leaner than the Mt Emmons deposit (Fig. 3.23). The detailed structural control for the occurrence of these deposits is unknown but they do not appear related to the same structural lineaments as the Climax and Henderson deposits which lie over 80 km to the north east (Fig. 3.18).

(a) Geology

The rocks of Mt Emmons are Cretaceous, dark carbonaceous shales and sandstones, overlain by

Tertiary sandstones. The stocks which intrude these sediments are fine grained to coarsely porphyritic granites dated as Miocene–Pliocene. A rhyolite pipe and intrusive breccia followed the emplacement of the stocks. This breccia contains rounded fragments of rhyolite, shale, hornfels and less commonly rhyolite porphyry fragments.

Upper and lower molybdenum deposits occur in the Redwell Basin area, northwest of the summit of Mt Emmons. The upper molybdenum deposit is a stockwork of veinlets in brown hornfels and rhyolite porphyry. Using a 0.1% MoS_2 assay cut-off this upper deposit contains about 20 million tonnes averaging 0.18% MoS_2. Neither the upper nor the lower deposit appears to have an inverted cup shape which is so characteristic of the other Climax-type porphyry molybdenum ore bodies (Thomas and Galey, 1982).

The molybdenite is fine grained and occurs chiefly on the borders of the fine-grained light-grey quartz veinlets. The lower deposit lies entirely within the Redwell stock and is a widely spaced network of sharp-walled quartz veinlets. These are relatively rich in molybdenite and contain trace amounts of K-feldspar, fluorite and pyrite.

The phase of the stock which contains the lower deposit, unlike the porphyry that hosts part of the upper deposit, is depleted in fluorine, weakly enriched in K_2O and strongly enriched in Na_2O.

(b) Mt Emmons molybdenite deposit

The following description is based on Thomas and Galey (1982). This deposit lies south-east of the Redwell Basin deposits and is related to the upper contact of the Red Lady stock. This is a hydrothermally altered cylindrical-shaped composite intrusion. The crest of the stock is potassically altered but contains relict phenocrysts such as those in the porphyritic core of the intrusion. The potassic alteration zone overlaps the top of the stock and adjacent hornfels, shaped like a shallow inverted bowl. Little of this K-feldspathized zone remains as it has been extensively veined, flooded, and largely replaced by fine-grained magnetite, quartz, minor biotite and trace molybdenite. The magnetite content varies from 10 to 30%. The grade of molybdenite in the intense potassic zone averages less than 0.10% MoS_2, representing a low-grade footwall fringe to the molybdenum ore body.

In the Red Lady Basin the phyllic zone extends about 400 m above the upper limit of potassic alteration and at the surface lies somewhat east of the projection of the stock. From the surface downward the dominant mineral component of veinlets and pervasive alteration changes from sericite to pyrite and quartz. The pyrite content of the propylitic zone averages about 1% and locally reaches 6–8%.

Pyrrhotite occurs in a 500-m-thick zone which thins as it nears the zone of phyllically altered rocks above the stock (Fig. 3.23). The pyrrhotite content varies from about 3% in unaltered hornfels to less than 1% near the phyllic zone.

The Mt Emmons deposit is a contact-related stockwork of quartz veinlets containing molybdenite and variable amounts of fluorite, pyrite and very minor huebnerite. The thickest part of the ore shell and best grades of molybdenum are found on the steepest flanks of the stock. Estimates made by Amax indicate reserves of about 150 million tonnes of mineralized porphyry and hornfels with average grades of 0.44% MoS_2.

Molybdenite in the ore zone is generally fine grained and is concentrated principally in thin discontinuous ribbons in quartz veins containing variable but minor amounts of fluorite, pyrite, biotite and K-feldspar. Where the host rock is porphyritic, molybdenite also occurs as fine grains coating phenocrysts, as fine disseminations in the matrix and in short discontinuous gashes lacking quartz.

(c) Emplacement of the intrusions and genesis of the ore bodies

Although the structural controls which localized the stocks in the vicinity of Mt Emmons are unknown, the effects of the stocks themselves are well documented (Thomas and Galey, 1982). Contact metamorphism physically changed the rocks overlying the stocks and resulted in

Magmatic hydrothermal deposits

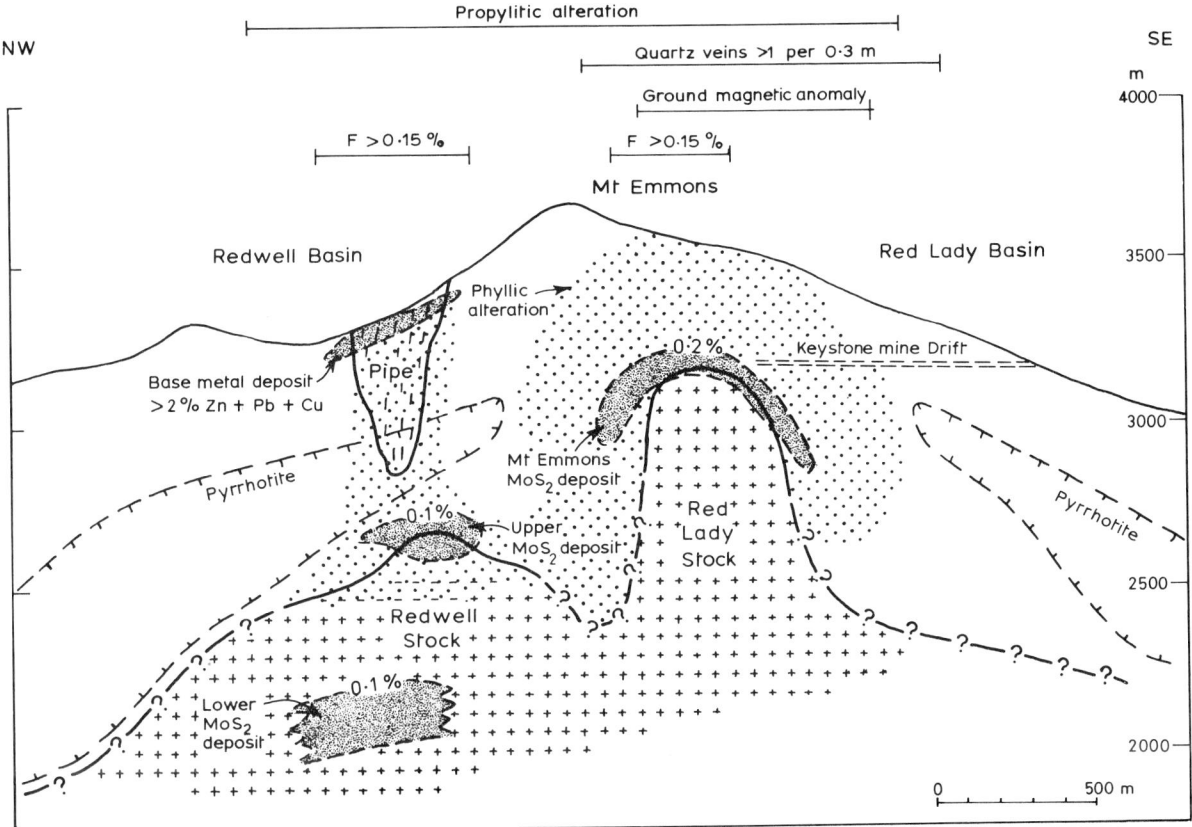

Fig. 3.23 Diagrammatic section showing the three molybdenite deposits at Mt Emmons, Colorado. Extent of propylitic alteration, quartz vein intensity >1 per 0.3 m, ground magnetics and fluorine anomalies also shown (after Thomas and Galey 1982).

increased brittleness and the ability to sustain fractures. The rhyolite of the pipe crystallized from a fluid enriched in tungsten, fluorine, rubidium and niobium that was 'partitioned' from the underlying stock. The fluid moved into the crackled hornfels overlying the rhyolite porphyry cupola and exploded upward from a point where impervious shales blocked further upward migration. Intrusive breccias in the Redwell and Red Lady Basins carried fragments of porphyry containing molybdenite at least 300 m upward from underlying deposits.

In comparing the geology of the Redwell and Red Lady Basins the upper Redwell deposit is over a small rhyolite porphyry cupola which apparently vented as a moderate-sized breccia pipe. On the other hand the Mt Emmons ore body is associated with a much larger stock which rose to a higher level and does not appear to have vented as a pipe. If a pipe formed well above the Red Lady stock as one formed above the Redwell Basin cupola all traces have been removed by erosion. Pervasive silicification, hydrothermal magnetite, and potassic alteration were significant features of the hydrothermal system associated with the Red Lady stock but were not major in the Redwell system.

The presence of orthoclase in quartz–molybdenite veins of the deposit and the fact that molybdenite-bearing veins cut the pervasively K-feldspathized zone suggests that the period of molybdenum mineralization was essentially contemporaneous with, or slightly later than, the period of K-feldspathization. Weak to moderate localized sericitic alteration of K-feldspar in the veins suggests that a period of weak phyllic alteration followed molybdenite mineralization.

(d) Exploration at Mt Emmons

Intrusive breccias, with fragments of porphyry containing molybdenite derived from the underlying deposits, are the only direct evidence at the surface of underlying porphyry molybdenum systems. The surface expression of the Mt Emmons deposit is a broad zone of phyllically altered rocks cut by a stockwork of quartz veins and veinlets. The area had been worked for base-metal vein mineralization. Extensive drilling in the Redwell Basin revealed extensions of this shallow base-metal mineralization and identified the two deep, low-grade molybdenum deposits.

The geochemical expression of the Mt Emmons molybdenum deposit is very subtle at the surface. Molybdenum is not anomalous in either stream sediments or waters; it is above background in only a few special samples of vein material and in the intrusive breccias referred to above. Lead, zinc, copper, tungsten and fluorine were anomalous in stream sediments. The last two elements generally corresponded to the zone of phyllically altered rocks when rock chip samples, collected on a 125 m grid, were examined.

Exploration for the Mt Emmons deposit involved surface drilling and reinvestigation of the Keystone Mine drift (Fig. 3.23). A regional aeromagnetic survey and a study of the down-the-hole physical properties were the only geophysical work carried out prior to the discovery. A hole was drilled north-east from underground in the Keystone Mine. This hole discovered a high-grade porphyry molybdenum system. After the discovery of magnetite in the footwall the regional aeromagnetic data were reinterpreted to show a small amplitude, closed magnetic high south of the surface projection of the Red Lady stock. A series of ground magnetic traverses confirmed the presence of this anomaly and sharpened the detail. Gravity surveys were also run but these proved inconclusive whereas complex resistivity responded well to the Keystone vein system and the high pyrite concentration in the phyllically altered rocks.

After the initial discovery the exploration programme involved about 15 000 m of drilling in a total of 20 holes to define the ore body. The drilling was guided, in part, by the magnetic surveys, rock chip sampling and additional detailed mapping. Figure 3.24 shows the definition of the target zone based upon these available data and the relationship of the ore body as outlined by drilling to these surface expressions. Although a full feasibility study and Environmental Impact Statement have been submitted for this project, the depressed World metal market has prevented the exploitation of this deposit at the time of writing.

3.5 EXPLORATION FOR PORPHYRY MOLYBDENUM DEPOSITS

The exploration philosophy for porphyry copper deposits, outlined earlier in this chapter, may be regarded as generally applicable to the search for porphyry molybdenum with some modifications. The first difference will be in the regional selection based upon the geotectonic setting of porphyry molybdenum compared to porphyry copper deposits.

The division of porphyry molybdenum deposits into back-arc rift-related and subduction-related deposits as described by Sillitoe (1980) may be used as the starting point for any exploration programme. Examples of suitable locations may be chosen from plate tectonic considerations. At present the highest grade and largest tonnage deposits are those belonging to the rift-related category of Sillitoe and further described as the Climax type (Fig. 3.19). Appreciable tonnage is available, albeit at a lower grade, in the subduction related category.

Magmatic hydrothermal deposits

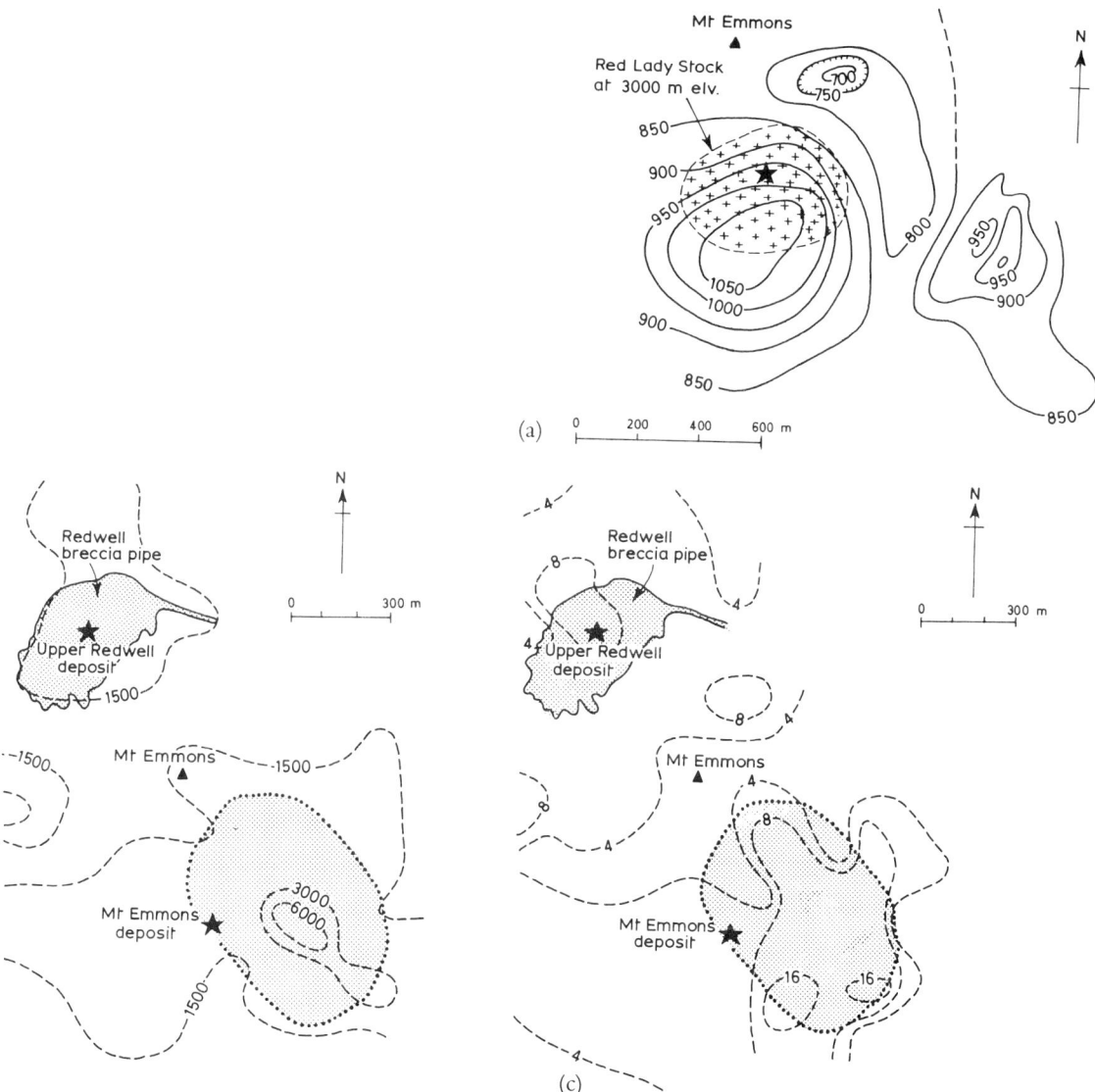

Fig. 3.24 (a) Results of ground magnetic survey over Red Lady Stock, Mt Emmons, Colorado, Contour values above 55 000 γ (total field measurements). Distribution of (b) fluorine and (c) tungsten in surface rock chip samples (values in ppm). Stars denote centre of molybdenite deposits projected to surface. Stippled areas in (b) and (c) show phyllically altered rocks at surface. (From Thomas and Galey 1982.)

Another method of defining areas with a high exploration potential, specifically for granite molybdenite systems, is by major element chemical 'fingerprints' (Mutschler et al., 1981). The following 'fingerprints' were defined: (a) SiO_2 >74.0 wt%; Na_2O <3.6 wt%; K_2O >4.5 wt% for fresh source rocks and altered rocks of potassic and quartz sericite assemblages. (b) SiO_2 >74.0 wt%; Na_2O <2.5wt%; K_2O > 4.5wt% which will reject fresh source rocks but will accept potassic and quartz–sericite assemblage rocks as well as some argillic altered samples. Having identified prospective areas it is still necessary to define the target more closely and sample and map it on the ground.

Within a chosen tectonic area remote sensing may be applied to define the target area more closely, and structural synthesis should be used. Despite the structural control evident in many deposits this control is not always evident, as in the case of Mt Emmons, and other methods of target definition are required. The percentages of magnetite within the deposits, as in the case of Henderson and Mt Emmons, means that regional aeromagnetic surveys are particularly relevant. The wide alteration halo surrounding the deposits enlarges the target considerably and the association of high fluorine and occasionally tungsten make soil geochemistry particularly attractive. The occurrence of high-level breccia pipes, often containing ore-bearing fragments from much deeper in the deposit, stresses the importance of careful geological mapping and lithogeochemistry. Ground magnetic surveys may be used to define more accurately the deposit and to guide drilling, as in the case of Mt Emmons.

3.6 PORPHYRY GOLD DEPOSITS

In describing porphyry deposits in this Chapter we have made a major subdivision into those producing copper and those producing molybdenum. But as we have seen, porphyry copper deposits can be made economically viable by their gold production. Others have appreciable by-product molybdenum, and Kesler (1973) divided porphyry copper deposits into molybdenum-rich and gold-rich categories. Since porphyry molybdenum deposits occur that carry minor copper the question arises do porphyry gold deposits occur that carry minor copper? We are not the first to ask this question. Sillitoe (1979) reviewed the evidence available from porphyry copper deposits with more than 0.4 g/t gold in an attempt to establish criteria for the occurrence of high gold content in porphyry deposits. He concluded that high gold content is not directly related to geotectonic setting, composition of the host intrusion, nature of the wall rocks, age of the mineralization, erosion level, size of the ore body or sericitic alteration.

Gold is closely associated with chalcopyrite (± bornite) mineralization. In the deposits studied by Sillitoe there was an abundance of magnetite, often accompanied by replacement quartz, in the gold-bearing feldspar-stable assemblage. Conditions conducive to this association would include abnormally high fO_2/fS_2 ratio in the magmatic environment the causes of which are unknown. The geochemical conditions for gold transport and deposition are the major factors for gold concentration and only porphyry systems which evolve these conditions will be gold-enriched. Why these conditions appear to be satisfied more commonly in island arc environments rather than in continental margins is, as yet, unknown.

Whatever the reason for the occurrence of gold-rich deposits the presence of a high magnetite content in many of them is of significance in exploration. When discussing exploration for porphyry copper deposits we suggested the use of airborne magnetic surveys as an integral part of the reconnaissance survey. If gold-rich porphyries are sought then the emphasis on magnetic surveys is even stronger, both airborne and as ground follow-up surveys. Drilling can be directed to the areas with the highest magnetic response if the gold-rich character of the deposit has been established. Sillitoe (1979) suggests that if leached capping appraisals (Section 3.3.e) indicate low-grade copper mineralization but a higher than normal magnetite content is shown in the feldspar-stable alteration zone more exploration may be warranted. If such high-gold low-copper deposits are discovered they may well be distinguished as porphyry gold deposits and these may become very important targets for exploration towards the end of this century.

3.7 PORPHYRY TIN DEPOSITS

Tin has been worked from stockworks in Germany, Cornwall, England and Australia.

Those in Cornwall, which occur as sheeted vein structures, have many characteristics in common with porphyry deposits. They appear to have been formed at a high level in the crust, and fluid inclusion studies show evidence of boiling. Hydrofracturing has been a major process in ground preparation for mineralization, and explosive brecciation has occurred. The mineralogy of these deposits is relatively simple with very few sulphides present. Perhaps the thermal system was insufficiently long-lived to allow meteoric waters to play a dominant role in the mineralization. Within the same metallogenic province there are greisen-bordered veins which have suffered argillic alteration and which have a more complex mineralogy including sulphides; here the hydrothermal system was longer-lasting with meteoric water becoming important; perhaps in these brecciation did not occur. (Thorne and Edwards, 1985.)

The tin in Cornwall belongs to a plutonic environment but Sillitoe *et al.* (1975) described porphyry tin associated with stocks in the subvolcanic portion of the Bolivian tin province south of Oruro. The stocks are considered to have been emplaced beneath stratovolcanoes and although these deposits have some fundamental differences from porphyry coppers, such as the lack of a potassic alteration zone, they have many features in common with them. There is pervasive sericitic alteration and propylitic alteration; also pyritic haloes are present. The grade of these deposits is between 0.2 and 0.3% Sn, which is higher than the majority of tin stockworks in the World but at present porphyry tin deposits are not economically viable.

3.8 VOLCANIC-ASSOCIATED MASSIVE SULPHIDE DEPOSITS

3.8.1 Introduction

Massive sulphide deposits were recognized as a genetically distinct group during the 1950s (Franklin *et al.*, 1981) and they occur associated with most upper crustal rock types. The most important massive sulphides are found with either pelites to semipelites (the 'shale-hosted' deposits described in Chapter 6) or associated with volcanic-dominated marine successions. The latter, often called volcanogenic or exhalative volcanogenic deposits, are major suppliers of copper, lead and zinc. They also have important by-product gold and silver.

3.8.2 Classification of volcanogenic sulphides

Attempts have been made to classify these deposits on various bases (Table 3.4). Solomon (1976) described a copper group, a zinc–copper group and a zinc–lead–copper group without any environmental criteria. A similar division is recognized by Franklin *et al.* (1981) who state that the subdivision into copper and copper–zinc may be artificial due to copper-producing deposits having zinc below cut-off value which is not reported (Fig. 3.25). Sawkins (1976) proposed a classification based upon the tectonic setting and associated volcanics:

(a) Cyprus type – associated with spreading ocean or back-arc spreading ridges and with basic volcanics usually ophiolites,

(b) Besshi type – associated with the early part of the main calc-alkaline stage of island arc formation, and

(c) Kuroko type – associated with the later stages of island arc formation and more felsic volcanics.

Stanton (1978) disagreed with these divisions, considering the volcanic ores to belong to one continuous spectrum showing geochemical evolution reflecting the evolution of the calc-alkaline rocks in island arcs. He suggested that the mainly pyritic copper deposits are spatially and temporally associated with tholeiitic and ultrabasic rocks. With the change to calc-alkaline volcanism the larger, zinc-rich deposits occur while the further development of the calc-alkaline trend to dacite and rhyolite is accompanied by the occurrence of lead-bearing deposits.

Hutchinson (1980) drew together the ore composition, the tectonic setting and the host

Table 3.4. Recent attempts to classify massive sulphide deposits

Hutchinson (1973)	Zn–Cu (principally Archaean)
	Pb–Zn–Cu–Ag (predominantly Proterozoic and Phanerozoic)
	Cupreous pyrite (principally Phanerozoic)
Sangster and Scott (1976)	Predominantly volcanic-associated
	Predominantly sedimentary-associated
	Mixed association of above
Sawkins (1976)	Kuroko type
	Cyprus type
	Besshi type
	Sullivan type
Solomon (1976)	Volcanic-hosted divided into:
	Zn–Pb–Cu
	Zn–Cu
	Cu
Klau and Large (1980)	Mafic volcanic-associated
	Felsic volcanic-associated
	Sedimentary types
Hutchinson (1980)	Primitive Zn–Cu
	Polymetallic Zn–Pb–Cu
	Kieslager Cu–Zn
	Cupreous Pyrite Cu
Franklin *et al.* (1981)	Cu
	Cu–Zn★
	Zn–Pb–Cu
	Pb–Zn

★Division into Cu and Zn may be artificial (see the text).

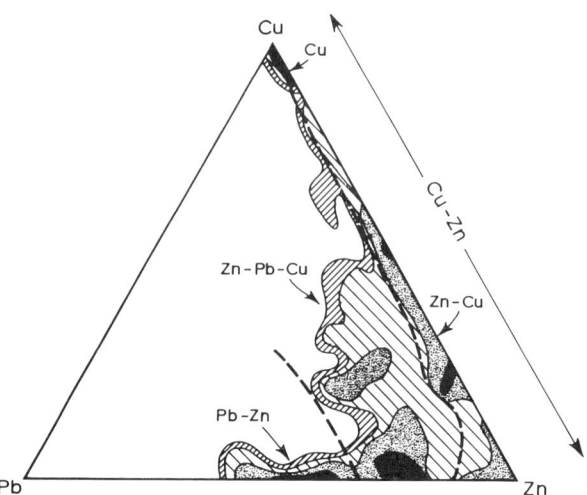

Fig. 3.25 Subdivision of volcanogenic sulphides (after Franklin *et al.* 1981).

Magmatic hydrothermal deposits

Table 3.5. Summary of volcanic-associated sulphide deposits (after Hutchinson, 1980).

Type	Volcanic rocks	Clastic sedimentary rocks	Tectonic environment			Approximate age range
			Depositional environment	General conditions	Plate tectonic setting	
Primitive Zn–Cu:Ag–Au	Fully differentiated suites Basaltic to rhyodacitic	Volcanoclastics Greywackes	Shallow to deep water Tholeiitic to calc-alkaline marine volcanism	Major subsidence compression	Subduction at consuming margin, island arc	Archaean–Early Phanerozoic
Polymetallic Pb–Zn–Cu:Ag–Au	Tholeiitic basalts calc-alkaline lavas Pyroclastics	Volcanoclastics increasing clastics minimum carbonates	Explosive shallow calc-alkaline-alkaline marine continental volcanism	Subsidence Regional compression but local tension	Back-arc or post-arc spreading; crustal rifting at consuming margin	Early Proterozoic–Phanerozoic
Cupreous pyrite Cu:Au	Ophiolitic suites Tholeiitic basalts	Minor to lacking	Deep tholeiitic marine volcanism	Minor subsidence tension	Oceanic rifting at accreting margin	Phanerozoic
Kieslager Cu–Zn:Au	Mafic: tholeiitic (?)	Greywacke, shale (?)	Deep marine sedimentation and tholeiitic (?) volcanism	Major subsidence Compression	Fore-arc trough or trench	Late Proterozoic Palaeozoic

rocks aspects and erected a fourfold division of the volcanogenic group (Table 3.5). Compared to the Sawkins classification mentioned above, Hutchinson has introduced an additional type – the Primitive (zinc-copper) type. This was mainly erected because of the size, frequency and significance of the deposits in the Canadian Archaean, deposits which have many differences from the lead-bearing, later, Kuroko type.

The Hutchinson classification has considerable advantages for the exploration geologist since it combines the plate tectonic setting, which will assist in selecting areas on a global scale, the localized tectonic and depositional environment, which will help on a regional scale, and the host rocks which will facilitate target selection. Although there is little doubt that this classification will be modified as more detailed evidence on the genesis of these deposits is forthcoming it is the most useful currently available and is the one we shall use.

3.8.3 General characteristics of volcanogenic sulphide deposits

(a) Distribution in time and space

Major economic deposits of volcanogenic sulphide types occur in Canada, Japan, USA, Australia and Europe (Fig. 3.26). They range in age from Archaean to Tertiary.

(i) Archaean

The volcanic rocks of the Canadian Shield have a great abundance of massive sulphide deposits; well-documented examples are those at Noranda and Kidd Creek. On the other hand Archaean rocks of generally similar age in the Yilgarn block of Australia, which are compositionally very similar although with somewhat different abundances of felsics, contain only two deposits of economic significance recognized to date. Similarly the large shield areas of South Africa and

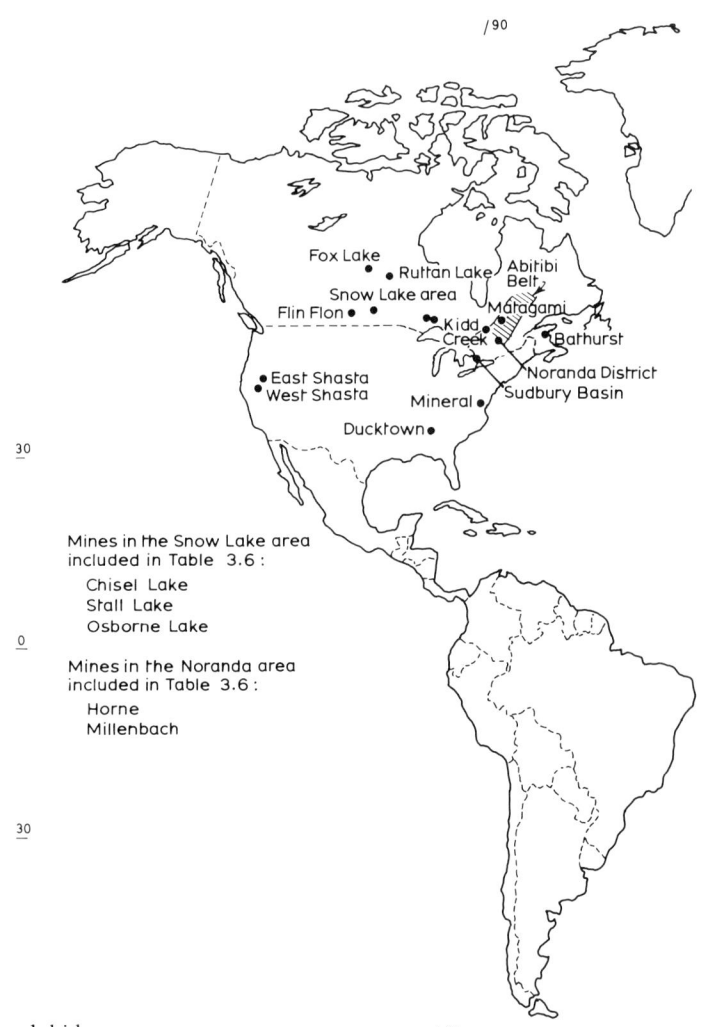

Fig. 3.26 Global distribution of volcanogenic sulphides referred to in text.

Zimbabwe have very few massive sulphide deposits. The same is true of the Brazilian Shield although this may be a reflection of the level of current exploration.

(ii) Proterozoic

Early Proterozoic deposits occur in the Sudbury Basin and at Flin Flon, Canada. There are no examples of volcanogenic deposits reported from the Middle Proterozoic (1200 million years to 1800 million years) but examples of Late Proterozoic deposits occur at Ducktown, Tennessee and at Matchless-Otjihase, Namibia.

(iii) Phanerozoic

Volcanogenic deposits occur at a number of stratigraphic levels in the Cambro-Ordovician succession of the Norwegian Caledonian Belt, one

Magmatic hydrothermal deposits

example being the Killingdal Mine (Fox, 1984). Cambro-Ordovician volcanogenic deposits are also found in Newfoundland and New Brunswick. Other early Phanerozoic ore bodies occur at West Shasta.

The volcanogenic deposits of the Pyrite Belt of Spain and Portugal are predominantly Lower Carboniferous although the first acid volcanic activity began in the Upper Devonian (Strauss *et al.*, 1977). Mesozoic deposits are recorded from Cyprus, Oman, Turkey and East Shasta. The Kuroko deposits of Japan are Tertiary with over 50% of the Kuroko ore in the Middle Miocene of the Hokuroko district.

(b) Size and grade of the deposits

Because of the postformation structural and metamorphic events which may affect deposits, it is difficult to assign accurate average dimensions to volcanogenic sulphide deposits. However,

Sangster (1980), in a study of seven volcanic districts in Canada and one in Japan, showed that certain semi-quantitative features were recognizable. Firstly within the districts concerned, clusters of volcanogenic massive sulphide deposits occurred with an 'average' cluster of about twelve per district. Each deposit cluster, and its associated ore rhyolite, had an average circle-equivalent radius of 16 km. Sangster also showed that in these eight districts the largest deposit of each district contained on average 67% of the total metal for the district. That this figure may be reasonable is supported by the sulphide occurrence in the Mount Lyell district of Australia where the largest of fifteen deposits in the field is the Prince Lyell which has 78% of the known reserves (Reid and Rowe, 1979).

Approximately 80% of all known volcanogenic deposits fall in the size range of 0.1 to 10 million tonnes (Fig. 3.27). About half of these contain less than 1 million tonnes (Sangster, 1977). Furthermore, Sangster states that 88% of Canadian deposits contain combined Cu + Pb + Zn grades of less than 10% irrespective of age. In these the most likely combined grade is about 6% as 4:1:1 for Zn:Cu:Pb. For example, the Noranda district has ore reserves of approximately 204 million tonnes (combined grade 2.4%) and the Bathurst district contains reserves around 278 million tonnes (combined grade 7%).

Larger deposits tend to be zinc-rich relative to copper; the lead grade varies inversely as copper and directly as zinc. Table 3.6 lists the reserve tonnages and grades for some volcanogenic deposits.

(c) Mineralogy

The mineralogy of these deposits is fairly straightforward: the main minerals are pyrite, pyrrhotite, sphalerite, galena and chalcopyrite. Occasionally chalcocite and bornite are important. Minor arsenopyrite, tetrahedrite and tennantite may be present. Traditionally it is stated that magnetite may occur as a minor component (Evans, 1980), but it can be significant in the stringer zones associated with these deposits, and in the Golden Grove Prospect in Western Australia abundant magnetite is present (Frater, 1983). The principal gangue mineral in base-metal sulphide deposits is quartz; carbonates may be developed although chlorite and sericite are more important. By-products of importance are gold and silver (Table 3.6). Cobalt reaches 0.35% in some of the Cyprus deposits (Constantinou, 1980), and nickel up to 220 ppm is recorded. In view of the crustal abundance of nickel and the geological setting of these Cupreous-Pyrite-type deposits it is surprising that they are not more enriched in nickel.

Most of these sulphide deposits are zoned, the

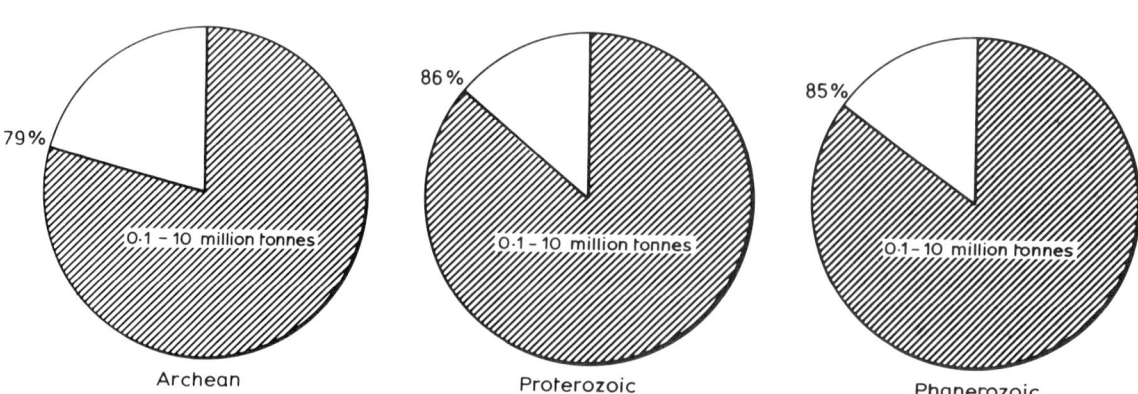

Fig. 3.27 Size range of volcanogenic sulphide deposits (data from Sangster, 1977).

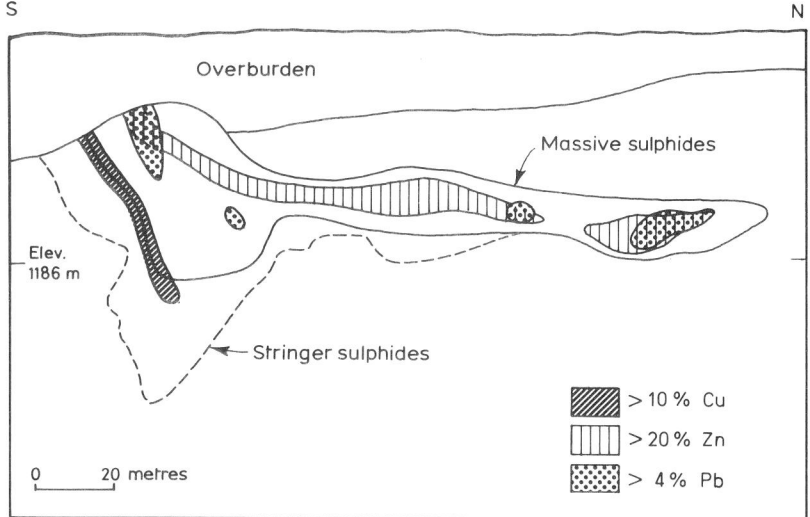

Fig. 3.28 Metal zoning of the Sturgeon Lake massive sulphide deposit, north-west Ontario (after Severin 1982).

zoning being best developed in the Polymetallic-type deposits. Lead and zinc (with possibly silver) occur principally in the upper zones, copper (and gold where present) increasing towards the footwall (Fig. 3.28). Stockworks of chalcopyrite are common below the footwall. A lateral zoning may also be recognized, with copper (plus gold) being developed in the proximal deposits and zinc, lead and silver in the distal deposits.

(d) Alteration

The alteration zones associated with these deposits are often much larger than the deposits themselves and if they can be correctly identified in the field they are very important in exploration. Four forms of alteration occur, a footwall pipe, a large semi-conformable alteration zone below the ore horizon, alteration associated with the ore horizon, and hangingwall alteration.

The footwall shows most evidence of alteration; chloritization and sericitization are the two most common forms. The alteration tends to be pipe-shaped and under Primitive-type deposits, such as Noranda, it may be well defined and may extend vertically to over 100 m. These pipes tend to have Mg-rich chlorite (or talc)-rich cores surrounded by a sericite halo with or without quartz. Furthermore there is pervasive Na_2O and CaO depletion with some K_2O addition. There is usually minor addition or redistribution of SiO_2. The alteration pipes associated with the Polymetallic types, such as Kuroko, are not as vertically extensive and show a zonation with a sericite–quartz core surrounded by an Mg-enriched chlorite halo. An idealized cross-section of the four principal alteration zones around the Kuroko deposits is shown in Fig. 3.29.

The alteration zones immediately below the Cupreous-Pyrite-type deposits show pervasive feldspar destruction and some introduction and redistribution of MgO and FeO to form chlorite. In these, sericite is a common accessory. Hydrothermal alteration including Ca depletion and Na, Mg and Fe enrichment in the footwall has been recognized in deposits of the Kieslager type. Large semi-conformable regional scale alteration zones underlie many deposits in the Precambrian Shield of Canada and in ophiolite terrains. These may enclose or lie below the alteration pipes beneath the individual deposits.

Hangingwall alteration zones are evident only in the least metamorphosed deposits. This hangingwall alteration may result from persistent reactivation of hydrothermal discharge activity or

Table 3.6. Reserve tonnages and grades for some volcanogenic deposits. Sources: Fox (1984), Franklin and Thorpe (1982), Frater (1983), Malone *et al.* (1981), Severin (1982), Schermerhorn (1980), Strauss *et al.* (1977).

Deposit	Reserves (million tonnes)	Cu%	Zn%	Pb%	Ag(g/t)	Au(g/t)
Canada						
Fox Lake	14.5	1.81	1.63			
Ruttan Lake	44.8	1.46	1.36		6.1	0.31
Osborne Lake	3.3	3.69	1.53			
Anderson Lake	3.1	3.79			6.4	0.4
Stall Lake	5.6	4.76	0.65		9.5	1.2
Chisel Lake	5.8	0.56	11.61	0.82	38.6	1.5
Flin Flon	63.4	2.31	4.25		38.6	2.5(?)
Centennial	1.4	2.06	2.60		21.5	1.23
South Bay	1.7	1.6	11.00		67	
Mattabi	11.5	0.83	7.98	0.76	97	0.031
Sturgeon Lake (Boundary)	2.2	2.74	10.01	1.41	175	0.61
Lyon Lake (Mattagami)	3.9	1.24	6.53	0.63	96	0.03
Geco	54.6	1.86	3.45		40.8	
Kidd Creek	155.4	2.46	6.00	0.2	63	
Corbet	3.1	2.99	2.13		19	0.75
Millenbach	3.5	3.43	4.23		42.5	0.7
Horne	61.3	2.18				4.6
Orchan	5.1	1.05	9.93		18	0.19
Mattagami Lake	21.2	0.65	8.40		29	0.48
Norita	4.5	2.16	3.57		22.3	0.56
Japan						
Mavovouni	15	4	0.5			
Skouriotissa	6					
Matsumine	60					
Europe						
Tharsis	110					
La Zarza	100					
Aznalcollar	45	0.44	3.33	1.77	67	1
Killingdale Mine						
Main Zone	2.2	1.7	3.5			
North Zone	1.1	1.1	7.0			
Australia						
Teutonic Bore	2.66	3.08	9.42		127	
Golden Grove						
Main Zone	15	3.4	0.1		14	0.1
Additive	1.7	0.4	14	1.6	87	2.2
Woodlawn	6.3	1.7	5.5	14.4	72	
Cu Zone	3.7	1.9	0.5	0.1		

Magmatic hydrothermal deposits

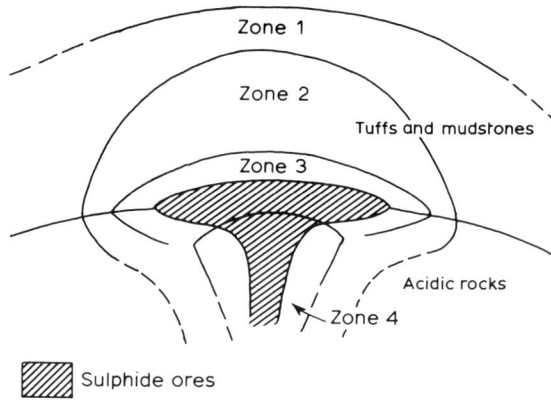

Fig. 3.29 Alteration zones around the Kuroko deposits (after Shirozo 1974) Zone 1, montmorillonite, zeolite and cristobalite; zone 2, sericite, interstratified sericite-montmorillonite Fe Mg-chlorite, albite, K-feldspar and quartz; zone 3, sericite, interstratified sericite-montmorillonite and Mg-chlorite; zone 4, quartz and sericite.

from postformation metamorphic effects or from decomposition of pyrite. The alteration around the essentially unmetamorphosed Kuroko deposits (Fig. 3.29) is as impressively developed above as below, suggesting that there were rocks above those containing the ores when the ores were introduced.

(e) Host rock lithology

Volcanogenic sulphides are hosted in volcanic-dominated marine successions. The relative amounts of the volcanic and sedimentary/metamorphic components vary considerably as do the types of volcanics and sediments (Table 3.5).

The Primitive-type deposits most commonly occur in sequences dominated by volcanic rocks and immature, volcanic-derived sedimentary rocks. Most of the Precambrian deposits of North America, for example, occur in volcanic-dominated portions of greenstone belts, over 90% of these volcanics being mafic. On the other hand the volcanic rocks associated with the Polymetallic type are largely felsic; there are virtually no mafic rocks in the footwall sequence in the Kuroko deposits. In the Cupreous Pyrite type, which are intimately associated with the tholeiitic basalts of ophiolite terrains, the clastic sediment component is very minor or lacking. Almost equal amounts of volcanic and sedimentary rocks occur in the Besshi deposits of Japan and the Kieslager type to which they belong is associated with tholeiitic mafic volcanism.

A factor which has been a very important key in volcanogenic exploration for deposits is the local prevalence of felsic rocks which may occur in the footwall of deposits, for example the Precambrian and Caledonide deposits of Scandinavia.

(f) Tectonic setting

The tectonic environments in which volcanogenic sulphide deposits are considered to have been formed are summarized in Table 3.5. The plate tectonic settings are shown in Fig. 3.30. According to this interpretation, the late Archaean deposits of the Primitive type arose in an environment of subsiding troughs suffering tholeiitic to calc-alkaline marine volcanism. The stress regime was dominantly compressive. This tectonic environment is compared by Hutchinson (1980) to island arc volcanism along Phanerozoic consuming plate boundaries, the environment in which the Phanerozoic examples of the Primitive-type deposits are thought to have formed.

Major subsidence and an overall compressive regime is thought to have been responsible for the Polymetallic and Kieslager types. In the case of the Polymetallic deposits, however, the local tectonic environment is considered to have been tensional, certainly during the early Proterozoic. At this time continental rifting with vertical displacement may have played the dominant tectonic role in the production of the lead–zinc–copper deposits that began to appear in major quantity for the first time. Phanerozoic Polymetallic-type deposits, such as those at Kuroko, are presumed to have been associated with back-arc volcanism.

Kieslager-type deposits are associated with tholeiitic mafic volcanism in a trench or fore-arc trough; the sort of environment we envisage for these Besshi type would be in a flysch formation

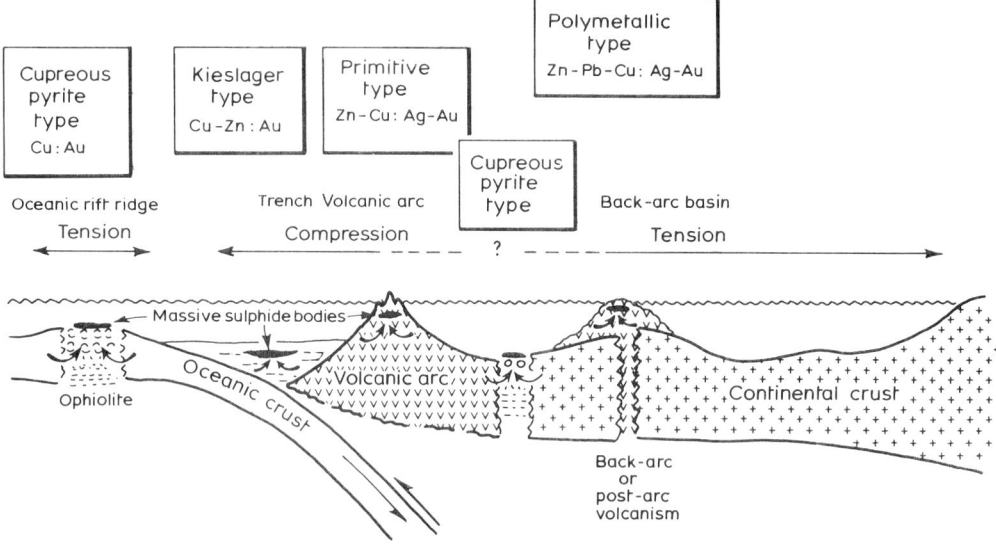

Fig. 3.30 Plate tectonic setting for principal types of massive sulphide deposits (modified from Hutchinson, 1980).

adjacent to a Cordilleran-style continental margin. The Cupreous Pyrite type of deposit is the only one of the volcanic-associated sulphide deposits which is associated with a strong tensional regime. These deposits are predominantly linked to the spreading centre environment of an accreting plate margin but they may also form in a back-arc basin where the continental crust is relatively thin. Deposits that are presumed to have formed at rift ridge environments include those in Cyprus and Saudi Arabia. The Red Sea and the East Pacific Rise may be current areas where such deposits are forming today (Edmond et al., 1979).

3.8.4 Examples of volcanogenic sulphides

Because of the wealth of published literature available and the enormous number of these deposits worldwide it is not possible to describe more than a few in a textbook of this size and nature. Readers who require details of other deposits are referred to the excellent paper by Franklin et al. (1981), from which the following description of the Abitibi belt largely originates.

(a) Primitive type

To illustrate some of the geological characteristics of these dominantly zinc–copper deposits which also may contain significant gold and silver we have chosen to describe three deposits from the economically very important, and well-documented, Abitibi Belt in Canada. In contrast to these we will describe the more recently discovered, relatively small, Teutonic Bore deposit in Western Australia.

(i) Abitibi Belt, Superior Province, Canada

This is an Archaean greenstone belt some 750 km long and 150 km wide which contains at least 350 million tonnes of ore. Among the major economic sulphide deposits in this Belt associated with volcanics are those in the Noranda, Timmins and Matagami mining districts.

(a) Noranda

Five major episodes of felsic volcanism have been recognized in this district, separated from each other by periods of mafic volcanism. Figure 3.31 shows the upper three felsic volcanic cycles; cycles

Magmatic hydrothermal deposits

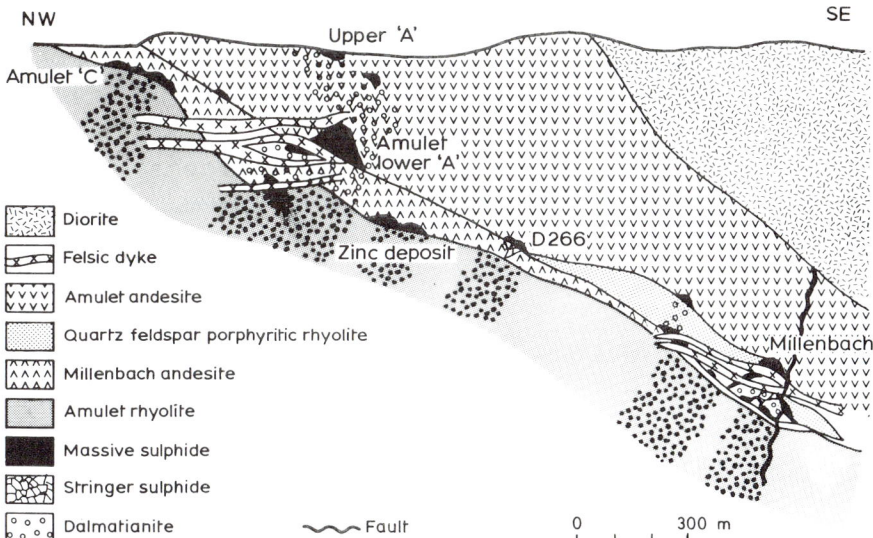

Fig. 3.31 Northwest-southeast section through the Millenbach and Amulet deposits, looking northeast (after Knuckey *et al*, 1982).

one and two contain over 16 000 m of felsic and mafic rock with no evidence of massive sulphide deposition. The third zone of felsic volcanism is about 3000 m thick containing at least three major felsic units. Most of the deposits occur in this third cycle with deposits at several levels within the cycle. A vertically extensive alteration and stringer copper zone occurs under most deposits, that beneath the Vauze deposit being over 1000 m deep (Spence, 1975).

The deposits occur in felsic and mafic rocks but felsic footwall rocks are more common. The Amulet 'A' ore bodies are in mafic flows and the origin of these particular bodies has been discussed by Beaty and Taylor (1982). Most of the ore bodies in the Noranda district are near small, sometimes steep-sided, felsic domes of rhyolite. These rhyolitic domes are usually underlain by feeders which have been controlled by synvolcanic fracture zones (Franklin *et al.*, 1981). Often the local occurrence of massive sulphides is the only evidence of a break in rhyolitic flow activity while in some cases the ore zones are capped by a thin pyroclastic and epiclastic tuff and chert horizon. Where the deposits are completely enclosed by flow rhyolite, the massive sulphides must have accumulated quite quickly, during a break in rhyolite flow activity, and the timespan involved here was insufficient to permit sedimentary material to accumulate. The Noranda deposits are quite undeformed large bulbous lenses of massive pyrite–sphalerite–chalcopyrite underlain by chalcopyrite–pyrite stringer zones with or without pyrrhotite.

(b) Matagami

Here the massive sulphide deposits occur on the two limbs of a large WNW trending anticline (Fig. 3.32). On the south limb the deposits all occur at the same stratigraphic level, underlain by a rhyolite – the Watson Lake rhyolite – and overlain by an extensive laminated pyritic tuffaceous and chert unit – the Key Tuffite. The deposits on the north limb are also associated with rhyolites; most, including the Norita deposit, lie on the Bell Channel rhyolite, while one, the Garon Lake deposit, is in the stratigraphically higher Garon Lake rhyolite.

(c) Timmins

This mining district includes the very large Kidd Creek deposit, with more than 150 million tonnes of massive sulphide ore, plus a few small deposits. The Kidd Creek deposit is typically about 46 m

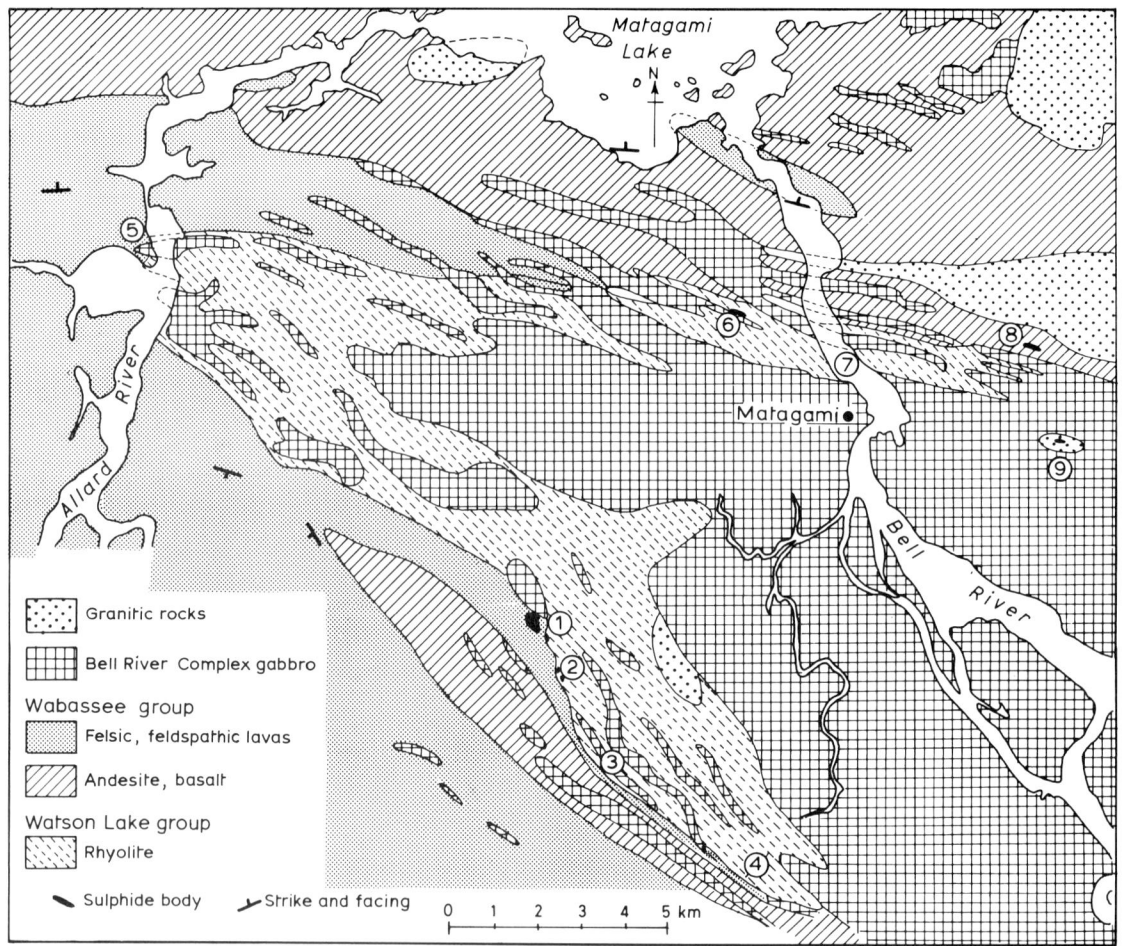

Fig. 3.32 Generalized geology of the Matagami, Quebec, area (from Franklin *et al.* 1981). The numbered ore bodies are: (1) Mattagami Lake; (2) Orchan; (3) Bell Allard; (4) Consolidated Mining and Smelting; (5) New Hosco; (6) Radiore A; (7) Bell Channel; (8) Garon Lake; (9) Radiore E.

thick and may be traced down dip for at least 1300 m (Fig. 3.33). The footwall of the deposit comprises both rhyolitic flow and epiclastic felsic rock, whereas the hangingwall has a distinctive carbonaceous argillite horizon. This argillite unit may be analogous to the tuff horizons at Noranda and the Key Tuffite at Matagami, representing a hiatus in volcanism (Franklin *et al.*, 1981). The mineralogy of the Kidd Creek ore body is interesting since in addition to the characteristic components of pyrite, sphalerite and chalcopyrite it also contains cassiterite and stannite. Tin locally reaches 3% and is recovered from the Kidd Creek ores. The stringer zone at Kidd Creek is also somewhat unusual since it contains a local bornite zone with a range of selenium minerals. Bornite is very rare in other Archaean deposits.

(ii) Teutonic Bore, Western Australia

This copper–zinc–silver deposit occurs in a volcanic and sedimentary sequence approximately

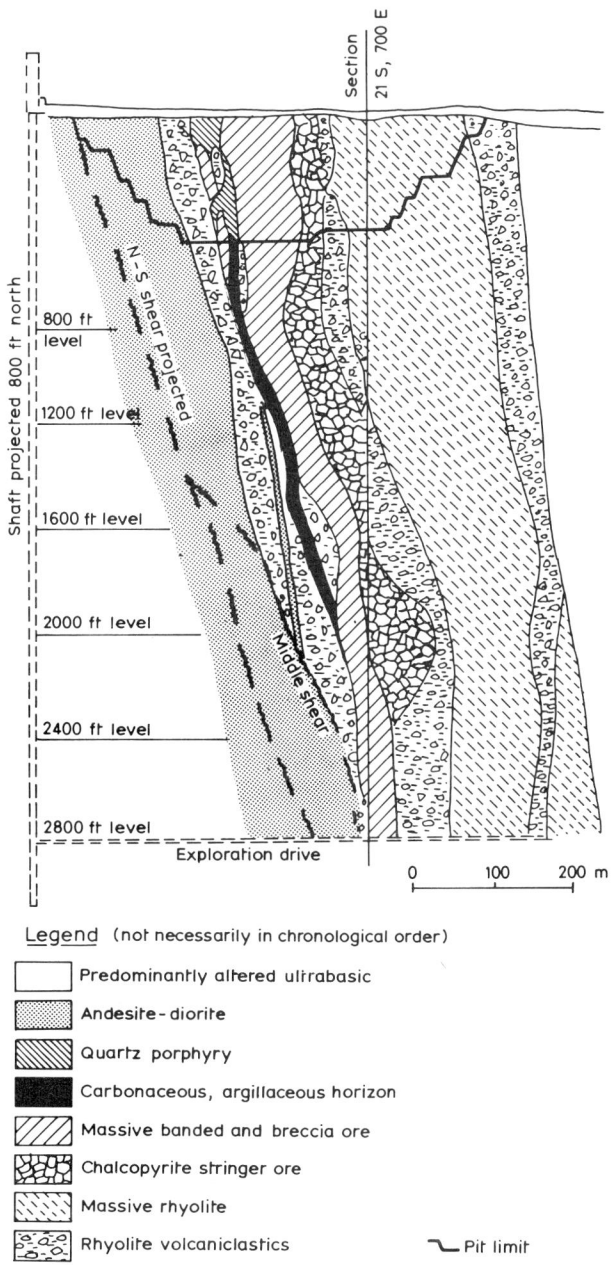

Fig. 3.33 Cross-section of the Kidd Creek Mine near Timmins, Ontario, (from Walker *et al*. 1975).

seventy kilometres northwest of Leonora, Western Australia (Fig. 3.34). Frater (1983) quotes reserves of 2.66 million tonnes for this deposit at 3.08% copper, 9.42% zinc and 127 g/t silver.

It lies in the Archaean greenstone Norseman-Wiluna belt in the Eastern Goldfields Province of the Yilgarn Shield. The volcanics strike NNW to N and are highly folded and weakly metamorphosed. The deposit is near the base of a predominantly mafic volcanic sequence underlain by

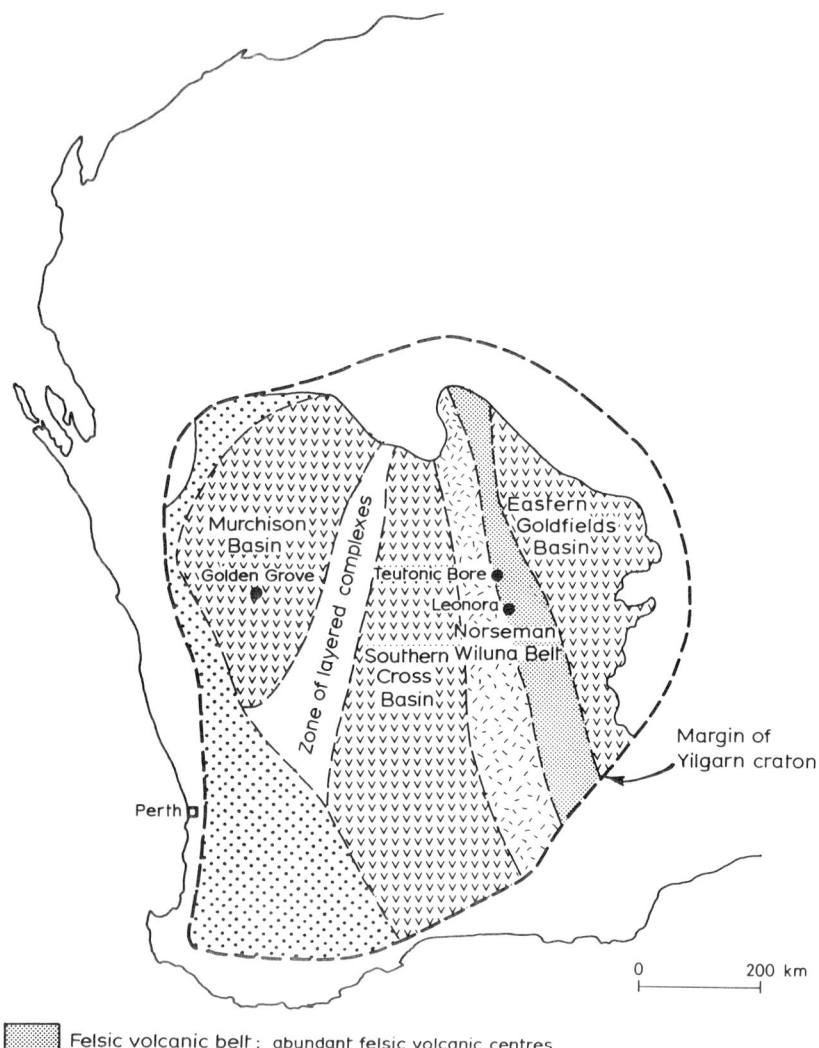

Fig. 3.34 Regional map of the Western Australian Shield showing the major tectonic units with Teutonic Bore and Golden Grove (modified from Groves and Batt, 1983).

felsic volcanics – subvolcanic rhyolite porphyries and associated pyroclastics (Fig. 3.35). The mafics are basaltic showing pillow structures and amygdales.

The massive sulphide lens is stratiform and is associated with mineralized, brecciated, bedded footwall cherts. The lens reaches a thickness of 25 m while barren pyritic sulphides persist away from the proven reserve area, thickening in depth at the northern end. Agnew and Harrison (1983) describe a zonation from footwall to hangingwall with chalcopyrite–pyrite more common near the footwall passing upward into a sphalerite–pyrite zone and then up into pyrite with minor sphalerite. Often occurring in cross-cutting veinlets, galena is a minor constituent. The silver present is mainly as tetrahedrite inclusions in chalcopyrite and pyrite. The sulphides are often well banded parallel to the lithologic layering, and contortions in the bedding of the sulphides represent soft sediment-type deformation. Although generally the copper grades increase towards the footwall, high copper grades occasionally appear near the top of the succession and these may be due to folding and remobilization.

The hangingwall basalt is tholeiitic and similar in primary composition to the footwall basalt but has undergone less alteration and has no base-metal sulphides. Sericitic alteration, associated with the waning stages of hydrothermal activity, is pervasive up to 15 m from the massive sulphides.

Economic stringer mineralization occurs in the footwall basalt. The mineralization here is in irregular veinlets and patches of chalcopyrite, pyrite and minor sphalerite. Distinct wall-rock alteration often accompanies these zones (Agnew and Harrison, 1983). This alteration zone beneath the massive sulphide lens extends for over 150 m through the footwall basalts. The zone is characterized by CaO, Na$_2$O, Sr depletions and SiO$_2$, FeO, MgO, Ba, F and base-metal enrichment. Agnew and Harrison recognize the Teutonic Bore deposit as a proximal volcanogenic massive sulphide deposit formed by the accumulation of sulphides precipitating from hydrothermal brines emanating from fumaroles on the sea floor. They were not able to deduce a paragenetic sequence for the multiple feeder zones from these fumaroles.

The Norseman–Wiluna Belt in which the Teutonic Bore is situated is described by Groves and Batt (1983) as a major rift zone superimposed on more extensive, relatively stable platforms. The belt developed rapidly under high total extension and appreciable crustal thinning. The development of volcanogenic sulphides was favoured here by the presence of synvolcanic faults which may have acted as conduits for hydro-

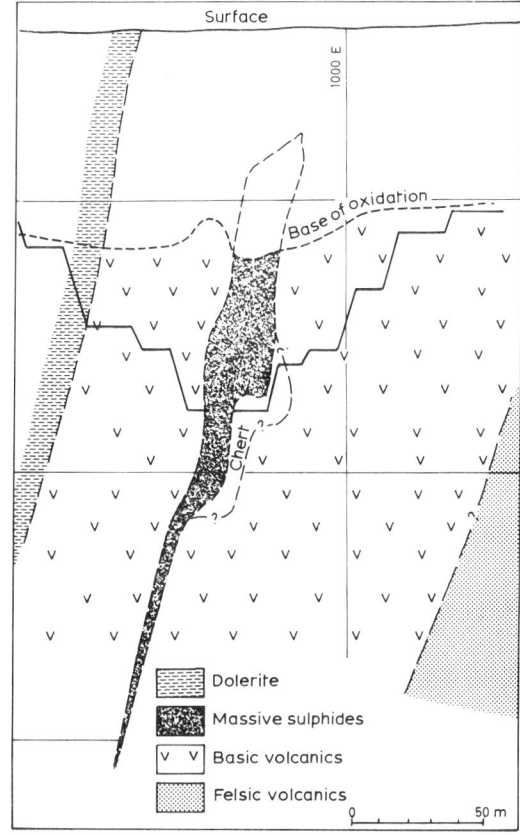

Fig. 3.35 Cross-section of the Teutonic Bore deposit, Western Australia (from Agnew and Harrison 1983).

thermal fluids. Convection cells generated under the prevailing high regional geothermal gradients would also have played an important role. The deep water covering would have inhibited subsurface boiling and permitted the preservation of sulphides without erosion or oxidation.

(b) Polymetallic type

The Kuroko deposits
The Japanese deposits are the standard volcanic-hosted massive sulphide deposits with which virtually all other deposits are compared – they are well preserved, show primary features and lack of deformation. These deposits occur in the Green Tuff Belt and are exposed over a length of 1500 km;

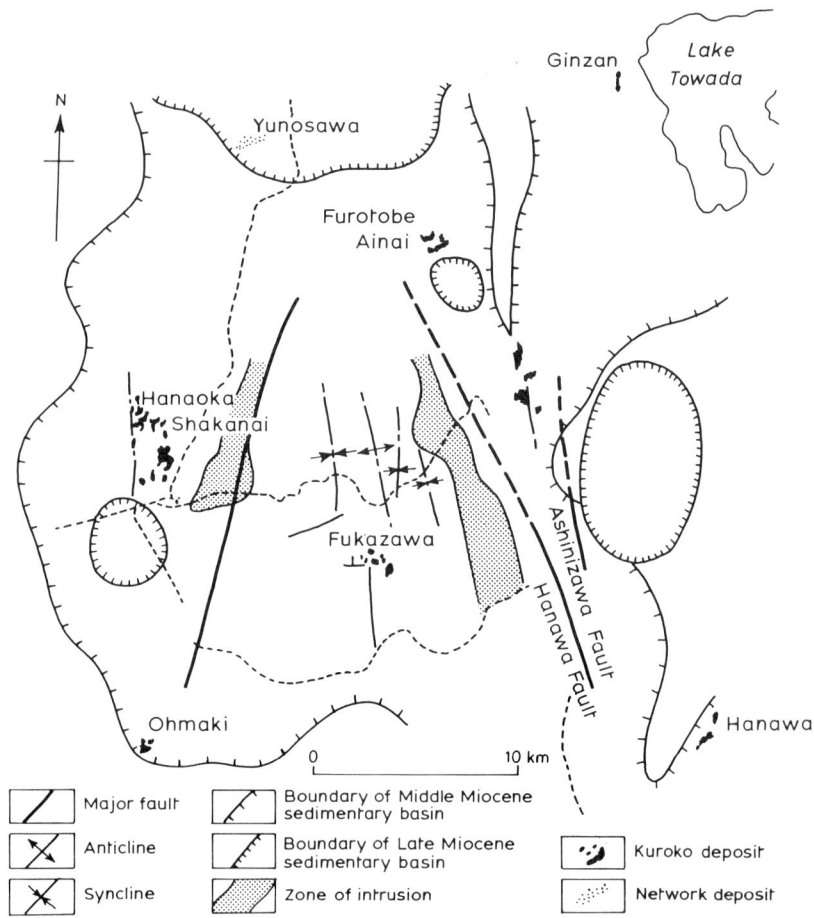

Fig. 3.36 Miocene sedimentary basins, major structures, and principal massive sulphide areas of the Hokuroku district, Japan (from Sato 1974).

the Belt has an average width of 100 km and contains up to 3000 m of volcanic and sedimentary rocks accumulated during fault-controlled subsidence during the Miocene. A major monograph devoted to the Kuroko and related deposits has been edited by Ohmoto and Skinner (1983); readers requiring detailed descriptions in this area are advised to read that publication. Over 50% of Kuroko ore is in the mines of the Hokuroku district. Situated primarily in the Middle Miocene basin, the deposits occur mainly in a broad ring towards the outer edge of the basin. The principal producers occur in distinct clusters on a conjugate set of NNW and NE striking linear features (Scott, 1980). Scott (1978) suggested these are basement fractures that controlled the emplacement of sulphide deposits and associated rhyolite domes and breccia piles (Fig. 3.36). Ohmoto (1978) postulated a spatial distribution controlled by caldera development or collapse.

The formation which contains the ore bodies has dacitic flows, including white rhyolitic domes and tongues of felsic lavas; well-sorted explosion breccias overlie the lavas and are close to many of the deposits. Tuff breccia forms the principal footwall sequence to the deposits where the white rhyolite domes are absent. During the deposition of this footwall tuff the basin wall subsided to 4000 m and was uplifted again to 3800 m during the

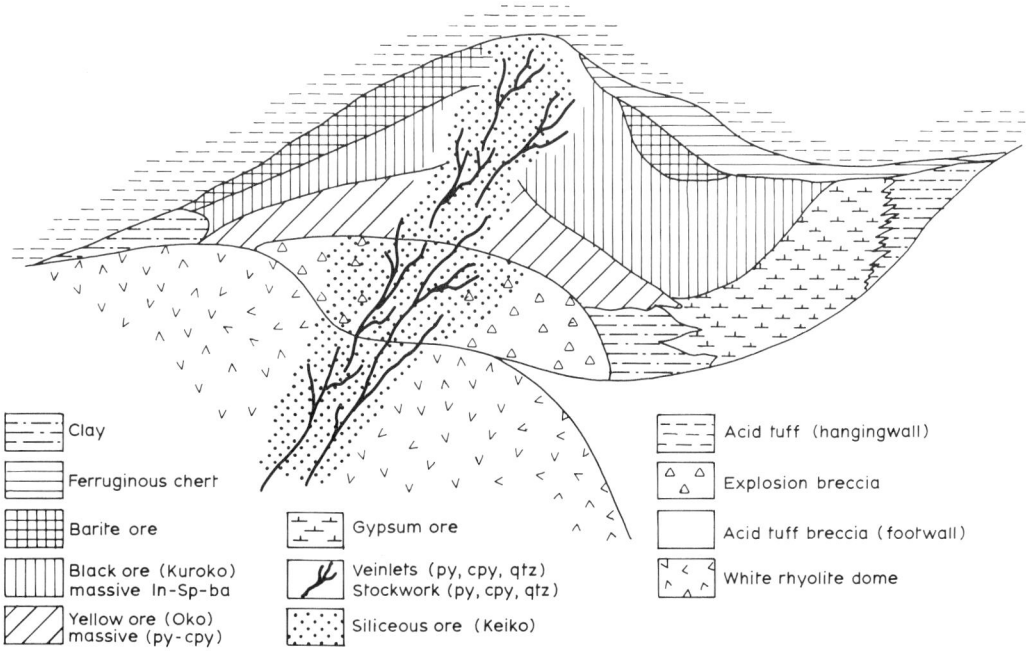

Fig. 3.37 Idealized cross-section of a typical Kuroko deposit (from Sato 1974 and Franklin *et al.* 1981).

later stages of the mudstone deposition which immediately followed ore deposition (Ohmoto, 1978). Sato (1974) on the other hand suggested water depths of 300–500 m for these deposits. The 'classic' Kuroko ore body (Fig 3.37) has seven zones (Lambert and Sato, 1974) in ascending order as follows, although it must be stressed that by no means all the deposits in the Japanese-type area have all of these zones preserved.

(1) Siliceous ore. This is the pyrite–chalcopyrite-quartz stockwork ore which commonly underlies only some of the stratiform part of the main massive black ore. The original volcanic rock structures may be still recognizable but these may be obscured by extensive overgrowth of quartz.

(2) Gypsum ore. The stratabound ore contains gypsum–anhydrite–pyrite–chalcopyrite–sphalerite–gelena–quartz and may occur beside the siliceous ore as well as on top of it. Less commonly this ore type is found in veins.

(3) Pyrite ore. As with the gypsum ore this may occur as veins or disseminations but more commonly it appears stratiform. Mineralogically it contains dominantly pyrite with subsidiary chalcopyrite and quartz.

(4) Yellow ore. This stratiform ore contains pyrite–chalcopyrite with sphalerite–barite and quartz. Yellow ore is considered largely as a replacement at the base of the previously deposited black ore. It may be deposited from a higher-temperature ore fluid than the black ore.

(5) Black ore. This 'Kuroko' ore is a sphalerite–galena–chalcopyrite–pyrite–barite stratiform ore. The ore may be compact and bedded or fragmental. Colloform banding is common in some parts of the bedded ore. Framboidal textures in pyrite are very common in this black ore. Towards the top of the zone there are significant amounts of tetrahedrite-tennantite. Bornite may also be present.

(6) Barite ore. This type is thin, well-bedded ore composed almost entirely of barite although

sometimes containing minor amounts of calcite–dolomite and siderite.

(7) Ferruginous chert is the stratigraphically highest zone.

In spite of the very low regional metamorphic grade evidenced by these deposits, the anhedral nature of some of the pyrite may indicate that some parts of the Kuroko deposits have been recrystallized. The sulphide mass might have undergone repeated recrystallization on the sea floor (Barton, 1978). Ridge (personal communication) believes that these deposits were not formed on the sea floor but at some little depth below. As evidence of rocks overlying the ore deposit at the time of formation of the ores, Ridge cites the extent of hangingwall alteration. In this latter case much of the deposit could form by replacement of already brecciated rock, and the anhedral nature of the pyrite would indicate that this replacement has occurred more than once.

Individual mines encompass a cluster of ore bodies, six in the case of the Kosaka Mine and seven in the case of the Shakanai Mine. The median size of the ore bodies in the Green Tuff Belt is 1.3 million tonnes with at least 13 of the ore bodies having more than 1 million tonnes. The Matsumine Mine, the largest ore body in the Green Tuff Belt, has approximately 60 million tonnes, and this has all the zones mentioned above.

(c) Cupreous Pyrite type

Cyprus deposits

Massive sulphide deposits may occur in the volcanic part of ophiolite complexes and these deposits are characterized by pyrite and chalcopyrite as the predominant sulphide minerals, with minor sphalerite in both the massive ores and the vertically extensive stringer ores. Deposits of this kind occur in the Troodos massif of Cyprus; this massif is interpreted by Gass (1980) as an off-axis portion of a mid-oceanic ridge system. The massive sulphide deposits lie in a pillow lava sequence which is divided into three: the Upper Pillow lavas are dominantly olivine basalts, the Lower Pillow lavas are oversaturated basalt while the Basal Group consists of altered basalts with large numbers of dykes. The majority of the deposits are at the contact between the Lower and Upper Pillow lavas and may lie adjacent to steep normal faults (Fig. 3.38). These faults probably controlled localized seabed subsidence and some of the footwall rocks show pre-ore brecciation.

Typically the deposits are of massive ore overlain by an ochre horizon and underlain by a basal siliceous ore; these are underlain by a stringer zone extending for hundreds of metres below the deposits. The ochre, where present, is interpreted as 'the accumulated product of submarine oxidative leaching of the sulphide ore that was exposed on the sea bed' (Constantinou and Govett, 1972). The massive ore is porous, colloform-banded blocks of pyrite and marcasite set in a sandy, friable sulphide matrix. The conglomeratic or brecciated structure of this ore increases downwards (Constantinou, 1977); this author has also suggested that the sandy texture of the sulphide results from intense leaching. Alteration of the sulphides may have formed secondary copper minerals such as covellite, chalcocite and bornite as well as producing the 'conglomeratic' texture.

The basal siliceous ore probably represents partial replacement of sulphides by silica (Constantinou, 1980). The ore here consists of pyrite and chalcopyrite in a quartz matrix; in some cases sphalerite is more common than chalcopyrite and some ores contain substantial amounts of zinc.

The stockwork is characterized by two types of ore: one is composed of sulphide (pyrite, chalcopyrite, sphalerite) and quartz veins cementing the basaltic breccia and fracture filling the pillow lavas while the other is disseminated sulphides (uniquely pyritic) impregnating altered lavas.

In general the deposits are small, mostly less than 1 million tonnes. The two major exceptions to this generalization are the Mavovouni and the Skouriotissa deposits. The former contains about 15 million tonnes of ore at 3.5 to 4.5% Cu and 0.5% Zn. At Skouriotissa the reserves are 6 million tonnes and here the cobalt reaches 0.35%

Magmatic hydrothermal deposits

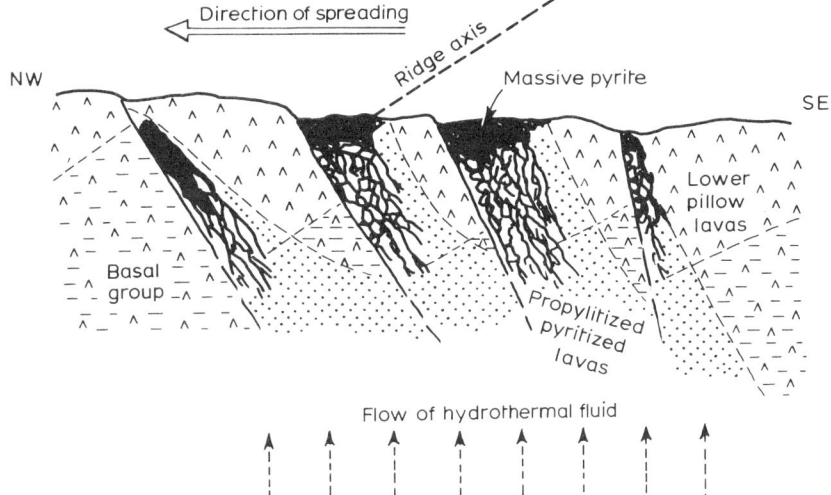

Fig. 3.38 Cross-section of Agrokipia ore deposits, Cyprus, illustrating the relationship of the massive and stringer sulphides to synvolcanic fault zones (after Adamides 1980).

(Constantinou, 1980) while nickel is variable, showing a general increase in the disseminated ores (from average 15 ppm in the massive ore to 220 ppm in the disseminated ores). In the other deposits the reserves are smaller but each district has several ore bodies. For example, the Kalavassos district has seven ore bodies which have been mined with an average size of 0.22 million tonnes (Adamides, 1980).

(d) Kieslager type

The Sanbagawa Belt, Japan

This belt of metasediments and metavolcanics in the southwest of Japan contains more than a hundred major copper–zinc deposits which are geologically and mineralogically similar to the biggest deposit, the Besshi Mine. This latter occurrence gave its name to the alternative name for the Kieslager type – the Besshi type. A review of Besshi-type volcanogenic sulphide deposits has been given by Fox (1984).

The deposits have a mafic volcanic rock (greenschist) beneath the ore and are overlain immediately by siliceous schist containing magnetite-rich layers (iron formation). The deposits consist of exceptionally elongated, to tabular, massive ore zones. The type Besshi deposit is 1800 m in strike length; individual ore beds attain a maximum thickness of only 10–20 m. The Motoyasu ore body in the Besshi district is 1400 m long, 100–180 m wide and 0.6–2.5 m thick. The Iimori deposit extends for more than 7000 m down its plunge and is 250–300 m wide and 0.2 to 2.8 m thick. Folding often locally increases the thickness of the ore.

The average copper content of the Besshi ores is 1.4%; the massive ores contain 3% Cu on average. The ores typically contain 0.3–0.9% Zn, 20–200 ppm Pb, and Zn/Cu ratios of 1:3. In general, the Besshi-type ores contain higher Co contents (1000 ± 200 ppm) than those of the Kuroko district and the cobalt is contained almost entirely in pyrite.

3.8.5 Genesis of volcanogenic massive sulphide deposits

Prior to about 1960 most of these massive sulphide bodies were considered to have formed by epigenetic replacement of chemically favourable rocks by some ascending hot metalliferous aqueous solution. The majority of the geological information collected since that time has pointed towards a broadly syn-depositional origin for these volcanic-associated deposits. We shall consider the various aspects that any genetic model must contain to explain the origin of these deposits.

(a) Source of the ore-forming fluids
In the Kuroko deposits, which have undergone minimal post-depositional deformation, the temperature and salinity of the fluid inclusions in the ore minerals and the oxygen isotopes indicate that the ore fluid was dominantly coeval seawater with a small meteoric and/or magmatic contribution (Ohmoto and Rye, 1974). Similar values for the Matagami deposit were also interpreted as recirculated seawater by Costa *et al.* (1980). As with Kuroko, the Cyprus deposits have ore fluids with isotopic compositions very close to modern seawater (Fig. 3.39). Heaton and Sheppard (1977) suggested a model involving deep circulation of seawater for this deposit. Hydrothermal activity within the modern oceanic crust particularly along mid-oceanic ridges indicates that seawater may circulate for depths up to 5 km. The Cyprus deposits indicate that similar extensive circulation of seawater operated in the ancient crust.

Therefore studies of stable isotopes of the fluids from modern hydrothermal systems and from the ancient ores indicate that much of the water is recirculated meteoric or marine with little, if any, of juvenile or magmatic origin.

(b) Nature of the ore-forming fluids
Active hydrothermal brine systems and fluid-inclusion studies indicate extreme chlorinity and low sulphide-ion activity suggesting that the base-metal transport was dominantly as chloride complexes (Hutchinson, 1982). Furthermore very low oxygen contents in these systems and the common association with carbonaceous strata suggest the fluids were strongly reducing. Fluid inclusions and stable isotopes in ores indicate temperatures up to 350°C (Heaton and Sheppard, 1977).

(c) Source of heat
The meteoric water and seawater circulated deep into the underlying sediments and volcanics, the circulation probably being driven by convection

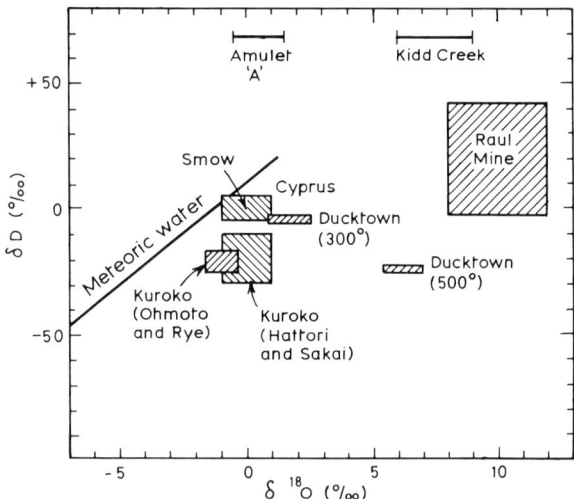

Fig. 3.39 Calculated ore fluids at six massive sulphide localities. Although the δD values of the Amulet and Kidd Creek ore fluids have not yet been obtained, their $\delta^{18}O$ values are shown at the top of the diagram for comparison. (From Beaty and Taylor 1982.)

due to igneous heating. The convection may have been helped by deep subsidence, compaction, lithification and low-grade metamorphism. Franklin and Thorpe (1982) cite the occurrence of subvolcanic comagmatic sills which may have been important in this convection in felsic volcanic environments. In ridge–rift environments magmatism would supply the necessary thermal drive.

(d) Source of the metals and sulphur
The ore-forming solutions may have been dominantly seawater but the bulk of the metals must be derived elsewhere. The metals may have been partially derived from the magma but many authorities suggest the major contribution is from metals leached from the rocks through which the deeply circulated, weakly acidic, heated solutions passed. As these solutions were also highly chlorinated and strongly reduced the conditions would allow the breakdown of the silicate mineral structures. This breakdown would release, probably only in trace amounts, the metals which

would be then transported by the solutions upwards towards the submarine vent. The deposition of the metals would result from E_h–pH and temperature changes as the ascending solutions mixed with the cooler oxygenated near-surface meteoric and sea waters.

Sulphur could, in part, be derived from seawater, sulphate being reduced to sulphide either during interaction with hydrothermal solutions and the ferrous components of the rocks through which the hydrothermal solutions passed, or by inorganic or organic reduction of coeval seawater sulphate at the site of deposition. Sulphur isotope studies indicate that much of the sulphur in these massive sulphide deposits is indeed derived from seawater by sulphate reduction (Ohmoto and Rye, 1979). On the other hand this is not universally true. Gregory and Robinson (1984) suggest a dominantly magmatic sulphur source for the Dianne and O.K. deposits in Northern Queensland, Australia, whereas the nearby Mt Molloy deposit shows mixing with seawater.

(e) Stratigraphic and structural control

In many of the deposits the massive sulphides are confined to a relatively narrow stratigraphic interval compared to the timespan occupied by volcanic activity. Also many submarine volcanic piles are barren, and the hydrothermal alteration of the hangingwall of some economic deposits indicates that barren hydrothermal activity occurred after the ore-forming event. Finlow-Bates (1980) and others suggest that the hydrothermal ore-forming solutions are trapped connate water that is suddenly released by the first movement of deep-seated faults. This would mean that the hydrothermal discharge would be more or less synchronous over large areas and might explain the location of the massive sulphides within defined stratigraphic intervals and why ore forming has taken place in some volcanic piles and not in others. It might also explain why post-ore barren hydrothermal activity may occur, the circulating meteoric waters having insufficient residence time, heat, or other requisite characteristic to remove the metals from the silicate minerals within the hydrothermal reservoir. Within cycles of volcanic activity in a particular district the most important massive sulphides are often associated with the lowermost cycle of activity.

The deposits that sit on the discharge vents are characteristically underlain by a footwall alteration pipe, often with economic quantities of sulphides with high Cu/Zn ratios. These ratios decrease upwards and outwards in the massive stratiform sulphide mound. This may be explained by re-mobilization of previously deposited sulphides by continued hydrothermal flow through the mound. An alternative explanation would be the gradual evolution of lower Cu/Zn ratios. However, evidence from deposits such as Millenbach, Noranda, Canada, shows that where stacked lenses are found above the same vent the older lenses were enriched in copper due to fracture filling and replacement by chalcopyrite (Knuckey et al., 1982). Some massive sulphide bodies show explosive brecciation of the initially deposited sulphides and this might aid the deposition of chalcopyrite in these deposits.

Sedimentary structures in some massive sulphide deposits indicate mechanical reworking and downslope transportation of the ores after deposition. If ore-forming solutions flow downslope, then ore deposition may take place in brine pools – ore bodies formed here are expected to be more tabular and should show regular banding or laminae.

3.8.6 A genetic model

The evidence we have presented above indicates that these massive sulphide bodies associated with volcanics were formed by hydrothermal fluids generated mainly by deep convective circulation of marine and meteoric waters rather than by magmatic differentiation. The introduction of the metalliferous hydrothermal fluids is considered to be syn- or slightly post-depositional. This would result in sulphide occurrence by chemical precipitation in mounds near vents on the seafloor, by diagenetic reaction with already lithified sediments and by subseafloor (footwall) pore space infill rather than by large-scale replacement of the rock. Early formed sulphides might be deformed or brecciated by subsequent hydrothermal activity. Downslope movement of the ore-forming solutions would result in the formation of distal deposits. The movement of the ore-forming solutions through the sulphide mound would result in metal zoning with dominantly copper towards the base and zinc higher up. This genetic model is summarized in Fig. 3.40.

3.9 EXPLORATION FOR VOLCANOGENIC SULPHIDE DEPOSITS

The currently accepted genetic model for these deposits such as the one we outlined in the previous section will form part of the basis upon which an exploration programme will be based. In addition to the model the exploration geologist will also be aware of the other features of these deposits which may be exploited for their discovery. The features to which we refer include the common host lithologies, the structures, the alteration phenomena and the physical and chemical properties of the deposits. The exploration geologist must also appreciate at what scale these features are evident and at what stage of the exploration programme they are best exploited. Further considerations must be the geographic/climatic setting, and the topography of the region; a method that has been very successful in Canada may result in less success in more arid terrains or in tropical climates.

3.9.1 Reconnaissance

In evaluating the potential of a completely new area to contain volcanogenic massive sulphides the two most important criteria are firstly that volcanic rocks should be present which are dominantly submarine in origin and secondly these volcanic rocks must be calc-alkaline or island-arc tholeiitic in composition (Sangster, 1980). There are, of course, many thousands of cubic kilometres of these rock types which are barren and therefore other criteria must be sought, and other techniques used, to narrow down the search within these favourable rock types.

With the availability of satellite imagery, remote sensing and the recognition of regional structural patterns will become increasingly used during the regional search. Very little published work appears to have been produced demonstrating the specific applications of remote-sensing techniques to massive sulphide deposits. Labovitz *et al.* (1983) describe experimental data for the application of certain remote-sensing techniques to geobotanical exploration for volcanogenic sulphides at Mineral, Virginia. They sampled oak trees, measuring the reflectance from leaves using a portable radiometer. They found the variation in leaf reflectance was associated with the change in concentration of trace metals in the soil, particularly Pb and Cu.

Sangster and Scott (1976) claimed that the massive sulphide deposits at Noranda were discovered in the 1960s and 1970s because exploration geologists were convinced they were volcanogenic in origin and associated with brecciated interflow surfaces. Ridge (1983) states, however, that the discoveries were actually made because the geologists observed that most deposits of the massive sulphide type in the area were located at the intersections of major north–south and east–west fractures. If this is the case then remote sensing may play a much more extensive role in the selection of areas for detailed exploration for volcanogenic massive sulphide deposits in the future since mapping of large lineaments is one of the main advantages of satellite imaging (Press, 1983).

As well as the possibility of being controlled by faults, the occurrence of volcanogenic massive sulphides in clusters, often associated with rhyolite domes, is well documented. Recognition of these factors is very important to the exploration geologist in deciding those areas for more detailed examination.

The presence of pyrrhotite in the massive sulphide lenses and magnetite in the underlying stringer zones should mean that airborne electro-

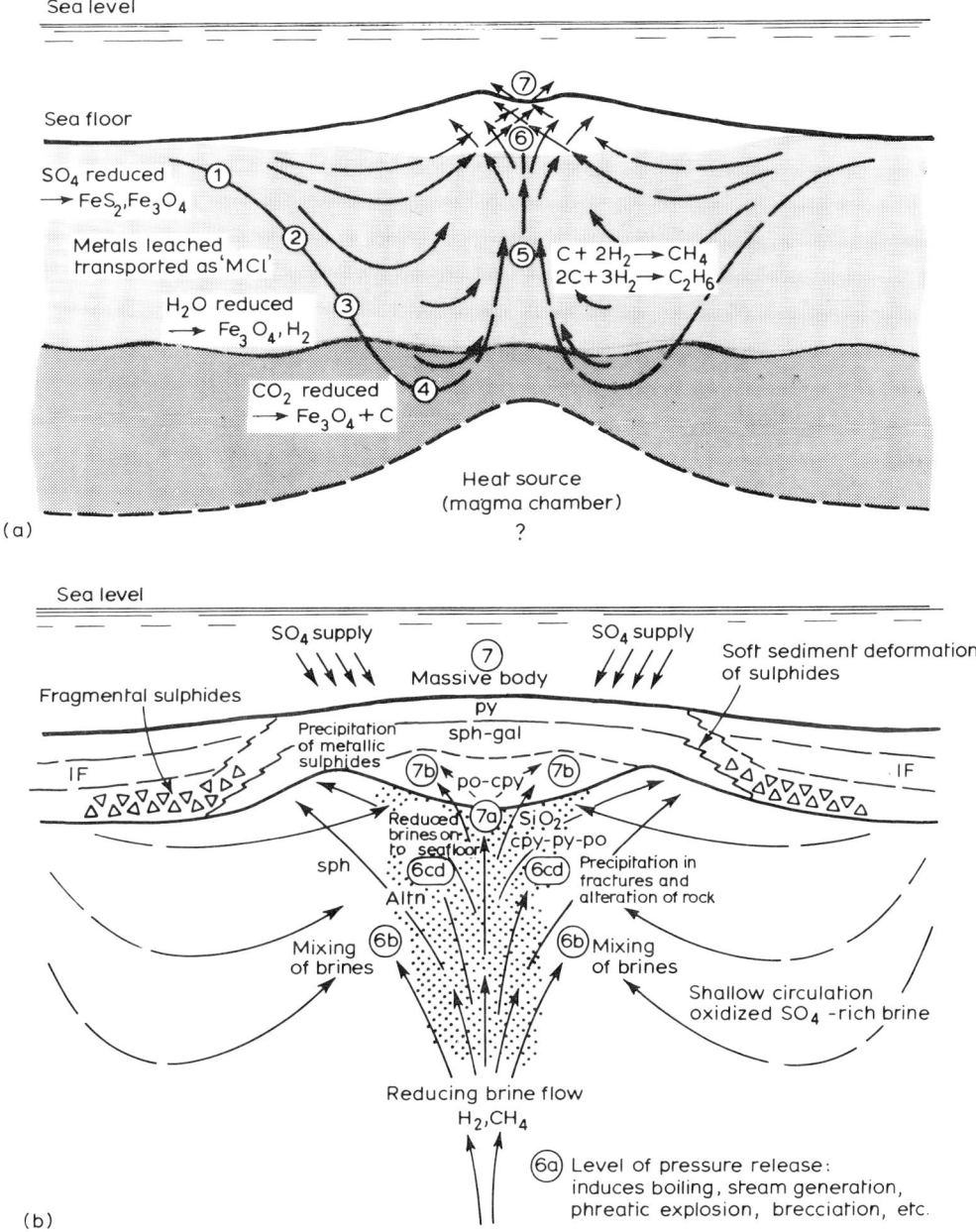

Fig. 3.40 Diagrammatic illustration of (a) synsedimentary subseafloor hydrothermal convective system (with formation of stringer sulphides below seafloor (6) and massive sulphide precipitation on seafloor (7)) and (b) detail of seafloor hydrothermal exhalative vent (from Hutchinson 1982).

magnetic (AEM) and airborne magnetic methods will be useful at the reconnaissance stage. This has proved to be the case in Canada, but these techniques have been less successful in more arid areas. Airborne magnetic surveys were used in the location of the Teutonic Bore deposit in Western Australia, the magnetic method being used to delimit the areas of ultramafics. In New South Wales, the Captains Flat and Woodlawn deposits were also discernible from aeromagnetic surveys (Malone and Whiteley, 1981). In the case of the latter deposits the regional gravity map revealed an apparent association of gravity highs with the major volcanogenic sulphide deposits. The regional gravity survey used an average station spacing of 11 km and was contoured at 5 mgal. As an alternative to regional geophysics, if the methods mentioned above indicate lack of response, or in addition to these methods, reconnaissance geochemical surveys should be undertaken. At this stage streamwater and sediment sampling are normally carried out. However, reconnaissance geochemical sampling of the B or C soil horizons adjacent to the drainage systems in the Woodlawn area disclosed anomalous values of copper, lead and zinc with higher than background silver and tin downstream from where the ore body was eventually discovered (Whiteley, 1981). Chip sampling from shallow pits gave anomalous values for all these elements and subsequent grid sampling outlined an extensive geochemical anomaly.

In addition to the use of water, stream sediment and soil geochemistry in the recognition of anomalous values of the metals associated with these deposits, lithogeochemistry has been used to recognize areas of depletion or addition of certain elements. The use of lithogeochemistry in exploration for volcanogenic massive sulphides has been described by Sopuk *et al.* (1980). The aim of these authors was to distinguish productive cycles (in their case Cu–Zn production) from their non-productive counterparts. They found that the elements most useful in this discrimination on a regional scale are also those whose concentrations vary intensely as a result of local metasomatism associated with mineral deposits. These are Fe_2O_3, Zn, MgO and Na_2O. On a mine scale Fe_2O_3 and MgO are enriched whereas Na_2O and CaO are depleted. Furthermore they found that the leaching and enrichment of certain major and trace elements may extend over areas of tens of square kilometres.

Although the work by Sopuk and co-workers (1980) was specific to one deposit type, this form of investigation may find increasing applicability and therefore we will provide brief details of their approach.

(a) Having located a suitable area, that is one having the correct volcanic sequence – submarine and felsic (Sangster, 1980) – attempts should be made to recognize the high-SiO_2 cyclic section of the felsic members. As an indication of the approximate level to be anticipated Sopuk and co-workers state that the SiO_2 level may be as high as 70%.

(b) To discriminate the potentially productive zones from the likely non-productive portions in the high-SiO_2 cyclic section samples should be taken on the basis of three or four per square kilometre. Two 1 kg samples should be collected at each site, one for slabbing and thin sectioning and the other for analysis. The samples should be analysed for SiO_2, Fe_2O_3 (total), MgO, CaO, Na_2O, Zn and Mn. Discriminant analysis is then applied, and felsic samples classified as 'mineralized' are separated from those classified as 'unmineralized'. Felsic areas with more than 75% of samples classified as 'mineralized' should be explored further.

3.9.2 Detailed survey and target definition

Once the regional survey is completed areas are selected for further detailed examination. These areas will have been selected as containing the correct host rocks, the calc-alkaline volcanic cycle most likely to contain massive sulphides, and will be those areas that show the most promising structures and those that appear anomalous on the regional geochemical and geophysical surveys. These chosen areas will be surveyed in more detail utilizing the likely physical and chemical features of the massive sulphide deposits and will be guided by the geological model derived for the deposit. The object of this stage is to define a target, or targets, for evaluation drilling. The location of this target relies on sound geological mapping, geochemistry, geophysics and a good appreciation of the regional structure and the possible localization structures.

Probably one of the most useful approaches to be adopted will be geochemistry, not only to outline the area likely to contain the deposit but also by recognition of the chemical signatures of the alteration phenomena surrounding these deposits to target in on the most likely site of the massive sulphide lens. In the Woodlawn area of New South Wales the soil geochemistry showed that the zinc anomalies extended over a much wider area than the lead peaks, and subsequent drilling showed that none of the zinc peaks coincided with the ore body (Malone, 1981). This may have resulted from zinc being a more mobile element. At this location the main area of the copper anomaly was located off the massive sulphide ore body, with only part of the anomalous copper area being underlain by the ore body, the rest coinciding with insignificant mineralization.

3.9.3 Alteration

One of the well-documented features of the volcanic-associated massive sulphides is the characteristic alteration phenomena that surround the ore bodies. Although this is dominantly in the footwall, hangingwall alteration does occur. Careful examination of the chemical and mineralogical characters of these alteration zones will indicate the proximity of the main ore zone. Using information from partial analyses of the Key Anacon deposit, Bathurst, New Brunswick, Boyle (1982) reported the chemical variation through the envelope of alteration zones marked by chloritization, sericitization and silicification which encloses the deposit. His data show a general and relatively consistent increase in the K_2O/Na_2O ratio towards the massive sulphide lenses. Marcotte and David (1981) have discussed geochemical data collected for target definition of Kuroko-type or Polymetallic-type deposits in the Abitibi Belt. They report an increase in the K_2O/Na_2O ratio, depletion in CaO and enrichment in MgO for a proximal group of samples relative to distal ones.

Boyle reports that the SiO_2/CO_2 ratio also shows a fairly consistent increase towards the massive sulphide bodies, reflecting increasing silicification towards the mineralization. Where silicification is a feature, a plot of SiO_2 content may indicate the proximity of ore – the SiO_2 content increasing as the ore is approached. Most primary halos associated with conformable massive sulphides are enriched in mainly chalcophile elements and principally Cu, Ag, Au and Zn. Boyle (1982) comments that when dealing with massive sulphides they are marked by a variety of leakage halos which may not be evident without considerable detailed field and laboratory work. Some may be characterized by albitized and/or tourmalinized zones as in the Bathurst Camp, New Brunswick. Goodfellow and Wahl (1976) utilized water-extractable analyses for Na, K, Ca, Mg, F and Cl in the halos of two massive sulphide deposits in New Brunswick. They identified halos up to 800 m radius in the volcanic rocks overlying the Brunswick deposits and concluded that the size and intensity of the anomalous zones associated with these deposits as well as the rapidity of the water-extractable analysis has much significance in the exploration for deeply buried massive sulphides in the Bathurst area.

3.9.4 Isotopes

When we discussed the genesis of these deposits we commented upon the isotopic evidence that has been used to indicate the likely source of the hydrothermal ore-forming fluids. These isotopes may have a further use, in exploration. At present their use in exploration is limited and may remain that way for some time because of the availability and cost of analysis. We can, however, illustrate how such data may be used if available.

The intensely altered stockwork chimneys below these deposits are lower in $\delta^{18}O$ values than the surrounding rocks. The almost concentric nature of the distribution of these values in the Amulet 'A' ore body, Amulet Mine, Noranda, gave rise to a 'bullseye' pattern as described by Beaty and Taylor (1982). The target is relatively small, however (Fig. 3.41) which means that random sampling of oxygen isotopes is likely to be of little or no use. However, if detailed geological mapping of the alteration zones is coupled with the ^{18}O sampling this might indicate the zone which has undergone the greatest water flux and therefore where the temperatures may have been most conducive to ore deposition. It is important to note that zones of ^{18}O depletion can survive relatively high metamorphism.

In the case of high ^{18}O the exploration potential is greater. The ore-forming fluid here is likely to imprint its distinctive isotopic value on the rocks through which it flows. For example the rocks altered by the high-^{18}O Kidd Creek ore solution have $\delta^{18}O$ values much higher than normal volcanic rocks elsewhere in the Abitibi Belt. Therefore in these circumstances if a high-$\delta^{18}O$ value is discovered, and weathering can be discounted, it is likely that a deep-seated high-^{18}O hydrothermal solution flowed through the rock.

In their work on the Amulet 'A' ore body Beaty and Taylor (1982) calculated $\delta^{18}O$ values indistinguishable from the ore fluids at Cyprus, Kuroko and at Ducktown (Fig. 3.39). As $\delta D°/_{oo}$ values are not available for the Amulet deposit the exact position relative to the other

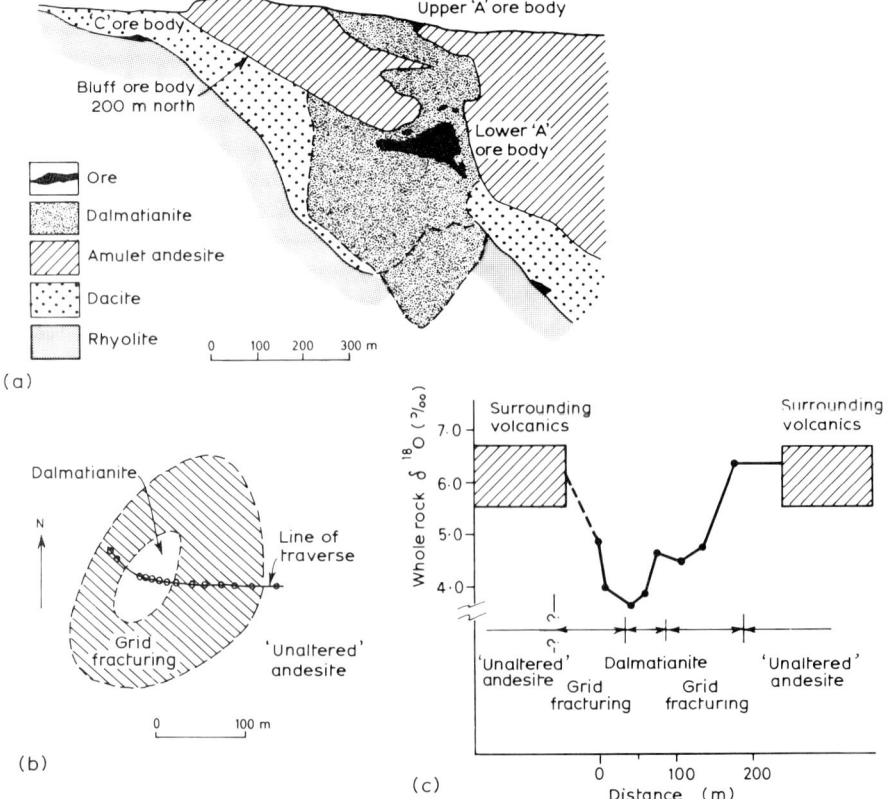

Fig. 3.41 Use of ^{18}O isotopes in exploration, Amulet 'A' Mine, Noranda district, Canada (after Beaty and Taylor 1982). (a) Generalized geological cross-section through the Amulet A mine. The dalmatianite chimney extends through to the surface, where it was sampled. (b) Sketch map of the Amulet 'A' alteration zone showing the sample localities. The unaltered andesite has been subjected to a regional hydrothermal flux and strictly speaking is not fresh. The Amulet pit lies to the northwest of the sampled traverse. (c) Variations in $\delta^{18}O$ along the Amulet traverse.

deposits could not be plotted on Fig. 3.39. In view of the age range of the deposits considered, Beaty and Taylor propose that the $\delta^{18}O$ of seawater has not changed significantly since the Archaean.

Also plotted on Fig. 3.39 are the isotopic values for the Raul and Kidd Creek deposits. It may be seen that these do not fall within the same domains as the other deposits. This variation is either due to evaporation in a closed basin or through exchange with high-^{18}O sediments during the fluid circulation. Therefore considering the deposit types included in Fig. 3.39 most are apparently derived from seawater circulating through oceanic crust while in at least two cases either the evolution of the ore fluid has been more complex or the ore fluid had a different origin altogether.

Those deposits with ^{18}O comparable to modern seawater – Amulet, Matagami, Ducktown, Cyprus and Kuroko – show age ranges from Archaean to Cenozoic; so do the deposits with abnormally high ^{18}O such as Kidd Creek and Raul. Beaty and Taylor raise the intriguing possibility that the size of the Kidd Creek deposit and its high ^{18}O may be genetically related. The Kidd Creek ore body is a giant massive sulphide deposit; it contains approximately 1% of the total known Canadian massive sulphide ore. Could this have arisen because the ore fluid was chemically different – an ore fluid that allowed more sulphides to be deposited? If more evidence becomes available that high ^{18}O can be equated with very large deposits, then measurement of $\delta^{18}O$ values could become an important factor in exploration programmes for these larger volcanogenic massive sulphide deposits.

3.9.5 Geophysics

The conductivity of these massive sulphide deposits means that they present very suitable targets for a variety of electrical and electromagnetic methods. We have already drawn attention to the use of airborne electromagnetic and magnetic methods. In exploration for shallow massive sulphide deposits, spontaneous potential must rank as one of the most *cost effective* methods.

Induced polarization is also very applicable to massive sulphide deposits and their associated stockworks and brecciated zones. At Woodlawn in New South Wales, this method defined an anomalous polarizable zone which partly coincided with the geochemical anomaly and from this the drilling targets were chosen. The resistivity response is often most marked over these large sulphide bodies and this may be very useful in delineating the ore body where the induced polarization response is complex.

In recent years the UTEM system has found greater use in Australia because of its ability to penetrate to deeper levels but in areas of conductive overburden the method has limited use depending on the thickness of such overburden. The magnetic induced polarization (MIP) technique may be useful in the search for this type of deposit but further work is required to prove the extent of its usefulness in this context.

3.9.6 Drilling

The optimum spacing of diamond drill holes to prove and evaluate any targets chosen will depend on the exploration budget and the irregularity of the deposit as interpreted from the data available. It is not possible for us to suggest an 'average' spacing. In the case of the Woodlawn ore body mentioned above the infill drilling was on a 30 m grid; this intensity would allow for the irregularity of the ore body and also supply representative samples for metallurgical study. The importance of good core recovery was highlighted by the case of the Teutonic Bore deposit in Western Australia where the original mineral reserve estimate undervalued the supergene copper because of poor recovery in the friable chalcocite zone. In addition to normal core logging and sampling techniques the characteristic features of the volcanogenic sulphides must also be looked for. Boyle (1982) suggests that during drilling for massive sulphides the K_2O/Na_2O ratio is the most useful indicator of the proximity of ore.

Diamond drilling is very expensive and consequently the maximum amount of information must be derived from the hole. To this end there will be an increasing use of down-the-hole geophysical techniques, particularly to detect 'off-hole' mineralization. Down-the-hole electromagnetic techniques, such as a SIROTEM probe, are particularly applicable to the search for massive sulphides. Depending on the conductive capacity of any massive sulphide body, such a probe should be able to locate sulphides within 100 m of the hole. This type of technique will also indicate the quadrant in which the conductive body lies relative to the hole and also the approximate distance from the hole. In connection with diamond drill holes the applied potential ('*mise-a-la-masse*') may be used to effect. Again off-hole mineralization may be detected; Templeton *et al.* (1981) quote distances of 60 m from the deposit to the hole in which the measurements were taken.

3.9.7 Mine-based exploration

Localized exploration, to extend the known limits of the deposit and to increase the ore reserves and to upgrade reserves from one category to another, will be achieved largely by diamond drilling. This drilling will be guided by the mine geologist's understanding of the structural and lithological controls on the ore body. The physical and chemical characteristics which we have previously described as useful during the larger, district-scale, exploration programme will also be utilized in the more local search. In addition, exploration will be aided by geological models based upon the perceived genesis of the ore body and the genetic model for the deposit type as a whole.

The occurrence of volcanogenic massive sulphides in clusters, often associated with rhyolite domes, close to hydrothermal vents, and possibly controlled by faults, has already been mentioned. At Millenbach, Noranda, Knuckey *et al.* (1982) showed that broad-scale metal zoning existed in a cluster at the same stratigraphic horizon, the average Cu/Zn ratio decreasing steadily with distance from the main hydrothermal vent. This occurrence of the deposits not only over the exhalative vent ('proximal' accumulations) but also away from this vent, usually down-palaeoslope from it ('distal' accumulations) will have major significance in mine-based exploration. The development of an exploration strategy to identify correctly these two types of accumulation relies on interpretation of thickness variations within each volcanic flow and in the recognition of palaeoslope conditions. MacGeehan and fellow workers used such features to advantage in mine-based exploration on the Norita deposit, an Archaean occurrence of distal type in the Mattagami mining district of Quebec (MacGeehan *et al.*, 1981). Although this deposit contains sulphide lenses that are now vertical because of folding (Fig. 3.42), it is possible to distinguish that the Main Zone ore body lies on the flanks of the rhyolite dome and the A-Zone ore body occurs in a basinal depression at the base of the dome. It is thought that the metal-bearing hydrothermal fluids discharged on to the seafloor (probably through the stringer zone shown in Fig. 3.42) reacted with the seawater, precipitated base-metal sulphide and generated a sulphide-rich density current. The flow of this current was controlled by the palaeotopography of the rhyolite surface. Correct identification of the characteristics of these deposits makes it possible to explore up palaeoslope from known deposits towards the mineralized hydrothermal feeder zone and possible proximal mineralization, and down palaeoslope for other distal deposits.

MacGeehan *et al.* also stress that too often the sulphide deposits at Mattagami, and possibly elsewhere, have been identified as proximal (Noranda)-type occurrences based on the fact that they are underlain by hydrothermally altered rocks. They contend that such pervasive alteration is a necessary feature of volcanogenic sulphide deposits, reflecting as it does the subseabottom geothermal activity of the area at the time the deposits were formed. We support this view and feel that district-wide footwall alteration associated with volcanogenic sulphides is becoming increasingly recognized. MacGeehan and co-workers recommend the best definitive method of recognizing the vent area, and the likely

Magmatic hydrothermal deposits

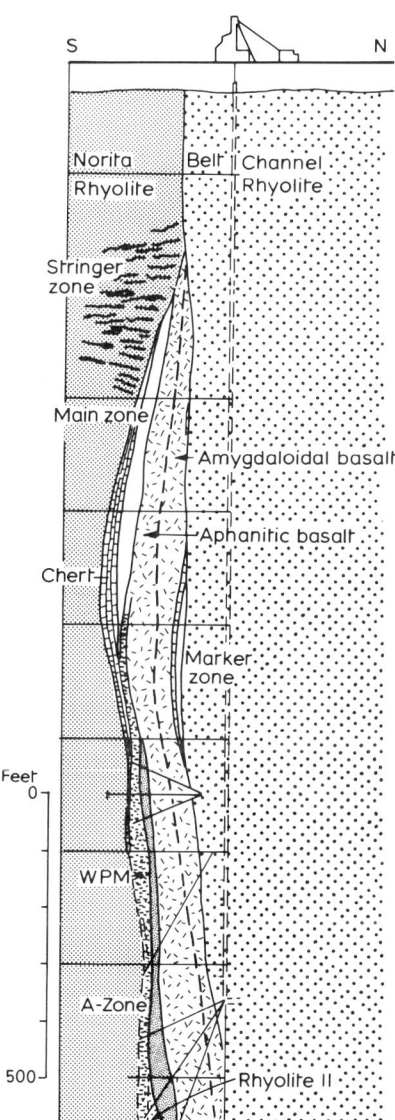

Fig. 3.42 Longitudinal vertical section along 9+OOE showing the stratigraphy and the relation between the main zone and underlying stringer sulphide mineralization, and the overlying A-zone and marker zone sulphide lenses Norita Deposit, Mattagami, Canada (from MacGeehan et al. 1981).

site for proximal deposits, is a vertical zone of intense stockwork mineralization terminating at the palaeosediment–seawater interface.

One further feature has a bearing on mine-based exploration. This is the observation on metal distribution in stacked lenses made by Knuckey et al. (1982). Although this observation was made specifically at Millenbach, Noranda, it may have much wider application. They showed that where stacked lenses formed over the same vent the older lenses were enriched in copper due to fracture filling and replacement by chalcopyrite. Therefore if stacked lenses are suspected the lower lenses are likely to be the highest grade.

3.9.8 Size characterization – a predictive grade/tonnage model

Some attempt has been made to characterize the grades and tonnages of volcanogenic massive sulphides to define predictive models for use in exploration (Sangster, 1980). If exploration has indicated the possible presence of a massive sulphide deposit within a district, Sangster's 'average' figure would suggest the likely grade to be about 6% combined base metal. Furthermore there is a more than 80% chance of any deposit discovered being less than 10 million tonnes. Sangster's work gives a 40% chance that the deposit will be of the order of 0.4 million tonnes.

If a volcanic centre can be recognized, then the possibility of a cluster of deposits occurring can be considered and in Sangster's opinion about five million tonnes will be in each cluster, distributed as shown in his size distribution curve (Fig. 3.43). An interesting proposition made by Sangster is that when more deposits are sought in areas where a few massive sulphide deposits are already known, a measure of the undiscovered potential can be made by comparison with the average total base-metal tonnage for such deposit clusters. This figure is calculated by Sangster as nearly 5 million tonnes, for the deposits studied in the Canadian base-metal sulphide province. The difference between this average and the known metal content would indicate the metal yet to be discovered.

Sangster (1980) used this model to analyse the known mineralization in the Snow Lake area and concluded that since the difference between the average and the known metal content was 3.2 million tonnes then this 'shortfall' indicated at least one other deposit remains to be discovered. If the total shortfall is in one large deposit then it will be bigger than any yet discovered in the area and the ratio between this 'new deposit' in Snow Lake, and the previous largest one in the area would be in line with the data from other deposit areas studied by Sangster (Fig. 3.43).

The type of statistical approach proposed by Sangster may not be readily acceptable by some authorities on mineral deposits but perhaps the unusual characteristics of volcanic-associated massive sulphide deposits may allow these deposits to be treated in this way. One of these features, to which we have already drawn attention, is that within a particular area or belt the massive sulphide deposits are confined to a particular stratigraphic horizon often representing a particularly short geological timespan. For example, the Kuroko deposits extend about 1500 km in a linear belt but are confined to one stage in the Middle Miocene and were formed in a 200 000 year timespan (Franklin et al., 1981). Other examples of the mineralizing process operating over a discrete, relatively short, period in a particular mineral belt include Noranda, Mattagami and Bathurst. If the hydrothermal event was more or less synchronous over a large area, then the deposits of this area may perhaps belong to one population and Sangster's approach may be acceptable. It is too early to tell if the Sangster method of predicting grade and tonnages will become widely used in exploration for volcanogenic massive sulphides. The Sangster model cannot readily accommodate the 'super-giant' deposits such as the Bathurst no. 1 or the Kidd Creek. These large deposits are exceptional, containing as they do about 2% of the total known

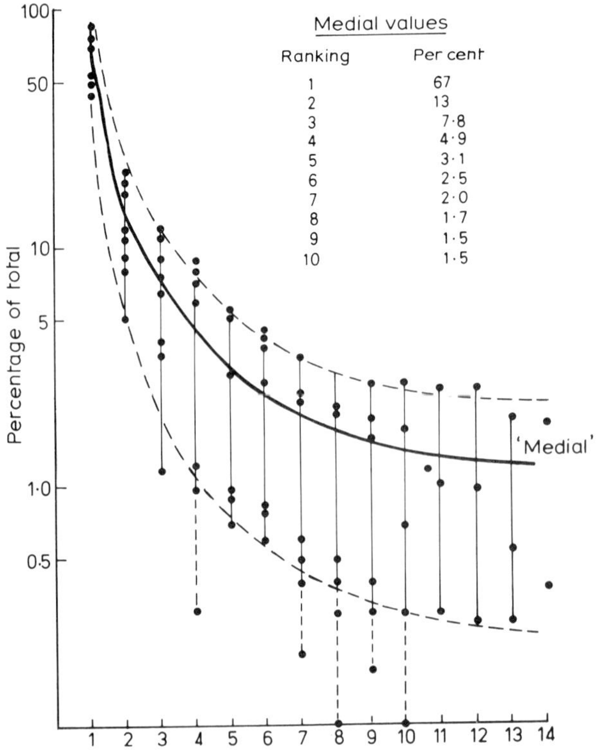

Medial values

Ranking	Per cent
1	67
2	13
3	7.8
4	4.9
5	3.1
6	2.5
7	2.0
8	1.7
9	1.5
10	1.5

(a)

Ranking	Present (%)	Predicted (%)	Medial (%)
1	51	68	67
2	21	18	13
3	12	7.8	7.8
4	8	3.4	4.9
5	3	2.7	3.1
6	2.7	1.1	2.5
7	0.5	0.85	2.0
8	0.5	0.15	1.7
9	0.4	0.15	1.5
10	0.3	0.13	1.5
11		0.11	

(b)

Fig. 3.43 (a) Percentage distribution of each deposit in a cluster. (b) Use of (a) in a prediction of size of undiscovered deposit in Snow Lake district. (From Sangster 1980.)

Canadian massive sulphide population. These supergiant deposits must represent a very efficient mineralizing process, the geological constraints for which are not yet fully understood.

3.10 CONCLUDING STATEMENT

The deposits we have considered in this chapter are currently the World's largest suppliers of copper, molybdenum and zinc. The porphyry deposits and volcanogenic massive sulphide deposits are also the two most economically important deposit types that are apparently directly related to plate tectonic processes. With the exception of the Cupreous Pyrite type all are correlated with destructive plate boundaries. The porphyries of South America are associated with an Andean-type boundary. Those of the Pacific, such as Bougainville and Ok Tedi, and the Polymetallic, Primitive and Kieslager-type volcanogenic sulphides are associated with island-arc-type boundaries. This similarity in tectonic environment between the two deposit types raises some interesting questions about their genesis and temporal distribution.

In the Tertiary there is strong evidence of this close spatial association between these deposits and destructive plate boundaries. In the Archaean and Proterozoic the volcanogenic sulphides are plentiful but there are no economic porphyry coppers. Our knowledge of the Precambrian crustal conditions is not yet sufficiently advanced to explain this absence. If subduction caused these two deposit types in the Tertiary it appears to have caused only one of them, the volcanogenic massive sulphides, in the Precambrian. The paucity of porphyry copper deposits cannot be explained by erosion and removal. If the current theories of genesis of these volcanogenic sulphides are correct they are produced higher in the crust than the porphyry deposits. Therefore they should have been removed first.

The apparent absence of the Cupreous Pyrite type in the Precambrian is also somewhat of an enigma. This type is thought to form at spreading centres. If subduction was operational in the Precambrian, to produce the other volcanogenic types, spreading centres also existed. Either different tectonic processes were involved or other crustal conditions existed which prevented the formation of the Cupreous Pyrite type and economic porphyry copper deposits. Continued research and exploration is necessary to provide answers to this problem.

Considering the characteristics of these deposits that are of exploration importance, firstly it is significant that they appear to be spatially related to plate boundaries. Global and regional search can be guided empirically by this observation without all the factors of genesis being explicable.

Once a search region has been chosen for these deposits, geological models have proved useful in exploration. The porphyry copper model was among the first geological models to be recognized and has been widely used in the search for new deposits. Possibly the decline in the metal markets in the early 1980s has prevented the development of the many deposits discovered by model-orientated exploration work carried out during the early 1970s. The model for porphyry, or stockwork, molybdenum deposits has developed from the porphyry copper model. The two models are thought to represent end members of a continuum, the intermediate members of which are those porphyry copper deposits with significant by-product molybdenum. The application of the Climax-type geological model resulted in the discovery of the Henderson ore body below the previously known Urad ore body. The success of this model, combined with data collected by indirect methods, helped in the discovery of the Mt Emmons ore bodies.

Although the recognition of the volcanogenic type as a distinct group was relatively recent, many models have been proposed for their formation. Any useful geological model must be flexible and constantly reviewed. The volcanogenic model is a good example of this. Currently the alteration phenomena and such features as the distribution of chalcopyrite are receiving considerable attention. The application of models, based upon the relatively undeformed and unmetamorphosed Tertiary deposits to the deformed Archaean and Proterozoic deposits, can be difficult but not impossible as seen at the Norita deposit.

The use of indirect methods such as geochemistry has been relatively successful in the search for the deposits covered in this chapter. Geophysics has been utilized to define targets more accurately. The deposits are characterized by large alteration patterns and these have been very important in their location. Lithogeochemistry and detailed geological mapping have been used to delimit these halos.

In these alteration halos detailed studies of isotopes could become important if the procedure is ever sufficiently cheap and routine to be included in a normal exploration programme. The approach offered by chemical fingerprinting, based upon whole-rock geochemistry, to predict the possible occurrence of productive intrusions appears a promising line for future research. The concept of predicting grade for undiscovered deposits may have merit when deciding whether it is worth continuing the search within a specific area.

REFERENCES

Abrams, M. J., Brown, D., Lepley, L. and Sadowski, R. (1983) Remote sensing for porphyry copper deposits in Southern Arizona. *Econ. Geol.*, **78**, 591–604.

Adamides, N. G. (1980) The form and environment of formation of the Kalavasos ore deposits, Cyprus. In *Ophiolites: Proc. Int. Ophiolite Symp., Cyprus 1977.* (ed. A. Panayiotou), Geol. Surv. Dept. pp. 117–178.

Agnew, P. and Harrison, S. (1983) Teutonic Bore Cu–Zn–Ag deposit. In *Eastern Goldfields Geological Field Conference 1983* (ed. P. C. Muhling). Geol. Soc. Aust. W.A. Div. and E. Goldfields Geol. Disc. Group. pp. 20–22.

Ambler, E. P. and Facer, R. A. (1975) A Silurian porphyry copper prospect near Yeoval, New South Wales. *J. Geol. Soc.*, **22(2)**, 229–241.

Anderson, J. A. (1982) Characteristics of leached capping and techniques of appraisal. In *Advances in Geology of the Porphyry Copper Deposits: Southwestern North America* (ed. S. R. Titley), University of Arizona Press, Tucson pp. 275–296.

Anthony, E. Y. (1983) The milieu of porphyry copper deposits. In *Revolution in the Earth Sciences Advances in the Past Half-Century* (ed. S. J. Boardman), Kendall/Hunt, Dubuque, USA, pp. 317–325.

Baldwin, J. A. and Pearce, J. A. (1982) Discrimination of productive and nonproductive porphyritic intrusions in the Chilean Andes. *Econ. Geol.*, **77**, 664–674.

Baldwin, J. T., Swain, H. D. and Clark, G. H. (1978) Geology and grade distribution of the Panguna porphyry copper deposit, Bougainville, Papua New Guinea. *Econ. Geol.*, **73**, 690–702.

Barton, P. B. Jr. (1978) Some ore textures involving sphalerite from the Furutobe mine, Akita Prefecture, Japan. *Min. Geol.*, **28**, 293–300.

Baumer, A. and Fraser, R. B. (1975) Panguna porphyry copper deposit, Bougainville. In *Economic Geology of Australia and Papua New Guinea*, Vol. 1 (ed. C. L. Knight). *Metals. Australas. Instn. Min. Metall. Monogr.* **5**, 855–866.

Bean, R. E. and Titley, S. R. (1981) Porphyry Copper Deposits Part II. Hydrothermal Alteration and Mineralization. *Econ. Geol. 75th Anniv. Vol.*, 235–269.

Beaty, D. W. and Taylor, H. P. (1982) Some petrologic and oxygen isotopic relationships in the Amulet mine, Noranda, Quebec and their bearing on the origin of Archean massive sulphide deposits. *Econ. Geol.*, **77**, 95–108.

Boyle, R. W. (1982) Geochemical methods for the discovery of blind mineral deposits. Part 1: *CIM Bull.*, **75 (844)**, 123–142; part 2 *CIM Bull.*, **75(845)**, 113–132.

Burnham, C. W. (1979) Magmas and hydrothermal fluids. In *Geochemistry of Hydrothermal Ore Deposits*, 2nd edn (ed. H. L. Barnes), Wiley-Interscience, New York, pp. 71–136.

Chaffee, M. A. (1982) Geochemical prospecting techniques for porphyry copper deposits in Southwestern US and North Mexico. In *Advances in Geology of the Porphyry Copper Deposits: Southwestern North America* (ed. S. R. Titley), University of Arizona Press, Tucson pp. 297–307.

Cole, M. M. (1971) The importance of environment in biogeographical, geobotanical and biogeochemical investigations. *Proc. Int. Geochem. Explor. Symp. 3rd, Toronto.* Can. Inst. Min. Metall., Spec. Vol. **11**, 414–425.

Constantinou, G. (1977) Hydrothermal alteration of basaltic lavas of the Troodos ophiolite complex associated with the formation of the massive sulphide ore of Cyprus. *Geol. Assoc. Can. Spec. Paper*, **14**, 187–210.

Constantinou, G. (1980) Metallogenesis associated with the Troodos ophiolite. In *Ophiolites: Proc. Int. Ophiol. Symp. Cyprus 1979* (ed. A. Panayiotou), Geol. Surv. Dept. Cyprus, pp. 663–674.

Constantinou, G. and Govett, G. J. S. (1972) Genesis of sulphide deposits, ochre and umber of Cyprus. *Trans. Inst. Min. Metall. Sect. B.*, **8**, B34–B46.

Coope, J. A. (1973) Geochemical prospecting for porphyry type mineralisation – a review. *J. Geochem. Explor.*, **2**, 81–102.

Costa, V. R., Fyfe, W. S., Nesbitt, H. W. and Kerrich, R. (1980) Archean sedimentary talc: evidence for an ancient greenhouse (abstr.). *EOS*, **61(17)**, 386.

Davies, H. L. (1978) History of Ok Tedi porphyry copper project, Papua New Guinea. Pt. 1. *Econ. Geol.*, **73**, 796–809.

Drummond, A. D. and Godwin, C. I. (1976) Hypogene mineralisation – an empirical evaluation of alteration zoning. In *Porphyry Deposits of the Canadian Cordillera. CIM Spec. Vol.*, **15**, 52–63.

Eastoe, C. J. (1982) Physics and chemistry of the hydrothermal system at the Panguna porphyry copper deposit, Bougainville, Papua New Guinea. *Econ. Geol.*, **77**, 127–153.

Edmond, J. M., Measures, C. et al. (1979) On the formation of metal-rich deposits at ridge crests. *Earth Planet Sci. Lett.*, **46**, 19–30.

Evans, A. M. (1980) *An Introduction to Ore Geology*. Blackwell Scientific, Oxford, 231 pp.

Finlow-Bates, T. (1980) The chemical and physical controls on the genesis of submarine exhalative orebodies and their implications for formulating exploration concepts. A review. *Geol. Jahrb.* **D40**, 131–168.

Fleming, A. W. and Neale, T. I. (1979) Geochemical exploration at Yandera porphyry copper prospect P.N.G. *J. Geochem. Explor.*, **11**, 33–55.

Fox, J. S. (1984) Besshi-type volcanogenic sulphide deposits – a review, *CIM Bull.*, **77** (**864**), 57–68.

Franklin, J. M., Lydon, J. W. and Sangster, D. F. (1981) Volcanic associated massive sulphide deposits. *Econ. Geol. 75th Anniv Vol.*, 485–627.

Franklin, J. M. and Thorpe, R. I. (1982) Comparative metallogeny of Superior, Slave and Churchill Provinces. *Geol. Assoc. Can. Spec. Pap.* **25**, 3–90.

Frater, K. M. (1983) Geology of the Golden Grove Prospect, Western Australia: a volcanogenic massive sulphide magnetite deposit. *Econ. Geol.*, **78(5)**, 875–919.

Gass, I. G. (1980) The Troodos massif: its role in the unravelling of the ophiolite problem and its significance in the understanding of constructive plate margin processes. In *Ophiolites: Proc. Int. Ophiol. Symp. Cyprus 1979* (ed. A. Panayiotou), Geol. Surv. Dept. Cyprus pp. 23–35.

Gilmour, P. (1982) Grades and tonnages of porphyry copper deposits. In *Advances in Geology of the Porphyry Copper Deposits Southwestern North America* (ed. S. R. Titley), University of Arizona Press, Tucson, pp. 7–36.

Goodfellow, W. D. and Wahl, J. L. (1976) Water extracts of volcanic rocks - detection of anomalous halos at Brunswick No. 12 and Heath Steele B-zone massive sulphide deposits. *J. Geochem. Expl.*, **6**, 35–59.

Gregory, P. W. and Robinson, B. W. (1984) Sulphur isotope studies of the Mt. Molloy Dianne and O.K. stratiform sulphide deposits, Hodgkinson Province, North Queensland, Australia. *Mineralium Deposita*, **19**, 36–43.

Groves, D. I. and Batt, W. D. (1983) Metallogeny of the Greenstone Belts in the Eastern Goldfields. In *Eastern Goldfields Geological Field Conference 1983* (ed. P. C. Muhling), Geol. Soc. Aust. W.A. Div. and Eastern Goldfields Disc. Group, pp. 9–11.

Guilbert, J. M. and Lowell, J. D. (1974) Variations in zoning patterns in porphyry ore deposits. *CIM Bull.*, **67**, 99–109.

Gustafson, L. B. (1979) Porphyry copper deposits and calc-alkaline volcanism. In *The Earth, its Origin, Structure and Evolution* (ed. M. W. McElhinny), Academic Press, London, pp. 427–468.

Gustafson, L. B. and Hunt, J. P. (1975) The porphyry copper deposit at El Salvador, Chile. *Econ. Geol.*, **70**, 857–912.

Haynes, F. M. and Titley, S. R. (1980) The evolution of fracture related permeability within the Ruby Star granodiorite, Sierrita porphyry copper deposit, Pima County, Arizona. *Econ. Geol.*, **75**, 673–683.

Heaton, T. H. E. and Sheppard, S. M. F. (1977) Hydrogen and oxygen isotope evidence for sea water hydrothermal alteration and ore deposition Troodos Complex Cyprus. *Geol. Soc. Lond. Spec. Publ.*, **7**, 42–57.

Heberlein, D. R., Fletcher, W. K. and Godwin, C. I. (1983) Lithogeochemistry of hypogene, supergene and leached cap samples, Berg Porphyry Copper Deposit, British Columbia. *J. Geochem. Explor.*, **19**, 595–609.

Heidrick, T. L. and Titley, S. R. (1976) Structural evolution of southwestern North American Laramide porphyry copper deposits and its relationship to the history of plate interactions (abs): *Int. Geol. Congr., 25th, Sidney*, **3**, 740.

Hollister, V. F. (1978) *Geology of the Porphyry Copper Deposits of the Western Hemisphere*. Society Mining Engineers AIME, New York, 219 pp.

Horton, D. J. (1978) Porphyry-type copper–molybdenum mineralisation belts in Eastern Queensland. *Econ. Geol.*, **73**, 904–921.

Hunt, J. P. (1977) Porphyry copper deposits. *Spec. Publ. Geol. Soc. Lond.*, 98 (abstr.).

Hutchinson, R. W. (1973) Volcanogenic sulphide deposits and their metallogenic significance. *Econ. Geol.*, **68**, 1223–1246.

Hutchinson, R. W. (1980) Massive base metal sulphide deposits as guides to tectonic evolution. In *The Continental Crust and its Mineral Deposits* (ed. D. W. Strangway). *Geol. Assoc. Canada Spec. Paper* **20**, 659–684.

Hutchinson, R. W. (1982) Syn-depositional hydrothermal processes and Precambrian Sulphide Deposits. *Geol. Assoc. Can. Spec. Pap.*, **25**, 3–90.

Kesler, S. E. (1973) Copper, molybdenum and gold abundances in porphyry copper deposits. *Econ. Geol.*, **68**, 106–112.

Kirkham, R. V. and Soregaroli, A. E. (1975) Preliminary assessment of porphyry deposits in the Canadian Appalachians. *Geol. Surv. Can. Paper* **75-1A**.

Klau, W. and Large, D. E. (1980) Submarine exhalative Cu–Pb–Zn deposits, a discussion of their classification and metallogenesis. *Geol. Jahrb. Sect. D*, **40**, 13–58.

Knuckey, M. J., Comba, C. D. A. and Riverin, G. (1982) Structure, metal zoning and alteration at the Millenbach Deposit, Noranda, Quebec. *Geol. Assoc. Can., Spec. Pap.* **25**, 255–295.

Labovitz, M. L., Masooka, E. J., Bell, R., Siegrista, A. W. and Nelson, R. F. (1983) The application of remote sensing to geobotanical exploration for metal sulphides – results from the 1980 field season at Mineral, Virginia. *Econ. Geol.*, **78**, 750–760.

Lambert, I. B. and Sato, T. (1974) The Kuroko and associated ore deposits of Japan: A review of their features and metallogenesis. *Econ. Geol.*, **69**, 1215–1236.

Learned, R. E. and Boissen, R. (1973) Gold – a sulphide pathfinder element in the search for porphyry copper exploration in Puerto Rico. In *Geochemical Exploration 1972 –Proc. Int. Geochem. Explor. Symp., 4th London, 1972.* (ed. M. J. Jones) *Instn. Min. Metall*, pp. 93–104.

Lowell, J. D. and Guilbert, J. M. (1970) Lateral and vertical alteration-mineralisation zoning in porphyry ore deposits. *Econ. Geol.*, **65**, 373–408.

MacGeehan, P. J., MacLean, W. H. and Bonenfant, A. J. (1981) Exploration significance of the emplacement and genesis of massive sulphides in the Main Zone at the Norita Mine, Matagami, Quebec. *CIM Bull.*, **74 (828)**, 59–75.

MacNamara, P. M. (1968) Rock types and mineralisation at Panguna porphyry copper prospect, Upper Kaverong Valley, Bougainville Island. *Proc. Aust. Inst. Min. Metall.*, **228**, 71–79.

McMillan, W. J. and Panteleyev, A. (1980) Ore deposit models – 1, Porphyry copper deposits, *Geoscience, Canada*, **7(2)**, 52–63.

Malone, E. J. (1981) Geology, geochemistry and setting of the Woodlawn deposit. In *Geophysical Case Study of the Woodlawn Orebody, New South Wales, Australia* (ed. R. J. Whiteley), Pergamon Press, Oxford, pp. 49–75.

Malone, E. J. and Whiteley, R. J. (1981) Regional mineralisation and Geophysics of the Eastern Lachlan Fold Belt. In *Geophysical Case Study of the Woodlawn Orebody, New South Wales, Australia* (ed. R. J. Whiteley), Pergamon Press, Oxford, pp. 23–47.

McMillan, W. J., Whiteley, R. J., Tyne, E. D. and Hawkins, L. V. (1981) Brief description of the Woodlawn Deposit and summary of geophysical responses. In *Geophysical Case Study of the Woodlawn Orebody, New South Wales, Australia* (ed. R. J. Whiteley), Pergamon Press, Oxford, pp. 3–11.

Marcotte, D. and David, M. (1981) Target definition of Kuroko-type deposits in Abitibi by discriminant analysis of geochemical data. *CIM Bull.*, **74 (828)**, 102–108.

Meyer, C. (1972) *Evolution of Ore-Forming Processes with Geologic Time*. Presidential address to the Society of Economic Geologists, San Francisco (unpublished).

Mutschler, F., Wright, E., Ludington, S. and Abbott, J. (1981) Granite molybdenum systems. *Econ. Geol.*, **76**, 874–897.

O'Driscoll, D. F. (1981) Exploration history of the Woodlawn orebody: Discovery to development. In *Geophysical Case Study of the Woodlawn Orebody, New South Wales, Australia* (ed. R. J. Whiteley), Pergamon Press, Oxford, pp. 15–20.

Ohmoto, H. (1978) Submarine calderas: A key to the formation of volcanogenic massive sulphide deposits? *Min. Geol.*, **28**, 219–231.

Ohmoto, H. and Rye, R. O. (1974) Hydrogen and oxygen isotopic compositions of fluid inclusions in the Kuroko deposits, Japan. *Econ. Geol.*, **69**, 947–953.

Ohmoto, H. and Rye, R. O. (1979) Isotopes of sulphur and carbon. In *Geochemistry of Hydrothermal Ore Deposits*, 2nd edn (ed. H. L. Barnes), Wiley, New York, pp. 509–567.

Ohmoto, H. and Skinner, B. J. (eds.) (1983) The Kuroko and related volcanogenic massive sulfide deposits. *Econ. Geol. Monogr.*, **5**.

Patton, T. C., Grant, A. R. and Cheney, E. S. (1973) Hydrothermal alteration at the Middle Fork copper prospect, Central Cascades, Washington. *Econ. Geol.*, **68**, 816–830.

Phillips, W. J. (1973) Mechanical effects of retrograde boiling and its probable importance in the formation of some porphyry ore deposits. *Trans. Instn. Min. Metall.*, **82**, B90–B98.

Pintz, W. S. (1984) *Ok Tedi. Evolution of a Third World Mining Project*. Mining Journal Books Ltd., London, 206 pp.

Press, N. (1983) Remote sensing in the extractive industries. In *Prospecting and Evaluation of Non-metallic Rocks and Minerals* (eds. K. Atkinson and R. Brassington), Institution of Geologists, London, pp. 15–20.

Reid, K. O. and Rowe, J. (1979) A case study in ore reserve procedures at Mount Lyell. In *Estimation and Statement of Mineral Reserves*, Australian Institute of Mining & Metallurgy, Victoria, pp. 241–254.

Ridge, J. D. (1983) Genetic concepts versus observational data in governing ore exploration. *CIM Bull.*, **76(852)**, 47–54.

Sangster, D. F. (1977) Some grade and tonnage relationships among Canadian volcanogenic massive sulphide deposits. *Geol. Surv. Can. Rep. Pap.* **77–1A**, 5–12.

Sangster, D. F. (1980) Quantitative characteristics of volcanogenic massive sulphide deposits. *CIM Bull.*, **73(814)**, 74–81.

Sangster, D. F. and Scott, S. D. (1976) Precambrian strata-bound massive Cu–Zn–Pb sulphide ore of North America. In *Handbook of Stratabound and Stratiform Ore Deposits* (ed. K. A. Wolf), Elsevier, Amsterdam, Vol. 6, pp. 129–222.

Sassos, M. P. (1983) Bougainville copper, *Eng. Min. Jrnl.*, October, 56

Sato, T. (1974) Distribution and geological setting of the Kuroko deposits. *Soc. Min. Geol. Jap. Spec. Iss.*, **6**, 1–9.

Sato, T., Tanimura, S. and Ohtagaki, T. (1974) Geology of the Aizu metalliferous district, Northeast Japan, *Soc. Min. Geol., Jap, Spec. Iss*, **6**, 11–19.

Sawkins, F. J. (1976) Massive sulphide deposits in relation to geotectonics. *Geol. Assoc. Can. Spec. Pap.*, **14**, 221–240.

Schermerhorn, L. J. G. (1980) Copper deposits of the Iberian Peninsula. *Proc. Int. Copper Symp. Bor.*, Yugoslavia.

Scott, S. D. (1978) Structural control of the Kuroko deposits of Hokuroko district Japan. *Min. Geol.*, **28**, 301–311.

Scott, S. D. (1980) Geology and structural control of Kuroko-type massive sulphide deposits. *Geol. Assoc. Can. Spec. Pap.*, **70**, 705–722.

Severin, P. W. A. (1982) Geology of the Sturgeon Lake copper–zinc–lead–silver–gold deposit. *CIM Bull.*, **75 (846)**, 107–123.

Shirozo, H. (1974) Clay minerals in altered wall rocks of the Kuroko-type deposits, *Soc. Min. Geol. Jap., Spec. Iss.*, **6**, 303–311.

Sillitoe, R. H. (1977) Permo-Carboniferous, upper Cretaceous and Miocene porphyry copper-type mineralisation in the Argentinian Andes, *Econ. Geol.*, **72**, 99–103.

Sillitoe, R. H. (1979) Some thoughts on gold-rich porphyry copper deposits. *Miner. Deposita*, **14**, 161–174.

Sillitoe, R. H. (1980) Types of porphyry molybdenum deposits. *Min. Mag.*, **142**, 550–553.

Sillitoe, R. H., Halls, C. and Grant, J. N. (1975) Porphyry tin deposits in Bolivia. *Econ. Geol.*, **70**, 913–927.

Singer, D. A., Cox, D. P. and Draw, L. J. (1975) Grade and tonnage relationships among copper deposits. *U.S. Geol. Surv., Prof. Pap.*, **907–A**, 11 pp.

Solomon, M. (1976) 'Volcanic' massive sulphide deposits and their host rocks – a review and an explanation. In *Handbook of Stratabound and Stratiform Ore Deposits. II. Regional Studies and Specific Deposits* (ed. K. A. Wolf), Elsevier, Amsterdam, pp. 21–50.

Sopuk, V. J., Lavin, O. P. and Nichol, I. (1980) Lithogeochemistry as a guide to identifying favourable areas for the discovery of volcanogenic massive sulphide deposits. *CIM Bull.*, **73(823)**, 152–166.

Soregaroli, A. E. and Sutherland Brown, A. (1976) Characteristics of Canadian Cordilleran molybdenum deposits. *Can. Instn. Min. Metall. Spec. Vol.* **15**, 417–431.

Spence, C. D. (1975) Volcanogenic features of the Vauze

sulphide deposit, Noranda Quebec. *Econ. Geol.*, **70**, 102–114.

Stanton, R. L. (1978) Mineralisation in Island Arcs with particular reference to the south-west Pacific Region. *Proc. Australas. Inst. Min. Metall.*, **268**, 9–19.

Strauss, G. K., Madel, J. and Alonzo, F. F. (1977) Exploration practice for strata-bound volcanogenic sulphide deposits in the Spanish-Portugese pyrite belt: geology geophysics and geochemistry. In *Time and Strata-bound Ore Deposits* (eds. D. D. Klemm and H. J. Schneider), Springer-Verlag, New York, pp. 55–93.

Templeton, R. J., Tyne, E. D. and Quick, K. P. (1981) Surface and downhole applied potential (mise-a-la-masse) surveys of the Woodlawn orebody. In *Geophysical Case Study of the Woodlawn Orebody, New South Wales, Australia* (ed. R. J. Whiteley), Pergamon Press, Oxford, pp. 509–518.

Thomas, J. A. and Galey, J. T. (1982) Exploration and Geology of the Mt Emmons Molybdenite Deposits, Gunnison County Colorado. *Econ. Geol.*, **77**, 1085–1104.

Thorne, M. G. and Edwards, R. P. (1985) Recent advances in concepts of ore genesis in South West England. *Trans. R. Geol. Soc. Corn.*, **21(3)**, 113–152.

Titley, S. R. (1982) The style and progress of mineralisation and alteration in porphyry copper systems. In *Advances in Geology of the Porphyry Copper Deposits. Southwestern North America* (ed. S. R. Titley), University of Arizona Press, Tucson, pp. 93–116.

Titley, S. R. and Beane, R. E. (1981) Porphyry Copper Deposits. Part 1 Geologic Settings, Petrology and Tectogenesis. *Econ. Geol. 75th Anniv. Vol.*, 214–235.

Tyne, E. D. and Whiteley, R. J. (1981) Electrical profiling with a number of arrays at Woodlawn. In *Geophysical Case Study of the Woodlawn Orebody, New South Wales, Australia* (ed. R. J. Whiteley), Pergamon Press, Oxford, pp. 349–374.

Walker, R. R., Matulich, A., Amos, A. C., Watkins, J. J. and Mannard, G. W. (1975) The geology of the Kidd Creek Mine, *Econ. Geol.*, **70**, 80–89.

Wallace, S. R., MacKenzie, W. B., Blair, R. G. and Muncaster, N. K. (1978) Geology of the Urad and Henderson Molybdenite Deposits, Clear Creek County, Colorado, with a section on a comparison of these deposits with those at Climax, Colorado. *Econ. Geol.*, **73**, 325–368.

Westra, G. and Keith, S. (1981) Classification and Genesis of Stockwork Molybdenum deposits. *Econ. Geol.*, **76**, 844–873.

White, W. H., Bookstrom, A. A., Kamilli, R. J., Ganster, M. W., Smith, R. P., Ranta, D. E. and Steininger, R. C. (1981) Character and origin of Climax-type molybdenum deposits. *Econ. Geol. 75th Anniv. Vol.*, 270–316.

Whiteley, R. J. (ed.) (1981) *Geophysical Case Study of the Woodlawn Orebody, New South Wales, Australia*, Pergamon Press, Oxford, 586 pp.

4

Hydrothermal vein deposits

4.1 INTRODUCTION

The term 'hydrothermal', when used in the context of mineral deposits, means that the minerals have formed from a hot aqueous fluid of unspecified origin. Many styles of ore deposits are considered to have been deposited from hydrothermal fluids, and in this chapter we restrict our attention to those deposits where the ore minerals are contained dominantly within veins. High-grade hydrothermal vein deposits, and their oxidation products, must have been one of the principal styles of mineralization worked by prehistoric man. Knowledge of this early activity is lost in antiquity, but the publication of *De Re Metallica* by Agricola in 1556 demonstrates that many of the characteristics of hydrothermal deposits were appreciated by the medieval miner. The most comprehensive review of current thinking on hydrothermal ore deposits is *The Geochemistry of Hydrothermal Ore Deposits* edited by Barnes (1979).

Hydrothermal veins remained a major source of metals until the end of the nineteenth century; but at present they are mined only for a restricted number of metals and industrial minerals, which are of sufficient value to offset the high mining costs which are associated with this style of deposit. The main metallic commodities which are mined from veins nowadays are tungsten, tin, gold, uranium, cobalt and silver. The vein association is of most significance for tungsten, because wolframite-bearing quartz veins associated with granitoids account for slightly more than half of the World's tungsten production (Bender, 1979). The mining of other metals from veins forms only a minor proportion of World production. Nevertheless production rates of gold from this style of deposit have increased dramatically in recent times. For example Australian gold production, which is mainly derived from veins, increased by 50% to 27 tonnes in 1982 (*Mining Annual Review*, 1983).

Uranium from classical vein-type deposits accounts for 5% of Reasonably Assured Western World Resources (Nash *et al.*, 1981). This is relatively unimportant on a global scale, but can be of very significant regional importance; for example it is predicted that France will generate 75% of its electricity from nuclear power in 1990 (*Mining Journal*, 22nd January, 1982). This reliance on nuclear power is based upon substantial local reserves of uranium, of which 50% are contained within veins.

4.2 CLASSIFICATION OF HYDROTHERMAL VEIN DEPOSITS

The classification of ore deposits has two main purposes: the systematization of knowledge, so that natural groupings become apparent, and the use of these natural groupings for the purpose of comparison and prediction. In considering the classification of hydrothermal vein deposits it is not immediately apparent which characteristic should be selected as being of prime importance. Should it be an aspect of fundamental genetic significance such as the nature of the hydrothermal fluid, or should it be more closely related to observations which can be made in the field such as morphology and mineralogy? The diversity of classification systems which have been proposed for hydrothermal deposits results from the selection of different criteria, which in turn stems from the differing cultural backgrounds and experience of geologists. We briefly consider some of the alternative systems which have been proposed, or might be utilized, for classifying hydrothermal deposits.

4.2.1 Classification based on the nature of the ore fluid

Geologists consider that most ore fluids are saline aqueous solutions. Hydrothermal fluids which have the potential to transport metals may be subdivided into four types (White, 1974): (i) surface water; (ii) connate and deeply penetrating groundwater; (iii) metamorphic water; (iv) magmatic water. From a genetic point of view the classification of hydrothermal deposits based upon the nature of the ore fluid is attractive. Unfortunately it is seldom possible to be certain which type of fluid has deposited ore, and there is evidence that in some cases hydrothermal deposits are the result of mixing of two separate fluids. In theory, the nature of a hydrothermal fluid can be determined from its isotopic signature using the isotopes D/H and $^{18}O/^{16}O$ (Table 1.7). However, only meteoric and ocean waters have distinctive isotopic signatures and other types of water overlap in their isotopic composition (Taylor, 1979). The nature of the ore fluid is therefore usually inferred from a range of evidence which includes the geological setting of the mineral deposit, and data from fluid inclusions, mineral parageneses, and wall-rock alteration. Table 4.1 shows some ore-deposit types and the ore fluids from which they are considered to have been derived. In view of the speculative nature of the ore fluid it is less suitable as a basis for classification than might appear to be the case.

4.2.2 Classification based upon the temperature and pressure conditions of ore deposition

The most famous classification of hydrothermal deposits is that proposed by Lindgren (1913). Lindgren recognized that hydrothermal deposits might be formed by meteoric water, but he considered that the bulk of deposits was formed by hot ascending waters of uncertain origin 'charged with igneous emanations'. Deposits were subdivided into hypothermal, mesothermal and epithermal types on the basis of the estimated temperature and pressure conditions prevailing at the time of mineral deposition (Table 1.1). Subsequently the term xenothermal was added by Buddington (1935). Ironically it is only within recent years that reliable data have become available on the pressure and temperature conditions under which hydrothermal minerals were deposited. In our experience these data often show that the Lindgren classification is too rigid. For example, the Sn/Cu mineralization in Cornwall is classed as hypothermal (>300°C) by Lindgren (1913) but fluid-inclusion studies indicate a temperature range which embraces both hypothermal and mesothermal subdivisions (Jackson et al., 1982). We have the impression that the Lindgren classification is little used by geologists nowadays. The reasons are partly due to the rigidity which has just been referred to, and are in part a tendency for geologists to prefer a classification which considers the broader geological context, in

Table 4.1. Differing hydrothermal fluids and some associated ore deposits.

Type of ore fluid	Deposit type	Location	Evidence	Source
(1) Surface water	Sandstone-hosted uranium (Wyoming-type roll deposit)	Wyoming, USA	Inferred from geochemistry of groundwater (see Section 9.4.3)	Rackley (1976)
(2) Connate and deeply penetrating groundwater	Hydrothermal tin	West Cornwall, UK	Isotopic evidence	Jackson et al. (1982)
(3) Metamorphic water	Hydrothermal auriferous veins	Eastern Goldfields Province, Western Australia	Inferred from tectonic setting and fluid inclusion data	Groves et al. (1984)
(4) Magmatic fluid	Hydrothermal tin deposits of variable style	Bushveld Complex, South Africa	Close spatial association between cassiterite and fluorite	Wilson (1979)

addition to the details of structure, mineralogy and geochemistry of the vein itself.

4.2.3 Classification based upon the geological setting of the deposits

R. L. Stanton (1972) has emphasized the close genetic connection between ores and the rocks with which they are associated. Stanton suggested that even hydrothermal vein deposits, which had previously been regarded as though they were divorced from their host rocks, might be associated with specific geological environments. Stanton proposed that veins might be assigned to different categories based in part on their mineralogy and in part on their associated lithology; for example 'precious metal telluride ores of volcanic association'. Burnham and Ohmoto (1981) have presented a more comprehensive subdivision of veins which is given with some modification below:

(1) Cu–Pb–Zn–Au–Ag subtype in veins associated with igneous intrusions (±Sb, Bi, As, Ga, Ge, In). Example: Butte, Montana (Meyer et al., 1968).

(2) W–Sn subtype in sheeted veins and stockworks associated with granitoids (As, B, F). Example: Xingluokeng deposit, Fujian Province, China (Wengzhang, 1982).

(3) In subtype in complex veins associated with granitoids (±B, Fe, F, As, Cu, W). Example: Cornwall, UK (Thorne and Edwards, 1985).

(4) Au subtype associated with mafic, ultramafic volcanics and banded iron formation (±As, Sb, Hg, Te, Cu, Pb, Zn, W). Example: Kalgoorlie, Eastern Goldfields Province, Western Australia (Groves et al., 1984).

(5) U subtype associated with granitoids, gneisses and schists (±Co, Ni, Bi, Ag, As). Example: Bois Noirs Limouzat, Massif Central, France (Cuney, 1978).

(6) Co–Ag–Ni subtype (±Fe, As, Cu, Pb, Bi). Example: East Arm, Great Slave Lake, Canada (Badham, 1978).

We do not have sufficient space to describe all of these ore types. The Au subtype has therefore been selected for further discussion. The reason for this choice is that gold has been the main focus of mineral exploration in the early 1980s and is likely to remain a major exploration target for the remainder of the decade.

4.3 CLASSIFICATION OF HYDROTHERMAL GOLD DEPOSITS

The classification of hydrothermal gold deposits is a difficult task and geologists are not in agreement about the best approach. A traditional subdivision is based upon the type of structure within which the valuable minerals are contained (Bateman, 1950). Boyle (1979) proposed a sixfold categorization of hydrothermal gold deposits based in part on structure and in part on their enclosing lithology. Stanton (1972) restricted his attention to those hydrothermal gold deposits which are associated with volcanics and identified two main categories: deposits of Tertiary age associated with the circum-Pacific volcanic belt, and deposits associated with Archaean greenstones. Stanton's contribution is important because it recognizes tectonic setting as the basis for classification. Bache (1981) has produced the most comprehensive classification of gold deposits which is based on a combination of tectonic setting, host rock lithology and mineral paragenesis (Table 4.2). Our main criticism of Bache's classification is his distinction between pre-orogenic and post-orogenic classes. In some Canadian mining districts the gold mineralization is interpreted as being partly pre-orogenic and partly syn-orogenic (Karvinen, 1981). Bache (1981) classified the auriferous veins of the Kalgoorlie district as pre-orogenic (Type 3) although field investigations indicate that they are mainly syn-orogenic or post-orogenic (Groves *et al.*, 1984).

We prefer Stanton's approach and we intend to concentrate on those gold deposits which are associated with Archaean greenstone belts. Further subdivision of the deposits is discussed in the context of the Porcupine Mining District, Canada (Section 4.4.2).

4.4 HYDROTHERMAL GOLD DEPOSITS IN ARCHAEAN TERRAIN

4.4.1 General characteristics

(a) Distribution in space and time
Gold has been mined from the Archaean shield areas of Canada, Australia, Zimbabwe, Brazil, India and South Africa (Fig. 4.1). Canada has been

Table 4.2. Classification of hydrothermal gold deposits (modified from Bache, 1981).

Group 1 – Pre-orogenic volcano–sedimentary deposits	
Type 1	Polymetallic massive sulphides
Type 2	'Itabirite' type in which the gold is closely associated with a ferruginous, siliceous or carbonate horizon
Type 3	Discordant deposits in a volcano-sedimentary environment
Group 2 – Post-orogenic plutonic–volcanic type	
Type 4	Porphyry copper deposits
Type 5	Massive ores formed by replacement of carbonates
Type 6	Vein deposits centred on an intrusion
Type 7	Vein deposits associated with Tertiary volcanics

the principal source of hydrothermal gold with an estimated total production of 5280 tonnes (Kavanagh, 1979). Australia and Zimbabwe are of secondary importance with similar production statistics of 2187 tonnes of gold (Kavanagh, 1979). The output of gold from hydrothermal deposits in Archaean terrain is of minor significance from other countries.

Gold reserves are unevenly distributed within the main producing countries. For example most of the gold production from the Canadian Shield is from the Superior Province. (Hodgson and MacGeehan, 1982). Gold is also of restricted occurrence in Western Australia. Little gold has been mined from the Pilbara block and the bulk of production in the Yilgarn block has been obtained from the Eastern Goldfields Province. Some 50% of gold production from Western Australia is from the Kalgoorlie district, known as the Golden Mile (Groves et al., 1984).

(b) Size and grade of deposits

Most published information on the size and grade of hydrothermal gold deposits in Archaean terrain refers to Canada and Australia.

The bulk of Canadian production has been mined from the Superior Province of the Canadian Shield. A detailed survey by Bertoni (1983) shows that nearly all the gold production in the Superior Province has come from 155 mines. Bertoni's analysis reveals that the mines are highly variable in size but have a relatively narrow range of grades. If we separate those mines which have a combination of reserves and past production in excess of 10×10^6 tonnes of ore it is evident from Bertoni's work that twelve Canadian mines fall into the 'giant' size category (Hollinger, Dome, Kerr-Addison, McIntyre, Pamour No. 1, Lamaque, East Malartic, Sigma, Lake Shore, Can Malartic, Beattie-Donchester and McCleod-Cockshutt). The newly discovered Hemlo deposit is also of giant size with a published reserve of 13.5×10^6 tonnes at 7.2 g/t (0.236 oz/ton) (Patterson, 1983). The grade of ore mined from the Superior Province varies from 3.0 to 9.2 g/t (0.1–0.3 oz/ton) with only a few mines having grades above 15 g/t. There is no linear relationship between grade and tonnage in the Canadian gold deposits (Bertoni, 1983).

The bulk of gold production mined from Archaean terrain in Australia is from the Yilgarn block. Only one mining district, the Golden Mile at Kalgoorlie, can be classed as a giant deposit by Canadian standards with a past production of 92×10^6 tonnes of ore (Woodall, 1975). A further twelve deposits have produced more than 1×10^6 tonnes of ore (Woodall, 1975). The average grade of ore mined from Western Australia has remained at about 8 g/t since 1940 (Woodall, 1975). However, in the case of small open-pit operations near existing mill facilities grades as low as 4 g/t may be worth exploiting.

Gold from Archaean greenstone belts in the Barberton district of South Africa and Zimbabwe has been extracted from many small mines and a few large ones. In the Barberton district the grade averages about 8.5 g/t (Anhaeusser, 1976).

(c) Mineralogy

The mineralogy of hydrothermal gold deposits in Archaean terrain is summarized in Table 4.3. Native gold is the dominant economic mineral but is commonly alloyed with variable but significant amounts of silver. Gold tellurides may be present but have a restricted distribution: they are mainly of significance in the Kalgoorlie district of Western Australia and in some of the Zimbabwean deposits (Fig. 4.2). Pyrite and pyrrhotite are the most common metallic minerals although other sulphide minerals may be locally important. Quartz and carbonates are the main gangue minerals.

The composition of the gold alloys and the mineralogical siting of the gold have an important effect on recovery. According to Gasparrini (1983), the main factors affecting gold recovery are the grain size of the gold-bearing minerals, the nature of the associated sulphides, and the manner of occurrence of gold in the host mineral. The problem of gold recovery from hydrothermal ores is also discussed by Anhaeusser (1976).

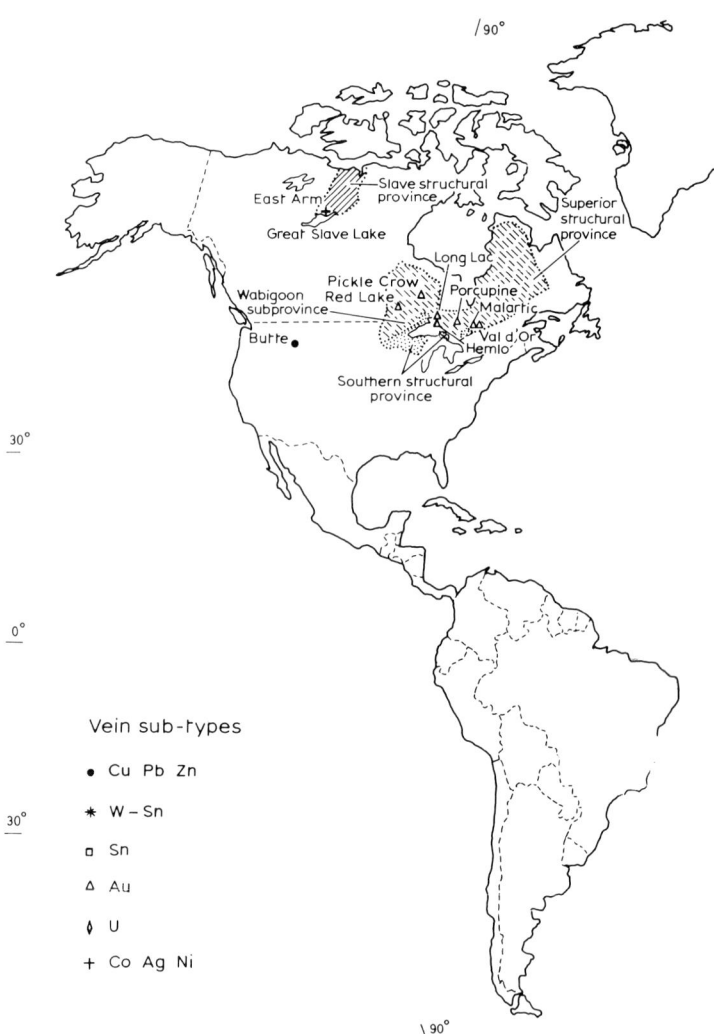

Fig. 4.1 Location map.

(d) Host lithology

One of the most consistent features of auriferous veins in Archaean terrain is the lithology of the country rock. In Canada, Western Australia, Zimbabwe, and India metavolcanics, normally of tholeiitic composition, are the predominant host rock. In the Kalgoorlie district of Western Australia the ore bodies are hosted by a tholeiitic dolerite (Woodall, 1965). Ultramafic lithologies have a less consistent association with gold deposits. Hodgson and MacGeehan (1982) noted that ultramafics are present within the host rocks of 80% of the main gold-mining camps of the Superior Province of Canada. However, ultramafics form only a minor component of the country rock in the gold-mining districts of Zimbabwe and Western Australia and only rarely form the host rocks for epigenetic gold deposits (Foster and Wilson, 1984; Groves *et al.*, 1984). Felsic volcanics are not prominent lithologies in the context of auriferous veins in Australia and Zimbabwe but they form a significant part of the volcanic sequences in the Red Lake, Val d'Or and Porcupine districts of Canada (Hodgson and

Hydrothermal vein deposits

MacGeehan, 1982). In addition, Hodgson and MacGeehan found in their survey of major gold-mining districts in the Superior Province that 90% of deposits were associated with felsic intrusions.

Sedimentary lithologies are common in all gold-mining districts in Archaean terrain. Hodgson and MacGeehan (1982) record that turbiditic greywackes and shales are dominant in the main gold-mining districts of the Superior Province. Iron-formation is commonly represented in sedimentary sequences which host auriferous veins. Foster and Wilson (1984) state that in Zimbabwe there is a strong correlation between the presence of auriferous iron-formation and gold-bearing quartz veins and shear zones within the same major lithostratigraphic units. Iron-formation is also present within the sedimentary sequences which host gold-bearing veins in the Pickle-Crow, Little Long Lac and Porcupine districts of Ontario (Hutchinson and Burlington, 1984). In the Southern Cross and Murchison Provinces of the Yilgarn block, Western Australia, **gold-bearing veins are found not only in the same sedimentary sequence as, but are commonly hosted**

Table 4.3. Summary of mineralogy of auriferous veins in Archaean terrain.

Country	Nature of gold	Principal metallic minerals	Principal gangue minerals	Wall-rock alteration
Canada (Superior Southern & Slave provinces)	Free or combined as[1] telluride, or antimonide. Some intimately associated with pyrite, arsenopyrite and sulpho-salts	Pyrite, arsenopyrite and pyrrhotite. Minor galena, sphalerite chalcopyrite, stibnite, molybdenite, scheelite and various sulpho-salts	Quartz, carbonates[1] (mainly ankerite, dolomite and calcite)	Chloritization, silicification[1] sericitization, carbonatization, tourmalinization, albitization, pyritization and arsenopyritization
Australia (Yilgarn block)	In Kalgoorlie area both native gold and gold tellurides[2]. Most of native gold in pyrite.[3] Other areas: mainly native gold	Kalgoorlie area:[3] pyrite, galena, chalcopyrite, sphalerite, stibnite, arsenopyrite, tetrahedrite–tennantite, realgar. In other mining districts: pyrite and pyrrhotite ubiquitous. Variable chalcopyrite, arsenopyrite	Kalgoorlie:[3] quartz, carbonates. Other mining districts: dominantly quartz with minor carbonates in some cases	Dominantly carbonatization,[6] sericitization, muscovitization and biotitization. Pyritization also common
Zimbabwe	Mainly native gold but tellurides significant in some cases.[4] Native gold commonly as inclusions in sulphides[4]	Pyrite normally[4,7] dominant. Pyrrhotite, chalcopyrite and galena usually present. Other sulphides may be present in small quantities. Scheelite is a common constituent[7]	Mainly quartz[7] and lesser amounts of carbonates	Non-stratabound:[7] propylitic alteration and sericitization. Strata-bound: variable styles of alteration; sodium-depletion in some cases
South Africa (Barberton)	Native gold[5] associated with pyrite and to a lesser extent arsenopyrite	Pyrite dominant.[5] Arsenopyrite and pyrrhotite important. 56 other minerals have been described	Dominantly quartz[5] with some ankerite	Not conspicuous. In some mines[8] pyritization and chloritization may be evident

Sources: 1, Boyle (1979); 2, Travis et al. (1971); 3, Woodall (1975); 4, Twemlow (1984); 5, Anhaeusser (1976); 6, Groves et al. (1984); 7, Foster and Wilson (1984); 8, Gribnitz (1964).

by banded iron-formation (Groves et al., 1984). Hutchinson and Burlington (1984) emphasize the importance of carbonate-rich and carbonaceous sediments in the country rock adjacent to gold-bearing quartz veins in some Canadian mining districts.

(e) Wall-rock alteration

Bates and Jackson (1980) define wall-rock alteration as the alteration of country rocks adjacent to hydrothermal ore deposits by the fluids responsible for, or derived during, the formation of the deposits. The ambiguity present in this definition nicely conveys the uncertainty of whether wall-rock alteration is caused by the same hydrothermal fluid which deposited the ore-forming minerals. The diversity of wall-rock alteration associated with auriferous vein deposits is partly a reflection of the wide range of lithologies which can host epigenetic gold deposits, an aspect which is discussed by Boyle (1979). However, differing styles of wall-rock alteration are also

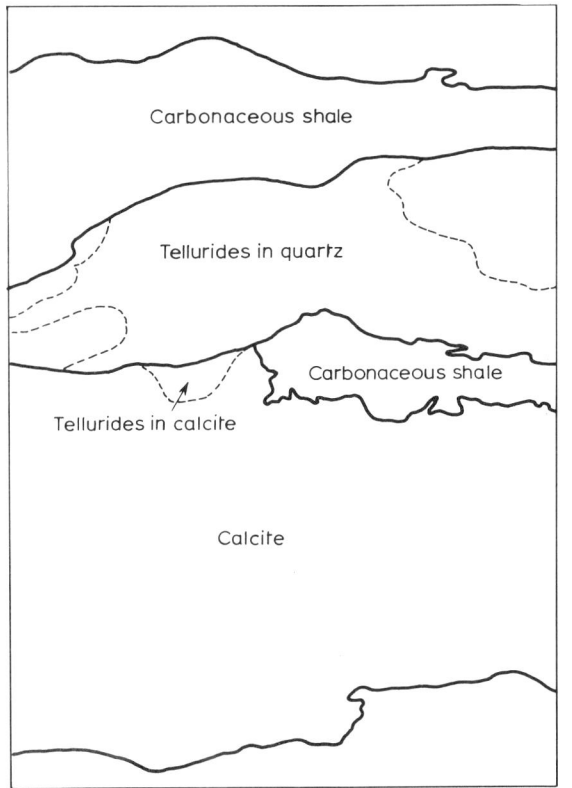 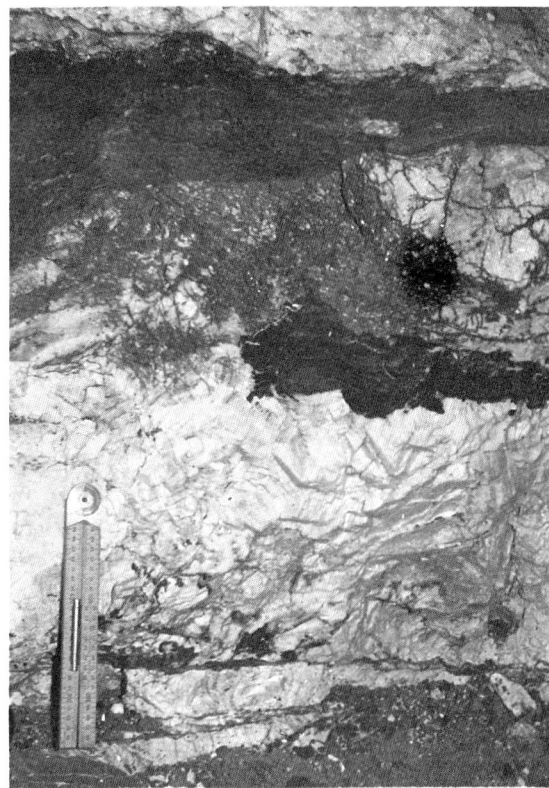

Fig. 4.2 Tellurides in the Commoner ore body, Commoner Mine, Zimbabwe (from Fernandes *et al.* 1979). Scale in centimetres.

caused by ore fluids of differing chemistry and it is this aspect which is of most interest to geologists.

Boyle (1979) reviewed the mineralogy and chemistry of wall-rock alteration associated with gold deposits and reported that sericitization, carbonatization and pyritization are the most consistent alteration processes. Boyle (1979) concluded that 'the constituents of most auriferous veins and lodes were deposited from an aqueous medium that was highly fluxed with CO_2 and S in most cases and with As, Sb, Se, Te, Bi, B, F and Cl in some cases. The gold was probably transported mainly as sulphide, arsenide, antimonide or telluride complexes.'

(f) Tectonic setting

Archaean gold mineralization is associated with thick, highly deformed sequences of mafic volcanics and metasediments which are referred to as greenstone belts. However, not all greenstone belts contain gold mineralization and those that are mineralized have a highly irregular distribution of gold. Recent studies of the tectonic setting of Archaean gold deposits have been confined to Australia and Canada.

Groves and Batt (1985) have discussed the structural evolution and metallogenesis of the Pilbara and Yilgarn blocks in Western Australia. The older Pilbara block ($3.5 \times 10^9 - 3.3 \times 10^9$ years) contains few gold deposits of significance. Groves and Batt (1985) classify the Pilbara block as a platform-phase greenstone belt. The main characteristics are a laterally continuous volcanic stratigraphy dominated by basalts, the presence of shallow-water sediment, and the general absence of faulting. There is a distinctive metallogenic

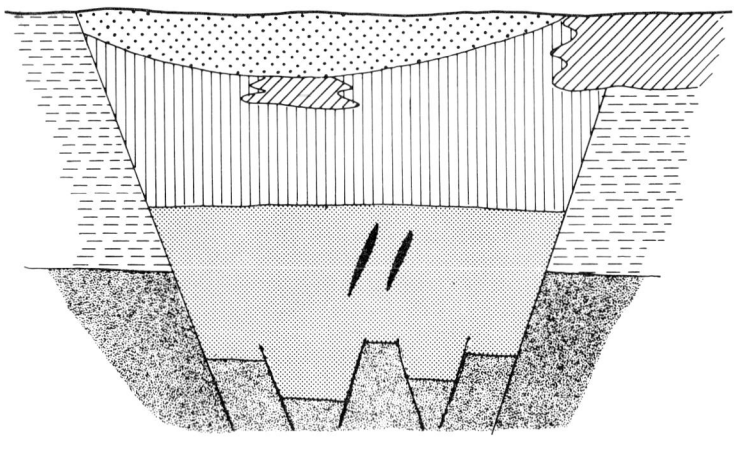

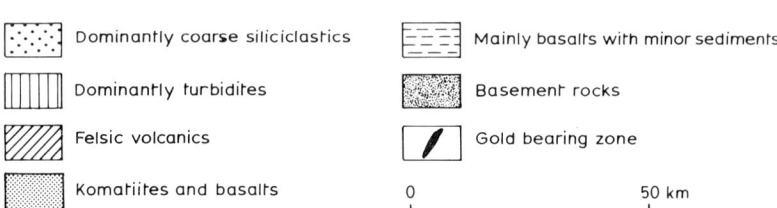

Fig. 4.3 Main rift zone, Eastern Goldfields Province, Western Australia (modified from Groves and Batt, 1985). Note the small rift and graben structures and the restriction of sediments to the upper part of the stratigraphy.

association characterized by volcanogenic sulphides, porphyry-style copper, and sedimentary barite. The younger greenstone belts of the eastern part of the Yilgarn block (2.8×10^9 $-2.7 + 10^9$ years) are classified by Groves and Batt (1985) as rift-phase greenstone belts. They have been the principal source of gold production in Western Australia. The rift-phase greenstones have a linear pattern and are characterized by a complex volcanic stratigraphy in which komatiites are abundant. Their main tectonic feature is extensive faulting which gave rise to small fault-bounded basins (Fig. 4.3).

Hodgson and MacGeehan (1982) have reviewed the results of recent studies concerned with the tectonic setting of gold mineralization in the Superior Province of Canada. Many of the gold-mining districts in Canada are associated with major zones of faulting. In some cases recent mapping has led to the reinterpretation of the fault zones as the sites of a major change in facies from volcanic to clastic sedimentary successions. It appears that the geological setting of many of the gold-mining districts in the Superior Province can be explained in terms of linear volcanic domes with flanking sedimentary troughs. The sedimentary succession is dominated by volcaniclastics but also contains significant proportions of chemical sediment, particularly carbonate units, and iron-formation. Fig. 4.4 is a section through the Campbell and Dickenson Mines, Red Lake, Ontario, which shows the lateral transition from volcanic to sedimentary facies.

There are thus clear differences in the tectonic setting of Archaean gold in Canada and Australia. In Canada sedimentation is coeval with volcanism and there is good evidence that part of the gold mineralization is synvolcanic. Faulting is in some cases synvolcanic but in other cases the structural disturbance is focused at the interface of major facies. In the Eastern Goldfields Province, Australia, volcanism was dominant in the lower part of the stratigraphy and was confined to narrow linear graben. Sedimentation post-dated

Hydrothermal vein deposits

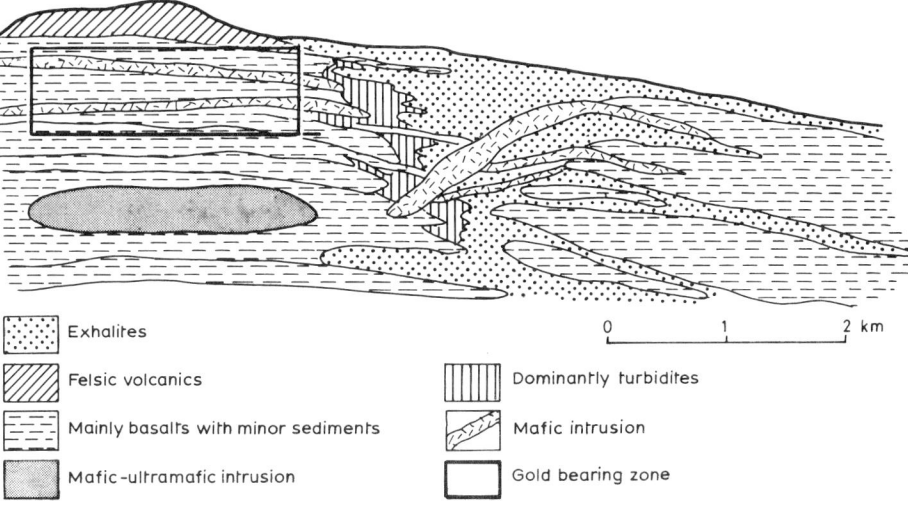

Fig. 4.4 Cross-section through volcanic complex and adjacent sedimentary basin, Campbell and Dickenson Mines, Red Lake Camp, Ontario. (From MacGeehan and Hodgson 1981.) Note the lateral facies change from volcanics to sediments.

the volcanism and there is therefore only a minor content of carbonates and iron-formation and little evidence of synsedimentary gold.

4.4.2 The Porcupine mining district (Timmins area), Ontario, Canada

(a) Introduction

The Porcupine mining district (or camp in Canadian terminology) is located at the western end of the Archaean Abitibi greenstone belt. The Porcupine district has been the most prolific centre of gold production in North America with an estimated production of 1.4×10^3 tonnes (Karvinen, 1980). Gold has been produced from over 24 deposits but today there are only four underground mines (Dome, Pamour No. 1, Pamour No. 3, and Schumacher) and one open pit (Hollinger) in operation (see Fig. 4.6).

(b) Stratigraphy and principal lithologies

The geology of the Porcupine district is dominated by mafic and ultramafic volcanics and clastic metasedimentary rocks. The volcanics are subdivided into the Deloro and Tisdale Groups and metasediments are assigned to the Porcupine Group. According to Pyke (1981), the Porcupine Group is the lateral equivalent of the upper part of the Deloro Group and the Tisdale Group (Fig. 4.5). However, Hodgson (1983a) considers that some of Pyke's correlations are contentious.

The Deloro Group comprises 4500–5000 m of calc-alkaline volcanics which are mainly restricted to the Shaw Dome, to the south of the mining district (Pyke, 1981). The Tisdale Group is about 5000 m in thickness. The lower part of the Tisdale

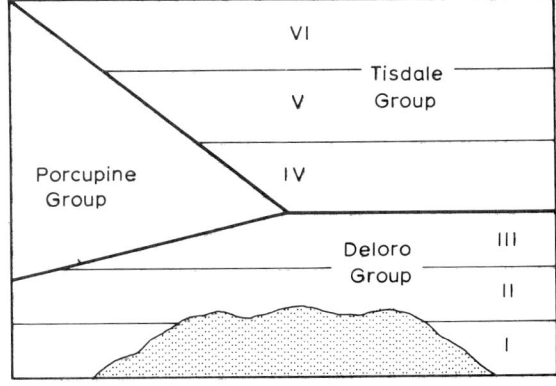

Fig. 4.5 Diagramatic illustration of the time equivalence of sedimentation (Porcupine Group) and volcanism (Deloro and Tisdale Groups) in the Timmins area (from Pyke 1981).

153

Group is largely komatiitic and is overlain by a thick sequence of tholeiitic basalt. The upper part of the Tisdale Group consists of volcaniclastics of calc-alkaline composition (Pyke, 1981). The Tisdale Group includes many minor carbonate zones, and Karvinen (1981) has traced two of the larger carbonate zones over a strike length of 20 km. The carbonate zones are mainly hydrothermally altered volcanics although the larger zones may have a sedimentary component.

Quartz feldspar porphyry bodies of varying size occur in close association with faults and carbonate units in the Tisdale Group. Karvinen (1981) believes that textural evidence and the presence of compositional layering indicate that many of the quartz–feldspar porphyries are of extrusive origin.

The Porcupine Group is about 3000 m in thickness. It consists mainly of interlayered greywackes and siltstones with subsidiary conglomerates (Pyke, 1981).

(c) Tectonic setting and structural style

Roberts (1981) interpreted the geological setting of the Porcupine district as a volcanic trough which received volcanic detritus from two linear domes which flanked the trough on its northern and southern margins. The dominant structure in the area is the Porcupine syncline. It is flanked on the north-west by the Central Tisdale anticline and on the south-west by the South Tisdale anticline (Fig. 4.6).

ENE-trending faults play an important role in the geology of the Porcupine district. The Destor–Porcupine fault (Fig. 4.6) separates gold-bearing volcanic rocks of the Tisdale Group in the north from predominantly barren rocks of the Deloro Group to the south. Pyke (1981) noted that the Destor–Porcupine fault forms part of a major fracture which extends for a distance of about 440 km. Faults of similar trend within the Porcupine syncline (Dome fault, Hollinger fault) are closely associated with feldspar porphyries and gold mineralization. Major faults of NW trend such as the Burrows–Benedict fault do not appear to have either a spatial or genetic relationship to gold mineralization.

(d) Mineralogy

Gold in the Porcupine district occurs principally as the native metal alloyed with a variable amount of silver. The gold may be free or have differing associations with pyrite. Gold–silver tellurides are of only minor significance. The principal metallic minerals associated with gold are pyrite, scheelite, arsenopyrite, pyrrhotite, chalcopyrite, galena and sphalerite. The gangue minerals are quartz, tourmaline, albite and fuchsite.

(e) Classification of veins

The gold-bearing veins of the Porcupine district have traditionally been regarded as epigenetic in origin and their subdivision has been based mainly on their geometry. For example, at the Hollinger mine the ore bodies were classified as follows: well-defined continuous veins, sinuous veins, tabular zones consisting of parallel quartz stringers, and *en echelon* vein zones (Lang et al., 1976).

Karvinen (1981) has suggested that the ores should be classified on a genetic basis as follows:

Synvolcanic ores

(1) Continuous stratabound bodies of quartz, ankerite and tourmaline characterized by cross-cutting 'ladder veins'.

(2) Massive siliceous carbonate rock with only minor quartz-carbonate stringers.

(3) Breccia bodies within fossil vents where mineralization occurs as disseminated pyrite in matrix material or as pyritic quartz-carbonate stringer-vein systems.

Metamorphogenic ores

Hodgson (1983a) does not accept that the ores are synvolcanic. He states that one of the key items of evidence are those carbonate rocks at the Dome Mine which Karvinen interprets as a chemical sediment of volcanogenic type. According to Hodgson (1983a) the carbonate zone is represented by anastomosing veinlets which cut across well-defined flow units.

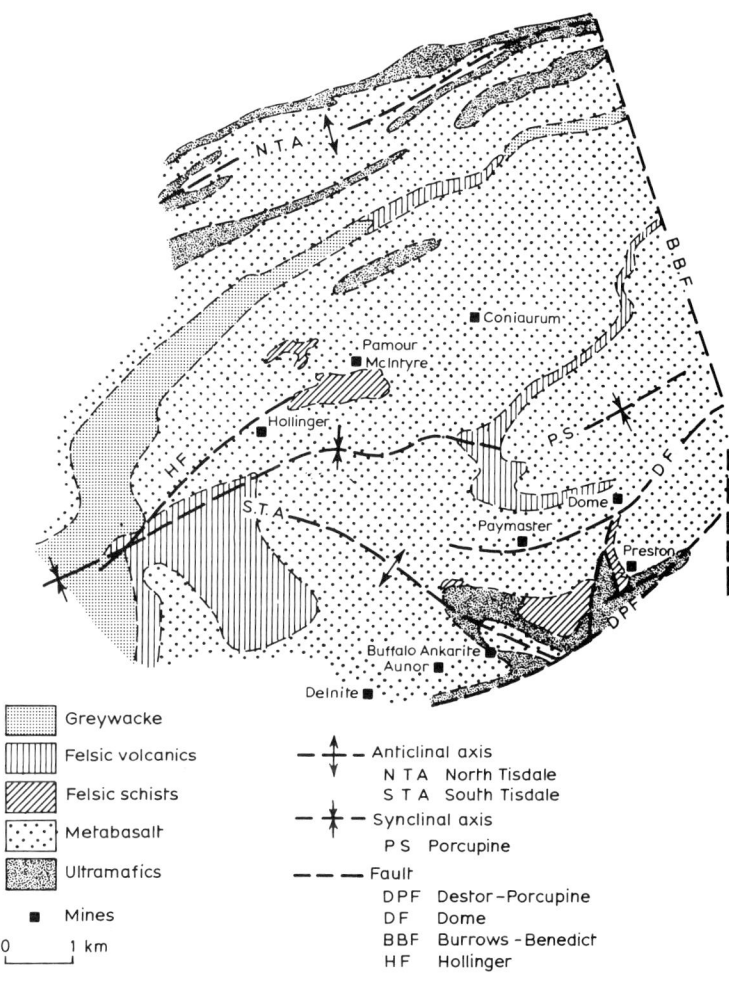

Fig. 4.6 Generalized geological map of the Porcupine mining district (from Fryer *et al.* 1979).

(f) Alteration

Wall-rock alteration in the Porcupine mining district is of several ages and types. An early stage of carbonatization was contemporaneous with the volcanic evolution of the Tisdale Group (Fyon and Crocket, 1981) and may sometimes be associated with gold mineralization. The nature of wall-rock alteration associated with discordant veins varies according to the lithology of the host rock and commonly exhibits zoning parallel to the vein. At the McIntyre mine, discordant auriferous quartz veins are hosted mainly by the Pearl Lake porphyry. Carbonatization is widespread. Zones of pyritization and silicification occur with increasing proximity to the vein, and pyrite and chalcopyrite occur adjacent to the vein contact (Griffis, 1962).

(g) Timing of mineralization

The timing of gold mineralization in the Porcupine district is controversial and will remain so until a systematic programme of radiometric dating is undertaken. Karvinen (1981) has proposed that there are two main stages of gold deposition: firstly a synvolcanic stage which is associated with submarine hydrothermal activity during the formation of the Tisdale volcanics; secondly a period of remobilization of gold from carbonate source rocks into veins of metamorphic origin during deformation of the Porcupine syncline.

Karvinen's sequence of gold deposition is based upon Pyke's (1981) interpretation of the stratigraphy of the Timmins area and utilizes the observations of both Karvinen (1981) and other workers (Fryer et al., 1979; Fyon and Crocket, 1981) on the volcanogenic nature of the carbonate rocks. Hodgson (1983a) disagrees with Karvinen's interpretation of the gold mineralization. He concludes, mainly from structural evidence, that the bulk of gold mineralization post-dates the Tisdale volcanics, quartz–feldspar porphyries, and major ENE trending faults.

4.4.3 Genesis of hydrothermal gold deposits in Archaean terrain

(a) Source of gold

There are two lines of approach which are usually taken in attempting to establish the source of metal for hydrothermal ore deposits. Firstly the recognition of a persistent association of ore deposits with a particular lithology. Secondly the anomalous enrichment or depletion of metal in such a lithology is usually interpreted as being indicative of its potential as a source rock.

Gold deposits in Archaean terrain are nearly always associated with mafic volcanics and may also occur in close proximity to clastic or chemical sediments and felsic igneous rocks. Several workers (Pyke, 1975; Viljoen and Viljoen, 1969) have emphasized the close spatial relationship between gold deposits and komatiites in Archaean terrain. Others have preferred to draw attention to the association of gold with chemical sediment and the sites of hydrothermal alteration (Karvinen, 1981). Cherry (1983) has emphasized the proximity of gold deposits to felsic intrusives in some of the mining camps within the Abitibi greenstone belt.

In recent years the development of analytical techniques with very low detection limits, such as neutron activation analysis, has resulted in a good data base on gold distribution in unmineralized lithologies.

Kwong and Crocket (1978) investigated the gold content of rocks in the western part of the Wabigoon volcanic–plutonic complex in N. W. Ontario. The area was selected because of suitability for sampling appropriate rocks but does not contain any commercial gold deposits. Table 4.4 shows that the average gold content of the five major rock types in the study area is very similar. However, carbonated rocks at the base of an ultramafic sill were enriched at least twofold with respect to background. Kwong and Crocket (1978) concluded that process-related factors such as alteration are more significant than the background content of host rocks in the generation of anomalous gold concentrations.

Keays (1984) has investigated the gold content of ultramafic volcanics as these are often prominent in Archaean greenstone belts which contain gold deposits of large size. In an earlier study of the geochemistry of ocean-ridge basalts Keays and Scott (1976) concluded that a significant amount of gold was lost at an early stage due to basalt/seawater interaction. This means that the original gold content of regionally metamorphosed extrusives such as komatiites must be determined indirectly. Table 4.5 compares the gold content of komatiites from Archaean terrain with more recent extrusions of similar composition. It is evident that in most cases appreciable depletion of gold from komatiites has taken place. Keays concludes that komatiites may be the ultimate source of gold in hydrothermal deposits but suggests that the gold is either retained in subvolcanic peridotite lenses or released to interflow sediments soon after extrusion. Keays also emphasizes that the availability of gold may be more important than its total concentration. For example gold associated with sulphides is more likely to be available for leaching by hydrothermal fluids than gold locked in the lattice of silicate minerals.

(b) The ore fluid

What was the nature of the ore fluid which

Table 4.4. Gold contents of various rock types in the Wabigoon volcanic–plutonic complex (from Kwong and Crocket, 1978).

Rock type	No. of analyses	Gold contents (ppb)				
		Range	Mean	Standard deviation	Median	Geometric mean
Major lithologies						
Mafic volcanics						
All	27	0.33–8.14	1.75	1.67	1.27	1.28
Tholeiitic rocks	17	0.33–8.14	1.97	1.91	1.31	1.44
Felsic volcanics	35	0.09–6.78	1.47	1.55	0.80	0.92
Sedimentary rocks	24	0.14–8.84	1.12	1.95	0.51	0.56
Felsic intrusions						
Stephen Lake stock	9	0.40–2.74	1.05	0.74	0.77	0.88
Regina Bay stock	9	0.10–4.88	1.18	1.60	0.55	0.57
Mafic/ultramafic intrusions						
All	18	0.11–3.25	0.78	0.78	0.64	0.55
Emm Bay sill	12	0.18–3.25	1.01	0.88	0.74	0.76
Minor lithologies						
Carbonates	9	0.90–17.75	4.26	5.29	2.41	2.29
Diabase dikes	3	0.45–0.81	0.63	0.18	0.62	—
Qtz–feld. porphyries	3	0.43–0.72	0.61	0.16	0.68	—

Table 4.5. Average gold content of Tertiary picrite and Archaean komatiites (from Keays, 1984).

Location	Lithology	Average gold content (ppb)
Disko Island, West Greenland	Fresh picrite (Tertiary)	4.17
Gorgona Island, Colombia	Picrite (Cretaceous)	
Munro Township, Canada	Komatiite	2.9
Mt Clifford, Western Australia	Serpentinized and chloritized komatiite	0.49
Barberton district, S. Africa	Highly altered komatiite	0.43
Kambalda, W. Australia	Komatiite	log average 4.43

transported gold to its site of deposition? At what temperature was the gold deposited? In what form was the gold transported? The answers to these questions cannot be provided by any single technique but by combining information from fluid inclusion studies, from experimental investigations of gold solubility and from the geochemistry of wall-rock alteration, it is possible to build up a comprehensive data base from which the main parameters of the ore fluid can be determined with some confidence.

In the case of hydrothermal gold deposits in

Archaean terrain the problem is further complicated by the evidence in Canada (Karvinen, 1981), Zimbabwe (Foster and Wilson, 1984) and South Africa (Anhaeusser, 1976) that part of the gold may be syngenetic in origin. In these cases there have been two ore fluids: an initial fluid related to submarine volcanic activity and a later fluid responsible for the remobilization of gold during deformation and metamorphism. In this section we will restrict discussion to those deposits in which the gold is associated with discordant veins.

(i) Evidence from fluid inclusions

Small quantities of liquid, sometimes associated with solid and vapour phases, which are trapped during crystal growth have proved to be a valuable source of information about the nature of the ore fluid. Roedder (1967, 1984) has explained the general principles involved in fluid inclusion work and has reviewed the application of fluid inclusion studies to hydrothermal gold deposits.

There are three types of information which may be obtained from fluid inclusions: the temperature of formation of minerals and the salinity and chemical composition of the fluid. These aspects are now described in the context of Archaean gold deposits.

Temperature of formation
The temperature of formation of minerals can be determined from the temperature at which the differing phases present in inclusions homogenize. The experimental technique is described by Roedder (1967). In order to provide meaningful geological information, the homogenization temperature must be corrected for pressure. The pressure is normally inferred from the geological setting of the mineralization. Roedder (1984) provides an example of a pressure correction applied to a homogenization temperature of 250°C determined from a gold-bearing quartz vein formed during greenschist facies metamorphism. The pressure in this geological environment is thought to be at least 2 kilobars. Assuming a pure 5

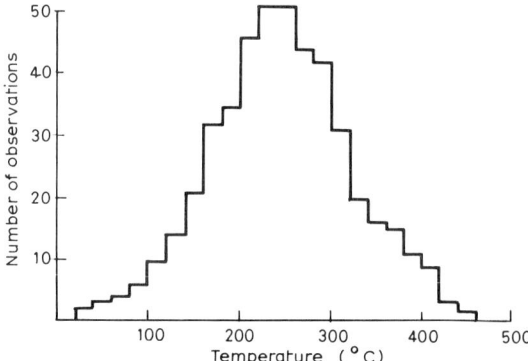

Fig. 4.7 Histogram of fluid inclusion data from auriferous quartz veins in Archaean and younger terrains (from Foster 1980).

wt% NaCl solution the pressure correction would be 170°C giving a formation temperature of 420°C.

Foster (1980) has compiled data from the literature on the temperature of formation of gold deposits in Archaean and other terrains obtained from fluid inclusion studies (Fig. 4.7). Most of the temperatures fall within the range 100–400°C with a mode of 240°C. More recent fluid inclusion data, exclusively from Archaean gold deposits in Western Australia, indicate temperatures of formation which range from 285 to 400°C (Groves *et al.*, 1984). The latter data are in accord with temperatures of 380–480°C determined from oxygen isotopes on Archaean gold deposits in Canada (Kerrich, 1981).

Salinity
Minerals such as halite, sylvite or gypsum can sometimes be observed within fluid inclusions and provide one of the most commonly used guides to the composition of the ore fluid. However, daughter minerals are seldom seen in fluid inclusions in quartz obtained from Archaean gold deposits, which suggests that the fluids have a low salinity. This inference is supported by investigations on fluid inclusions from auriferous quartz veins from Western Australia which show salinities of less than 4 wt% equivalent NaCl (Groves *et al.*, 1984).

Chemical composition

It is extremely difficult to obtain reliable information about the chemical composition of fluid inclusions. This is partly because the mineral has to be crushed to liberate the fluid and there is a strong possibility of contamination from inclusions of secondary origin. The very low concentration at which most elements are present in fluid inclusions also poses a major problem. The components which are determined routinely are CO_2 and H_2O because their approximate relative concentrations can be determined optically (Roedder, 1984). CO_2 appears to be a ubiquitous constituent of fluid inclusions within quartz obtained from Archaean gold deposits (Roedder, 1984).

(ii) Evidence from experimental studies

The form in which gold is transported in hydrothermal solutions is a matter for speculation. However, experimental studies carried out within the assumed pressure conditions of greenschist facies metamorphism (up to 2 kilobars), and within the temperature range indicated by fluid inclusion studies, can provide evidence about the varying concentration of gold in hydrothermal solutions. From a combination of experimental work and calculation, Seward (1984) has deduced that gold is mainly transported as Au^+ complexes. Seward also observes that the electronic structure of Au^+ is such that it will tend to form covalent bonds, with ligands such as CN^- and HS^- forming the most stable complexes. Ligand availability in the ore fluid is clearly a limiting factor. Seward (1984) suggests, from the analysis of deep geothermal water, that chlorides, iodides, ammonia, sulphides and bisulphides are amongst the most probable candidates. Alternatively, elements, such as arsenic, which are consistently associated with gold in veins, may have formed an appropriate ligand.

The bulk of experimental work at elevated temperatures and pressures has been carried out using chloride and sulphide solutions. Figure 4.8 shows the results of selected experimental work.

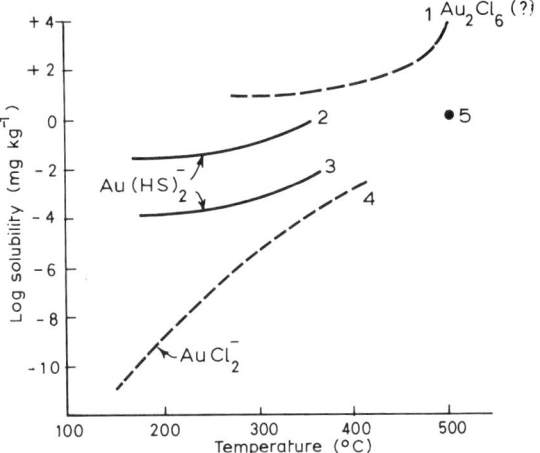

Fig. 4.8 Experimental and theoretical studies on the solubility of gold due to the formation of different complexes (modified from Seward, 1984)
1. Experiments at pressures of 1 and 2 kb in 1 and 2 M KCl solution in equilibrium with potassium feldspar–muscovite–quartz–magnetite–hematite, (Henley 1973) 2, 3. Experiments at 1 kb in equilibrium with pyrite and pyrrhotite at pH = 5 and total reduced sulphur ($H_2S + HS$) = 0.05 M (curve 2) and 0.001 M (curve 3). (Seward 1973) 4. Calculation of stochiometric ion activity coefficient in 1 M NaCl solution. (Helgeson 1969.) 5. Experiments at 1 kb in equilibrium with pyrite and pyrrhotite in 1 M NaCl solution. (Ryfuba and Dickson 1977.)

Henley's research (1973) suggests that gold has the highest solubility when forming a complex anion with chlorine under relatively oxidizing conditions (Fig. 4.8, curve 1). However, the data of Rytuba and Dickson (1977) and Helgeson (1969) suggest that gold solubility in chloride solutions is significantly reduced in the presence of pyrite and pyrrhotite (curve 4 and point 5). The experimental work of Seward (1973) suggests that gold in the form of $Au(HS)_2^-$ has appreciable solubility at elevated temperatures in the presence of excess sulphur.

At the present time, experimental data on gold-complexing with chloride and sulphide ligands is incomplete and there are few data on complex formation with ammonia or arsenic. Therefore

Ore deposit geology and its influence on mineral exploration

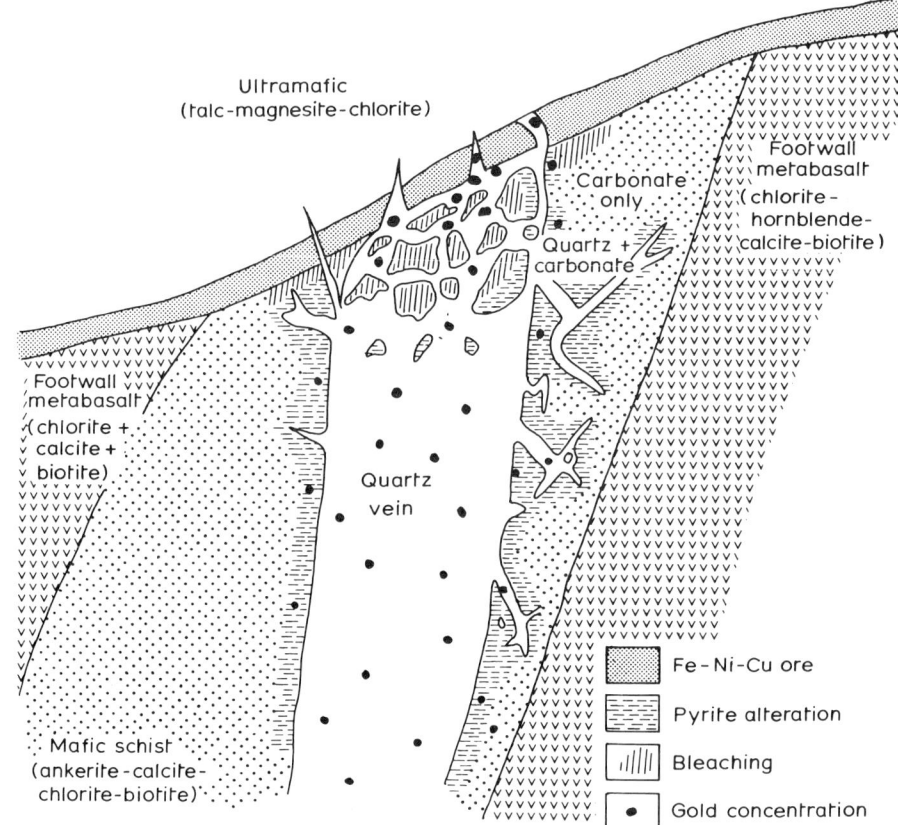

Fig. 4.9 Schematic diagram showing auriferous quartz vein and associated wall-rock alteration, Hunt Mine, Kambalda, Western Australia (from Phillips and Groves 1984).

only very tentative conclusions can be drawn from this discipline.

(iii) Evidence from wall-rock alteration

The types of wall-rock alteration which are associated with auriferous veins in Archaean terrain are summarized in Table 4.3. It is clear that there is considerable regional variation in the products of alteration, and thus generalization is not warranted. We have therefore selected the Eastern Goldfields Province of the Yilgarn block, Western Australia, to demonstrate the type of information which can be determined about the ore fluid from wall-rock-alteration studies.

The regional setting of gold deposits in the Eastern Goldfields Province has been described by Groves *et al.* (1984). Detailed studies of wall-rock alteration have been described from Kalgoorlie (Bartram and McCall, 1971), from Mount Charlotte Mine (Phillips *et al.*, 1983) and from the Hunt Mine, Kambalda (Phillips and Groves, 1984).

In the famous Golden Mile, adjacent to Kalgoorlie, carbonate and sericite alteration zones are dominant. Bartram and McCall (1971) considered that the bulk of the carbonate alteration precedes lode formation. Phillips *et al.* (1983) have focused their attention on one mine, Mount Charlotte, within the Golden Mile. The deposit is stratabound and is confined to an iron-rich granophyric zone in the Golden Mile dolerite. Alteration consists of an inner bleached zone comprising sericite + pyrite + pyrrhotite + siderite and an outer unbleached zone containing sulphides. Phillips and Groves (1984) have described wall-rock alteration associated with the Hunt Mine, Kambalda. The gold mineralization in the Hunt Mine has a strong lithological control and is confined to a metabasalt unit. The pattern of alteration consists of a broad zone of carbonate, a less extensive zone of biotite and a very restricted zone of pyrite (Fig. 4.9).

In both mining districts, the dominant alteration process is the hydrolysis of silicates. In the case of the Hunt Mine, extensive carbonatization suggests a CO_2-rich ore fluid. The presence of micas and iron sulphides in the wall rock adjacent to the lodes at both mines suggests that potassium and sulphur were significant components of the ore fluid.

(c) Factors controlling deposition

The controls of ore deposition are of great practical consequence because an understanding of this aspect of ore genesis has immediate relevance to mineral exploration. For example the identification of lithology as a significant depositional control allows the geologist to focus the detailed stages of exploration within a more limited area.

One of the most characteristic features of gold-mining districts in Archaean terrain is their close proximity to major zones of faulting. This association has already been referred to in Section 4.4.1 and has been described more specifically in the context of the Porcupine mining district where the Destor–Porcupine fault has an important control on the distribution of gold deposits (Section 4.4.2).

The structural control of gold mineralization has been recognized for many years by geologists working in Archaean terrain (Horwood, 1948; McKinstry, 1942; Goldberg, 1964), and major fault zones have been interpreted as the channelways along which the ore fluid has migrated. The more difficult problem is to ascertain those influences which have caused the gold to be deposited at particular sites along the conduit. The most probable causes of deposition are cooling of the ore fluid during movement along a thermal gradient, cooling or dilution of the ore fluid by mixing with meteoric water, adiabatic expansion, and reaction between the ore fluid and the wall rocks. These differing influences on ore deposition are now reviewed in the context of Archaean gold deposits.

(i) Cooling of the fluid along a thermal gradient

Evidence for a thermal control of gold deposition can best be demonstrated on a regional scale. Archaean terrain normally comprises greenstone belts, high-grade gneiss zones and sedimentary basins of Proterozoic style (thick sequences of clastic sediments which are relatively little deformed). Within greenstone belts greenschist and amphibolite facies of metamorphism predominate.

Anhaeusser (1976) has observed that in Zimbabwe the high-grade metamorphic belts which flank the gold-producing greenstone belts are virtually barren of gold mineralization. Anhaeusser has also examined the distribution of gold deposits within the Zimbabwe greenstone belts and concluded that there was an 'optimal thermal zone' in which gold tends to be concentrated. Groves et al. (1984) compared the distribution of gold deposits and metamorphic grade in the Eastern Goldfields Province of the Yilgarn block, Western Australia. They found that the bulk of the gold deposits is concentrated near to the transition from greenschist to amphibolite facies of metamorphism. The common association between gold deposits and rocks of greenschist facies has also been recognized in the Canadian Shield (Poulsen, 1983; Blackburn and Janes, 1983). However, Muir (1983) has described the geological setting of the newly discovered Hemlo deposit in Ontario where pelitic sediments adjacent to the deposit have a mineral assemblage characteristic of middle to upper amphibolite facies. It would therefore be prudent not to emphasize the significance of metamorphic facies in designing an exploration model for Archaean gold deposits.

There is little evidence for a thermal gradient on the scale of a mining district or mine. Boyle (1979) commented on the absence of vertical zoning in gold deposits in Archaean terrain and quoted the example of the Kolar goldfield, India, where there is little change in either grade or mineralogy although some of the mines extend in depth to more than 3000 m. The main exception appears to be Zimbabwe where a number of deposits show telescoped zoning and where gold and sulphides sometimes decrease dramatically with increasing depth (R. Foster, personal communication).

(ii) Cooling or dilution of the ore fluid by mixing with meteoric water

This mechanism of ore deposition can be evaluated by considering the isotopic signature of quartz which, from paragenetic studies, can be inferred to have formed at the same time as gold. In Section 4.2.1 we explained that minerals deposited from either meteoric or magmatic water were characterized by their oxygen isotope ratios. Values for meteoric water generally range from $\delta^{18}O$ of -10 to $-4^0/_{00}$ (Taylor, 1979). Kerrich (1981) has investigated the oxygen isotope ratios from auriferous quartz veins in the Canadian Shield. He calculated that $\delta^{18}O$ of fluids from which the minerals were precipitated were in the range of $7.5-11.6^0/_{00}$. Kerrich concluded that the results were compatible with a fluid derived from a metamorphic hydrothermal reservoir. There is no evidence of mixing with meteoric water.

(iii) Adiabatic expansion

As the ore fluid rises from depth towards the surface, pressure conditions will change and might exert some influence on mineral deposition. This aspect of ore deposition has been reviewed by Toulmin and Clark (1967) who have suggested that when an ore fluid passes through the restricted part of a vein system an irreversible adiabatic expansion (throttling) may occur. Furthermore, they concluded that this process could have important consequences in cooling the ore fluid and mineral deposition. Toulmin and Clark (1967) described two features which might be regarded as indicative of throttling. Firstly zones of relatively high temperature separated from regions of low temperature by narrow zones of erratically variable temperature; secondly, the evidence of well-developed mineral zoning.

The study of fluid inclusions from Archaean hydrothermal gold deposits suggests a remarkable uniformity of both salinity and temperature within individual mining districts. Neither is there much evidence for mineral zoning. We may therefore conclude that throttling is not an important control of gold deposition in Archaean terrain.

Phillips (1972) has proposed that hydraulic fracturing is a significant mechanism of vein formation and suggested that the rapid drop of pressure may be attended by ore deposition. Groves et al. (1984) suggested that hydraulic fracturing may have been significant in the genesis of gold deposits in the Eastern Goldfields Province, W. Australia. However, they interpreted the role of hydraulic fracturing as a mechanism for facilitating fluid access to reactive wall rocks rather than directly causing ore deposition. In this context it is pertinent to mention Seward's (1973) experimental work which indicated that, in the case of $Au(HS)_2^-$, a decrease of pressure above 250°C will lead to an increase in gold solubility.

(iv) Reaction between ore fluid and wall rock

Mafic volcanics are the predominant host lithology of auriferous quartz veins in Archaean terrain (Section 4.4.1). The reason for this common association may be the tendency for competent volcanic units to fracture and provide structural sites for ingress of the ore fluid. The common development of alteration assemblages adjacent to the veins suggests that extensive interaction between the ore fluid and the wall rock may be an additional control of deposition. Possible effects of wall-rock alteration include changes of E_h and pH in the ore fluid.

Iron is an important component of tholeiitic basalts where it occurs mainly within silicates. The breakdown of silicate minerals during wall-rock alteration results in the release of iron which

Hydrothermal vein deposits

becomes available to participate in iron-related redox reactions. Phillips and Groves (1984) carried out a detailed mineralogical and chemical study of wall-rock alteration at the Hunt Mine, Western Australia, and suggested that sulphidization of ferromagesian minerals could be considered as a redox reaction:

$$Fe_6Si_4O_{10}(OH)_8 + 12H_2S + 3O_2$$
$$\text{(chlorite)} \quad (S^{2-})$$
$$= 6FeS_2 + 4SiO_2 + 16H_2O$$
$$(S^{1-})$$

Groves *et al.* (1984) and Foster and Wilson (1984) suggest that wall-rock reactions of this type have been the principal cause of gold deposition in discordant veins in Western Australia and Zimbabwe respectively.

4.4.4 Genetic models

Many genetic models have been proposed for Archaean gold-bearing quartz veins (Fripp, 1976; Karvinen, 1981; Groves *et al.*, 1984; Keays, 1984). Most genetic models are variants of two types which may be termed exogenous/epigenetic and indigenous (including both syngenetic and epigenetic styles).

(a) Type 1 – exogenous/epigenetic (Figs 1.6 and 4.10)

This genetic model does not require the presence of specific source rocks which are enriched in gold (Kwong and Crocket, 1978), although the availability of gold in a form which can be readily leached is desirable. The more important aspect is the sweeping of very large volumes of rock by metamorphic water, expelled from predominantly mafic volcanics by dehydration reactions during prograde regional metamorphism. Fyfe and Kerrich (1984) have carried out a number of geochemical calculations to determine the volume of source rocks required. One of their calculations is based on the Porcupine district which is estimated to have produced 1.7×10^9g of gold. They estimated that this mass of gold could be produced from a source-rock volume of 1200 km^3 assuming an average gold concentration of 2 ppb, and 50% efficiencies of leaching and deposition. The large volumes of water required in a model of this type must be channelled along zones of high permeability, and the major fault structures associated with some greenstone belts are considered to represent major conduits for the ore fluid. Deposition is in part related to the geothermal gradient but more particularly to

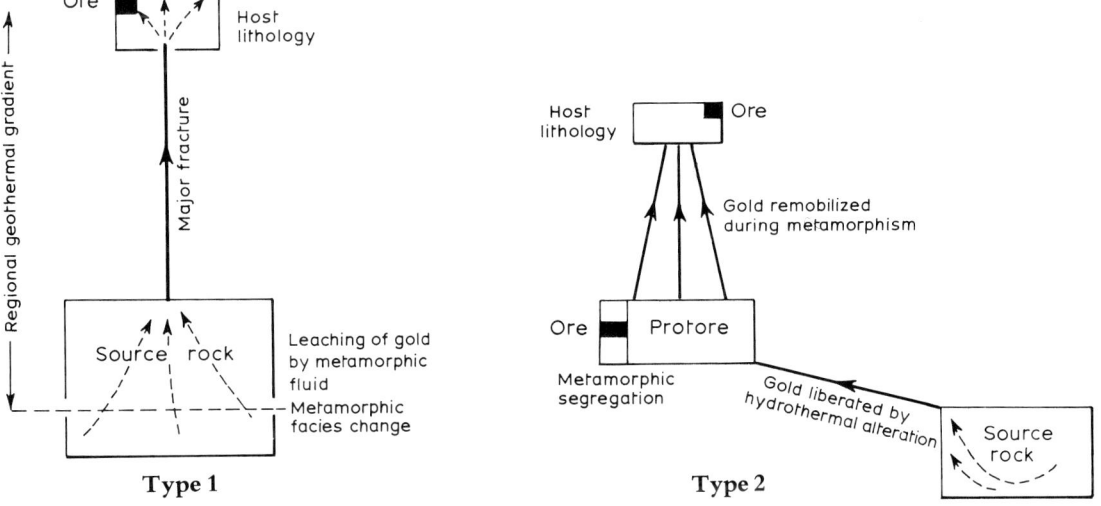

Fig. 4.10 Two genetic models of gold concentration for hydrothermal deposits in Archaean terrain.

fluid/wall-rock interaction with favourable lithologies.

Many examples of this genetic model are evident in the Eastern Goldfields Province of Western Australia (Groves et al., 1984).

(b) Type 2 – indigenous/syngenetic and epigenetic (Figs. 1.6 and 4.10)

In this genetic model, the gold may be derived in the first instance from rocks containing average crustal concentrations of the metal, although Keays (1984) points out that metal availability is an important factor. A critical aspect of this model is that gold is leached soon after lithification, either by rock/seawater interaction or by hydrothermal alteration, and is redeposited in the form of a chemical sediment. The concentration of gold present in the chemical sediment may be very variable. Foster and Wilson (1984) quote average grades of 0.3 and 1.6 g/t for iron-formation from different parts of Zimbabwe. Bavinton and Keays (1978) reported an average value of 0.146 g/t from interflow metasediments at Kambalda, Western Australia. Kerrich (1981) interpreted stratiform siliceous carbonates from the Timmins area as being synvolcanic and reported high gold concentrations. In some cases the stratabound auriferous chemical sediment is mined as ore, albeit of a low tenor. More commonly some degree of remobilization has taken place. However, the protore and discordant vein mineralization usually occur sufficiently close together for the former to be identified. The type example of this genetic model is the stratabound gold ores of Zimbabwe (Fripp, 1976).

4.5 EXPLORATION FOR GOLD IN ARCHAEAN TERRAIN

Gold has been the glittering prize sought by prospectors from the earliest days of mining and it continues to be a prime exploration target in the 1980s. At the present time there are two reasons why many mining companies are restricting their attention to gold. Firstly, it is one of the few metals which has increased significantly in price in real terms since 1970 (Hargreaves and Williamson, 1984). Secondly, the rate of return on investment is very much higher for gold than for most other mineral commodities (Mackenzie, 1984).

Our discussion of exploration is confined to the Canadian Shield and Western Australia. These are the foremost producers of gold from hydrothermal vein deposits in Archaean terrain and both regions have experienced a surge in exploration effort since 1980. However, there are several interesting points of contrast between the two regions which result in differing approaches to exploration. In Section 4.4.1 we outlined the evidence of contrasting tectonic settings and styles of mineralization which have led to the dissimilar genetic models described in Section 4.4.4. In addition, the extensive lateritization of Western Australia and the widespread cover of glacial sediments in the Canadian Shield present particular problems for the exploration geologist (Table 4.6).

In both Canada and Western Australia most of the major ore discoveries were made in the late 19th or early 20th centuries by conventional prospecting. However, modern exploration methods have also been successful in the search for gold. We therefore propose to review the application of remote sensing, geochemistry, geophysics and new geological concepts in the search for auriferous veins.

(a) Remote sensing

In Western Australia, reconnaissance exploration is directed towards mapping basalts, gabbros, and iron-formations within the greenstone belts, as these lithologies commonly host gold mineralization. The effectiveness of geological mapping is constrained by paucity of outcrop but is aided by photointerpretation. False colour photography permits the distinction of lateritic terrain and moreover is able to distinguish ultrabasic and felsic lithologies using their distinctive tonal characteristics (D. Hanley, personal communication). At the reconnaissance stage, photointerpretation is used in conjunction with airborne magnetometry

Hydrothermal vein deposits

Table 4.6. Contrasting features of the Canadian Shield and Western Australia of relevance to mineral exploration.

	Western Australia (Yilgarn block)[1]	Canadian Shield[2]
Geomorphology	Peneplain with elevations of 300–600 m	Peneplain with elevations rarely exceeding 760 m
Past climatic conditions of relevance to exploration	Temperate or warm climate in Mesozoic becoming humid-tropical or subtropical in Oligocene–Miocene	The mainland of the Canadian Shield was covered by glacier ice for most of the past 100 000 years
Nature and extent of overburden	Deep weathering profiles with indurated upper horizons – formerly continuous but now dissected. Kaolinitic profiles vary from 10 to 20 m in thickness	Glacial drift of varying type: lodgement and ablation till, gravels, and sands of variable association with ice sheets, marine and lake sediments
Present climate	Semiarid Rainfall 200–250 mm Temperature range 0–45°C	Arctic zone: long cold winter and short cool summer Boreal zone: long cold winter and short warm summer South zone: temperate climate
Soil type	Ferruginous and siliceous duricrusts predominate	North: regosolic and brunisolic soils South: podzolic and organic soil
Vegetation	Sparsely vegetated with trees of the genus Eucalyptus and Acacia dominant	Tundra: cold climate flora Boreal zone: partly coniferous forest South zone: mainly mixed conifer and hardwood

Sources: 1, Butt and Smith (1980); 2, Boyle *et al.* (1975).

to establish the geological base for more detailed exploration. At this stage data are normally compiled at a scale of 1:25 000.

The extensive cover of glacial sediments has precluded the application of remote sensing to mineral exploration in the Canadian Shield.

(b) Geophysical exploration

Geophysical techniques are valuable at both the reconnaissance and detailed stages of exploration for gold deposits. The objectives may vary and include the following aspects:

(i) Provision of data to assist geological mapping

Airborne magnetic methods are of importance for this purpose and have the advantage of depth penetration below both glacial sediments and zones of deep weathering. In Western Australia, airborne magnetic surveys have been flown over the greenstone belts by the Australian Bureau of Mineral Resources using a sensor height of 150 m and a flight line spacing of 1.5 km. More detailed surveys are sometimes commissioned by mining companies when the sensor height may be reduced to 30 m and flight line spacings narrowed to 100 m.

In Canada, airborne magnetometry is also used as a mapping tool but gamma-ray spectrometry and conductivity are increasingly used to provide supporting data. In the latter case, several airborne electromagnetic devices are available and multiple frequency measurements are made in order to cover a range of earth conductivities (Siegel, 1979).

(ii) Determination of overburden thickness

This approach is mainly relevant to the Canadian Shield and is used in conjunction with geochemical surveys. Killeen and Hobson (1974) carried out a shallow seismic refraction survey in the Timmins area, Ontario, and found distinct changes in seismic velocity when passing from overburden to bedrock. The bedrock topography was found to be highly irregular, varying in depth from 7.6 m to 45.7 m over a horizontal distance of 300 m.

(iii) Detection of geophysical signatures associated with gold mineralization

The application of detailed ground geophysical surveys to the search for gold deposits has been more successful in the Canadian Shield than in the deeply weathered terrain of Western Australia. Hodgson (1983b) reported that drilling and trenching on geophysical targets accounted for 8.6% of discoveries (63 deposits) in the Canadian Shield. At the detailed stage of exploration a variety of geophysical techniques may be used. Fortescue (1983) reported the results of a successful integrated survey for gold in Hoyle Township, near Timmins, Ontario. The geophysical part of the investigation entailed defining a zone of carbonatization using induced polarization and resistivity. Until the early 1970s the conductive laterite cover was found to preclude electromagnetic methods in Western Australia but during the last decade the transient electromagnetic method (SIROTEM) has been extensively used. The method was used successfully at the Water Tank Hill prospect (Mount Magnet) to define gold mineralization associated with pyrite and pyrrhotite (Western Mining, 1979).

(c) Geochemical exploration

Glaciated and deeply weathered terrain both present challenging environments for the exploration geochemist. Nonetheless, geochemical methods have become an important part of modern exploration programmes for gold in both Western Australia and the Canadian Shield. The traditional geochemical approach in both regions has been to use pathfinder elements and in principle the following elements may be appropriate: As, Sb, Hg, Te, Cu, Pb, Zn, W, Bi, B, F, V and Ag. The introduction of analytical techniques with low detection limits, such as flameless atomic absorption and neutron activation, has increased the trend towards the direct analysis for gold to define targets for trenching and drilling.

(i) Reconnaissance exploration

In Western Australia, geochemistry is used both as a mapping tool and as a direct guide to gold. In areas where deep weathering inhibits the identification of lithology the Zr/TiO_2 ratio of residual soil has proved to be a reliable discriminator between acid, intermediate and basic units (Hallberg, 1982). Mazzuchelli and James (1980) have established that arsenic is the most appropriate pathfinder element for gold in the Yilgarn block and this element is now determined routinely to establish large-scale patterns indicative of hydrothermal mineralization. In reconnaissance stream sediment sampling, a density of 1–2 samples/km^2 is commonly used. The effectiveness of the geochemical approach is enhanced by the separation of heavy mineral concentrates for analysis (Watters, 1983).

In the Canadian Shield, stream sediments, lacustrine sediment and water have been investigated as sampling media, mainly for base metals. The effectiveness of stream sediments depends upon topography and the thickness of glacial overburden. In areas of moderate relief covered by a thin veneer of glacial material, regional stream sediment sampling can provide a rapid evaluation of the mineral potential of an area (Wolfe, 1976). Research indicates that stream waters and lake sediments are of limited value in reconnaissance work for gold but may play a significant role at more detailed stages of exploration (Fortescue, 1983).

(ii) Detailed exploration

In Western Australia, detailed geochemical surveys for gold mainly use soil samples. The appropriate sample depth and size fraction for analysis must be based on an orientation survey as both factors are of critical importance. In areas which are masked by laterite or transported soil, samples for geochemical analysis may be collected using a light portable drill.

Detailed primary geochemical studies have been carried out at several mines with particular reference to arsenic (Mazzuchelli and James, 1980). In the Kalgoorlie area peak arsenic values were found to coincide with or lie adjacent to gold mineralization and there appears to be a quantitative relationship between the abundance of gold and arsenic in the ore. Primary dispersion halos associated with deposits from the Golden Mile were found to vary from 5 ft (1.52 m) to 50 ft (15.24 m) in width. At the Norseman Mine, anomalous arsenic values are confined to the zone of alteration and shearing in which the ore bodies were emplaced. In the Crown Shear, the arsenic anomaly extends for 2000 ft (609.6 m) up dip from the ore body and could be detected in the soil over the suboutcrop although the ore body itself is blind. The greater persistence of arsenic within the shear zone compared with gold suggests that it could prove a useful aid in exploration.

In the Canadian Shield, glacial till has been the most widely used sampling medium for detailed geochemical surveys. The method is based on the observation that clastic dispersion patterns are often produced by the ice sheet as it plucks out portions of bedrock and smears them in the direction of ice movement. When the till occurs at the surface the mapping of mineralized boulders is an effective technique which can be enhanced by the use of geochemistry and the study of the heavy mineral fraction in the till. The discovery of gold at the Lamaque mine, Val d'Or, and the Malartic gold field can be attributed to the tracing of mineralized boulders to their source (Dreimanis, 1958).

More commonly the till is obscured by a cover of glaciolacustrine or glaciofluvial deposits, and sampling of the till requires the use of a light portable drill. The general principles of using lodgement till as a sampling medium are described by Levinson (1980). Descriptions of the application of the technique to gold mineralization in Canada are provided by Fortescue and Lourim (1982) and Brown (1982). Dispersion trains detected in lodgement till have been recorded up to 8 km from the source of mineralization (Smee and Sinha, 1979). An essential aspect of this approach to exploration is a good understanding of Quaternary geology and some careful drilling may be required to establish the local stratigraphy of Quaternary sediments to enhance the value of available geological maps (Fortescue, 1983).

Lithogeochemical surveys are proving to be of value in the detailed stages of exploration for gold in the Canadian Shield. Research has shown that some gold deposits are associated with very extensive alteration zones which can best be defined using geochemical criteria. Durocher (1983) described an alteration halo which has a strike length of 9 km associated with the Madsen and Starratt-Olsen gold deposits in the Red Lake area, Ontario. Major element patterns involved depletion in Na_2O and enrichment in K_2O. Arsenic and antimony were found to have a high correlation with gold and could be used as indicators of proximity to an ore body. Carbonate alteration zones are common in volcanic rocks within Archaean greenstone belts and are sometimes associated with gold mineralization. Fyon and Crocket (1983) studied barren and gold-associated carbonate zones in the Timmins area, Ontario, and reported enhanced values of Sb, B, Li and Au in those alteration zones which were related to gold mineralization.

(d) Geological exploration

The bulk of the gold deposits in both Western Australia and the Canadian Shield were found at a relatively early stage. In most cases the deposits were found by prospectors either at outcrop or under a shallow cover of soil. However, a significant number of suboutcropping deposits have been found by geological methods. Hodgson (1983b) noted that 13.7% of gold deposits in the Abitibi Belt of the Canadian Shield were found by trenching or drilling a geological target. It is

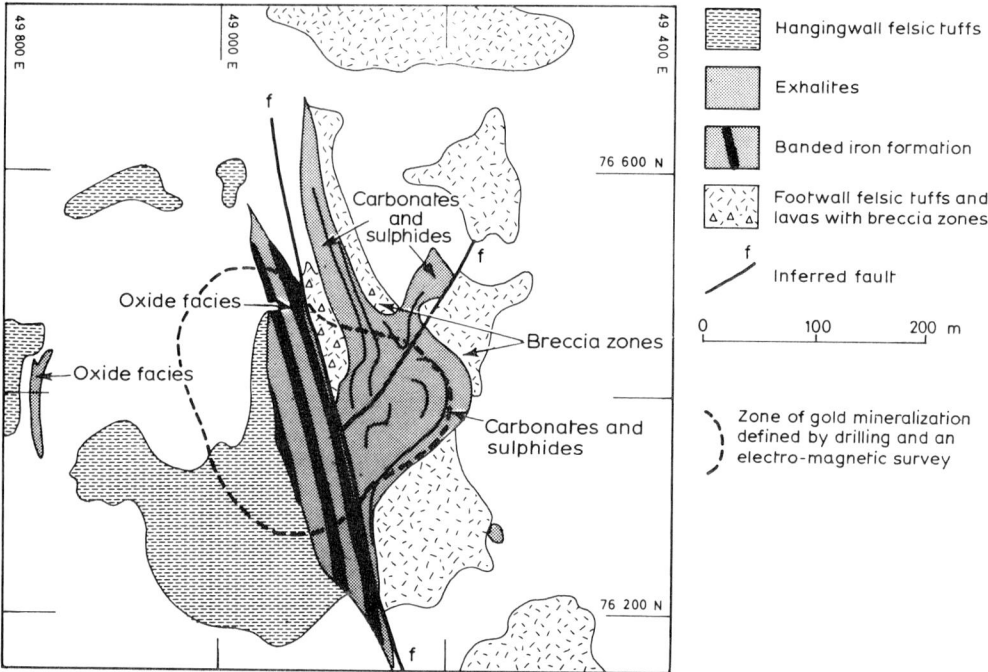

Fig. 4.11 Geological map of the Water Tank Hill deposit, Western Australia, showing features indicative of gold mineralization (from Western Mining 1979).

probable that in many cases the target was the extension of favourable geological features from known mineralization.

Differing geological criteria are valuable at differing stages of an exploration programme. Hodgson *et al.* (1982) stated that, at the scale of a mining district (mining camp) in the Superior Province of the Canadian Shield the most important geological factors are the contact between mafic ± ultramafic volcanics and metasediments, the presence of major faults and the evidence of indicator minerals. In the Eastern Goldfields province of the Yilgarn block the regional controls of gold mineralization are metamorphic grade (greenschist facies), regional structure (evidence of uplift) and the presence of mafic lithologies (Groves *et al.*, 1984).

At the more detailed stages of exploration other criteria assume greater prominence. In the Superior Province, the presence of felsic intrusives or extrusives, major carbonate alteration zones and local structures are significant indicators of ore (Hodgson *et al.*, 1982). In the Yilgarn block, the key factors to be observed during the detailed stage of exploration are evidence of metasomatic alteration, sulphide mineralization, veins, shears, faults and breccias (Western Mining, 1979) (Fig. 4.11).

It must be emphasized that detailed geological, geophysical and geochemical studies are normally integrated at this stage of exploration. The data obtained are commonly plotted at a scale of 1:5000.

(e) Drilling

In both Western Australia and the Canadian Shield drilling is used at an early stage in the exploration programme. We have mentioned that light portable drills are used in geochemical programmes in both regions. In Western

Australia, reverse-circulation drilling techniques are preferred because the sample does not come into contact with the walls of the hole. This reduces sample loss and also minimizes contamination. Techniques for drilling overburden in glaciated terrain have been described by Wennervirta (1973). Skinner (1972) discussed a drilling programme in lodgement till in the Abitibi greenstone belt.

Diamond drilling is used for hard-rock exploration and constitutes the most expensive stage of the exploration programme. The irregular nature of gold mineralization requires that a greater emphasis is placed on drilling than would be necessary in a larger, more uniform style of mineralization. This requires continuing confidence by management in their exploration geologists. The point is well made by the discovery of the Hemlo deposit in Ontario (Patterson, 1983). The Hemlo area was originally investigated by mapping and drilling between 1947 and 1951. Several changes of ownership of the option area occurred during the ensuing thirty years with episodic periods of drilling which defined a small reserve of gold mineralization. In 1980 a new programme of exploration was undertaken under the geological direction of D. Bell. A preliminary electromagnetic and magnetic survey was followed by a systematic programme of diamond drilling. However, it was the 75th diamond drill hole which intersected the major eastern zone of the ore body!

Drilling provides information about the persistence of the vein structure and gives some idea of the grade of gold. However, underground development is the only conclusive way of confirming the economic viability of a vein-gold deposit.

4.6 CONCLUDING STATEMENT

At the beginning of this chapter we commented on the wide variety of metals which is produced from veins. In most cases veins represent a minor source of production and we have therefore made the decision to concentrate on gold deposits. This was a difficult choice as we are more closely familiar with the tin-bearing veins of south-west England. However, the fact that hydrothermal auriferous veins are one of the principal exploration targets in the 1980s dictated our decision to 'go for gold'.

The classification of gold deposits is discussed in Section 4.3. It is evident that most hydrothermal gold deposits are concentrated in either Archaean greenstone belts or Tertiary volcanics. We have focused our attention entirely on the former environment and our approach may have been too narrowly defined. However, there has been an enormous amount of research published on gold in Archaean greenstones in recent years which is difficult to condense into a single chapter. More importantly Archaean greenstones represent a specific exploration environment for gold. We have therefore reviewed the geological setting of Archaean greenstone belts and examined the evidence for the genesis of gold deposits associated with them. How does such an understanding help the exploration effort in the 1980s? To what extent have recent ideas on ore genesis and modern exploration techniques improved on the time-honoured approach of the prospector with a nose for gold and a prospecting pan?

Firstly, let us recall the history of exploration for gold in Archaean terrain. In all countries the bulk of the gold was found during the late 19th and early 20th centuries by prospectors. In many cases the evidence for gold was found at outcrop or by panning for heavy minerals in streams and areas of thin residual soil. This phase of activity continued to the late 1930s. During World War II and the ensuing two decades there was little interest in exploration for gold. However, in 1968 the price of gold which had previously been held at US$35 per oz was decontrolled. This action led to rapid fluctuations in price but a gradual appreciation from US$34–39 per oz in 1970 to US$600–850 per oz in 1980 (*Mining Annual Review*, 1971, 1981). As the price of gold steadily rose the precious metal became an exploration target of increasing importance. This resurgence in interest has been reflected in recent reinvestigations of the genesis of gold in Archaean terrain. At the same time, methods of exploration developed for base metals in the post-war years have been applied to the search for gold.

One of the most fundamental changes in our understanding of Archaean geology has come from the appreciation that greenstone belts do not represent a single geological entity but can develop in several contrasting tectonic environments (Section 4.4.1). Some tectonic environments appear to be particularly favourable for gold mineralization. A further development in our understanding has come from the recognition that some gold deposits are stratabound and are best considered as chemical sediments rather than veins. This interpretation appears to be valid for Zimbabwe and Canada but does not seem to be the case in Australia. Clearly this is an advance in our theoretical understanding of gold mineralization which has assisted in the planning of exploration programmes. In this case, the genetic model for gold has changed and as a result exploration models have also been modified.

In most cases modern exploration programmes have been carried out in areas with poor exposure. Geochemical and geophysical techniques have therefore been vital in new ore discoveries. It is interesting to note that indirect methods have been valuable both in helping to define the detailed geology of prospective areas and in providing direct guides to gold.

REFERENCES

Anhaeusser, C. R. (1976) The nature and distribution of Archaean gold mineralisation in southern Africa. *Miner. Sci. Eng.*, **8**, 46–84.

Bache, J. J. (1981) A tentative quantitative classification of World gold deposits. *Chron. Rech. Min.*, **459**, 43–50.

Badham, J. P. N. (1978) Magnetite–apatite–amphibole–uranium and silver–arsenide mineralisations in Lower Proterozoic igneous rocks, East Arm, Great Slave Lake, Canada. *Econ. Geol.*, **73**, 1474–1491.

Barnes, H. L. (ed.) (1979) *Geochemistry of Hydrothermal Ore Deposits*, Wiley Interscience, New York, 798 pp.

Bartram, G. D. and McCall, G. J. H. (1971) Wall-rock alteration associated with auriferous lodes in the Golden Mile, Kalgoorlie. *Geol. Soc. Austr. Spec. Publ. No. 3*, 191–199.

Bateman, A. M. (1950) *Economic Mineral Deposits*, John Wiley & Sons, New York, 916 pp.

Bates, R. L. and Jackson, J. A. (1980) *Glossary of Geology*, 2nd edn, American Geological Institute, Falls Church, Virginia, 749 pp.

Bavinton, O. A. and Keays, R. R. (1978) Precious metal values from inter-flow sedimentary rocks from the komatiite sequence at Kambalda, Western Australia. *Geochim. Cosmochim. Acta*, **42**, 1151–1163.

Bender, F. (1979) The tungsten situation: supply and demand, present and future. In *Tungsten. Proceedings of the First International Tungsten Symposium, Stockholm 1979*, pp. 2–17.

Bertoni, C. H. (1983) Gold production in the Superior Province of the Canadian Shield. *CIM Bull.*, **76(857)**, 62–69.

Blackburn, C. E. and Janes, D. A. (1983) Gold deposits in N.W. Ontario. In *The Geology of Gold in Ontario, Ontario Geological Survey Miscellaneous Paper 110* (ed. A. C. Colvine), pp. 194–210.

Boyle, R. W. (1979) The geochemistry of gold and its deposits. *Geol. Surv. Can. Bull.*, **280**, 584 pp.

Boyle, R. W., Bradshaw, P. M. D., Clews, D. R., Fortescue, J. A. C., Gleeson, C. F., Hornbrook, E. H. W., Shilts, W., Tauchid, M. and Wolfe, W. (1975) The Canadian Shield. In Conceptual Models in Exploration Geochemistry (ed. P. M. D. Bradshaw), *Geochem. Expl.*, **4(1)**, 109–199.

Brown, M. R. (1982) Asarco's Nighthawk Lake Gold project reaches a most interesting stage. *The Northern Miner*, **68(23)**, pp. 1, 20.

Buddington, A. F. (1935) High temperature mineral associations at shallow to moderate depths. *Econ. Geol.*, **30**, 205–222.

Burnham, C. W. and Ohmoto, H. (1981) Late magmatic and hydrothermal processes in ore formation. In *Mineral Resources: Genetic Understanding for Practical Applications*, National Academy Press, Washington, DC, pp. 62–72.

Butt, C. R. M. and Smith, R. E. (eds) (1980) Conceptual models in exploration geochemistry: Australia. *J. Geochem. Explor.*, **12(2/3)**, 365.

Cherry, M. E. (1983) The association of gold and felsic intrusions – examples from the Abitibi Belt. In *The Geology of Gold in Ontario, Ontario Geological Survey Miscellaneous Paper 110* (ed. A. C. Colvine), pp. 48–55.

Cuney, M. (1978) Geologic environment, mineralogy, and fluid inclusions of the Bois Noirs-Limouzat uranium vein, Forez, France. *Econ. Geol.*, **73**, 1567–1610.

Dreimanis, A. (1958) Tracing ore boulders as a prospecting method in Canada. *CIM Bull.*, **51**, 73–80.

Durocher, M. E. (1983) The nature of hydrothermal alteration associated with the Madsen and Starratt-Olsen gold deposits, Red Lake area. In *The Geology of Gold in Ontario, Ontario Geological Survey Miscellaneous Paper 110* (ed. A. C. Colvine), pp. 123–140.

Fernandes, T. R. C., Foster, R. P., Storey, M. J. and Twemlow, S. G. (1979) Telluride mineralisation at the Commoner gold mine, *Chamber of Mines (Rhodesia) Journal*, **21**, 34–37.

Fortescue, J. A. C. (1983) Geochemical prospecting for gold in Ontario. In *The Geology of Gold in Ontario, Ontario Geological Survey Miscellaneous Paper 110* (ed. A. C. Colvine), pp. 251–271.

Fortescue, J. A. C. and Lourim, J. (1982) Descriptive geochemistry and descriptive mineralogy of the basal till in the Kirkland Lake area, districts of Timiskaming and Cochrane. In *Summary of Fieldwork 1982 by the Ontario Geological Survey, Ontario Geological Survey Miscellaneous Paper 106*, 235 pp.

Foster, R. P. (1980) Gold in Archaean greenstone belts: a review of depositional controls. *Min. Eng. (Zimb.)*, **45(6)**, 21–25 and **45(8)**, 31.

Foster, R. P. and Wilson, J. F. (1984) Geological setting of Archaean gold deposits in Zimbabwe. In *Gold '82: the Geology, Geochemistry and Genesis of Gold Deposits, Geological Society of Zimbabwe* (ed. R. P. Foster), A. A. Balkema, Rotterdam, pp. 521–552.

Fripp, R. E. P. (1976) Stratabound gold deposits in Archaean banded iron-formation, Rhodesia. *Econ. Geol.*, **71**, 58–75.

Fryer, B. J., Kerrich, R., Hutchinson, R. W., Pierce, M. G. and Rogers, D. S. (1979) Archaean precious metal hydrothermal systems, Dome Mine, Abitibi greenstone belt. 1. Patterns of alterations and metal distribution. *Can. J. Earth Sci.*, **16**, 421–439.

Fyfe, W. S. and Kerrich, R. (1984) Gold: natural concentration processes. In *Gold '82: The Geology, Geochemistry and Genesis of Gold Deposits, Geological Society of Zimbabwe* (ed. R. P. Foster), A. A. Balkema, Rotterdam, pp. 99–128.

Fyon, J. A. and Crocket, J. H. (1981) Volcanic environment of carbonate alteration and stratiform gold mineralisation, Timmins area. In *Genesis of Archaean, Volcanic Hosted Gold Deposits*. Symposium held at the University of Waterloo March 7 1980 Ontario Geological Survey MP 97, pp. 47–58.

Fyon, J. A. and Crocket, J. H. (1983) Gold exploration in the Timmins area – using field and lithogeochemical characteristics of carbonate alteration zones. *Study Geol. Surv. Ont.*, **26**, 56 pp.

Gasparrini, C. (1983) The mineralogy of gold and its significance in metal extraction. *CIM Bull.*, **76(851)**, 144–153.

Goldberg, I. (1964) Notes on the relationship between gold deposits and structure in Southern Rhodesia. In *The Geology of Some Ore Deposits in Southern Africa*, Vol. II (ed. S. H. Haughton), Geological Society of South Africa, Johannesburg, pp. 9–14.

Gribnitz, K. H. (1964) Notes on the Barberton goldfields. *The Geology of Some Ore Deposits in Southern Africa*, Vol. II, (ed. S. H. Haughton), Geological Society of South Africa, pp. 77–90.

Griffis, A. T. (1962) A geological study of the McIntyre Mine. *Trans. Can. Inst. Min. Metall. Min. Soc. Nova Scotia.*, **65**, 47–54.

Groves, D. I. and Batt, W. D. (1985) Spatial and temporal variations of Archaean metallogenic associations in terms of evolution of granitoid-greenstone terrains. In *Archaean Geochemistry* (ed. A. Kroner), Springer-Verlag, Berlin.

Groves, D. I., Phillips, G. N., Ho, S. E., Henderson, C. C., Clark, M. E. and Woad, G. M. (1984) Controls on distribution of Archaean hydrothermal gold deposits in Western Australia. In *Gold '82: The Geology, Geochemistry and Genesis of Gold Deposits, Geological Society of Zimbabwe* (ed. R. P. Foster), A. A. Balkema, Rotterdam, pp. 689–712.

Hallberg, J. A. (1982) An aid to rock-type discrimination in deeply-weathered terrain. In *Geochemical Exploration in Deeply Weathered Terrain* (ed. R. E. Smith), CSIRO Institute of Energy & Earth Resources, Division of Mineralogy, Floreat Park, pp. 29–32.

Hargreaves, D. and Williamson, D. R. (1984) *The Annual Review of the Metal Markets 1983–4*, Shearson/American Express, 262 pp.

Helgeson, H. C. (1969) Thermodynamics of hydrothermal systems at elevated temperatures and pressures. *Am. J. Sci.*, **267**, 729–804.

Henley, R. W. (1973) Solubility of gold in hydrothermal chloride solutions. *Chem. Geol.* **11**, 73–87.

Hodgson, C. J. (1983a) The structure and geological development of the Porcupine Camp – a re-evaluation. In *The Geology of Gold in Ontario,*

Ontario Geological Survey Miscellaneous Paper 110 (ed. A. C. Colvine), pp. 211–225.

Hodgson, C. J. (1983b) Preliminary report on a computer file of gold deposits of the Abitibi belt, Ontario. In *The Geology of Gold in Ontario, Ontario Geological Survey Miscellaneous Paper 110* (ed. A. C. Colvine), pp. 11–37.

Hodgson, C. J., Chapman, R. S. G. and MacGeehan, P. J. (1982) Application of exploration criteria for gold deposits in the Superior Province of the Canadian Shield to gold exploration in the Cordillera. In *Precious Metals in the Northern Cordillera*. Proceedings of a symposium held on April 13–15 1981 in Vancouver, British Columbia (ed. A. A. Levinson), pp. 173–206.

Hodgson, C. J. and MacGeehan, P. J. (1982) A review of the geological characteristics of 'gold-only' deposits in the Superior Province of the Canadian Shield. *Can. Inst. Min. Metall. Spec. Vol.*, 211–229.

Horwood, H. C. (1948) General structural relationships of ore deposits in the Little Long Lac-Sturgeon River Area. In *Structural Geology of Canadian Ore Deposits*, Vol. 1, Canadian Institute of Mining and Metallurgy, pp. 377–384.

Hutchinson, R. W. and Burlington, J. L. (1984) Some broad characteristics of greenstone belt gold lodes. In *Gold '82: The Geology, Geochemistry and Genesis of Gold Deposits, Geological Society of Zimbabwe* (ed. R. P. Foster), A. A. Balkema, Rotterdam, pp. 339–369.

Jackson, N. J., Halliday, A. N., Sheppard, S. M. F. and Mitchell, J. G. (1982) Hydrothermal activity in the St. Just mining district, Cornwall, England. In *Metallization Associated with Acid Magmatism* (ed. A. M. Evans), John Wiley & Sons, Chichester, pp. 137–179.

Karvinen, W. O. (1981) Geology and evolution of gold deposits, Timmins area, Ontario. In *Genesis of Archaean, Volcanic-Hosted Gold Deposits*, Symposium held at the University of Waterloo March 7th 1980 Ontario Geological Survey, MP 97 pp. 29–46.

Kavanagh, P. M. (1979) Precambrian gold deposits. In *Proceedings of the Gold Workshop, Yellowknife N.W.T.* (ed. R. D. Morton, University of Alberta), pp. 47–62.

Keays, R. R. (1984) Archaean gold deposits and their source rocks: the upper mantle connection. In *Gold '82: The Geology, Geochemistry and Genesis of Gold Deposits, Geological Society of Zimbabwe* (ed. R. P. Foster), A. A. Balkema, Rotterdam, pp. 17–52.

Keays, R. R. and Scott, R. (1976) Precious metals in ocean ridge basalts as source rocks for gold mineralisation. *Econ. Geol.*, **71**, pp. 705–720.

Kerrich, R. (1981) Archaean gold-bearing chemical sedimentary rocks and veins: a synthesis of stable isotope and geochemical relations. In *Genesis of Archaean, Volcanic-Hosted Gold Deposits*. Symposium held at the University of Waterloo March 7 1980, Ontario Geological Survey MP 97 pp. 144–167.

Killeen, P. G. and Hobson, G. D. (1974) Project EGMA seismic survey – Timmins Ontario to Val d'Or, Quebec. *Geol. Surv. Can. Paper*, **74–44**, 33 pp.

Kwong, Y. T. J. and Crocket, J. H. (1978) Background and anomalous gold in rocks of an Archaean greenstone assemblage, Kakagi Lake area, N.W. Ontario. *Econ. Geol.*, **73**, 50–63.

Lang, A. H., Goodwin, A. M., Mulligan, R., Whitmore, D. R. E., Gross, G. A., Boyle, R. W., Johnston, A. G., Chamberlain, J. A. and Rose, E. R. (1976) Economic minerals of the Canadian Shield. In *Geology and Economic Minerals of Canada, Geological Survey of Canada, Economic Geology Report No. 1* (ed. R. J. W. Douglas), pp. 151–226.

Levinson, A. A. (1980) *Introduction to Exploration Geochemistry* 2nd edn., Applied Publishing, Wilmette, Illinois, 924 pp.

Lindgren, W. (1913) *Mineral Deposits*, McGraw-Hill, New York, 930 pp.

MacGeehan, P. J. and Hodgson, C. J. (1981) The relationship of gold mineralization to volcanic and alteration features in the area of the Campbell Red Lake and Dickenson Mines, Red Lake area, North-western Ontario. In *Genesis of Archaean, Volcanic-Hosted Gold Deposits*. Symposium held at the University of Waterloo, March 7, 1980, Ontario Geological Survey, MP 97, pp. 94–110.

Mackenzie, B. W. (1984) Economic mineral exploration targets. Centre for Resource Studies. Queens University, Ontario, Working paper no. 28, 54 pp.

Conceptual Models in Exploration Geochemistry (eds C.

Mazzuchelli, R. H. and James, C. H. (1980) Yilgarn gold deposits, Eastern Goldfields province. In *Conceptual Models in Exploration Geochemistry* (eds. C. R. M. Butt and R. E. Smith), Elsevier, Amsterdam, pp. 100–104.

McKinstry, H. E. (1942) Norseman Mine, Western Australia. In *Ore Deposits as Related to Structural Features* (ed. W. H. Newhouse), reprinted by Hafner (1969), New York, 224 pp.

Meyer, G., Shea, E. P. and Goddard, C. C. (1968) Ore deposits at Butte, Montana. In *Ore Deposits of the United States 1933–1967, The Graton-Sales Volume* (ed. J. D. Ridge), American Institute of Mining,

Muir, T. L. (1983) Geology of the Hemlo-Heron Bay area. In *The Geology of Gold in Ontario, Ontario Geological Survey, Miscellaneous Paper 110* (ed. A. C. Colvine), pp. 230–239.

Nash, G. T., Granger, H. C. and Adams, S. S. (1981) Geology and concepts of genesis of important types of uranium deposits. *Econ. Geol. 75th Anniv. Vol.*, 63–116.

Patterson, G. C. (1983) Exploration history in the Hemlo area. In *The Geology of Gold in Ontario, Ontario Geological Survey Miscellaneous Paper 110* (ed. A. C. Colvine), pp. 226–229.

Phillips, G. N. and Groves, D. I. (1984) Fluid access and fluid-wall rock interaction in the genesis of the Archaean gold-quartz vein deposit at Hunt mine, Kambalda, Western Australia, In *Gold '82: The Geology, Geochemistry and Genesis of Gold Deposits, Geological Society of Zimbabwe* (ed. R. P. Foster), A. A. Balkema, Rotterdam, pp. 389–416.

Phillips, G. N., Groves, D. I. and Clark, M. E. (1983) The importance of host rock mineralogy in the location of Archaean epigenetic gold deposits. *International Conference Applied Mineralogy, Johannesburg* 1981, Geological Society of South Africa Special Publication no. 7.

Phillips, W. J. (1972) Hydraulic fracturing and mineralization. *J. Geol. Soc. Lond.*, **128**, 337–359.

Poulsen, K. H. (1983) Structural setting of vein-type gold mineralisation in the Mine Centre – Fort Frances area: implications for the Wabigoon Subprovince. In *The Geology of Gold in Ontario, Ontario Geological Survey Miscellaneous Paper 110* (ed. A. C. Colvine), pp. 174–180.

Pyke, D. R. (1975) On the relationship of gold mineralisation and ultramafic volcanic rocks in the Timmins area. *Ontario Division of Mines Miscellaneous Paper* 62, 23 pp.

Pyke, D. R. (1981) Relationship of gold mineralisation to stratigraphy and structure in Timmins and surrounding areas. In *Genesis of Archaean, Volcanic-Hosted Gold Deposits*, Symposium held at the University of Waterloo, March 7th 1980 Ontario Geological Survey MP 97 pp. 1–15.

Rackley, R. I. (1976) Origin of Western-states type uranium mineralisation. In *Handbook of Stratabound and Stratiform Ore Deposits* Vol. 7, (ed. K. H. Wolf), Elsevier, Amsterdam, pp. 89–152.

Roberts, R. G. (1981) The volcanic-tectonic setting of gold deposits in the Timmins area, Ontario. In *Genesis of Archaean, Volcanic-Hosted Gold Deposits*, Symposium held at the University of Waterloo, March 7 1980, Ontario Geological Survey MP 97 pp. 16–28.

Roedder, E. (1967) Fluid inclusions as samples of ore fluids. In *Geochemistry of Hydrothermal Ore Deposits* (ed. H. L. Barnes), Holt, Rinehart and Winston, New York, pp. 515–567.

Roedder, E. (1984) Fluid-inclusion evidence bearing on the environment of gold deposition. In *Gold '82: The Geology, Geochemistry and Genesis of Gold Deposits, Geological Society of Zimbabwe* (ed. R. P. Foster), A. A. Balkema, Rotterdam, pp. 129–164.

Rytuba, J. J. and Dickson, F. W. (1977) Reaction of pyrite + pyrrhotite + quartz + gold with NaCl–H_2O solutions, 300–500°C, 500 to 1500 bars and genetic implications. In *Problems of Ore Deposition*, 4th IAGOD Symposium, Varna 1974, vol. II, Bulgarian Academy of Science, Sofia, pp. 312–313.

Seward, T. M. (1973) Thio complexes of gold in hydrothermal ore solutions. *Geochim. Cosmochim. Acta*, **37**, 379–399.

Seward, T. M. (1984) The transport and deposition of gold in hydrothermal systems. In *Gold '82: the Geology, Geochemistry and Genesis of Gold Deposits, Geological Society of Zimbabwe* (ed. R. P. Foster), A. A. Balkema, Rotterdam, pp. 165–182.

Siegel, H. O. (1979) An overview of mining geophysics. In *Geophysics and Geochemistry in the Search for Metallic Ores* (ed. P. J. Hood), Geological Survey of Canada, Economic Geology Report 31, pp. 7–24, Proceedings of Exploration 77.

Skinner, R. G. (1972) Drift prospecting in the Abitibi clay belt; overburden drilling programme – methods and costs. *Geol. Surv. Can. Open File* 116.

Smee, B. W. and Sinha, A. K. (1979) Geological, geophysical and geochemical considerations for exploration in clay-covered areas: a review. *CIM Bull.*, **72**, 67–82.

Stanton, R. L. (1972) *Ore Petrology*, McGraw-Hill, New York, 713 pp.

Taylor, H. P. Jr. (1979) Oxygen and hydrogen isotope relationships in hydrothermal mineral deposits. In *Geochemistry of Hydrothermal Ore Deposits* 2nd edn. (ed. H. L. Barnes), John Wiley & Sons, Chichester, pp. 236–272.

Thorne, M. G. and Edwards, R. P. (1985) Recent advances in concepts of ore genesis in Southwest England, *Trans. R. Geol. Soc. Corn.* **21 (3)**, pp. 113–152.

Toulmin, P., III and Clark, S. P. Jr. (1967) Thermal aspects of ore formation. In *Geochemistry of Hydrothermal Ore Deposits* (ed. H. L. Barnes), Holt, Rinehart & Winston, New York, pp. 440–462.

Travis, G. A., Woodall, R. and Bartram, G. D. (1971) The geology of the Kalgoorlie gold field. *Spec. Publ. Geol. Soc. Austr.*, **3**, 175–190.

Twemlow, S. G. (1984) Archaean gold-telluride mineralisation of the Commoner mine, Zimbabwe. In *Gold '82: The Geology, Geochemistry and Genesis of Gold Deposits, Geological Society of Zimbabwe* (ed. R. P. Foster), A. A. Balkema, Rotterdam, pp. 469–492.

Viljoen, M. J. and Viljoen, R. P. (1969) The geology and geochemistry of the Lower Ultramafic unit of the Onverwacht Group and a proposed new class of igneous rock. *Geol. Soc. S. Afr. Spec. Publ.*, **2**, 113–152.

Watters, R. A. (1983) Geochemical exploration for uranium and other metals in tropical and sub-tropical environments using heavy mineral concentrates. In *Geochemical Exploration 1982* (ed. G. R. Parslow), Association of Exploration Geochemists Special Publication no. 10, pp. 103–124.

Wengzhang, L. (1982) Geological features of mineralisation of the Xingluokeng tungsten (molybdenum) deposit, Fujian Province. In *Tungsten Geology, Jiangxi, China* (eds J. V. Hepworth and Y. H. Zhang), ESCAP/RMRDC, Bandung, Indonesia and Geological Publishing House, Beijung, China, pp. 339–348.

Wennervirta, H. (1973) Sampling of the bedrock/till interface in geochemical exploration. In *Prospecting in Areas of Glaciated Terrain* (ed. M. J. Jones), Institute of Mining and Metallurgy, London, pp. 67–71.

Western Mining (1979) Gold: a review of the technology of exploration and mining. Vol. I: Exploration and evaluation. Report prepared by the staff of Western Mining Corporation Ltd., 360 Collins Street, Melbourne, Victoria, Australia, 158 pp.

White, D. E. (1974) Diverse origins of hydrothermal ore fluids. *Econ. Geol.*, **69**, 954–973.

Wilson, J. G. (1979) The major controls of tin mineralisation in the Bushveld Igneous Complex, South Africa. *Bull. Geol. Soc. Malaysia*, **11**, 239–252.

Wolfe, W. J. (1976) Regional geochemical reconnaissance of Archaean metavolcanic metasedimentary belts in the Pukaskwa Region. Ontario Geological Survey Geological Report, 158, 54 pp.

Woodall, R. (1965) Structure of the Kalgoorlie Goldfield In *Geology of Australian Ore Deposits* 2nd edn. (ed. J. McAndrew), 8th Comm. Min. Metall. Congr. Melbourne, pp. 71–79.

Woodall, R. (1975) Gold in the Precambrian shield of Western Australia. In *Economic Geology of Australia and Papua New Guinea* (ed. C. L. Knight), Australasian Institute of Mining & Metallurgy, pp. 175–184.

5

Placers and palaeo-placers

5.1 INTRODUCTION

Placer deposits are formed by the mechanical concentration of resistate minerals, which are released by weathering from source rocks in which their distribution is normally subeconomic. The main commodities derived from placer deposits are tin, gold, platinum, niobium, tantalum, titanium, zircon and diamonds. Some metals, such as gold, occur within the entire spectrum of placer deposits, but it is more common for a mineral to be concentrated in a specific placer environment, which is ultimately controlled by tectonic setting. In this chapter, the main environments of placer formation are reviewed with particular emphasis on the genesis of the deposits and the appropriate exploration techniques.

Palaeo-placers may be defined as the lithified equivalents of placer deposits. However, heavy mineral concentrations preserved within sedimentary sequences are usually only of academic interest (Zimmerle, 1973). Palaeo-placer deposits of economic importance are of limited geographical distribution (Fig. 5.1) and are mainly confined to rocks of Archaean and Proterozoic age. They therefore represent a rather specific type of preserved placer. Palaeo-placers are a major source of uranium in Canada, and gold and uranium in South Africa. Tin-bearing palaeo-placers of minor economic importance are recorded from Nigeria, Brazil and Malaysia.

5.2 PLACER DEPOSITS

5.2.1 Classification of placer deposits

Various classification systems have been proposed for placer deposits (Wells, 1969; Smirnov, 1976). We prefer to use a simplified version of a classification proposed by Macdonald (1983) which is based primarily on geological environment (Table 5.1). Figure 5.2 is a schematic cross-section which shows the distribution of the sub-environments of placers.

A more detailed discussion of these differing placer environments will take place in later sections. At present it is sufficient to draw attention to the importance of correctly identifying the genetic type, because each has its characteristic distribution of valuable heavy minerals, which must be borne in mind during exploration and evaluation.

Fig. 5.1 Location map.

5.2.2 General characteristics of placer deposits

(a) Distribution in space and time

Placer deposits are formed by normal surface processes, and therefore they have a wide geographical distribution. Sometimes the size of placer deposits is anomalously large. Anomalous size is difficult to quantify, but it is possible to suggest the Bakwanga eluvial diamond field in Zaire, the Otago fluvial gold deposits in New Zealand, and the titanium-bearing beach sands of Eastern Australia as examples (Fig. 5.1). Henley and Adams (1979) used the term 'giant placer' to describe those auriferous placers from which more than 148 tonnes of gold have been extracted. If we restrict our attention to placer deposits of very large size, it is possible to recognize two patterns of distribution which are discussed in the section concerned with tectonic setting.

The bulk of the World's placer deposits is of

Placers and palaeo-placers

Tertiary and Quaternary age. There are several reasons for this restriction. In part, it is probably a question of preservation with older placer deposits having been destroyed. However, it is also a reflection of geological events in late Tertiary and Quaternary times which have favoured the accumulation of placers in some environments. Many large fluvial and eluvial gold placers can be attributed to rapid uplift of mountainous terrain during the Alpine orogeny, with the subsequent shedding of detrital gold. Figure 5.3 shows the episodic supply of gold to Tertiary sediments in New Zealand from lode gold in the Mesozoic basement.

An additional factor which has favoured placer accumulation is the change in sea levels during Quaternary times.

(b) Size and grade of deposits

The diversity of geological environments in which placer deposits may form, and their variation in mineralogy, makes it very difficult to generalize

Table 5.1 Classification of placer deposits (modified from Macdonald, 1983)

Environment	Subenvironment	Main products	Environmental processes
Continental	Eluvial (sometimes referred to as residual)	Au, Pt, Sn, WO$_3$, Ta, Nb, gemstones (all varieties)	Percolating waters, chemical and biological reactions, heat, wind and rain
	Colluvial	Au, Pt, Sn, WO$_3$, Ta, Nb, gemstones (all varieties)	Surface creep, wind, rainwash, frost, elutriation
	Fluvial (or alluvial)	Au, Pt, Sn: rarely Ta, Nb, WO$_3$ Diamonds	Flowing streams of water
Transitional	Strand-line	Ti, Zr, Fe, ReO, Au, Pt, Sn	Waves, currents, wind, tides
Marine	Drowned placers	Au, Pt, Sn, diamonds minor Ti, Zr, Fe, ReO industrial sand and gravel	Eustatic, isostatic and tectonic movements – net rise in sea level

about either size or grade. However, the following points can be made:

(a) In nearly all cases, the average grade of ore which is mined from placers is considerably lower than from hard-rock mining. We will consider tin mining in Cornwall as an example, using information obtained from personnel within the industry. Narrow, steeply dipping quartz–tourmaline lodes, which are mined by a variety of underground mining techniques, have average grades within mineralized structures of about 1.0–1.5% Sn. There is usually some dilution during mining, so that the feed to the mill is in the order of 0.8% Sn. Large stockwork deposits which are amenable to open pit mining, but are not yet in production, have grades of about 0.20% Sn and WO$_3$. There are no land-based placer deposits being mined in Cornwall at present, apart from mine dumps, but grades of 0.25–0.40% Sn would be economic, assuming that recovery was reasonable and environmental constraints not too excessive. At the present time plans are at an advanced stage to mine tin from deposits off the north coast of Cornwall. The average grade of ore to be mined is 0.15% Sn. The grades of placer tin which are mined in Cornwall are rather higher than from other parts of the World, because a significant proportion of the cassiterite is locked with quartz, chlorite and tourmaline. This is due to the fact that much of the tin is derived from active and abandoned mining operations.

(b) From general principles one would anticipate that, in most cases, placer deposits formed in eluvial or colluvial subenvironments would tend to be smaller, and have lower grades, than in the fluvial subenvironment. Similarly, deposits formed in a beach environment tend to be much larger than deposits formed in a continental environment. Few data are available in the literature to substantiate these points. Boyle (1979) stated that most eluvial gold placers have lower

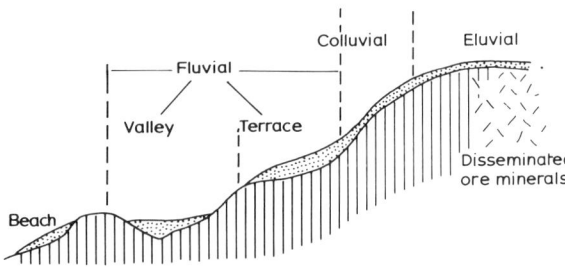

Fig. 5.2 Cross-section showing the subdivision of environments of placer formation (modified from Smirnov 1976).

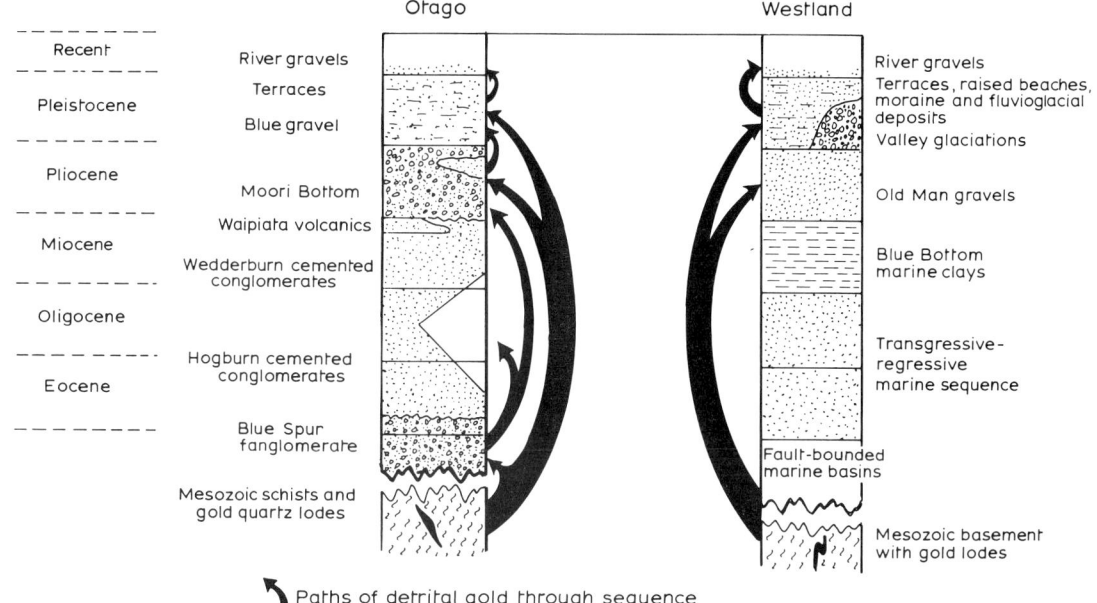

Fig. 5.3 Stratigraphy of auriferous gravels of Otago and Westland, New Zealand, showing the movement of placer gold (modified from Henley and Adams 1979).

Table 5.2 Selected physical properties and environments of deposition of the economic minerals found in placer deposits

	Mineral	Formula	Specific gravity	Hardness	Principal placer environment
Heavy heavy minerals	Gold	Au	15.5–19.4	2.5–3	Fluvial, eluvial (beach)
	Platinum	Pt	14–19	4–4.5	Fluvial
	Cassiterite	SnO_2	6.8–7.1	6–7	Eluvial, fluvial, marine
	Wolframite	$(FeMn)(WO_4)$	7.0–7.5	5–5.5	Eluvial, colluvial
Light heavy minerals	Magnetite	Fe_3O_4	5.2	5.5–6.5	Beach sand
	Ilmenite	$FeTiO_3$	4.5–5.0	5–6	Beach sand
	Rutile	TiO_2	4.2	6–6.5	Beach sand
	Columbite–tantalite	$(FeMn)(NbTa)_2O_6$	5.3–7.3	6	Fluvial
	Pyrochlore	$(NaCa)_2(NbTi)_2(O,F)_7$	4.2–4.4	5–5.5	Eluvial
	Xenotime	YPO_4	4.5	4–5	Beach sand
	Monazite	$(CeLaDi)PO_4$	4.9–5.3	5–5.5	Beach sand
	Bastnaesite	$CeFCO_3$	4.9	4.5	Eluvial
	Baddeleyite	ZrO_2	5.5–6.0	6.5	Eluvial
	Zircon	$ZrSiO_4$	4.6–4.7	7.5	Beach sand
	Diamond	C	3.5	10	Beach, fluvial, eluvial

grades than fluvial deposits, although some are noted for their large nuggets and some are very large in size. The size and grade of some selected beach sand deposits are provided in Table 5.4.

(c) The cut-off grade and the size of placer deposits are dependent upon the mining method used. This point may be illustrated by the example of tin mining in Indonesia. In 1983 the average grade of tin ore obtained from placer deposits was 0.41 kg/m^3 for open cast mining, 0.18 kg/m^3 for gravel pumping and 0.14 kg/m^3 for dredging (Williamson and Fromson, 1984).

(c) Mineralogy

Table 5.2 lists the placer minerals of principal economic importance. Placer minerals are often typified by very high specific gravities, and are usually hard. These are important characteristics but it is the fact that the mineral is physically durable and chemically inert which ensures its concentration in the placer environment. Table 5.2 shows that placer minerals may be classified on the basis of their specific gravity (S.G.). Heavy heavy minerals have an S.G. in excess of 7.0, and are mainly restricted to a continental environment. Most of the light heavy minerals are confined to either the eluvial or beach environments.

Commonly the placer minerals which occur in fluvial and beach environments have undergone several cycles of erosion, transport and deposition. As a consequence, the mineral grains are usually liberated from their associated gangue, and a minimal amount of comminution is required before the ore is processed.

(d) Tectonic setting

In discussing the classification of placer deposits (Table 5.1) we have seen that they may be subdivided into two types: those of continental setting which are characterized particularly by gold and tin, and placers occurring within a transitional environment, which contain mainly titanium and zircon minerals. This subdivision reflects two contrasting tectonic settings:

(i) Transitional environment (beach deposits)

The location of the major deposits belonging to this class is shown in Fig. 5.1 and their main characteristics are summarized in Table 5.4. They occur on passive continental margins (Mitchell and Garson, 1981), where the predominant influences are appropriate source materials, the fluctuation of sea level in Quaternary times and the

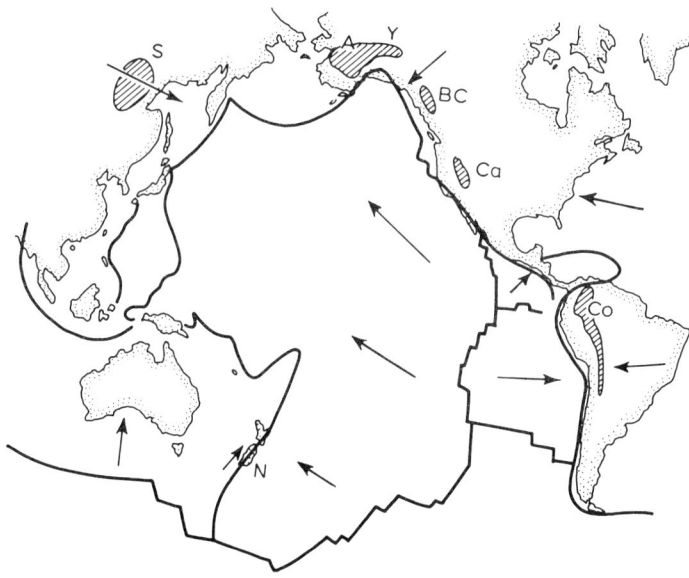

Fig. 5.4 Tectonic structure of circum-Pacific area, showing giant placer regions. A, Alaska; Y, Yukon; BC, British Columbia; Ca, California; Co, South America (Colombia, etc.); N, New Zealand (Otago, Westland); S, USSR (Siberia). (From Henley and Adams 1979.)

mechanical concentration of heavy minerals by wave and current activity. Although the presence of suitable source rocks ultimately depends upon tectonic environment, the role of tectonic activity appears to be of minor importance for this class of deposit.

(ii) Continental environment (eluvial, colluvial and fluvial deposits)

The tectonic setting of large fluvial gold placers has been discussed by Henley and Adams (1979). Their study was mainly confined to Otago and Westland, New Zealand, California, USA, and Columbia. They demonstrated that giant gold placers are restricted to regions of tectonic instability, associated with plate margins of the circum-Pacific (Fig. 5.4). Two aspects of the tectonic setting are of fundamental importance for placer formation. Firstly, rejuvenation of an older orogenic belt provides the source of gold. Secondly, the high rates of uplift associated with this tectonic setting provide multiple episodes of reworking of sediment, which leads to the concentration of gold.

5.3 ELUVIAL (RESIDUAL), COLLUVIAL AND FLUVIAL (ALLUVIAL) DEPOSITS

5.3.1 Introduction

The eluvial deposits of most significance economically are probably the Bakwanga diamond deposits in Zaire, which are mainly eluvial in origin (Bates, 1969). The Ghanaian diamond deposits are also considered to be partly eluvial in nature (Applin, 1974). Eluvial gold deposits have been mined from many countries in the past, but most are now exhausted. Important exceptions are the deposits in the Quadrilatero Ferrifero, Minas Gerais, Brazil where eluvial gold is associated with auriferous banded iron-formation (Boyle, 1979). Brazil also contains important tin and niobium deposits of eluvial origin. The most important tin deposits are those in the Rondonia district, from which more than 90% of Brazilian tin is produced (Beurlen and Cassadene, 1981).

Gold and cassiterite are the main ore minerals produced from fluvial placers. In the Western World, many of the mining districts which were famous for their production of alluvial gold are now exhausted or in decline. In contrast it is estimated that about 60% of Soviet gold production is of placer origin (Potts, 1980), mainly from the Lena and Amur Rivers, Siberia. Approximately 65% of the World's tin is produced from placer deposits in Malaysia, Thailand and Indonesia where fluvial deposits predominate.

5.3.2 Genesis of eluvial, colluvial and fluvial placers

Eluvial deposits overlie, or are located very near, their source rocks. The constant reworking of resistate minerals, which is a characteristic feature of fluvial and beach deposits, does not affect eluvial deposits so their grade is normally lower than equivalent ores in other placer environments. The degree to which the resistate minerals are up-graded depends largely on topography and climate. According to Macdonald (1983), the most favourable conditions are elevated ground surfaces, where local depressions cover the mineralized zone. Level surfaces covered by dense vegetation, in regions of heavy subtropical rainfall, are also suitable. The enrichment in placer minerals is partly caused by the removal of soluble minerals by groundwater, and partly by the transport of the lighter minerals by running water and wind action.

Fluvial placers can form at any stage within a river system although the stages of youth and old age, in which deposition is dominant compared with erosion, are likely to contain the most favourable sites for the accumulation of large deposits of significant grade. Within the fluvial environment river terraces and valley-fill are the most important locations for the concentration of valuable heavy minerals.

One of the requirements for the formation of a placer is an appropriate source rock, which contains anomalously high concentrations of the relevant heavy mineral. In most cases the valuable

Table 5.3 The provenance of economic minerals in placer deposits from (Macdonald, 1983)

Provenance	Economic mineral	Associated minerals
Ultramafic and mafic terrains including pyroxenites and norites	The platinoids	Olivine, enstatite, basic plagioclase, chromite, titano-magnetite, ilmenite, chrome spinel, diopsidic augite
Granitoid terrains and related pegmatites and greisens	Cassiterite, monazite, zircon, rutile, gold	Wolframite, potash feldspar, quartz, topaz, beryl, spodumene, petalite, tourmaline, tantalite, columbite, monazite, fluorite, sphene
Plateau basalts	Magnetite, ilmenite	Pyriboles, basic plagioclase, apatite
Syenitic rocks and related pegmatites	Zircon, rare earth minerals including uranium- and thorium-bearing minerals	Ilmenite, magnetite, fluorite, pyriboles, potash feldspar, apatite, feldspathoids, zircon
Contact metamorphic aureoles – skarns	Scheelite, rutile, occasionally corundum	Diopside, grossularite, wollastonite, calcite, basic plagioclase, epidote
Kimberlites	Diamonds	Ilmenite, magnetite, pyrope garnet, pyroxene including diopside, kyanite, sphene, apatite
High-grade metamorphic terrain	Gold, rutile, zircon, gemstones	Kyanite, pyriboles, quartz, sillimanite, almandine garnet, feldspars, apatite
Serpentine belts	Platinoids, chromite, magnetite	Chrome garnet, pyroxenes, olivine
Carbonatites including associated rare basic igneous rocks	Rutile, ilmenite, magnetite, rare-earth minerals, uranium, niobium, thorium and zirconium minerals	Potash feldspars, calcite, pyriboles, garnets, apatite

mineral is dispersed within the host rock at grades which are not of economic significance. Table 5.3 shows the principal lithologies from which most placer minerals are derived. Evans (1981) has suggested that the normal processes of lateritization may be important in upgrading the gold content of ultrabasic rocks and therefore increase their suitability as a source for placer deposits. The distance between the bedrock source and the site of deposition is also an important factor. For example, Emery and Noakes (1968) suggested that the median distance from the primary source at which economic placer cassiterite deposits can develop is about 8 km. Placers are not always derived from subeconomic sources; in many cases primary ore deposits have been discovered by tracing the source of placer minerals.

In order for a heavy mineral to become sufficiently concentrated to form a deposit of economic significance it is normally found that extensive reworking of the sediment has taken place. This aspect is discussed by Schumm (1977) in an excellent account of placer formation. Schumm attributed the reworking of alluvial sediment to three mechanisms: (i) changes in the erosional history of the valley, for example by a change in the base level of erosion; (ii) climatic changes; (iii) a change of relief in the source area.

Climate plays an important role in the development of placer deposits. Under dry conditions the valley system can be fed from the

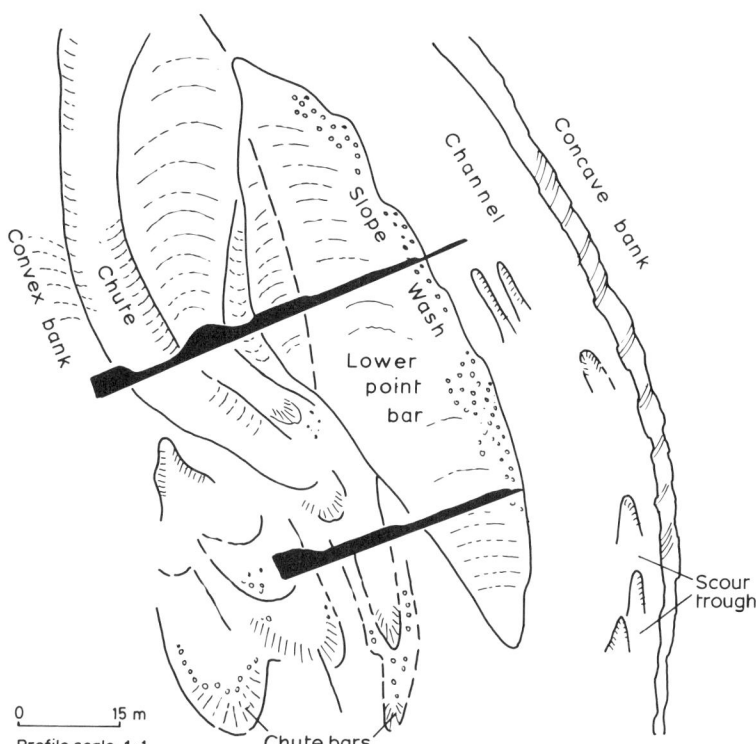

Fig. 5.5 Plan view of a coarse-grained point bar including profiles of sediment thickness (from McGowan and Garner 1970).

products of rock wastage, whereas in a wet climate the feed to the valley is virtually restricted to what the river system itself can erode, although landslides may contribute. In a hot/wet climate the land is protected by a cover of vegetation which will restrict the movement of soil and rock towards the drainage system (K. F. G. Hosking, personal communication).

Even without major changes in climate or base level there may be preferred sites for heavy mineral deposition. Thus in a meandering river valley, heavy minerals tend to be concentrated at point bars and chute bars (Fig. 5.5). As the meander pattern evolves, the isolated concentrations of heavy minerals become elongated paystreaks within the flood plain sediments. Experimental work cited by Schumm (1977) suggests that even where the river channel maintains a preferred position within a valley, the concentration of valuable heavy minerals may take place.

Tributaries play an important part in controlling the stability of channel position and their intersection with the main channel is often a site for heavy mineral deposition. The importance of bedrock gradient in controlling the formation of stanniferous placer deposits in Nigeria has been described by Mackay et al. (1949). Figure 5.6 shows the roles of tributary intersection and bedrock gradient in controlling the concentration of gold in the Klondike area, Yukon. The gold is concentrated along the valley trough and the change in the valley slope is an additional factor causing an increase in gold values.

In addition to their association with river gravels, valuable heavy minerals are often concentrated at the sediment/bedrock interface, and in many cases are found within the weathered bedrock (Applin, 1974). Schistose rocks, clays and carbonates are found to be particularly favourable as repositories for the infiltration of heavy

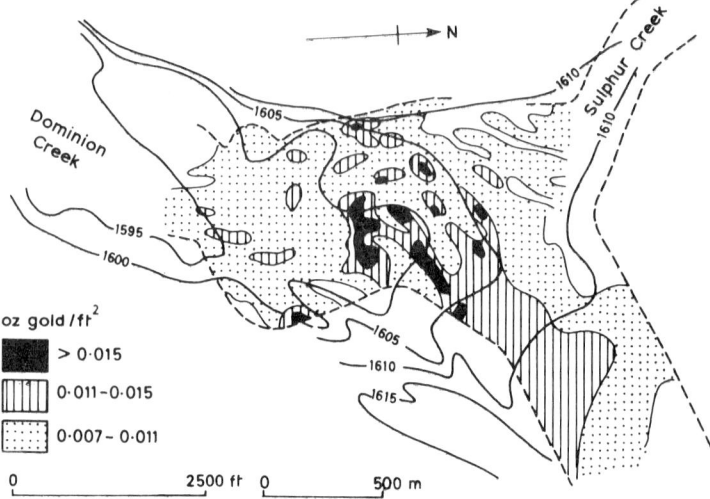

Fig. 5.6 Plan of part of Lower Dominion Creek, near confluence with Sulphur Creek, Klondike area, showing distribution of gold values revealed by dredging in relation to contours on bedrock. Contours at 5 ft (1.52 m) intervals. (From Hester 1970.)

minerals, and as a result exploration drill holes are usually continued for a short distance into the bedrock. Caution must be exercised in distinguishing 'false bottoms' caused by barren sediment or volcanic ash which may conceal valuable concentrations of heavy minerals. According to A. D. Francis (personal communication) the tin-bearing alluvial gravels in Nigeria are often multi-filled, with concentrations of cassiterite sometimes occurring in the same channel at three separate levels, separated by barren sediment.

5.3.3 The Kinta Valley tin field, Malaysia

One of the most significant fluvial deposits in the World is the Kinta tin field of Malaysia, which has been the largest tin producer since 1890.

At the present time it accounts for about 10% of World output with an annual average production of about 20 000 tonnes of metallic tin. At the end of 1977 there were 396 tin mines operating in the Kinta Valley. The main memoir on the area is by Ingham and Bradford (1960), and recent development in the interpretation of the geology and tin deposits has been reviewed by Rajah (1979).

The rocks which lie beneath the Kinta Valley are dominantly limestones, which range in age from Silurian to Permian. Granitoid rocks of Lower Triassic age encircle the sedimentary strata and occupy about half the Kinta area (Fig. 5.7). The granitoids are predominantly medium to coarse grained, and no marked differences in the chemistry or mineralogy of the plutons can be discerned. Most of the Kinta Valley is covered by alluvium which varies in thickness from 6 to 30 m and is considered to be Lower to Middle Pleistocene in age. Walker (1956) recognized the following four types of alluvium:

(1) Boulder Beds, which consist of subangular to rounded boulders of vein quartz, granitoid and schist within a sandy clay and clay matrix.

(2) Old Alluvium, which represent the thickest alluvial deposits and consist of gravel, sand, silt and clay in all possible mixtures with peaty sediments and lignitized wood.

(3) Young Alluvium, which consists of unconsolidated sand and gravel with some clay and peat.

(4) Organic mud and peat.

The tin deposits of the Kinta Valley are predominantly alluvial, but eluvial deposits are found in association with them. The alluvial tin is

Placers and palaeo-placers

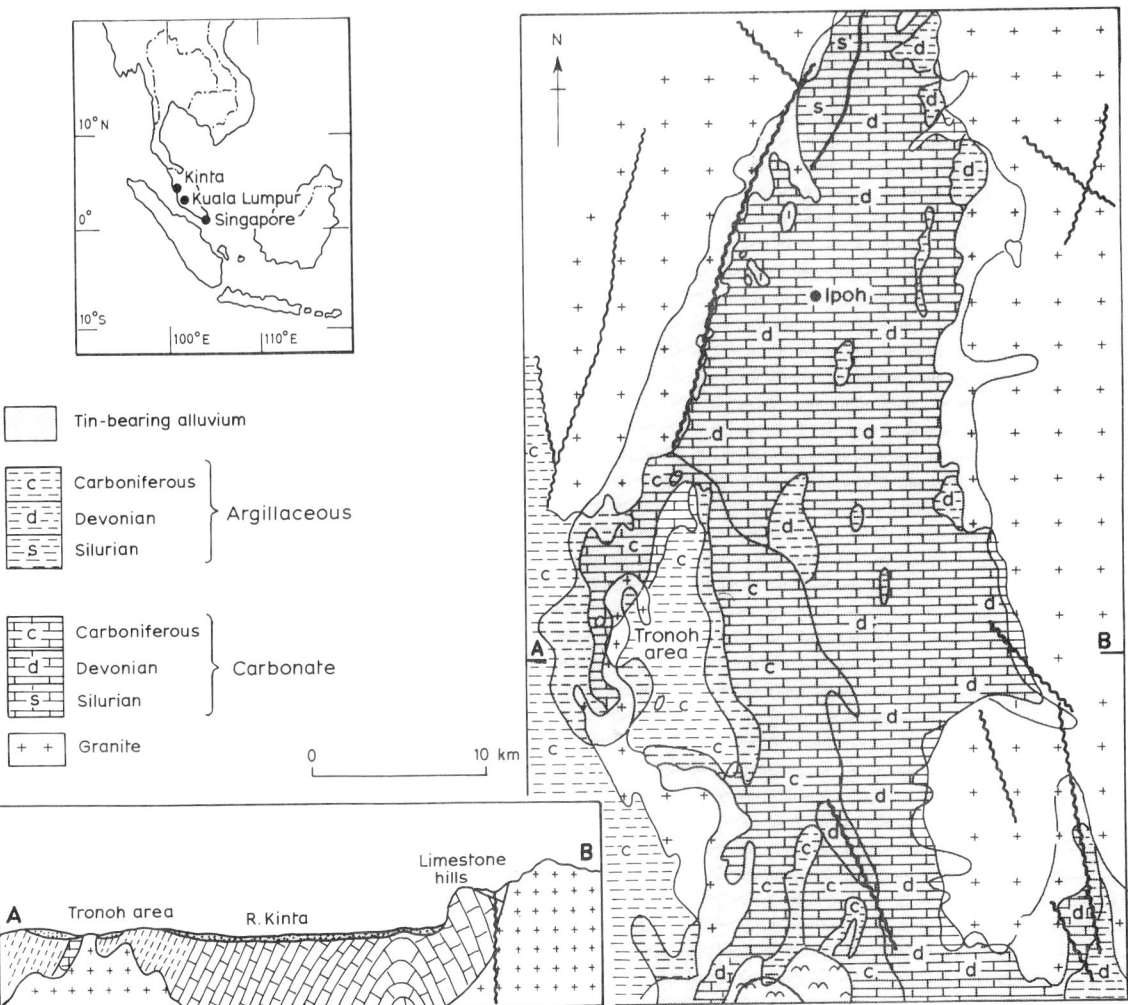

Fig. 5.7 Simplified geological map and section of the Kinta Valley tin field (from Dixon 1979, based on Ingham and Bradford 1960).

believed to be derived from the erosion of tin-bearing veins, pipes, stockworks, greisen and stringers cutting granitoids or sedimentary rocks in contact with them. However, tin production from primary ore deposits is negligible.

Cassiterite is the ore mineral of predominant importance but ilmenite, monazite, zircon, scheelite, wolframite, xenotime, columbite, gold and fluorite are recovered as by-product minerals.

In general, the richest alluvial deposits are found when the underlying limestone bedrock forms a trough-and-pinnacle topography. The deep solution channels form a series of natural riffles which concentrate the heavy grains of cassiterite (see Fig. 5.8). Rich deposits are also found in the granite contact zone where slumped eluvial and alluvial material is concentrated in deep alluvial channels.

Fig. 5.8 Gravel-pump mining alluvial tin in the Kinta Valley, Malaysia. Note the karstic topography of the limestone below the alluvium. High tin values commonly occur near the limestone/alluvium interface, tin here can be recovered most effectively by reducing the alluvium to a slurry. The slurry is then pumped to a gravity separation plant.

The tin deposits of the Kinta Valley are mined by gravel pump and dredging. In many dredged-out areas gravel pumping has been introduced to recover the cassiterite lodged in the crevices between limestone pinnacles (Figs. 5.8 and 5.9).

5.3.4 Exploration for eluvial, colluvial and fluvial placer deposits

Exploration for placer deposits may take place in virgin terrain, or may be carried out in an area adjacent to known mineralization. Macdonald (1983) suggests that in the former case one should be looking for evidence for the following:

(a) a substantial provenance in an appropriate geological setting;
(b) a prolonged weathering regime;
(c) a dynamic transport system with favourable conditions for deposition;
(d) a favourable sequence of land and sea movements.

In reconnaissance exploration, data obtained from remote sensing and aerial photography may be valuable. For example, in the Rondonia district of Brazil, side-scanning radar delineates the ring structures with which cassiterite-bearing granitoids are associated (K. F. G. Hosking, personal communication). In the Orange River Valley, Namibia, the interpretation of aerial

Fig. 5.9 Bucket-dredge operating in the Kinta Valley, Malaysia. The line of buckets conveying alluvium to the dredge can be seen on the left-hand side. Dredges of this type can work placer deposits to a depth of about 50 metres.

photographs aids in the delineation of abandoned river meanders (J. Fowler, personal communication).

The main basis for exploration for placers is the preparation of maps which provide details of the solid geology, Quaternary cover and geomorphology. Goosens (1980) suggested that the best approach for large areas is to prepare a preliminary topographic map with contour levels and drainage patterns, and then superimpose the information from remote sensing. Kuzvart and Bohmer (1978) emphasized the importance of very detailed geomorphological mapping, and suggested that for maps at scales of 1:5000 to 1:10 000, contours of 2–5 m intervals are required, whereas for maps at scales of 1:1000, contours at 1–2 m intervals are more appropriate.

An important component of any exploration programme for placers is the study of heavy-mineral assemblages obtained from panning soils and sediments. The main objective at this stage is to identify valuable heavy minerals but the total assemblage of resistate minerals may be a guide to favourable areas (Table 5.3).

The evaluation of an alluvial or eluvial deposit normally takes place in several successive stages. In preliminary evaluation, use is sometimes made of geophysical techniques. In principle seismic, electrical logging, resistivity and magnetic techniques are of potential value, but in practice geologists report varying degrees of success. Seismic techniques appear to be the most valuable. Petersen *et al.* (1968) determined the configuration of a large, gravel-filled channel of Tertiary age in northern Nevada, using a combination of shallow seismic refraction and gravity methods. Singh

(1983) describes the use of shallow reflection seismics in the Kinta Valley, Malaysia, and reports that the method helped in the delineation of bedrock, and in determining the thickness of alluvium. However, in the evaluation of diamondiferous gravels in the Orange River Valley, Namibia, it was found that neither seismic, resistivity nor magnetic methods provided useful results (J. Fowler, personal communication). Similarly S. Camm (personal communication) attempted to use electrical resistivity and shallow refractive seismic profiling in the evaluation of tin placer deposits on Goss Moor, Cornwall, UK. However, subsequent drilling found that the channels suggested by the geophysical work were spurious.

The detailed evaluation of placer deposits invariably requires the analysis of samples collected by pitting, trenching or drilling. Normally pitting is restricted to countries with an abundant supply of cheap labour, but mechanical pitting and trenching are also used. The advantages of pits and trenches are that they permit the accurate sampling of mineralized horizons and they facilitate the collection of very large samples which is particularly important in the evaluation of diamondiferous deposits.

If the terrain is unfavourable for trenching, or if greater depth of penetration is required, drilling techniques are employed. The different drilling methods which are appropriate to placer deposits are discussed by Macdonald (1983).

One of the most common problems encountered in the valuation of alluvial deposits is an undervaluation of the heavy minerals present. In many cases this has been attributed to errors in the determination of volume. Hester (1970) has demonstrated that in the case of the Klondike area in the Yukon, where estimates from drill holes consistently undervalued low-grade deposits and overvalued high-grade deposits, errors could be attributed principally to the particle size of the gold. The larger-sized gold particles were rarely intersected by drill holes, whereas these particles made a significant contribution to the overall value of the deposit.

5.4 BEACH SAND DEPOSITS

5.4.1 Introduction

Placer deposits may be formed in a beach environment by the natural sorting action of the surf, which concentrates heavy minerals at the high-water mark, and removes the lighter fraction of minerals. In some cases the transport of sediment by wind also plays a part in the formation of beach sand deposits. The most important deposits are the titanium-bearing 'black sands', and their main characteristics are summarized in Table 5.4. The unique diamondiferous gravels of the Namibian coast are also of very considerable economic importance.

5.4.2 Genesis of beach sand deposits

Precambrian metamorphic and igneous complexes are regarded as the ultimate sources of the valuable heavy minerals found in beach sand deposits. In some cases mineralogical studies have indicated the nature of the source rocks more precisely. Puffer and Cousminer (1982) carried out a detailed textural study of ilmenites from beach sands in the Lakehurst area, New Jersey, and compared them with ilmenite associated with possible source rocks. Ilmenite in the granite and gneisses was found to be characterized by haematite exsolution blebs, and the weathered equivalents of the exsolution textures could also be identified in ilmenite from the beach sand.

In a study of the Capel beach sand deposits, Western Australia, Welch *et al.* (1975) suggested that an essential part was played by Mesozoic sediments in collecting, but not necessarily concentrating, the heavy minerals liberated during millions of years of erosion of Archaean rocks.

Welch *et al.* also suggested that rapid uplift of lateritized terrain, under relatively wet climatic conditions, is an important prerequisite to the accumulation of heavy mineral sands. Puffer and Cousminer (1982) utilized palynological evidence from a lignite band, interbedded with the titanium ore in the Lakehurst area, to show that at the time

Table 5.4 Summary of characteristics of beach sands from main producing countries

Country and average annual production[4] (tonnes per year)	Location and size of deposits	Principal valuable heavy minerals (VHM)	Grade	Associated minerals
Australia (1.5×10^6)	(a) Eastern Australia rutile[1] province 1000 km × 3 km	(a) Rutile, zircon and ilmenite	(a) About 3% VHM	(a) Chromite, monazite
	(b) Eneabba, W. Australia[2] 20 km × 1 km	(b) Ilmenite, leucoxene, zircon, rutile, kyanite	(b) 5–10% VHM	(b) Monazite, staurolite
	(c) Capel area, W. Australia[3] 40 km	(c) Ilmenite and leucoxene	(c) About 10% VHM	(c) Zircon, monazite, rutile, xenotime
S. Africa (470 000)	Richards Bay[7] 17 × 2 km	Ilmenite, zircon, rutile	7% VHM	Magnetite, augite, hornblende, garnet, leucoxene, monazite
USA (390 000)	Trail Ridge, Florida[8] 27 × 1.6–3.2 km	Ilmenite, leucoxene, rutile	About 2% VHM	Staurolite, zircon, kyanite, sillimanite, tourmaline, spinel, topaz, corundum
India (235 000)	Tamil Nadu and Kerala States[4]	Ilmenite[5]	40–80% VHM[5]	Garnet, monazite[5]
Sri Lanka (100 000)	Pulmoddai area, 73 km north of Trincomalee 7.2 km × 228 m[6]	Ilmenite, rutile, zircon, monazite	80% VHM	Sillimanite

Source: 1, McKellar (1975); 2, Lissiman and Oxenford, (1975); 3, Welch et al. (1975); 4, Coope (1982); 5, Lynd (1960); 6, Clark (1983); 7, R. P. Edwards technical visit 1980; 8, Pirkle and Yoho (1970).

of formation of the mineral sands the area had a subtropical climate.

In all cases the main concentration process of valuable heavy minerals is the natural winnowing away of lighter minerals by the sea, particularly in storm conditions. However, it is important to emphasize the importance of climatic changes in Plio-Pleistocene times, during which changes of sea level have resulted in several periods of marine transgression and regression, which have reworked an extensive zone of clastic sediment.

Sometimes onshore winds may cause the transport of heavy mineral sands inland. In some cases, as at Richards Bay, Natal, S. Africa, this produces a dissemination of the valuable heavy minerals. However, in Eastern Australia wind action tends to produce a zoning of the valuable heavy minerals downwind from the ocean.

5.4.3 The Eneabba rutile–zircon–ilmenite sand deposit, Western Australia

The Eneabba deposit was discovered and pegged out by a local farmer and prospector in 1970, and subsequent exploration has revealed that it is one of the largest beach sand deposits in the World. Details of the size, grade and mineralogy of the deposit are provided in Table 5.4.

The main part of the Eneabba deposit forms a linear feature, about 20 km in length, parallel to the

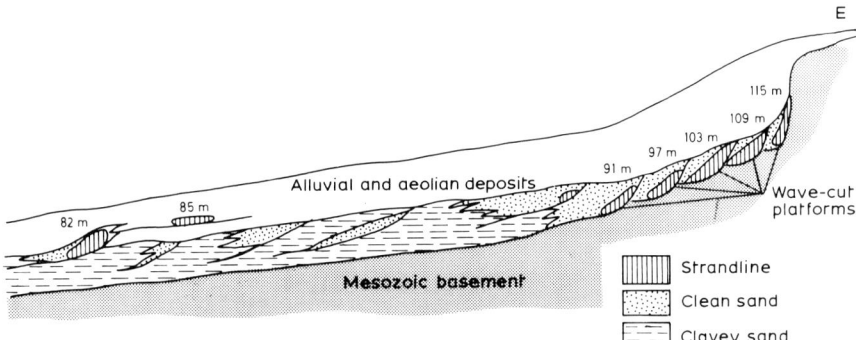

Fig. 5.10 Diagrammatic geological section of the Eneabba deposit (from Lissiman and Oxenford 1975).

Pleistocene shoreline. The sediments consist of an association of beach and dune sand with fluviatile, lacustrine and dunal sediments deposited in a paralic sequence (Baxter, 1977). Concentrations of valuable heavy minerals occur in lensoid beach sediments within the lower marine-estuarine succession, and in the unconformably overlying beach sand deposits. The bulk of the sands is poorly consolidated but about 15% of the deposit is lithified by an argillaceous cement. The valuable heavy minerals are contained in seven parallel zones which represent the positions of old strand lines, the bases of which occur on wave cut platforms at successively lower elevations to the west (Fig. 5.10).

Mineral assemblages tend to vary both between and within individual strands. Lissiman and Oxenford (1975) considered that the variation in mineralogy reflects differences in mode of deposition. The two older strands, which have sands characterized by a high zircon content, derived the bulk of their heavy mineral content by direct erosion of the existing cliff face. Subsequent strands, which have less zircon and higher rutile and ilmenite contents, were formed by the concentration of heavy minerals brought into 'Eneabba Bay' by various streams.

According to Lissiman and Oxenford (1975) the heavy mineral sands were deposited and concentrated in a favourable coastline environment during a period of high sea levels, probably in late Tertiary or early Pleistocene times. Comparison of the palaeo shoreline, and particularly the 'Eneabba Bay', with the present-day coastline, where beach sands are being formed at the present time, suggests that a combination of coastal configuration and a prevailing north-westerly wind/wave action are important controls in the concentration of heavy minerals (Fig. 5.11).

5.4.4 Diamondiferous beach sands of the south-western coast of Africa

(a) Distribution

Diamondiferous beach placers have been exploited along approximately 1600 km of the S.W. coastline of Africa and extend southwards from the Kunene River in Namibia to the Olifants River in Namaqualand, South Africa (Fig. 5.1). At the present time the most important diamond-mining operations are those owned by CDM, along an 80 km coastal strip, north of Oranjemund in Namibia.

(b) The CDM operations north of Oranjemund, Namibia

The diamondiferous deposits north of the Orange River consist of a series of raised beaches (Fig. 5.12) resting on a wave-cut platform, which varies from over 3000 m in width immediately north of the Orange River, to 200 m wide in the north of the CDM mining lease. The abrasion platform is

Placers and palaeo-placers

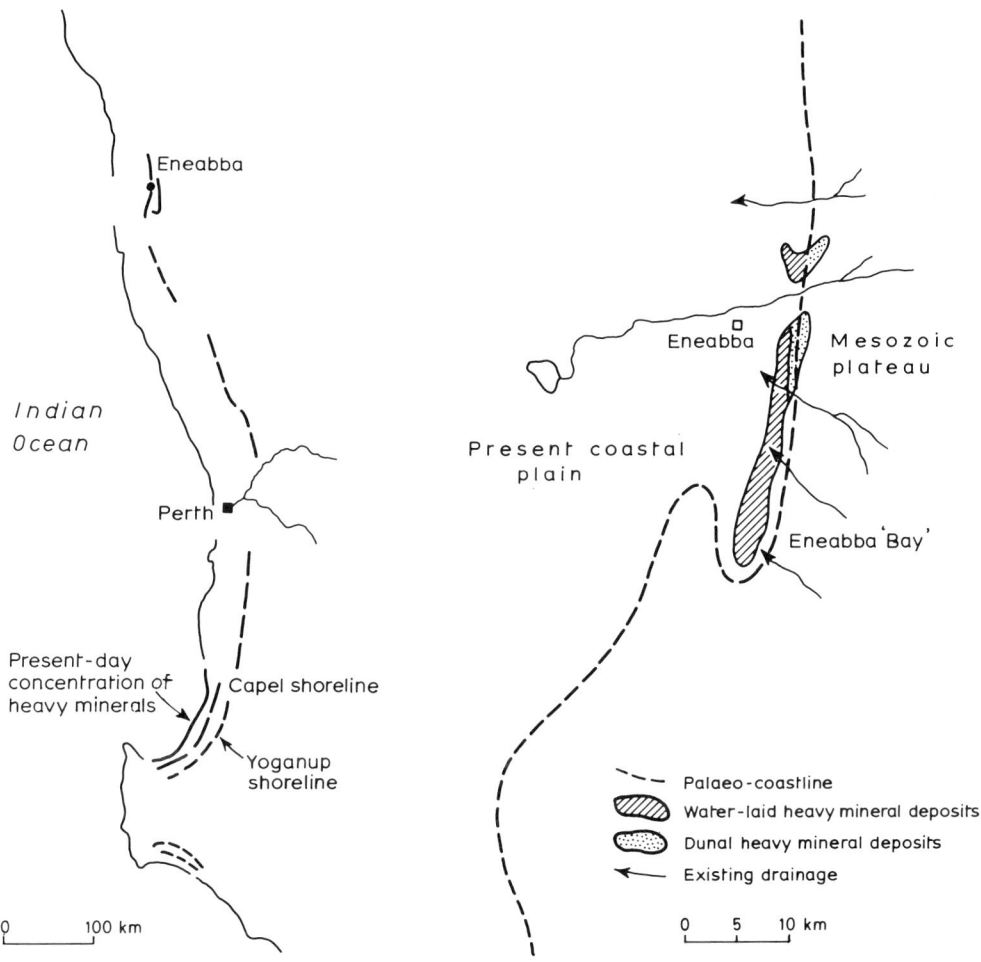

Fig. 5.11 Comparison of the configuration of the palaeo-coastline of Eneabba Bay with the present-day shoreline at Bunbury, south of Perth, where heavy minerals are concentrated (from Baxter 1976 and Lissiman and Oxenford 1975)

cut into Precambrian schists, and has been intensely gullied by wave action (Fig. 5.13). The irregular surface precludes the use of mechanization to remove all the diamondiferous gravel, and teams of sweepers are employed to clear thoroughly the exposed bedrock surface.

Six distinct beaches are recognized in the CDM area, and these are designated A to F (Fig. 5.14). The beaches are thought to be mainly Upper Pleistocene in age.

The beaches may be divided into two groups. An upper, older group represented by beaches D, E and F, and a lower younger group represented by beaches A, B and C. The upper group of beaches terminates against a cliff, cut into the bedrock by the marine transgression responsible for the formation of beach F. All the beaches comprising this group have suffered tilting to the north, and are characterized by the formation of a well-developed calcrete layer. The younger raised beaches were emplaced subsequent to the down-warping which affected the older terraces.

Fig. 5.12 Diamondiferous gravel. At this locality 'A' beach is resting disconformably on '−5 metre' beach. CDM operations, Oranjemund, Namibia.

Fig. 5.13 Diamond mining and processing at CDM, Oranjemund, Namibia. Foreground: intensely gullied wave-cut platform. Background: diamond recovery plant.

Placers and palaeo-placers

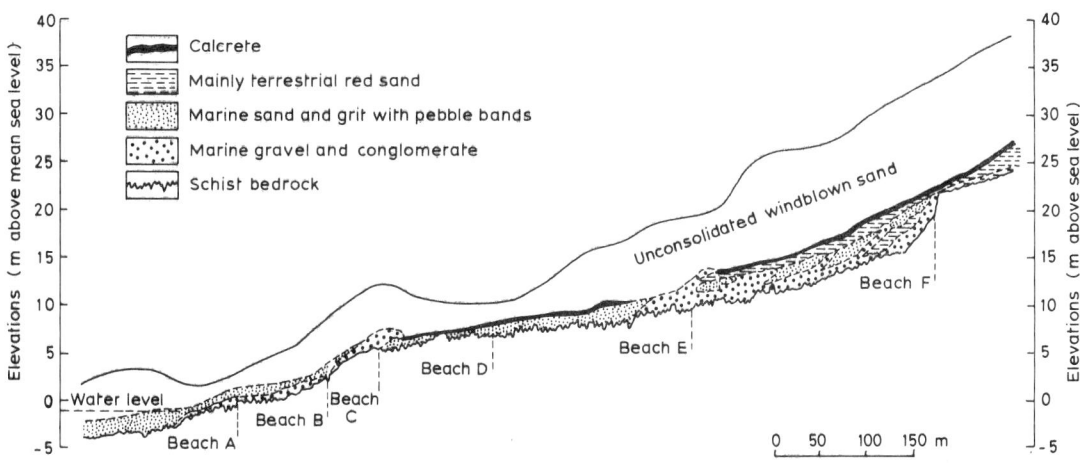

Fig. 5.14 Cross-section through raised beaches at CDM operations, Namibia (modified from CDM Internal Report, 1980).

The beaches are overlain by an overburden of unconsolidated wind-blown sand, which has to be removed before the diamondiferous gravels can be exploited. Overburden removal by large, bucket-wheel excavators is therefore one of the more important aspects of mining the deposits, and in 1979 more than 30 million m^3 of sand was stripped during mining operations.

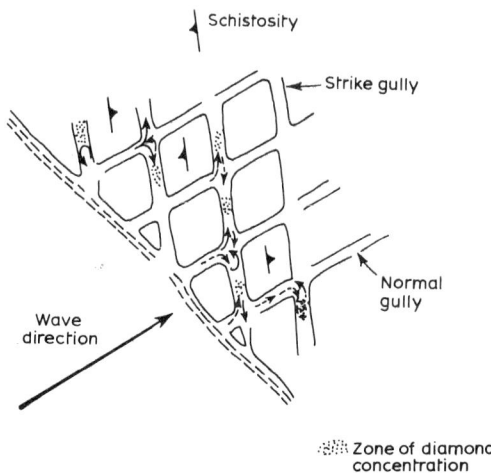

Fig. 5.15 Plan of wave-cut platform showing mechanism of diamond concentration, Namibia.

The diamonds are believed to have been transported from kimberlitic parent rock by the Orange River. Their concentration within the gravels has been mainly the result of wave action. Hallam (1964) pointed out that the pattern of gullies within the wave-cut platform probably facilitated the concentration of the diamonds (Fig. 5.15). The waves break at right-angles to the beach, and the swash flows up the normal gully, the water returning partly by way of the strike gully. The loss of carrying power means that some of the diamonds are left behind in the strike gullies.

Wave action is considered to be the main mechanism of diamond concentration but there is also a generalized northward movement of material, partly caused by northward-flowing currents. As a result the best concentrations of diamonds tend to be on the north side of a bay.

5.4.5 Exploration for beach sand deposits

The most comprehensive accounts of the exploration for and evaluation of beach sand deposits are by Macdonald (1973, 1983) and Baxter (1977) and much of the following account is based on their publications.

Exploration and evaluation is usually

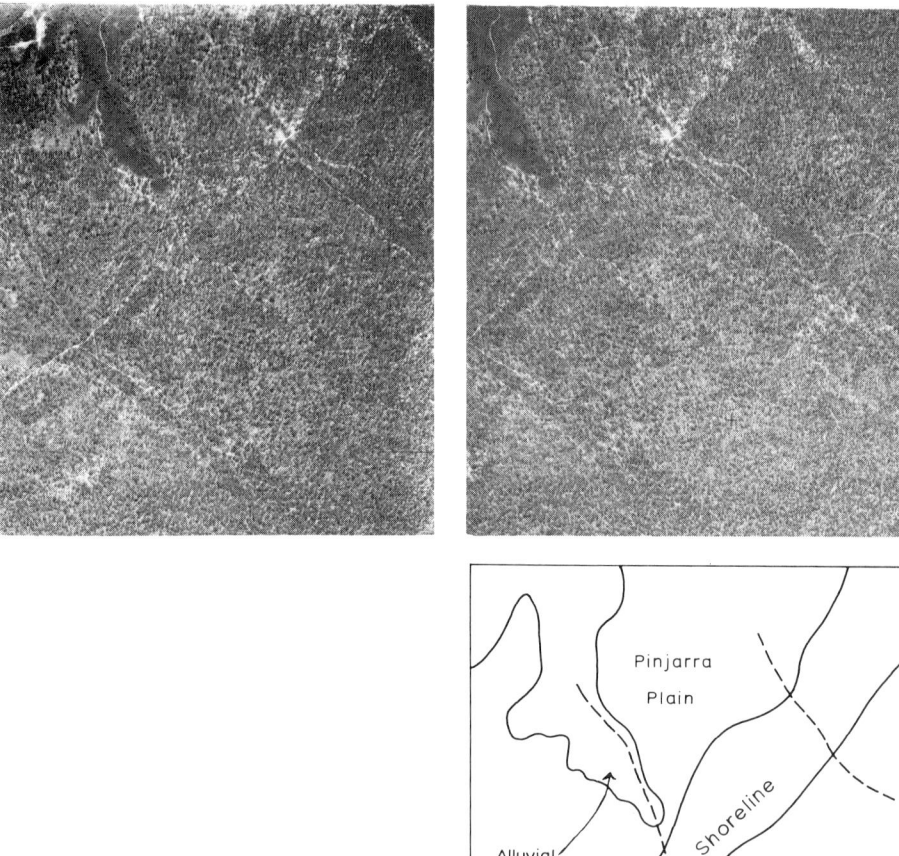

Fig. 5.16 Stereoscopic pair of photographs of the Yoganup Central Mine area prior to mining showing the appearance of the shoreline deposits and associated sediments (from Baxter 1977 with permission of the Director, Geological Survey of Western Australia).

undertaken in three stages: (1) reconnaissance; (2) scout testing; (3) close boring and computation of reserves. At the reconnaissance stage of exploration maximum use should be made of aerial photography. Aerial photographs are of assistance in identifying possible deposits (Fig. 5.16), and may also be used in planning the subsequent drilling campaign, and eventually in designing the infrastructure of the mine.

Baxter (1977) stated that colour aerial photography assisted in the discovery of a new beach sand deposit in the Capel area of Western Australia. Deposits in this area are commonly covered with a thin veneer of yellow sand, which contrasts clearly with adjoining white–grey sand.

Shepherd (1979) is of the opinion that geophysical methods are unsuitable in exploration for beach sand deposits. However, Baxter (1977) stated that airborne geophysical surveys can be effective. In Western Australia, airborne magnetic

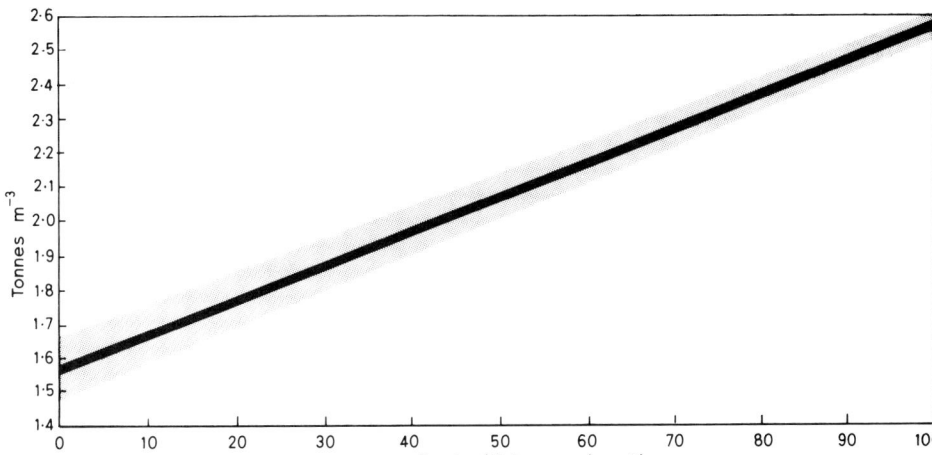

Fig. 5.17 Recommended conversion factors from volume to tonnage for beach sands (from Baxter 1977).

surveys at heights below 60 m have delineated lines of mineral sands, and airborne radiometric surveys have detected beach sands containing monazite and zircon.

Scout boring of beach sands is the next phase of exploration. Macdonald suggests that initially scout holes may be placed 400 to 800 m apart for square grids. Beach sands often have a linear form. If this is apparent from the interpretation of aerial photographs, or from the study of adjacent deposits, it is more appropriate to drill at intervals of 20–100 m along lines which are 200 m apart or more. At Eneabba the drilling was carried out using a 152.4 × 15.2 m grid (500 × 50 ft).

The use of open-hole rotary drilling at Eneabba occasionally caused contamination due to the collapse of the poorly consolidated sand. The content of heavy minerals was determined using a γ-ray scintillometer probe. Orientation studies confirmed that γ-emission from the monazite and zircon is proportional to the heavy mineral content of the sand (J. C. Lissiman, personal communication).

When this phase of exploration has been completed it should be possible to state the following:

(1) the approximate size and grade of the deposit;
(2) the difficulties likely to affect mining and milling;
(3) the most suitable aspect and spacing of holes for the final grid;
(4) the most suitable boring and sampling equipment for complete evaluation.

Close boring is required to increase the accuracy of the reserve estimate and for purposes of mine control. Macdonald (1973) suggested that drilling at this stage may be reduced to lines 40 m apart with holes drilled at 10 m intervals so that narrow pay streaks will not be missed. This approach may not always be appropriate, for example, the beach sand deposits at Richards Bay, S. Africa. In this locality the sands are of aeolian origin, and the heavy minerals have an irregular distribution. In this context drilling is largely directed to defining the base of the deposit in advance of mining operations (see Fig. 5.19).

Shepherd (1979) noted that the conversion of sand volume to tonnage is the weakest link in the calculation of ore reserves. This is because the degree of compaction, slime content, grade of heavy minerals and mixture of grain sizes all have an effect on the bulk density. Figure 5.17 shows the relationship between the grade of heavy minerals and density which is used in Western Australia. The shaded area shows the range that can be expected due to variation in clay content and compaction.

The methods of mining beach sands are surprisingly diverse and are usually related to the

Fig. 5.18 A scraper mining unconsolidated sand at Eneabba, Western Australia.

Fig. 5.19 A concentrator separating valuable heavy minerals from an unconsolidated aeolian sand. Richard's Bay, Natal, South Africa.

nature of the distribution of valuable heavy minerals within the sand. In the case of the Eneabba deposit there is a vertical variation in the content of heavy minerals, but the ore can be blended by mining with scrapers using a 15° cut (Fig. 5.18). In contrast, the heavy minerals in the aeolian beach sands at Richards Bay, Natal, South Africa, have a random distribution. The beach sands are therefore mined by processing the unconsolidated sand through a concentrator (Fig. 5.19) which floats on a man-made pond. The concentrator is moved systematically along the strike of the dunes, and the tailings are landscaped and revegetated.

5.5 MARINE PLACERS

5.5.1 Introduction

The ore minerals which have recently been, or are actively being, mined from this type of deposit include cassiterite, gold, diamonds and titaniferous magnetite. Chromite, fergusonite, ilmenite, monazite, platinum, rutile, wolframite and zircon are known to occur in marine placers and may be exploited in the future.

The principal valuable mineral extracted from marine placers is cassiterite. The main producing zones are off the coast of Thailand and Indonesia. An increasing proportion of tin production from these countries comes from coastal waters, and 50% of Indonesia's tin resources is believed to lie in offshore areas (Batchelor, 1979). Offshore tin deposits have the advantage of lower dredging costs, and higher grades, than land-based deposits. Batchelor (1979) pointed out that, whilst the grade of tin mined from Indonesian land-based deposits remained stable at 0.27 kg/m^3 from 1971 to 1975, during the same period the grade of tin from offshore dredging rose from 0.36 kg/m^3 to 0.73 kg/m^3.

Cassiterite was extracted from marine placers in St. Ives Bay, Cornwall, during 1966 but the operation proved to be unprofitable. Renewed mining activity is planned to commence in 1986, with an anticipated annual production of 750 tonnes/year of tin metal (M. Wolle, personal communication).

Diamonds were obtained by dredging off the coast of Namibia during the 1960s, but at the present time mining operations below low-water mark are confined to those areas which can be protected by man-made sea walls.

Gold has been exploited from marine placers at South Primor'ye in the Laptev Sea, USSR, and around the Seward peninsula, Alaska, particularly in the Nome area (Boyle, 1979). Ilmenite, rutile, zircon and monazite are reported to occur off the east coast of Australia, but are of lower grade than the land-based deposits (Burns, 1979). Magnetite deposits have been mined off the coast of Japan (Kyushu Island) and the Philippines (Luzon Island). At the present time 1.8 million tonnes/year of titanomagnetite sand is mined off the west coast of North Island, New Zealand (*Mining Magazine*, May 1982).

5.5.2 Genesis of marine placers

Marine placers occur on the continental shelf, and are generally within 5 km of the coast. They have mainly formed by the submergence of heavy mineral accumulations, which were originally concentrated in a continental or littoral environment. In some cases offshore current activity has played a significant role in concentrating or redistributing the heavy minerals. For example, Schofield (1976) considered that the 'drowned beaches' off the coast of Otago, New Zealand, are in the process of formation under present-day hydraulic conditions.

Yim (1978) described an investigation of cassiterite distribution in an area selected for mining off the north coast of Cornwall, UK. His results indicated a complex interplay of alluvial and marine influences. The cassiterite is partly derived from the erosion of hydrothermal vein deposits. However, the main concentration of tin can be related to streams discharging from areas of previous mining activity. Part of the zone of investigation is covered by a well-sorted sand containing coarse cassiterite, which Yim (1978)

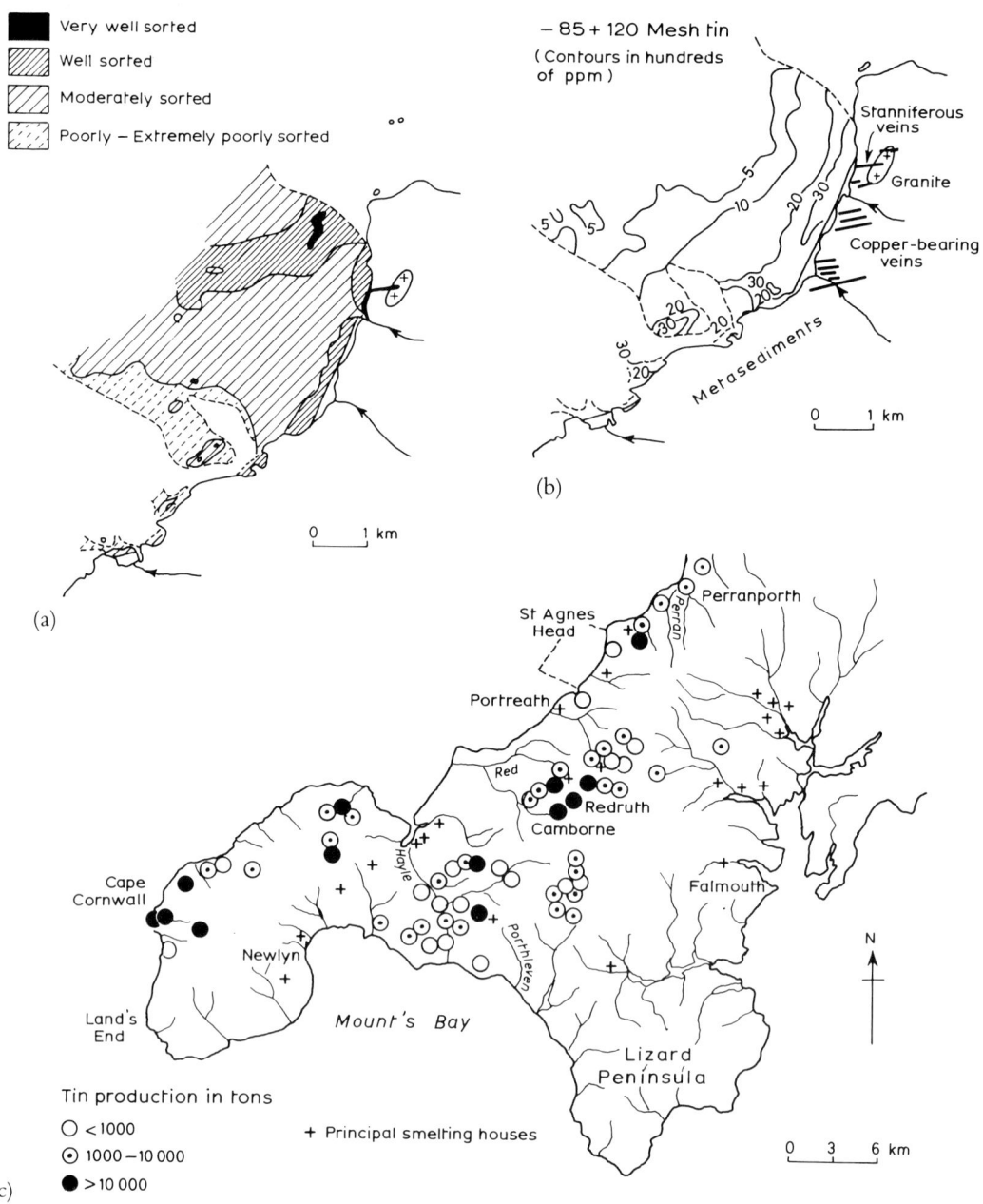

Fig. 5.20 Sediment sorting distribution (a) and tin content of sediment (b) off the coast of St Agnes (c), Cornwall, UK (from Yim 1978).

attributed to reworking by strong bottom currents (Fig. 5.20).

5.5.3 The marine placers of Indonesia

The Indonesian marine placers surround the islands which occur between Sumatra and Borneo, and rest on a submerged platform which is referred to as Sundaland (Fig. 5.21). Originally the geology of the Sunda Shelf was inferred from the islands which emerge above the southern part of the South China Sea. However, from a combination of seismic data, offshore drilling and the investigation of adjacent coastal mines, Batchelor (1979) has reinterpreted the stratigraphy of the Late Cainozoic sediments. He considers that from the Late Miocene to recent times there have been three major phases of sedimentation, characterized by distinctive climatic regimes, and accompanied by a progressively rising sea level (Fig. 5.22).

In Late Miocene–Early Pliocene times semi-arid conditions prevailed on the Sundaland continent when sea levels were as much as 230 m below their present level. Deep lateritic weathering of greisen-bordered vein swarms led to the development of eluvial deposits with colluvial placers forming by the downslope creep of eluvium. In Late Pliocene times intermittent intense rainfall caused debris flows of sodden stanniferous regolith to form piedmont fan placers.

Towards the end of Early Pleistocene times, and subsequently in Middle Pleistocene times, periods of increased precipitation resulted in cassiterite being released by the headward erosion of stanniferous bedrock and the reworking of the piedmont placers.

Two major marine transgressions at 120 000 and less than 10 000 years BP have modified the placers; in some cases elutriation by waves and tides has increased the tenor of the ore. Placer deposits have in some places been destroyed by the incision of meandering rivers during an emergent phase between the two marine transgressions. Sediment subsequently deposited in these valleys

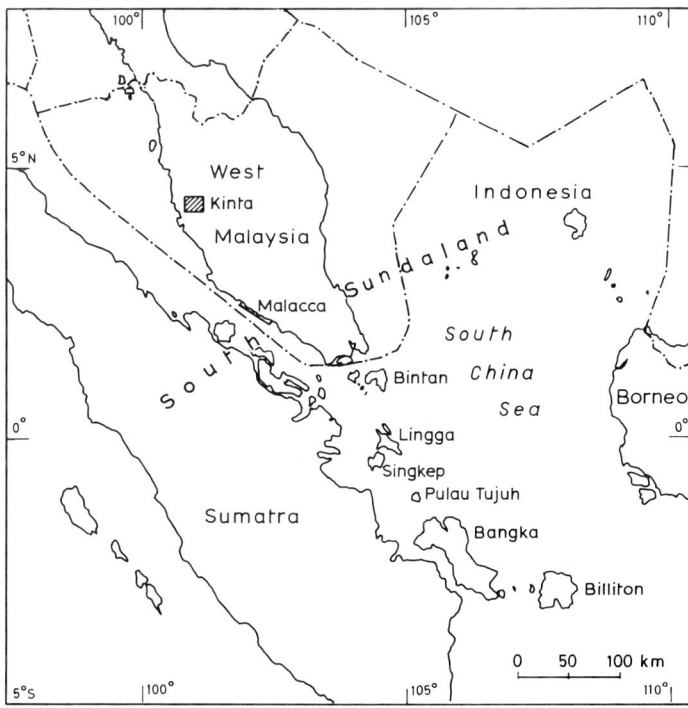

Fig. 5.21 Location map of south Sundaland, Southeast Asia (modified from Batchelor, 1979).

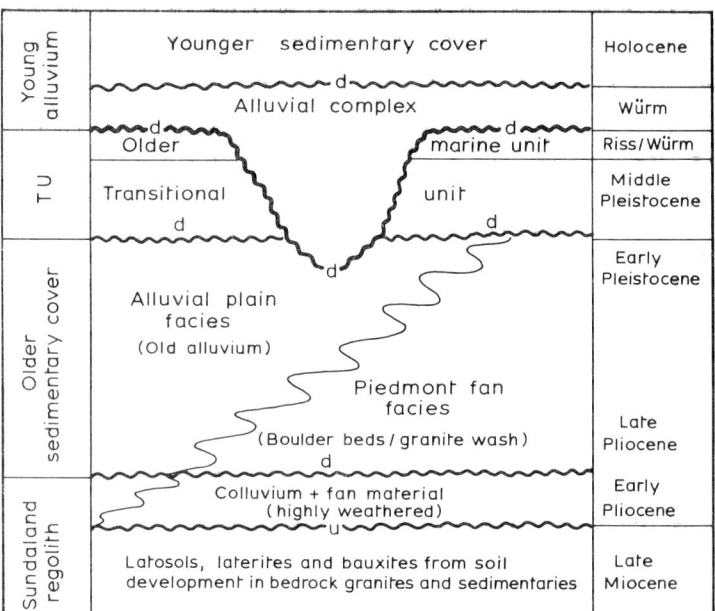

Fig. 5.22 Regional Late Cainozoic stratigraphic relationships in Sundaland, Southeast Asia. d, disconformity; u, unconformity. (From Batchelor 1979.)

has generally proved to be barren with respect to cassiterite.

Batchelor's interpretation of the Cainozoic history of the Sunda Shelf, and his recognition of the principal environments of placer formation, make an important contribution to mineral exploration for offshore deposits in this area. Batchelor pointed out that exploration should be restricted to the flanks of positive granite features, contact zones, and adjacent shallow bedrock troughs. He believes that the steeply incised valleys filled with Würm and Holocene alluvium are of little mineral potential.

5.5.4 Exploration for marine placers

It is vital that exploration for marine placers should be based on a thorough understanding of the stratigraphy of the enclosing sediments and the processes which control the transport, deposition and enrichment of valuable minerals. Area selection based upon careful geological studies is the most important phase of exploration for marine placers.

Detailed investigations of selected areas are normally based on a combination of geophysical surveys and sediment sampling. Moore (1978) mentioned that magnetic data may provide initial clues for locating prospects. However, seismic profiling is the most widely used technique in exploration for marine placers. Bon (1979) described the use of a 'Sonia MK 1' acoustic reflection profiler in an exploration programme for offshore tin deposits in the Pulau Tujuh area, Indonesia (Fig. 5.21). The echo traces produced by the profiler must initially be correlated with information from boreholes, but once this has been carried out the seismic reflections can be utilized to map areas between drill sites. In the Pulau Tujuh area it was possible to distinguish granite with a weathering mantle, sedimentary basement rocks, gravelly sediments and clayey sediments using the technique of seismic profiling.

Moore (1978) noted that since 1972 it has become common practice to conduct grab sampling at the same time as the geophysical survey in order to aid interpretation of seismic data. However, heavy pipe-dredge sampling and coring must be carried out after the completion of the geophysical survey.

Description of the drilling techniques used in offshore exploration for placers is provided by van Overeem (1970) and Bon (1979).

Moore (1978) advised that in view of the high cost of coring this technique should be used only when all other survey information has proved positive and all prospects in the area have been explored using alternative methods.

The sampling grid to be utilized during exploration for marine placers has been discussed by Moore (1978). He suggested that the higher the S.G. of the metal, the closer the sampling stations must be. His recommendations for gold placers are a maximum spacing of 150 m extending to 600 m centres for light heavy minerals, such as rutile and zircon.

5.6 PALAEO-PLACER DEPOSITS

5.6.1 General characteristics of palaeo-placer deposits

(a) Distribution in space and time

Conglomerates have been a common component of sedimentary assemblages throughout geological time but they are only rarely found to contain economically interesting gold and uranium concentrations. The principal deposits are located in the Witwatersrand Basin, South Africa (sometimes referred to as the Rand) and in the Blind River–Elliot Lake district, Ontario, Canada (Fig. 5.1). Gold is mined from similar deposits near the town of Tarkwa in south west Ghana, although it should be noted that the principal gold producer in Ghana is the Ashanti Mine where gold-bearing quartz lodes occur in schists of Birrimian age (1915–2210 million years). The gold- and uranium-bearing deposits of the Jacobina district, Bahia, Brazil, have been of importance in the past, but the bulk of Brazilian gold production is from placer deposits at present. However, a new hardrock gold mine is currently being developed in the Jacobina district (Govett and Harrowell, 1982) which will re-establish the importance of the quartz-pebble conglomerate as a source of Brazilian gold.

The Canadian and Brazilian deposits occur in Lower Proterozoic sediments, and the Ghanaian auriferous conglomerates are of Middle Proterozoic age (1900 million years). The gold- and uranium-bearing conglomerates of South Africa are Archaean in age (2500–2600 million years) but have a tectonic setting which is more typical of Proterozoic basins in other continents. Pretorius (1981) points out that the boundary between Archaean and Proterozoic sediments is generally assumed to be about 2500 million years, whereas in South Africa there is a clear change from an Archaean to a Proterozoic style of crustal development at about 3100 million years. Pretorius suggests that the greater time during which the sedimentary basins have developed in South Africa may be of significance in determining their abnormally high gold and uranium contents.

(b) Size and grade of deposits

Gold- and uranium-bearing palaeo-placers are of limited distribution, but are usually of very large size. The Witwatersrand Basin in South Africa is the largest repository of gold in the World. Since its discovery in 1886 more than 35 000 tonnes of gold have been produced, with an average grade of 10 g/t (Pretorius ref. in Viewing, 1984). According to Glyn (ref. in Bache, 1981) the combination of tonnage exploited and existing reserves in the Witwatersrand Basin is 50 000 tonnes of gold. The reserves of ore in the Blind River–Elliot Lake district are approximately 300×10^6 tonnes at a grade of 0.1% U_3O_8 and 0.05% ThO_2 (Boyle, 1979). The grade of ore in the Jacobina deposit, Brazil, varies from 7.1 to 12.7 g/t (Boyle, 1979).

(c) Mineralogy

The valuable ore minerals in quartz-pebble conglomerates are few in number, highly variable in their concentration, and often have complex relationships with associated minerals. Native gold is the principal ore mineral in the South African, Ghanaian and Brazilian deposits.

Uranium is the main economic product of the Blind River–Elliot Lake district where uraninite and brannerite are the principal ore minerals. Uranium is also a significant by-product from some South African mines and this is discussed in more detail in Section 5.6.2.

The mineralogy of all the auriferous and uraniferous palaeo-placers is dominated by quartz, although they contain a diverse assemblage of accessory resistate and sulphide minerals. Pyrite is the dominant sulphide, and in the Witwatersrand and Blind River region has a high correlation with economic ore minerals, although this relationship was not found to exist in the Jacobina deposits (Pretorius, 1981). Pyrite in the Blind River district is considered to be detrital, but a much more complex classification is required for the Witwatersrand pyrite which has been studied in detail by several workers. Ramdohr (1958) has suggested that part of the pyrite is detrital and part formed by the sulphidization of magnetite, banded iron-formation clasts and limonite. Pyrite is absent from the Ghanaian auriferous conglomerates where haematite, ilmenite and magnetite constitute a more typical placer mineral assemblage.

(d) Host rock lithology

One of the most striking features of this class of ore deposit is the close similarity in appearance between the conglomerates from the principal mining districts. In all cases the dominant lithology which hosts the mineralization is an oligomict conglomerate (Figs. 5.23 and 5.24), in which vein quartz is the main pebble type, usually constituting more than 90% of the clasts. Schist, chert and quartzite are usually reported as pebbles of minor importance. In addition the matrices of the conglomerates are usually very similar with quartz being the major component.

The position of the ore-bearing conglomerates within the sedimentary sequence has similarities in the Brazilian, Canadian and Ghanaian deposits. In all three cases there are three mineralized conglomerate horizons which are found near the base of the sequence. The Blind River deposits are developed in conglomerates of rather variable thickness, which fill palaeo-valleys cut into the Archaean basement (Theis, 1979). Individual lenses have widths of 120 m and thicknesses of 5 m. In the Tarkwa area of Ghana there are three mineralized conglomerate zones in the lower part

Fig. 5.23 Auriferous conglomerate, Ventersdorp Contact Reef, Western Deep Levels, South Africa.

Fig. 5.24 Auriferous conglomerate, Kimberley Reef, Winkelhaak Mine, South Africa.

of the Tarkwaian Series. Most of the gold is in the basal conglomerates, which are discontinuous lenses 600–1000 m long in the direction of current flow and 100–150 m wide (Sestini, 1973). In the Witwatersrand Basin mineralized sediments occur throughout the succession from base to top, although the main proportion of ore-bearing horizons occurs in the upper part of the Witwatersrand Supergroup. Conglomerates are of dominant importance, but banded pyritic quartzites and orthoquartzites are also exploited for gold and uranium (Pretorius, 1981).

In all the deposits mentioned there appears to be a close correlation between the grade of ore, sedimentary textures and structures. This has obvious economic implications which will be discussed more fully in the context of the Witwatersrand Basin.

(e) Tectonic setting

Geologists are uncertain about the tectonic setting of palaeo-placer deposits. Mitchell and Garson (1981) suggest that the sedimentary basins which contain palaeo-placers may have formed in a tectonic regime which existed only in Proterozoic times. The characteristic features of these sedimentary basins are that they contain thick successions of dominantly shallow-water, terriginous sediment and rest unconformably on a basement of metamorphosed continental rocks. From this it is possible to conclude that they have an intracontinental setting. It is probable that the sedimentary basins have formed as a result of rifting, but at present there is little evidence to suggest that they have formed in an aulacogen.

5.6.2 The gold and uranium deposits of the Witwatersrand Basin

The gold and uranium deposits of the Witwatersrand Basin have been selected for more detailed discussion, because in terms of value it represents the most important mining district in the World. The mineral deposits have been the subject of debate for many years between 'hydrothermal' and 'placer' schools of thought, and whilst the conflict over genesis has now largely been resolved there is still a lively discussion concerning whether the uranium and gold are entirely detrital or partly the products of mobilization during diagenesis. The evolution of ideas concerning the genesis of the Witwatersrand

Ore deposit geology and its influence on mineral exploration

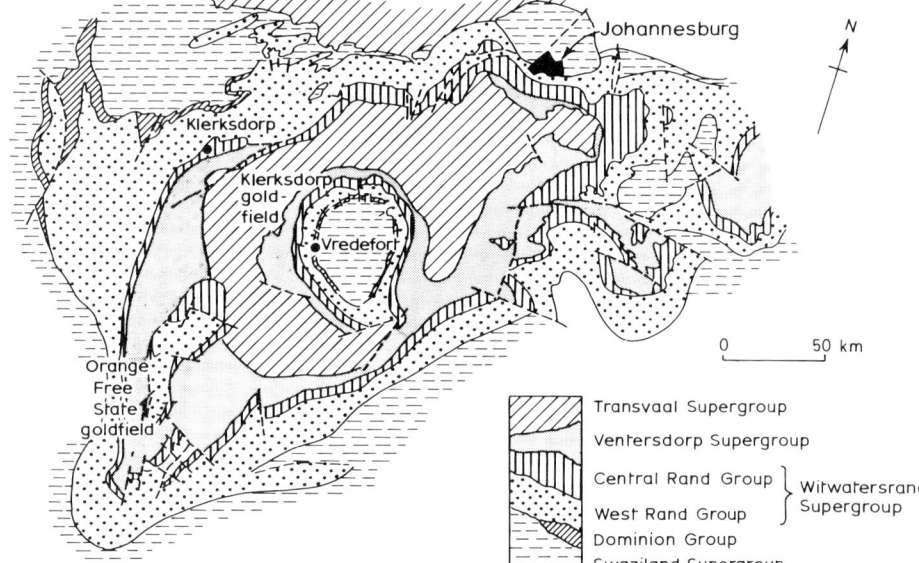

Fig. 5.25 Map of the distribution of the Witwatersrand Supergroup beneath the cover (modified from Park and MacDiarmid, 1975, after Borchers, 1961).

deposits has been described by Pretorius (1975), and his account will be of interest to those who wish to understand the historical development of concepts in ore genesis.

The Witwatersrand Basin is located in the Transvaal and Orange Free State provinces of South Africa and contains six major gold fields. The basin has a north-easterly trend with its axes being approximately 320 km and 160 km in length (Fig. 5.25).

Table 5.5 shows a generalized lithostratigraphic column of the Witwatersrand Basin. Only those

Table 5.5 Generalized lithostratigraphic column of the Witwatersrand Basin (from Tankard et al., 1982)

Supergroup	Group	Subgroup
Transvaal	Pretoria Chuniespoort Wolkberg	
Ventersdorp	Priel Sequence Platberg Klipriviersberg	
Witwatersrand	Central Rand West Rand Dominion	Turffontein Johannesburg Jeppestown Government Hospital Hill
Swaziland	Moodies Fig tree Onverwacht (Barberton area)	

points of relevance to the economic mineralization will be mentioned here, and those seeking a fuller description of the stratigraphy are referred to Tankard et al. (1982).

The Dominion Group and Witwatersrand and Ventersdorp Supergroups rest with profound unconformity on rocks of the Swaziland Supergroup, which are considered to be the primary source of gold. The Dominion Group comprises a basal layer of coarse, clastic sediments overlain by lavas of andesitic and felsic composition. Two conglomerate bands at the base of the succession carry economic gold concentrations, and have been mined in the Klerksdorp area. Extensive drilling has shown that the main potential of these conglomerates is for uranium. The original mines which worked the Dominion Group closed for some years but are now reopened and new extensions are being developed (D. Pretorius, personal communication).

The Witwatersrand Supergroup rests unconformably upon the Dominion Group and is subdivided into lower, West Rand and upper, Central Rand Groups. The West Rand Group comprises mainly barren shales and quartzites, although a few auriferous conglomerates occur.

The Central Rand Group is of major economic significance, and comprises mainly quartzites and conglomerates with minor shales. Although most of the conglomerates contain some gold and uranium, only about twelve are rich enough to repay the cost of mining.

The Ventersdorp Supergroup consists mainly of a thick sequence of basaltic and andesitic lavas, but near the base of the succession lenticular conglomerates occur, which contain the erosion products of the Central Rand Group. The principal mineralized horizon is the Ventersdorp Contact Reef (Fig. 5.23) which contains concentrations of both gold and uranium. Similarly the mineralized Black Reef occurs at the base of the Transvaal Supergroup which is dominated by a succession of clastic sediments.

It is worth reflecting on the time interval during which the sediments of Archaean and Lower Proterozoic age formed within the Witwatersrand Supergroup. Age determinations indicate that 500 million years elapsed during the deposition of 14 000 m of clastic sediments and lavas. The very slow rate of deposition and subsidence explains the mineralogical maturity of the sediments and must be borne in mind in discussing the genesis of the mineralization.

(a) Mineralogy

The mineralogy of the auriferous conglomerates of the Witwatersrand Basin has been reviewed by Feather and Koen (1975). In this brief account the emphasis will be placed on those minerals which are of economic value. Historically the average grade of gold mined in the Witwatersrand Basin has been about 10 g/t, although in recent years there has been a dramatic decline in the average grade from 12.7 g/t in 1972 to 6.9 g/t in 1981 (*Mining Annual Review*, 1982). Commonly lower-grade ores are mined during periods of high metal price. For example, some of the mines in the Evander area were mining ore with an average grade of 4 g/t in 1980 when the average price of gold was US$612/oz. However, the main reason for the decline in grade is because the ores which are being mined are of poorer quality. In many mines the grade is highly variable, a point which is well demonstrated by the Vaal Reef in the Klerksdorp area, where values of 1500 g/t may be attained near the base of the conglomerate although the average grade is 15 g/t (Minter, 1976). Very little of the gold is considered to be primary in origin; the bulk of the gold has recrystallized under the influence of diagenesis or metamorphism, and has numerous associations with the fine siliceous minerals, with sulphides and with carbonaceous material.

Uranium is being mined as a by-product metal from the following reefs: Carbon Leader, Main Reef, Vaal Reef, Basal Reef, Elsburg Reef, Birds Reef and Steyn Reef, from which the average grade is reported to be 500–700 g/t. Uranium also occurs in the Ventersdorp Contact Reef, but the calcitic nature of the tuffaceous matrix would make extraction of the ore by conventional acid leaching prohibitively expensive, a point which

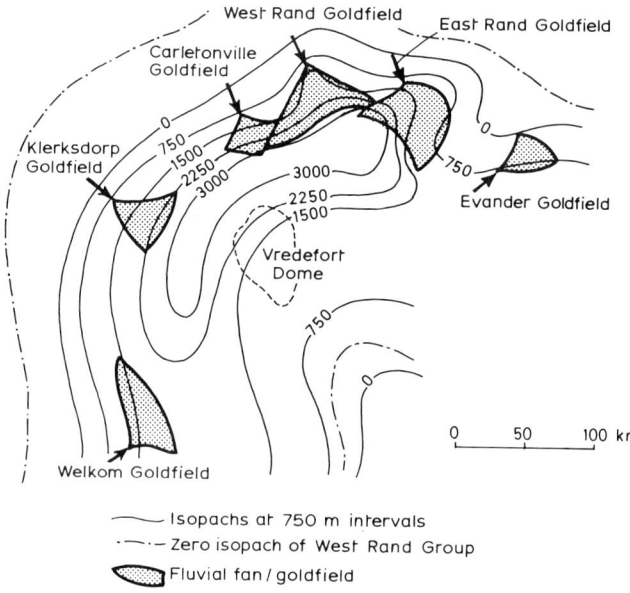

Fig. 5.26 Geometry of the Witwatersrand Basin as revealed by the depositional isopachs of the Central Rand Group. The asymmetry of the basin is shown by the distances between the zero isopachs and the depositional axis. Six fluvial fans, hosting the major gold fields, are all located on the short, shrinking side of the depository (from Pretorius 1981).

demonstrates that grade is not the sole criterion in evaluating the economic value of a conglomerate. The uranium content of the Witwatersrand ore is accounted for by uraninite (30%), uraniferous carbon (20%) and finely disseminated material (50%) which Thiel and co-workers (1979) have demonstrated is mainly associated with leucoxene and carbon.

Pyrite is the dominant sulphide in the auriferous conglomerates, and constitutes from 2 to 20% of the matrix. There is frequently a strong positive correlation between the gold content of the reef and the amount of pyrite present, and the latter mineral has been described in considerable detail (Saager, 1970). The results of studies by different workers (Ramdohr, 1958; Clemmey, 1981) have led to the conclusion that the pyrite is in part detrital and in part formed by the sulphidization of magnetite, banded iron-formation clasts, limonite and ferric clay.

Carbon is a major constituent of some ore-bearing conglomerates and is particularly abundant in the Carbon Leader, Vaal and Basal Reefs. In most cases it is found at or near the footwall of reefs, frequently forming a continuous mat over a few metres. Gold and uranium are commonly associated with the carbon. Hallbauer (1975) has provided convincing evidence, using the scanning electron microscope, that the carbon is biogenic in origin.

(b) Structure

The Witwatersrand Basin is an arcuate structure, bounded on the north-west side by faults, and defined by a major structural flexure on its south-eastern side. Palaeocurrent directions suggest that sediment was introduced into the basin by five major rivers. Isopachyte maps of the West and Central Rand Groups indicate that the areal extent of the basin decreased with the passage of time. The isopachyte maps also suggest that the basin was asymmetrical with respect to the depositional axis (Fig. 5.26).

From the rim of the basin the economic reefs dip towards the centre. The average dip on the West and Central Rand is 30–40°, and whilst dips elsewhere tend to be lower, there are cases when the dip is locally increased by faulting. Near the centre of the basin the strata are updomed, punctured and locally overturned by the Vredefort Dome.

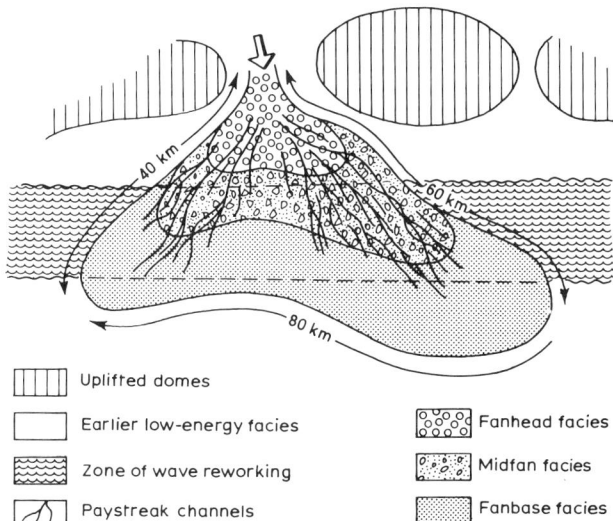

Fig. 5.27 Conceptual model of a typical prograding fluvial fan in the Witwatersrand Basin, showing the location of the fanhead, midfan and fanbase facies and the zone of reworking by wave action in transgressive depository waters (from Pretorius 1981).

The Witwatersrand Basin is a synclinorium and the geometry of the basin reflects a pattern of superimposed interference folding. According to Pretorius (1981) structural domes and depressions occur where major fold structures intersect. He also suggests that as the fold movements were episodically active during sedimentation, their distribution played a considerable role in localizing gold fields and in controlling facies variation.

Extensive major faulting has taken place tangential to the depositional axis of the basin. For example, the Buffelsdoorn fault in the Klerksdorp area is a normal strike fault, which displaces strata over 1300 m. The economic reefs have also been displaced by numerous relatively minor strike and transverse faults.

5.6.3 The genesis of the gold and uranium deposits in the Witwatersrand Basin

The genesis of gold- and uranium-bearing conglomerates has been the subject of a long and continuing debate amongst geologists. The Witwatersrand deposits have been the main focus of attention and many detailed studies have been carried out, particularly in the field of mineralogy. Far fewer investigations have been carried out on the Canadian, Ghanaian and Brazilian deposits, but as in South Africa the discussion has frequently been polarized between a 'placer' and 'hydrothermal' viewpoint. In this account the main emphasis is placed on the Witwatersrand Basin, but reference will be made to other localities where they amplify a particular point, or provide conflicting evidence.

South African geologists have consistently argued in favour of a placer, or modified placer, mode of origin for the genesis of economic ore minerals in the Witwatersrand Basin. The evidence to support the placer theory will be reviewed and those areas which appear to be in conflict with the theory will be highlighted.

The evidence for a placer origin is based mainly upon the consistent association of gold and uranium with particular sedimentary forms and facies within the Witwatersrand Basin. On a regional scale, the major gold fields are associated with fluvial fans which were extensively reworked during successive transgressions and regressions of a shallow inland sea (Fig. 5.27). On the scale of an individual mine, the distribution of gold and uranium can often be related to a distinctive sedimentary environment, within which the concentration of valuable ore minerals can sometimes be related to sedimentary structures and textures. For example, detailed studies by Minter (1976) in

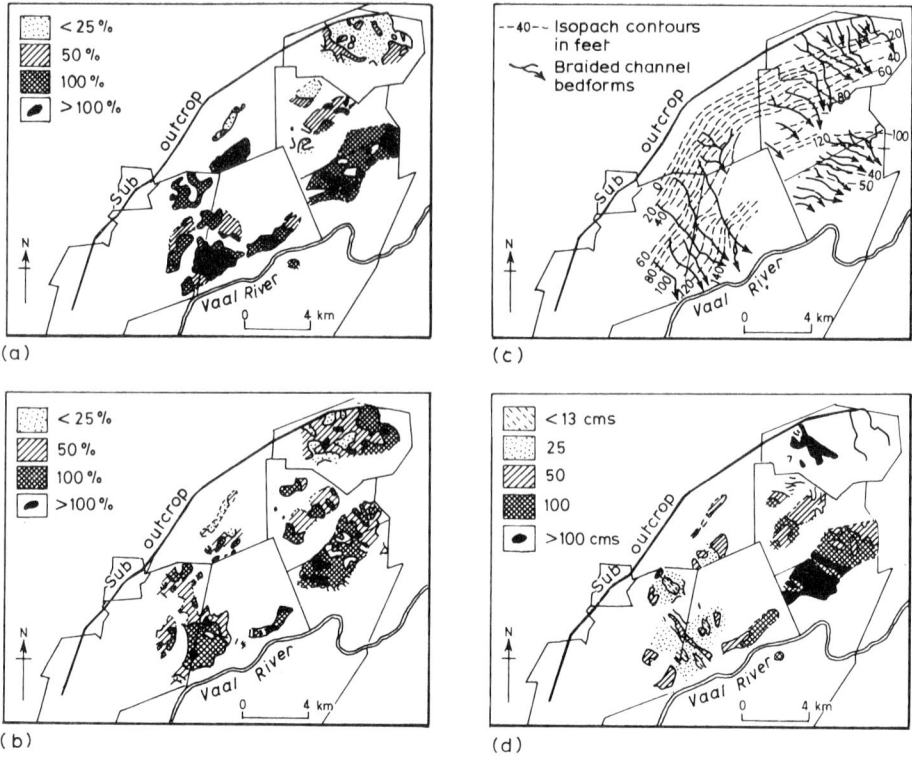

Fig. 5.28 Main characteristics of the Vaal Reef, Klerksdorp area (from Minter 1976) (a) Moving-average isocons of the uranium content expressed relative to the highest contour. (b) Moving-average isocons of the gold content expressed relative to the highest contour (c) Isopachs of truncated footwall MB5 formation with braided channels of Vaal Reef superimposed on the south-east paleoslope. (d) Moving-average isopachs of the Vaal Reef thickness.

the Klerksdorp area have shown that the sediment which encloses the gold and uranium in the Vaal Reef was deposited in a series of dendroidal shallow channels, incised into the truncated surface of underlying gravels. High uranium and gold concentrations are confined to the complex branching channelways (Fig. 5.28).

Smith and Minter (1979) have also carried out a careful examination of the Leader Reef, Welkom, and the number 5 reef, Klerksdorp, and concluded that the placer minerals are concentrated on well-worked scour surfaces and in pebble-supported conglomerates. Similar sedimentological studies have been carried out by Sestini (1973) in the Tarkwaian of Ghana where the highest gold values are found in well-sorted, well-packed, haematite-rich conglomerates.

Further support for a placer theory of origin is provided by the hydraulic equivalence of gold and uraninite with detrital minerals such as garnet, chromite and zircon. Also, Hallbauer (1977) has compared the morphology of gold from Witwatersrand conglomerates with gold particles from recent alluvial placers, and the evidence from scanning electron photo-micrographs suggests a similar mode of transport.

Most advocates of the placer theory accept that the form of the gold, and its distribution within the conglomerates, is not always consistent with deposition in a fluvial environment. They accept that some remobilization of both gold and uranium may occur during diagenesis and regional metamorphism and a 'modified placer' mode of origin seems to be widely accepted.

However, there are a number of mineralogical observations which suggest that only part of the gold and uranium is detrital in origin. The mode of origin of uranium has been a consistent problem in the elucidation of the Witwatersrand deposits. Early advocates of a hydrothermal origin for the uraninite suggested that it would be unstable during transport and deposition as detrital grains, but this suggestion was countered by the argument that the Early Proterozoic atmosphere was oxygen-deficient, thus allowing the possibility of detrital uraninite. Recently, a number of workers (Dimroth and Kimberley, 1976; Simpson and Bowles, 1977) have questioned the evidence for a non-oxygenic atmosphere, and have suggested that, whilst part of the uraninite may be detrital, the substantial component of uranium, which is associated with titanium-bearing minerals and carbon, may have been leached from detrital uraninite by intrastratal solution, and redeposited in the distal facies of the fluvial fan.

The source of the gold and uranium in the Witwatersrand Basin is generally attributed to rocks belonging to the Swaziland Supergroup. Most writers have considered gold-bearing quartz veins or ultramafic volcanics as appropriate source rocks, although iron-formation has also recently been suggested as an alternative.

5.6.4 Exploration for gold in the Witwatersrand Basin

Whilst there is a voluminous bibliography concerning the genesis of uraniferous and auriferous conglomerates, the industrial geologist is usually preoccupied with more immediate problems of exploration, reserve evaluation and grade control. In this section emphasis is placed on the role of the geologist on the Rand, due to the scale of the mining operations and the complexity of the mineralization.

The Witwatersrand deposits provide a difficult challenge to the exploration geologist, because the main proportion of economic ore occurs at great depth (the shallowest mine in the Orange Free State starts at 230 m) and because the ore-bearing conglomerates do not have a distinctive geochemical or geophysical signature, which can be utilized at surface. However, a successful programme of exploration has been carried out since the 1930s using airborne and ground magnetics and gravity methods (Roux, 1967). The magnetic method utilizes the presence of three shale bands in the West Rand Group, which are sufficiently magnetic to be detected at depths in excess of 1000 m. Geologists are able to interpret the magnetic data in terms of the presence and structure of the West Rand Group, and therefore the position of the ore-bearing conglomerates in the Central Rand Group can be inferred with sufficient accuracy to justify a programme of diamond drilling.

The Evander gold field was identified by a preliminary airborne survey in 1949, and a subsequent diamond drilling programme led to the intersection of the Kimberley Reef at a depth of 1041 m (Tweedie, 1978).

The gravity method utilizes density differences between the Central Rand Group (mean 2.63–2.66 g/cm^3) and the West Rand Group (mean 2.75–2.83 g/cm^3). The usefulness of the gravity method is twofold: firstly, to establish the existence of a basin beneath a cover of younger rocks and secondly to detail the structure within the basin. In particular, changes of gradient and gravity lows can indicate the presence of lighter Central Rand Group quartzites between the heavier West Rand Group and Ventersdorp rocks. Drilling a gravity low on the farm of St. Helena in the Orange Free State in 1938 led to the discovery of the Basal Reef, and the subsequent development of the Welkom gold field.

Magnetic and gravity methods are able to define the broad structure of the Witwatersrand Basin, but the evaluation of potential gold and uranium mineralization is dependent upon diamond drilling, often to depths of several thousand metres. The problem of correlation between rather uniform sequences of clastic sediments is partly resolved using down-the-hole geophysics, and in particular γ-ray logging. Drilling at these depths requires sophisticated directional control of

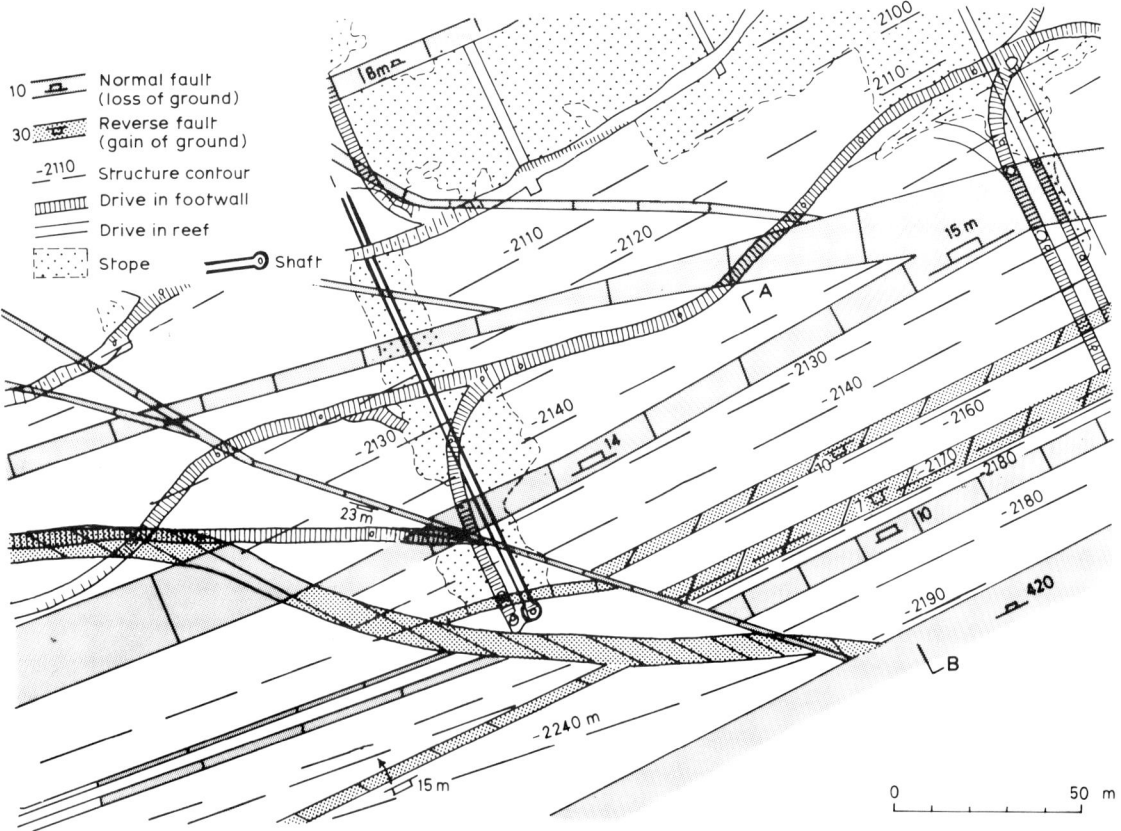

Fig. 5.29 Part of a plan of a South African gold mine.

the drill bit to ensure that any deviation can be either compensated for or at least recognized.

The Witwatersrand conglomerates are characterized by irregular concentrations of ore minerals and structural complexity. The role of the mine geologist is therefore partly to monitor fluctuations within the concentrations of valuable minerals, and to predict the trend of gold and uranium values beyond the working face of the mine. In recent years an attempt has been made on some mines to utilize sedimentological studies, in an attempt to assist the interpretation of these trends to aid mine planning. Studies of this nature have recently been described from the Evander area by Tweedie (1978), from the Welkom and Klerksdorp gold fields by Smith and Minter (1979), and from the Hartebeestfontein gold mine by Magri (1978). The conclusions drawn by these workers are rather tentative, but they suggest that although more work is required, there are discernible linear trends which can be of assistance in mine planning.

Many of the mines in the Witwatersrand have a complex structure, with the ore-bearing conglomerate being displaced by both normal and reverse faulting. Normal faults result in 'loss of ground', whilst reverse faults result in 'gain of ground', both of which have important economic implications and require the advice of a geologist to ensure that mining can proceed with the minimum of development in barren ground.

Part of the plan of a South African gold mine is

shown in Fig. 5.29. Unlike conventional (surface) geological maps, on mine plans faults are shown as two lines to indicate the amount of 'loss' or 'gain' of ground mentioned above. The loss or gain of ground is measured with reference to the reef being worked. On the mine plan, Fig. 5.29, the stoped areas of the reef are shown stippled; the elevation of the reef within the mine area shown is marked by structure contours and these are negative indicating the depth, in metres, of the reef below mine datum.

The gain of ground in a reverse fault is due to the continuation of the reef in the footwall of the fault below the *same reef* in the hangingwall of the reverse fault. This gain is sometimes referred to as 'fault overlap', while the loss of ground encountered in a normal fault is referred to as the 'fault gap'. Using the structure contours a short section has been constructed for part of the mine plan (Fig. 5.30). This shows the amount of displacement of the reef by faulting and illustrates the concept of 'gain' and 'loss' of ground. The relative positions of the 'loss' and 'gain' of ground have been shown by −ve and +ve signs respectively on this section.

5.7 CONCLUDING STATEMENT

Placers and palaeo-placers have in common the role of mechanical concentration as the principal mechanism of ore formation. In nearly every other respect they represent distinct styles of mineralization, and therefore we will continue to discuss them separately in this section.

Placer deposits are a major source of World production of tin, diamonds, niobium, zircon and titanium. They also represent an important source of gold and platinum. The economic importance of their minerals, and the relative ease with which they can be exploited, makes the placer deposit an attractive exploration target. Which techniques have proved most effective in the discovery of new deposits?

At the preliminary stage of area selection it is important to be aware that heavy resistate minerals are concentrated in specific geological environments. For example, wolframite and scheelite occur most commonly in the eluvial and colluvial subenvironments, and only rarely in the fluvial subenvironment or beach environment. At the reconnaissance stage of exploration geological studies are very important. In part this involves the identification of source areas, but more particularly the recognition of zones where minerals have been concentrated by mechanical processes. This aspect of an exploration geologist's work is aided by knowledge obtained from modern sedimentary environments and laboratory studies. For this reason the geologist should be able to base his exploration procedure on a well-researched genetic model. Geological mapping is an important part of exploration for placer deposits, and in many cases this is aided by data from satellites, and from the interpretation of aerial photographs. Indirect methods of exploration have not been so widely used in the search for placer deposits compared with other styles of mineralization. This is partly because the application of geophysics has commonly produced ambiguous results. Also the resistate nature of the placer mineralogy makes the use of geochemical pathfinder techniques of minor

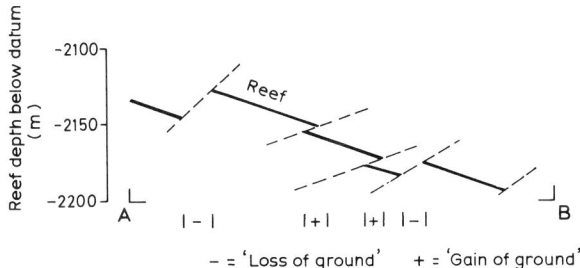

Fig. 5.30 Section taken along line A–B on gold mine plan (Fig. 5.29).

significance. Drilling is vital in both the later stages of exploration and in the evaluation of placer deposits.

Palaeo-placers are primarily of importance for uranium and gold, which are associated with sedimentary basins of Proterozoic style. At the present time they are of restricted geographical distribution, but exploration geologists should not exclude the possibility of finding new economic concentrations of gold in sedimentary sequences with evidence of prolonged reworking of terriginous sediment. Within the context of South Africa it is interesting to reflect on the important role that geophysics has played in identifying extensions to the Witwatersrand Basin – probably the single most important factor in the discovery of new gold fields. In mine exploration increasing use is being made of models to identify trends in gold concentration and therefore allow a more scientific basis for mine development. The models are genetically based in that they assume a placer environment of gold concentration, although the ultimate source of gold is not of great consequence.

REFERENCES

Applin, K. E. S. (1974) Sampling of alluvial diamond deposits in West Africa. In *Geological, Mining and Metallurgical Sampling* (ed. M. J. Jones), Institution of Mining & Metallurgy, London, pp. 120–135.

Bache, J. J. (1981) A tentative quantitative classification of World gold deposits. *Chron. Rech. Min.*, **459**, 43–50.

Batchelor, B. C. (1979) Geological characteristics of certain coastal and off-shore placers as essential guides for tin exploration in Sundaland, S.E. Asia. In *Geology of Tin Deposits* (ed. C. H. Yeap), *Bull. Geol. Soc. Malaysia*, **11**, 283–314.

Bates, R. L. (1969) *Geology of the Industrial Rocks and Minerals*, Dover Publications, New York, 459 pp.

Baxter, J. L. (1977) Heavy mineral sand deposits of Western Australia. Geological Survey of Western Australia. *Miner. Resour. Bull.*, **10**, 148 pp.

Beurlen, H. and Cassadene, J. P. (1981) The Brazilian mineral resources. *Earth Sci. Rev.*, **17**, 177–206.

Bon, E. H. (1979) Exploration techniques employed in the Pulau Tujuh tin discovery. *Trans. Instn. Min. Metall. (Sect. A, Min.)*, **88**, 13–22.

Borchers, R. (1961) Exploration of the Witwatersrand System and its extensions. *7th Commonw. Min. Metall. Congr. Trans. Pap. Discuss.*, **2**, 489–512.

Boyle, R. W. (1979) The geochemistry of gold and its deposits. *Can. Geol. Surv. Bull.*, **280**, 584 pp.

Burns, V. M. (1979) Marine placer minerals. In *Marine Minerals: Mineralogical Society of America Short Course Notes* (ed. R. G. Burns), vol. 6, pp. 347–380.

Clark, G. (1983) Sri Lanka's industrial minerals. *Ind. Miner.*, **193**, 61–72.

Clemmey, H. (1981) Some aspects of the genesis of heavy mineral assemblages in Lower Proterozoic uranium–gold conglomerates. *Miner. Mag.*, **44**, 399–408.

Coope, B. (1982) Titanium minerals – focus on production. *Ind. Miner.*, **178**, 27–36.

Dimroth, E. and Kimberley, M. M. (1976) Precambrian atmospheric oxygen: evidence in the sedimentary distribution of carbon, sulphur, uranium and iron. *Can. J. Earth Sci.*, **13**, 1161–85.

Dixon, C. J. (1979) *Atlas of Economic Mineral Deposits*, Chapman & Hall, London, 143 pp.

Emery, K. O. and Noakes, L. C. (1968) Economic placer deposits of the continental shelf. *UN ECAFE CCOP Tech. Bull.*, **1**, 95 111.

Evans, le Count D. (1981) Laterisation as a possible contributor to gold placers. *Eng. Min. J.*, **182(8)**, 86–91.

Feather, C. E. and Koen, G. M. (1975) The mineralogy of the Witwatersrand Reefs. *Miner. Sci. Eng.*, **7**, 189–224.

Goosens, P. J. (1980) Programming modern mineral exploration survey, a field geology-oriented approach. *Memoires of the Institute of Geology*, University of Louvain, Belgium.

Govett, M. H. and Harrowell, M. R. (1982) Gold – World supply and demand. *Australian Mineral Economics*, Sydney, 455 pp.

Hallam, C. D. (1964) The geology of the coastal diamond deposits of Southern Africa. In *The Geology of Some Ore Deposits of Southern Africa* (ed. S. H. Haughton), Geology Society of S. Africa, Johannesburg, pp. 671–728.

Hallbauer, D. K. (1975) The plant origin of the Witwatersrand carbon. *Miner. Sci. Eng.*, **7(2)**, 111–31.

Hallbauer, D. K. (1977) Geochemical and morphological characteristics of gold particles from recent river deposits and the fossil placers of the Witwatersrand. *Miner. Deposita*, **12(3)**, 293–306.

Henley, R. W. and Adams, J. (1979) On the evolution of giant gold placers. *Trans. Instn. Min. Metall. (Sect. B: Appl. Earth Sci.)*, **88**, B41–50.

Hester, B. W. (1970) Geology and evaluation of placer gold deposits in the Klondike area, Yukon Territory. *Trans. Instn. Min. Metall. (Sect. B: Appl. Earth Sci.)*, **79**, B60–67.

Ingham, F. P. and Bradford, E. F. (1960) The geology and mineral resources of the Kinta Valley, Perak Malaya. *Geol. Surv. Mem.*, **9**, 347 pp.

Kuzvart, M. and Bohmer, M. (1978) *Prospecting and Exploration of Mineral Deposits*, Elsevier, Amsterdam, 431 pp.

Lissiman, J. C. and Oxenford, R. J. (1975) Eneabba rutile–zircon–ilmenite sand deposit, Western Australia. In *Economic Geology of Australia and Papua New Guinea Monograph Series* 5 (ed. C. L. Knight), Australasian Institute of Mining & Metallurgy, Parkville, Victoria, pp. 1062–1070.

Lynd, L. E. (1960) Titanium. In *Industrial Minerals and Rocks* 5th edn (ed. J. L. Gillson) American Institute of Mining Metallurgical and Petroleum Engineers Inc., New York, pp. 851–880.

Macdonald, E. H. (1973) *Manual of Beach Mining Practice*, 2nd ed. Australian Government Publishing Service, Canberra, 120 pp.

Macdonald, E. H. (1983) *Alluvial mining. The Geology, Technology and Economics of Placers*, Chapman & Hall, London, 508 pp.

Mackay, R. A., Greenwood, R. and Rockingham, J. E. (1949) The geology of the Plateau tin fields – resurvey 1945–48. *Geol. Surv. Nigeria Bull.* No. **19**.

Magri, E. J. (1978) A comparison between geostatistical analyses and sedimentological studies at the Hartebeestfontein gold mine. *J. S. Afr. Inst. Min. Metall.*, pp. 218–223.

McGowen, J. H. and Garner, L. E. (1970) Physiographic features and stratification types of coarse-grained point bars: modern and ancient examples. *Sedimentology*, **14**, 77–111.

McKellar, J. B. (1975) The Eastern Australian rutile province. In *Economic Geology of Australia and Papua New Guinea Monograph Series* 5 (ed. C. L. Knight), Australasian Institute of Mining & Metallurgy, Parkville, Victoria, pp. 1055–1062.

Minter, W. E. L. (1976) Detrital gold, uranium and pyrite concentrations related to sedimentology in the Precambrian Vaal Reef placers. *Econ. Geol.*, **71**, 157–176.

Mitchell, A. H. G. and Garson, M. S. (1981) *Mineral Deposits and Global Tectonic Settings*, Academic Press, London, 405 pp.

Moore, J. R. (1978) Marine placers: exploration problems and sites for new discoveries. In *Off-Shore Mineral Resources*, Bureau de Recherches Géologiques et Minières, document no. 7, pp. 130–163.

Park, C. F. and MacDiarmid, R. A. (1975) *Ore Deposits*, 3rd edn, W. H. Freeman & Co., San Francisco, 530 pp.

Petersen, D. W., Yeend, W. E., Oliver, H. W. and Mattick, R. E. (1968) Tertiary gold channel gravel in northern Nevada County, California. *U.S. Geol. Surv. Circ.*, **566**, 22 pp.

Pirkle, E. C. and Yoho, W. H. (1970) The heavy mineral ore body of Trail Ridge, Florida. *Econ. Geol.*, **65**, 17–30.

Potts, D. (1980) *Gold 1980*. Consolidated Gold Fields Ltd., London, 66 pp.

Pretorius, D. A. (1975) The depositional environment of the Witwatersrand gold fields: a chronological review of speculations and observations. *Min. Sci. Eng.*, **7(1)**, 18–47.

Pretorius, D. A. (1981) Gold and uranium in quartz-pebble conglomerates. *Econ. Geol. 75th Anniv. Vol.*, 117–138.

Puffer, J. H. and Cousminer, H. L. (1982) Factors controlling the accumulation of titanium-iron oxide rich sands in the Cohansey Formation, Lakehurst area, New Jersey. *Econ. Geol.*, **77**, 379–391.

Rajah, S. S. (1979) The Kinta Tinfield, Malaysia. In *Geology of Tin Deposits* (ed. C. H. Yeap), *Bull. Geol. Soc. Malaysia*, **11**, 111–136.

Ramdohr, P. (1958) New observations on the ores of the Witwatersrand and their genetic significance. *Geol. Soc. S. Afr. Trans.*, **61**, 50 pp.

Roux, A. T. (1967) The application of geophysics to gold exploration in South Africa. *Mining & Groundwater Geophysics*, Canadian Geological Survey Economic Geology Report no. 26, pp. 425–437.

Saager, R. (1970) Structures in pyrite from the Basal Reef in the Orange Free State Goldfield. *Geol. Soc. S. Afr. Trans.*, **73(1)**, 29–46.

Schofield, J. C. (1976) Sediment transport on the continental shelf, east of Otago – a re-interpretation of so-called relict features. *N. Z. J. Geol. Geophys.*, **19(4)**, 513–526.

Schumm, S. A. (1977) *The Fluvial System*, Wiley, New York, 338 pp.

Sestini, G. (1973) Sedimentology of a palaeo-placer; the gold bearing Tarkwaian of Ghana. In *Ores in Sediments* (eds G. C. Amstutz and A. J. Bernard), Springer-Verlag, Heidelberg, pp. 275–305.

Shepherd, M. S. (1979) Ore reserves in mineral sands. In *Estimation and Statement of Mineral Reserves*, Australasian Institute of Mining & Metallurgy Symposium no. 22, pp. 167–174.

Simpson, P. R. and Bowles, J. E. W. (1977) Uranium mineralisation of the Witwatersrand and Dominion Reef Systems. *Philos. Trans. R. Soc. London A*, **286**, 527–548.

Singh, S. (1983) A study of shallow reflection seismics for placer tin reserve evaluation and mining. *Geoexploration*, **21(2)**, 105–135.

Smirnov, V. I. (1976) *Geology of Mineral Deposits*, MIR Publishers, Moscow, 520 pp.

Smith, N. P. and Minter, W. E. L. (1979) Sedimentological controls of gold and uranium in local developments of the Leader Reef, Welkom goldfield and Elsburg no. 5 Reef, Klerksdorp goldfield, Witwatersrand Basin. University of Witwatersrand, Economic Geology Research Unit Information Circular no. 137, 21 pp.

Tankard, A. J., Erikson, K. A., Hunter, D. R., Jackson, M. P. A., Hobday, D. K. and Minter, W. E. L. (1982) *Crustal Evolution of Southern Africa, 3.8 Billion Years of Earth History*, Springer-Verlag, Heidelberg and Berlin, 523 pp.

Theis, N. J. (1979) Uranium-bearing and associated minerals in their geochemical and sedimentological context, Elliot Lake, Ontario. *Geol. Surv. Can. Bull.*, **304**, 50 pp.

Thiel, K., Saager, R. and Muff, R. (1979) Distribution of uranium in early Precambrian gold bearing conglomerates of the Kaapvaal craton, S. Africa. *Miner. Sci. Eng.*, **11**, 225–245.

Tweedie, E. B. (1978) History, geology and value distribution of the Evander gold field, Eastern Transvaal, South Africa. In *Proceedings of the 11th Commonwealth Mining & Metallurgical Congress, Hong Kong, 1978* (ed. M. J. Jones), pp. 489–496.

van Overeem, A. J. A. (1970) Offshore tin exploration in Indonesia. *Trans. Instn. Min. Metall. (Sect. A Min.)*, **79**, A81–85.

Viewing, K. A. (1984) A summary of the technical sessions. In *Gold 82: The Geology, Geochemistry and Genesis of Gold Deposits, Geological Society of Zimbabwe* (ed. R. P. Foster), A. A. Balkema, Rotterdam, pp. 1–9.

Walker, D. (1956) Studies on the Quaternary of the Peninsula pt. 1 Alluvial deposits of Perak and the relative levels of land and sea. *Fed. Mus. J.*, **1–2**, 19–34.

Welch, B. K., Sofoulis, J. and Fitzgerald, A. C. (1975) Mineral sands of the Capel area W.A. In *Economic Geology of Australia and Papua New Guinea Monograph Series* 5 (ed. C. L. Knight), Australasian Institute of Mining & Metallurgy, Parkville, Victoria, pp. 1070–1088.

Wells, J. H. (1969) *Placer Examination, Principles and Practice, Technical Bulletin* no. 4, U.S. Dept. Int. Bur. Land Management, Washington, D.C.

Williamson, D. and Fromson, S. (1984) *Annual Review of the World Tin Industry 1984/85*, Shearson Lehman/American Express, 107 pp.

Yim, W. W. S. (1978) Geochemical exploration for off-shore tin deposits in Cornwall. In *Proceedings of the 11th Commonwealth Mining & Metallurgical Congress Hong Kong 1978* (ed. M. J. Jones), pp. 67–78.

Zimmerle, W. (1973) Fossil heavy mineral concentrations. *Geol. Rundschau*, **62**, 536–548.

6

Sediment-hosted Copper–Lead–Zinc deposits

6.1 INTRODUCTION

The recognition of sedimentary processes in the genesis of base-metal deposits can be traced back to Werner's observations on the German Kupferschiefer in the late eighteenth century. Werner's belief that all rocks and ores were of sedimentary origin led eventually to the discrediting of his views on ore genesis. In the mid-nineteenth century, Hunt (1873) expressed a more balanced and sophisticated sedimentological explanation for the genesis of some base-metal sulphides, and foreshadowed much contemporary thinking on the subject. However, during the first fifty years of this century concepts of ore genesis were strongly influenced by American geologists, who observed the proximity of many ore deposits in the USA to igneous intrusions, and considered that igneous hydrothermal solutions were of dominant importance in ore formation. Lindgren and Bateman were amongst the leading advocates of the 'hydrothermal' school, and in their opinion ore deposits associated with sediments were largely the products of replacement by solutions emanating from a cooling magma. As a result the proximity of igneous intrusions was seen as a favourable indication of mineralization, and structure was regarded as a major control of mineral deposition. The role of igneous activity in the formation of base-metal deposits became increasingly questioned during the 1940s and 1950s, because in many newly discovered mining districts igneous rocks were found to be either absent, or to have age relationships with the adjacent ores which were inconsistent with a common mode of genesis. As it became evident that igneous rocks might have no part to play in the formation of sediment-hosted ore deposits geologists began to consider a variety of alternative mechanisms. In some cases volcanically derived gases were seen to be significant, in other areas the leaching of metals from adjacent sediments or the erosion of land masses appeared to be more appropriate mechanisms. Whilst the source of metals was debated geologists began to accept that in many cases the timing of mineral deposition was essentially contemporaneous with sedimentation.

Two papers by Stanton (1955a,b) were of seminal influence on ideas concerning the timing of mineralization, and the publication of Stanton's *Ore Petrology* (1972) had a most important impact on the thinking of geologists during the 1970s. Stanton classified sediment-hosted base-metal deposits into stratiform and stratabound types, on the basis of the geometry of the ore body.

Stratiform ores were in turn subdivided, partly on the relative importance of volcanism, and partly on the environment of sedimentation. Stratabound ores were classified according to the dominant lithology of the host sediment. Thus the following stratabound ore types were described: limestone–lead–zinc association; sandstone–uranium–vanadium–copper association; conglomerate–gold–uranium–pyrite association. As a consequence, at least in part, of Stanton's influence sedimentary ore deposits have in recent years been classified on the nature of the enclosing sedimentary rock.

Two separate developments within the 1970s have led to a reappraisal of both the classification of sediment-hosted base-metal deposits, and to a reinterpretation of the timing of ore deposition within sediments. In the first case we now recognize that in many mining districts the ores are hosted by a diversity of sediments. This will become evident in reading the section on the Zambian copperbelt (Section 6.2.2). In addition ores within dissimilar lithologies are seen to have many common characteristics. For example, there are more common features between carbonate-hosted Pb–Zn deposits in Ireland, and shale-hosted Pb–Zn deposits in Northern Australia, than there are with carbonate-hosted Pb–Zn deposits of Mississippi Valley type. There has been perhaps an over-preoccupation with the nature of the sediment hosting the ore body.

Detailed mineralogical studies of many mineralized districts during the last decade have changed our perception of the timing of the ore-forming events within sedimentary sequences. Whilst in some cases the ore-bearing fluid has issued on to the seafloor to form mineral deposits contemporaneously with sedimentation (syngenetic), there is increasing evidence that many deposits are diagenetic in origin. In other cases the ore fluid has invaded the rocks after lithification (epigenetic). Does this mean that the pendulum is swinging back towards an epigenetic view of sediment-hosted metal deposits? Much depends on our understanding of the terms 'syngenetic' and 'epigenetic'. Gustafson and Williams (1981) provide a valuable discussion of these terms, and suggest that although in the context of a specific mineral deposit the age relationship of ore minerals and sediments may be epigenetic, when broadened to include the whole sedimentary basin or sequence then the deposits may be described as diagenetic.

Sediment-hosted copper deposits appear to have sufficiently distinctive characteristics to separate them from sediment-hosted lead–zinc deposits. However, lead–zinc mineralization is hosted by a considerable diversity of associated lithologies and exhibits a range of timing of the ore fluid with respect to the enclosing sediment. It therefore seems appropriate to separate those lead-zinc ores which are either syngenetic or diagenetic from those where the age relationship is epigenetic. The term sedimentary-exhalative (Section 6.3) is sometimes used for those ores of syngenetic or diagenetic type whereas epigenetic ores are commonly described as being of Mississippi Valley-type (Section 6.4).

Many geologists are inclined to emphasize the convergence of these apparently dissimilar styles of deposit. However, there are significant points of contrast which are of importance for both the explorationist and mine geologist.

6.2 SEDIMENT-HOSTED COPPER DEPOSITS

Sediment-hosted copper deposits account for 27% of the World's copper reserves (Comrate, 1975). About one-third of the deposits are hosted by sandstones and the remaining two-thirds by calcareous shales. The publication *Gisements Stratiformes et Provinces Cupriferes* produced in 1974 by the Société Géologique de Belgique contains an excellent series of papers on this style of mineralization, and the subject has more recently been reviewed by Gustafson and Williams (1981).

6.2.1 General characteristics

(a) Distribution in space and time

The spatial and temporal aspects of the principal ore deposits are summarized in Table 6.1 which

Sediment-hosted copper-lead-zinc deposits

Table 6.1 Age, size and grade of selected sediment-hosted copper deposits

Location	Age	Size and grade	Source
Tenke-Fungurume, Shaba, Zaire	Upper/Middle Proterozoic (750–1300 million years)	350×10^6 t at 4.5% Cu	1
Zambia	Upper/Middle Proterozoic (840–1320 million years)		2
Chingola		281×10^6 t at 3.14% Cu	
Konkola		186×10^6 t at 3.68% Cu	
Rokana		118×10^6 t at 2.37% Cu	
Bwana Mkubwa		0.72×10^6 t at 3.6% Cu	
Luanshya		128×10^6 t at 2.44% Cu	
Mufulira		117×10^6 t at 3.11% Cu	
Chibuluma		6.8×10^6 t at 4.68% Cu	
White pine, Michigan, USA	Middle Proterozoic (1050 million years)	550×10^6 t at 1.2% Cu	
Creta, USA	Middle Permian	1.5×10^6 t at >2% Cu	
Nacimiento, USA	Lower Trias	11×10^6 t at 0.65% Cu	
Corocoro, Bolivia	Tertiary (Oligocene)	7×10^6 t at 5% Cu	1
Mansfeld, DDR	Lower Permian	75×10^6 t at 2.9% Cu	
Lubin, Poland	Permo-Triassic	1000×10^6 t at 2% Cu	
Dzhezkazgan, USSR	Devonian–Carboniferous	10^9 t (estimate) at 1.5% Cu	
Udokan E. Siberia, USSR	Upper Proterozoic	10^9 t (estimate)	
Mount Isa,* Queensland, Australia	Middle Proterozoic (1650 million years)	143×10^6 t at 3.1% Cu	4
Morocco	Lower Adoudonian (Infra-Cambrian)		3
Talaat N Oumane		0.58×10^6 t at 2.7% Cu	
Tizert		2.0×10^6 t at 1.93% Cu	
Amadous		0.53×10^6 t at 2.04% Cu	
Tiferki		0.25×10^6 t at 1.88% Cu	

*See Section 6.3.2.
Sources: 1, Gustafson and Williams (1981); 2, *Zambia Mining Yearbook* 1979; 3, Holman (1982); 4, MIM Holdings Annual Report 1983.

shows that the deposits are of Upper/Middle Proterozoic or Phanerozoic age. There are two major cupriferous provinces: the Zambian–Zairean copperbelt of Upper Proterozoic age and the German–Polish Kupferschiefer of Lower Permian age (Fig. 6.1). The other deposits have more isolated occurrences, and ages ranging from Mid-Proterozoic to Oligocene.

(b) Size and grade of deposit

Table 6.1 summarizes the size and grade of sediment-hosted copper deposits. It can be seen that whereas the size of the deposits varies from $< 1.0 \times 10^6$ tonnes to $\sim 1000 \times 10^6$ tonnes the grade is comparatively uniform. Guilloux and Pelissonier (1974) stated that the average grade from this style of mineralization is 2.4% Cu. The most important

Ore deposit geology and its influence on mineral exploration

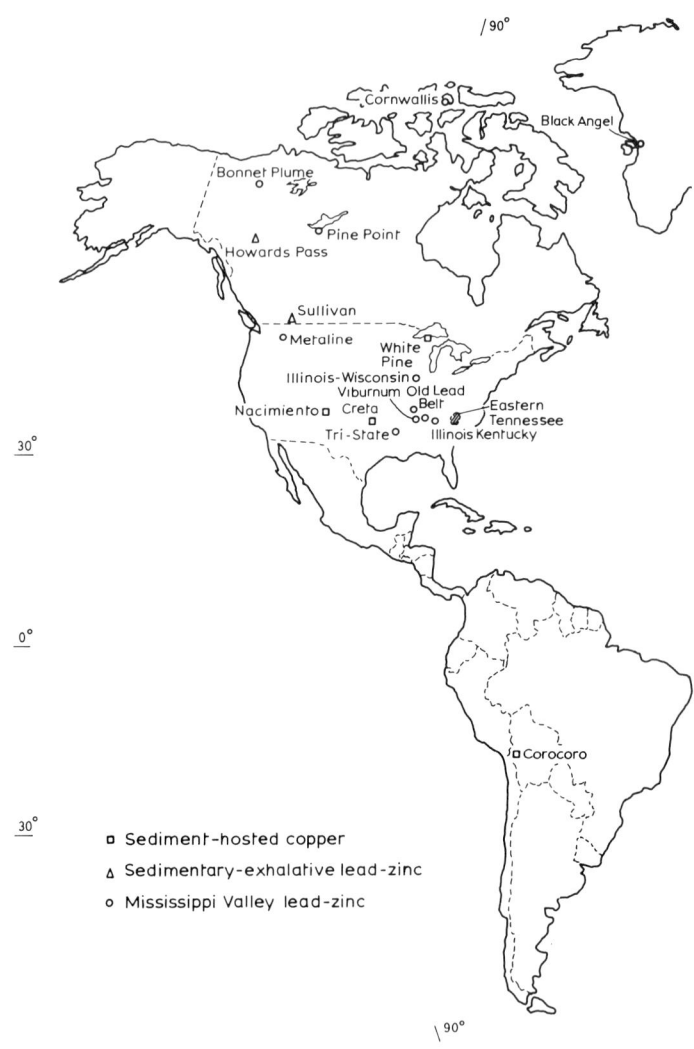

Fig. 6.1 Location map.

by-product is cobalt, which is produced mainly from Zaire, but some of the Zambian mines are also cobaltiferous.

(c) Mineralogy

Chalcopyrite, bornite and chalcocite are ubiquitous ore minerals in sediment-hosted copper deposits. Tennantite, galena, sphalerite and carrollite may locally be of economic significance. Apart from pyrite the most common gangue minerals are the constituents of the enclosing sediments: quartz, feldspar, clay minerals and carbonates.

The Cu:Pb:Zn ratio of ores from sediment-hosted copper deposits produces a distinctive grouping, which distinguishes them from lead and zinc deposits of sedimentary affiliation (Fig. 6.2). The relative deficiency of sediment-hosted copper deposits in pyrite separates them mineralogically from base-metal sulphides of volcanic association.

Both lateral and vertical mineral zoning is characteristic of all the districts which have been

218

studied. Horizontally the classical pattern of zonation is: pyrite, chalcopyrite, bornite and chalcocite but it is more difficult to generalize about the vertical mineral zoning.

The age relationships between the ore minerals and the enclosing sediment are variable. In some cases an early phase of detrital copper minerals can be distinguished. However, an increasing number of mineralogical studies suggests that the main copper ore minerals are diagenetic. The evidence comes partly from studies of mineral paragenesis (Bartholome, 1974) and partly from textural relationships between the ore minerals and the enclosing rock (Guilloux and Pelissonier, 1974) (see Fig. 6.7).

(d) Host lithology

Routhier (1963) suggested the following classification of copper deposits based upon the lithology of their enclosing sediments:

(1) Ores associated with coarse-grained detrital sandstones which are mainly grey or red in colour

Ore deposit geology and its influence on mineral exploration

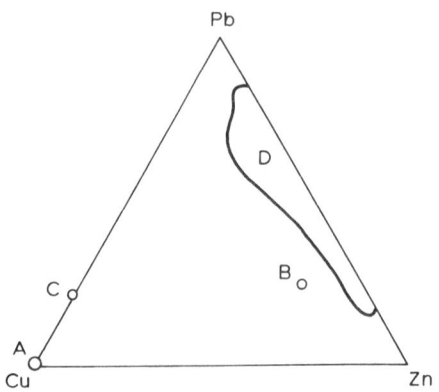

Fig. 6.2 Metal ratios of major sediment-hosted copper and lead–zinc deposits (modified from Gustafson and Williams 1981). *Sediment-hosted copper deposits*: A, Zambian–Zairean copperbelt, White Pine Spar Lake, Creta and Nacimiento, USA, and Corocoro, Bolivia; B, Kupferschiefer regional ratio (apparently includes much lead and zinc which is not mined). C, Djezkazgan, USSR. *Sediment-hosted lead–zinc deposits*: D, McArthur River (HYC), Rammelsberg, Meggen, Lady Loretta Broken Hill, Howards Pass, Laisvall, Largentiere, Aggeneys, Faro.

(red beds), e.g. Corocoro, Bolivia.

(2) Ores associated with shales which have a varying amount of calcareous and carbonaceous material, e.g. Kupferschiefer, Mansfeld, E. Germany.

(3) Ores associated with carbonates or dolomitic, psammitic schists, e.g. Katanga, Zaire.

Recent authors have concentrated on the relationship between the first two categories. Guilloux and Pelissonier (1974) noted that types (1) and (2) normally have a close spatial association which is particularly evident in Zambia and Lubin, Poland.

In spite of their lithological diversity the ore deposits have several similar factors which derive from their common palaeogeographical setting.

(a) Mineralization tends to be associated with a major marine transgression, is commonly diachronous and frequently stratiform. The sequence usually changes from coarse clastics through shales to dolomitic rocks, which represents a gradual deepening of the sedimentary basin.

(b) Palaeohighs in the basement affect the facies distribution in the lower part of the transgressive sedimentary sequence (Mansfeld, Zambia, Lubin and Moroccan Anti-Atlas). This in turn can be related to the distribution of copper, for example siliciclastic facies above palaeohighs are normally barren whereas high-grade copper is commonly found in shales at the periphery.

(c) Evaporites, particularly gypsum and anhydrite, are frequently found in close proximity to the ore-bearing horizon. They commonly, but not always, occur in the hangingwall of the mineralization. Sedimentary structures indicative of emergence such as desiccation cracks are also found in the ore horizons (Mansfeld, Chibuluma). Guilloux and Pelissonier (1974) suggested on this evidence that the sedimentary basins had only intermittent connection with the open sea.

The sediments which host the copper ore sometimes have a significant carbon content. Annels (1979) stated that the graphitic carbon content of the Mufulira greywackes varies from 1 to 2% but he concluded that there is no relationship between carbon and copper on a local scale. In the Kupferschiefer the carbon content varies from 2.7 to 9.0% (Rentzsch, 1974). Haranczyk (1961) found that there was a high correlation coefficient of 0.73 between copper and organic carbon in the Lubin–Sieroszowice region of Poland. He observed that most of the organic carbon consisted of algae, foraminifera and bacterial forms but left unanswered the possibility of a genetic link between the organic component and copper concentration.

(e) Tectonic setting

The most common regional tectonic setting is one of rifting (Zambia, Anti-Atlas). However sediment-hosted copper deposits are located in other tectonic environments. For example, the Corocoro deposit is associated with the Andean

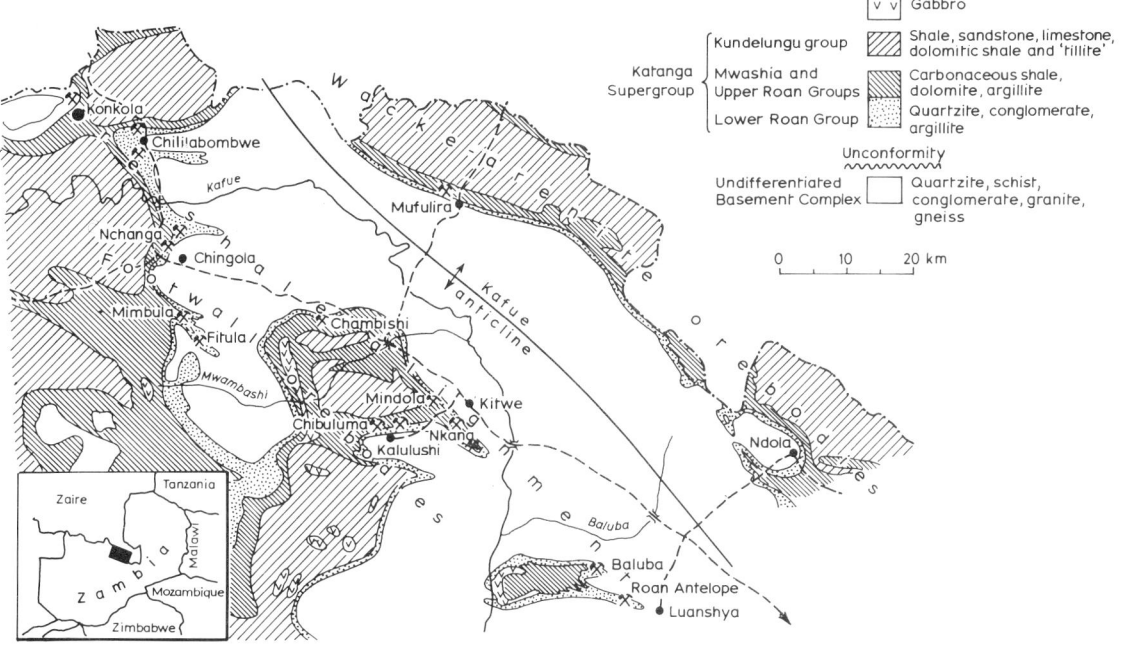

Fig. 6.3 Geological map of the Zambian copperbelt (from Diederix 1977).

subduction zone, whereas the plate tectonic environment of the Polish and German copper deposits is ambiguous (Guilloux and Pelissonier, 1974).

The structural style of sediment-hosted copper deposits is rather variable. Normal faulting is the main type of deformation. The tenor of mineralization is not usually affected by faulting apart from some of the red bed deposits in the Western USA. In the copperbelt, sediments are deformed by an open style of folding in Zambia but are subjected to intense folding and thrusting in Zaire (Section 6.2.2).

6.2.2 Geology of the Zambian copperbelt

(a) Geological setting

The Zambian copperbelt occupies the southern part of the Lufilian arc which extends for 800 km into Zaire and Angola (see Fig. 6.1).

The major copper–cobalt ore bodies are hosted by sediments of the Katanga Supergroup, which were deposited in a north-westerly trending intracratonic basin (Fig. 6.3). Isotopic age determinations indicate that the Katanga sequence was deposited before 620 million years and that the Lower Roan is between 840 and 1300 million years.

The Lower Roan sediments are dominantly siliciclastic, and were deposited unconformably on a basement complex which had an irregular topography, with differences of elevation of several hundred metres. Conglomerates, arenites and feldspathic arenites of continental facies pass upwards into sediments of more varied lithology, indicative of an incursion of marine conditions which transgressed in an easterly direction. In the Nchanga area, this marine transgression is manifested by a black carbonaceous shale which hosts many of the major ore bodies. Subsequent oscillations of the shoreline are expressed by alternations of shale with thinly banded sandstone and quartzite. The major ore bodies at Mufulira are associated with carbonaceous arenites, referred to as greywackes. The sedimentary environment in which the Mufulira 'greywackes' were deposited has been the subject of different interpretations (Brandt, 1961; Eden, 1974). Annels (1979) considered that the 'greywackes' are petrographically and texturally similar to adjacent massive and well-bedded siliciclastics. He considered that they differ only in their content of organic matter which he suggested is diagenetic in origin.

The Lower Roan sediments were subjected to local uplift and emergence prior to the resumption of marine conditions. The Upper Roan has a basal succession of clastic sediments which pass upwards into a predominantly carbonate sequence.

(b) Structure and metamorphism

The Katangan sediments were deformed as a consequence of the northern movement of the Kibaran massif, and their subsequent compression against the Bangweulu massif. In the Shaba province of Zaire the copper–cobalt deposits occur in diapiric anticlines, and thrust faulting is the dominant structural style (Fig. 6.4). In Zambia the main response to deformation is a series of open north-westerly trending folds (Fig. 6.5). The Kafue anticline is the main structure and later cross-folding has resulted in folds, the axial traces of which trend in a north-easterly direction. In general the Roan sediments have deformed plastically and faulting is uncommon.

Regional metamorphism has transformed the Katangan sediments to a sequence of argillites and quartzites, which range from greenschist facies on the Zairean border to garnet–amphibolite facies in the Luanshya area. More importantly metamorphism has resulted in recrystallization of the sulphide minerals, which must be taken into account in using textural criteria to interpret the genesis of the ores.

(c) Mineralogy

The relative abundance of the major ore minerals is shown in Table 6.3. Chalcopyrite and bornite are the main copper sulphides, both of which are

Table 6.2 Stratigraphy of the Katanga Supergroup, Zambia (from Mendelsohn, 1961)

Group	Subgroup	Rock types
Kundelungu	Upper Kundelungu	Shale, quartzite
	Middle Kundelungu	Shale, tillite
	Lower Kundelungu	Shale, dolomite and tillite
Mine	Mwashia	Carbonaceous shale, argillite (dolomite and quartzite)
	Upper Roan	Dolomite and argillite
		Argillite and quartzite
	Lower Roan	Quartzite, argillite and feldspathic quartzite
		Argillite, impure dolomite
		micaceous quartzite
		Argillaceous quartzite
		Feldspathic quartzite
		Aeolian quartzite
		Conglomerate

Basement complex.

Sediment-hosted copper–lead–zinc deposits

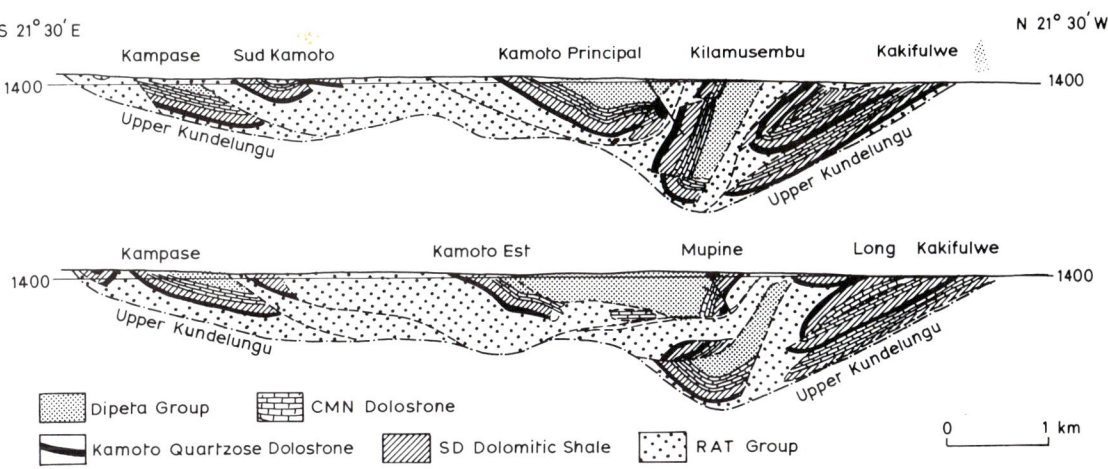

Fig. 6.4 Style of deformation in the Zairean part of the Copperbelt illustrated by cross-sections through the Kolwezi thrust plane (Kamoto Mine area) (from Bartholome *et al.* 1973).

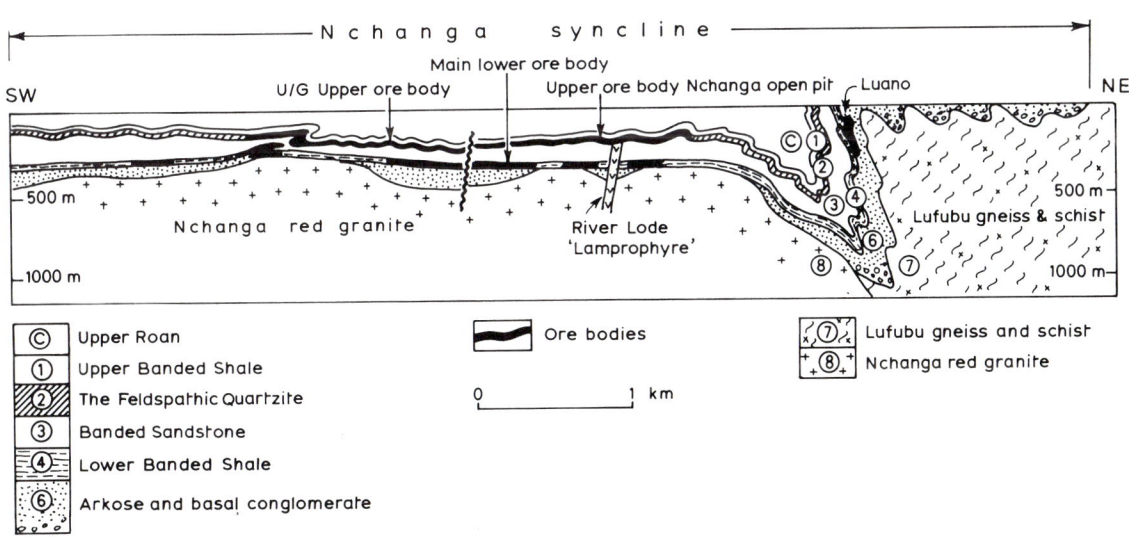

Fig. 6.5 Style of deformation in the Zambian part of the Copperbelt illustrated by a cross-section through the Nchanga area (modified from Diederix 1977).

normally disseminated in their host rocks. Near the surface the primary sulphides are oxidized, and whilst at many mines the oxide ores have now been exhausted, malachite, azurite and chrysocolla remain a substantial component of production from Chingola (Fig. 6.3). In addition, Chingola produces a refractory cupriferous mica which is stockpiled at present (1983) due to the very high cost of treatment. Most of the chalcocite in the copperbelt is of supergene origin and the most

Table 6.3 Relative abundances of the major copper and cobalt minerals in mines on the Zambian copperbelt (after Notebaart and Vink, 1972)

Mineral	Baluba	Bwana Mkubwa	Chambishi	Chibuluma	Chililabombwe	Mufulira	Nchanga/Chingola	Luanshya	Rokana Mindola
Chalcopyrite	xxxx	x	xx	xxxx	x	xxx	x	xxxx	xxx
Bornite	xx	xx	xxxxx	xx	xxx	xxxx	x	xx	xx
Chalcocite	x	xxx	xx	x	xxx	xx	xxx	xx	x
Cobaltiferous pyrite	xx	—	—	xx	—	—	—	—	xxx
Carrollite	xx	—	—	x	x	—	x	—	xx
Malachite	xx	xxxx	xx	—	xx	x	xxxx	xx	x
Cuprite	x	x	x	—	x	x	x	x	—
Native copper	trace	trace	—	—	x	x	x	trace	trace
Pseudo-malachite	trace	x	x	—	x	—	xx	x	x
Chrysocolla	trace	x	trace	—	x	—	xx	x	trace

x, 1–5%; xx, 5–20%; xxx, 20–40%; xxxx, 40–70%; xxxxx, >70%.

important concentrations are at Nchanga and Chililabombwe.

Carrollite and cobaltiferous pyrite are the principal cobalt-bearing minerals and are mined mainly from Rokana/Mindola, Konkola, Baluba and Chibuluma.

More detailed descriptions of the major and minor ore minerals in the Zambian copperbelt can be obtained from Notebaart and Vink (1972).

6.2.3 Genesis of the Zambian copperbelt

The evolution of ideas concerning the origin of the Zambian copperbelt has mirrored the changing pattern of geological thought concerning the genesis of sediment-hosted base-metal deposits during the last fifty years.

From their discovery in the 1920s, through to the 1950s, the main consensus of opinion favoured an igneous hydrothermal origin for the Zambian copper ores. Dissenting viewpoints were expressed originally by Schneiderhön (1932) and subsequently by Garlick who has written extensively about the genesis of the copper–cobalt mineralization and has argued forcibly in favour of a syngenetic origin (Garlick, 1961; Garlick and Fleischer, 1972).

Garlick's reinterpretation of the genesis of the copperbelt ores was based upon a study of the field relationships between the Lower Roan sediments and the Basement Complex, the relationship between ore minerals and sedimentary textures, the relationship between ore grade and folding, and mineral zoning within some of the ore bodies. In the following section we review the evidence used by Garlick in the formulation of a syngenetic theory for Zambian mineralization, and then re-examine his conclusions in the light of more recent mineralogical studies.

When Garlick began his investigations of the copperbelt it was assumed that the basement granites were intrusive into the Lower Roan sediments, and that the copper mineralization had been introduced epigenetically into the sediments by hydrothermal fluids emanating from the granite. Garlick was able to demonstrate, by careful geological mapping, that the Lower Roan sediments rest with a profound unconformity on an older Basement Complex.

In order to determine the age of the sulphide mineralization relative to the enclosing sediment Garlick examined the relationship of ore grade to folding. At Nkana, he carried out comparative studies of the variation in copper grade between the expanded axial region, and the squeezed limbs of a major dragfold. If the mineralization had been

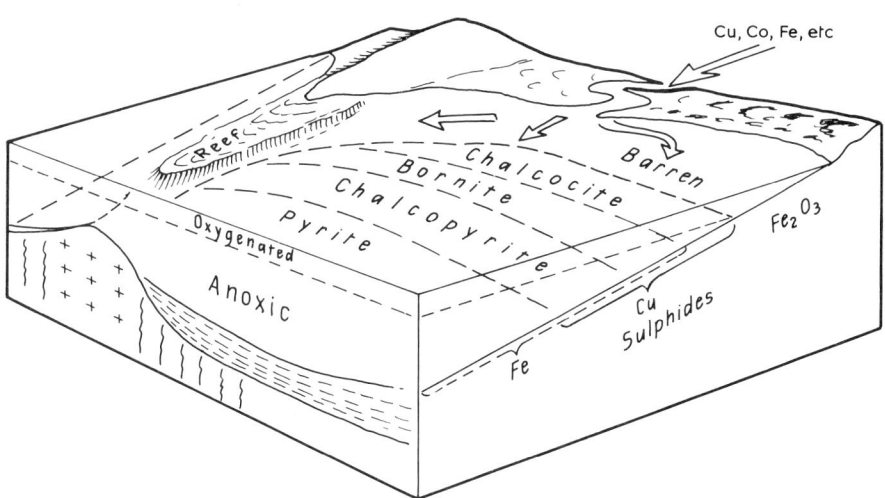

Fig. 6.6 Block diagram showing diffusion of copper, cobalt, iron and other metals in solution into a sea. The syngenetic viewpoint considers that metal deposition is controlled by depth of water which in turn determines the E_h conditions. (From Fleischer et al. 1976.)

post-tectonic, one would have expected it to be concentrated in the highly fractured and thickened fold crest. In practice Garlick found little variation in grade within the fold, which pointed to a pre-deformation age for the mineralization; an age relationship which has not been challenged.

Garlick concluded that the ore minerals were synsedimentary from a study of their relationship with sedimentary textures, and from the pattern of mineral zoning at a number of mines. At Mufulira the concentration of sulphides with zircon and tourmaline on foresets and bottomsets of cross-bedded arenites was noted, and sulphides were also found in the hollows of ripple marks at Mufulira. At Chibuluma pyrite and carrollite were found with quartz pebbles in the potholed surface of an arenite.

Garlick studied the distribution pattern of ore minerals at Mufulira, Chambishi and Nkana and concluded that both vertical and lateral mineral zoning were evident. Upward zoning is normally from chalcocite through bornite and chalcopyrite to pyrite. Lateral zoning is more complex but may be simplified as chalcocite; bornite; chalcopyrite; pyrite.

The model of ore genesis proposed by Garlick entailed the introduction of metals to a shallow marine environment by rivers draining a copper-rich source. The copper and cobalt are transported in part as detrital sulphide grains but are mainly present in solution. Metals introduced into the marine environment are believed to be precipitated by reaction with hydrogen sulphide formed by anaerobic bacteria. The pattern of mineral distribution is believed to be controlled by E_h/pH conditions which are in turn determined by depth of water (Fig. 6.6).

Garlick's model of ore genesis is not accepted by all geologists who have studied the copperbelt mineralization. Clemmey (1978) has suggested that the mineral zoning at Rokana can be explained by several cycles of submergence and emergence in a very quiet epeiric marine or lacustrine environment. Studies of comparable patterns of mineral zoning in the Kupferschiefer (Konstantynowiz, 1972; Rentzsch, 1974) concluded that they are not parallel to palaeoshorelines but reflect post-depositional features. It might be thought that the nature of the sulphur could at least be resolved from a study of sulphur isotopes. Dechow and Jensen (1965) investigated the sulphur isotopes from both Zairean and Zambian sulphide ores and found a similar pattern of results. However, they show a narrow range of values which probably reflects homogenization by metamorphism, and Dechow and Jensen con-

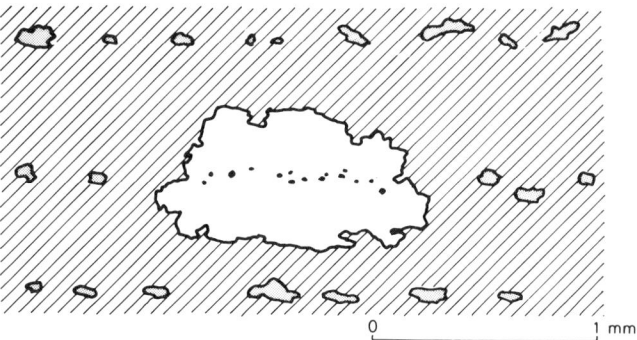

Fig. 6.7 Relationship between pyrite and other sulphides in and around a carrollite porphyroblast. The drawing shows a carrollite porphyroblast (white) in a dolomite matrix (shaded). This matrix consists of thin laminae rich in bornite blebs (stippled) and of thick barren laminae. Pyrite (black) is found only inside the carrollite porphyroblast at the approximate level of a bornite-rich lamina. (From Bartholome 1974.)

cluded tentatively that the stratiform copper deposits may possibly contain sulphur of biogenic origin.

In recent years several research workers have published evidence which points towards a late diagenetic origin for the Zambian ores. Textural studies indicate that pyrite was the only sulphide present at an early stage during diagenesis. For example Bartholome (1974) described small remnants of pyrite preserved within porphyroblasts of carrollite in obvious banded continuity with bands of copper sulphide outside the porphyroblasts (Fig. 6.7).

Annels (1979) has drawn attention to the antipathetic relationship between anhydrite and high-grade copper ores. Annels pointed out that, whilst anhydrite is a common mineral in barren and subeconomic ores, it is absent from horizons containing high grades of copper. In high-grade zones, pseudomorphs of calcite and quartz after anhydrite are commonly fringed by pyrite. Annels suggested that the main source of sulphur is nodular anhydrite, which formed diagenetically, and this points to a late diagenetic origin for the copper.

The absence of copper and cobalt ores from arenites over palaeohighs is difficult to explain in terms of the syngenetic model. However, a diagenetic mechanism involving the reduction of copper-bearing connate water by carbonaceous material implies that copper will be deposited mainly in carbonaceous argillites. In the carbon-deficient arenites no precipitation of copper or cobalt will occur, and therefore the pyrite remains unaltered.

Further research is required before a fully acceptable model for the genesis of the Zambian copperbelt is defined. At the present time the source of copper and the nature of the ore fluid are poorly understood. Many writers favour a thick sequence of tholeiitic basalts beneath the sedimentary pile as a source for sediment-hosted copper deposits. However, the Basement Complex in Zambia is dominated by granites and schistose rocks. It would therefore appear that the source of copper lies within the sediments themselves, either as detrital oxide minerals or weakly adsorbed metal. Little research has been carried out on the nature of the ore fluid responsible for the Zambian ores. However, Rose (1976) concluded that the Nacimiento and Creta deposits (Table 6.1) were formed by sulphate-rich chloride brines that were slightly alkaline. Rose suggested that similar brines might have been responsible for the formation of the Zambian deposits.

The mechanism of ore deposition appears to be better understood. Essentially a trap is required to remove copper from brines which are migrating up dip within the sedimentary basin due to compaction. The alternatives which have been suggested are syngenetic pyrite, the formation of which will be favoured by a high organic content in the sediment, and anhydrite associated with sabkha-type sediments within the sedimentary basin.

Two aspects of the debate on the genesis of

copperbelt ores are remarkable: firstly, the virtual exclusion of the Zairean ore bodies from most discussions and, secondly, the changing emphasis of ideas concerning the genesis of the ores has had relatively little impact on mineral exploration strategy. In contrast Brock (1961) has pointed out that all the major Zambian deposits, when plotted on a map, form two north-westerly-trending lineaments which extend into the Zairean part of the copperbelt. All successful exploration efforts have been concentrated either within or adjacent to the lineaments.

6.2.4 Exploration in the Zambian copperbelt

Copper minerals were recognized and exploited as shallow workings by indigenous tribesmen in northern Zambia before the advent of European prospectors in the early part of this century. The first European miners were aware of outcrops with extensive malachite staining, but were not at first attracted by the oxide nature of the ore. However, in 1925 the first hole was drilled at Roan Antelope, and intersected sulphide ore at a depth of 500 feet (152.1 m) with a grade of 4% Cu – this may be regarded as the birth of the modern copperbelt.

A comprehensive account of the exploration techniques used on the copperbelt is provided by Mendelsohn (1961). The most successful indirect method of exploration has been geochemistry. Exploration geophysics has been little used due mainly to the almost complete oxidation of sulphides to a depth of 60 m. Much of the pioneering work by British geochemists was carried out in Zambia (Webb and Tooms, 1959). For reconnaissance work the sampling of seepage areas or dambos has proved successful. However, soil sampling has been the most widely used and successful technique, and located the Fitula and Luano ore bodies at Nchanga (Diederix, 1977). Rose et al. (1979) described a typical exploration sequence in the copperbelt from initial soil sampling through to trenching and drilling. An interesting case history is provided by Ellis and McGregor (1967) in describing the discovery of the Kalengwa deposit, some 320 km to the west of the copperbelt in which the sub-outcropping mineralization was covered by about 2 m of windblown sand.

Geobotany has been successfully applied to the copperbelt, particularly using the copper flower, *Ocimum homblei de wild*. Diederix (1977) mentioned that abnormal 'clearings' in the typical Brachystegia woodland of the copperbelt were investigated using oblique aerial photography in 1927, and commented that it is one of the earliest examples of aerial photography applied to mineral exploration.

6.2.5 The role of the mine geologist in Zambia

Although the task of evaluating tonnage and grade is an important aspect of the mine geologist's responsibility, in Zambia there are four additional areas where the geologist's expertise may be required: geotechnics, mine planning, hydrogeology and the application of mineralogical techniques to problems of recovery.

During the early development of the copperbelt in the 1930s and 1940s the stability of the rock within a mine was considered to be the responsibility of the mining engineer. However, since the mid 1960s the problem of monitoring the stability of the hangingwall of the ore body has increased, mainly due to the increased size of the haulageways. The introduction of tyred vehicles (load-haul-dump units or LHDs) for the transport of ore from the stope to the ore pass is a technological change which has provided the mines with greater flexibility of operation, compared with earlier tracked vehicles. LHDs require wider drives with attendant support problems, particularly where there are many discontinuities in the hangingwall (A. Brooks, personal communication). Recent textbooks which discuss the geotechnical aspects of rock stability include Hoek and Bray (1981) and

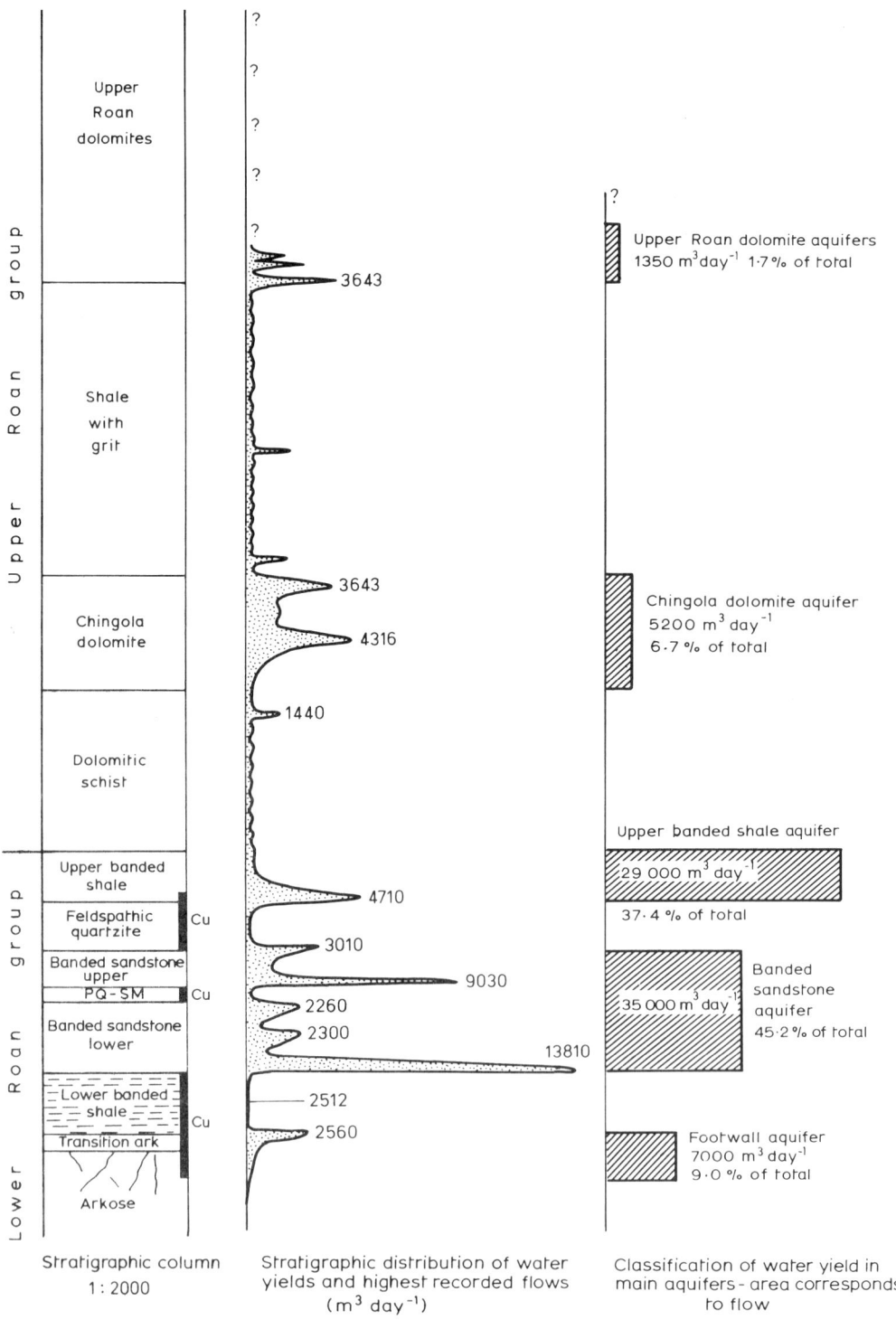

Fig. 6.8 Distribution of aquifers in the stratigraphic succession at Nchanga, Zambia (from Diederix 1977).

Hoek and Brown (1980). It is normal practice on the copperbelt for each mine to employ a geologist with special responsibility for geotechnics.

One of the duties of the Senior Mine Geologist is to liaise with the Planning Engineer in the design of future mine development. The extent to which the geologist is involved in this type of work depends upon the complexity of the ore body. For example, at Luanshya the early development of the eastern part of the ore body in a simple synclinal structure was undertaken with little need for geological advice. However, as the ore body has been developed in a westerly direction the folding has become more complex, with the ore being frequently deformed by minor folding. As a result stopes of irregular shape and size must be designed by the planning engineer based upon detailed information and advice from the geological department. The techniques of underground mapping, which forms the basis of the geologist's concept of the three-dimensional shape of the ore body, are described by McKinstry (1948) and Peters (1978).

In all mines it is the responsibility of management to prevent the sudden in-rush of water into the mine workings, and often this becomes one of the duties of the mine geologist. The Zambian mines are notoriously wet. Konkola, which is the wettest mine on the copperbelt, and perhaps in the World, has a daily pumping rate of 400 000 m^3 – a figure which may be difficult to visualize, but which results in water spouting from boreholes and pouring from joints and fractures. Readers who have already experienced an underground visit may recall a chilling trickle of water down the neck as they descend in the mine cage; at Konkola one descends through a waterfall!

In Zambia the problem of hydrogeology is heightened by the presence of numerous aquifers within both the Lower and Upper Roan Series. In this account discussion is restricted to Nchanga and Chingola underground workings where about 91 000 m^3 a day are pumped to surface, but it should be borne in mind that this is a general problem throughout the copperbelt.

At Nchanga and Chingola there are five main aquifers (Diederix, 1977). The Banded Sandstones and Upper Banded Shale aquifers account for 45% and 37% respectively of the total water yield within the mine. The Banded Sandstone comprises a basal sequence of barely coherent siltstones and sandstones, succeeded by a highly jointed shale and quartzite which pass upwards into dolomitic mica schists. The aquifers consist of lenses and strata of high permeability in a porous formation of low permeability. The Upper Banded Shale aquifer has isolated and interconnected weathered porous zones, possibly after dolomite, in a formation of laminated, micaceous shale (Fig. 6.8).

The general approach which is adopted to counteract the problem is to dewater the aquifer, so that the water table is lowered in advance of mining operations. At Nchanga this is done by drilling dewatering boreholes upwards from the mine workings into the aquifer – the flow of water then being controlled by valves on boreholes, and ultimately by the construction of watertight doors. The progress of dewatering relative to stoping is monitored by regular estimates of the water level in each main aquifer, by underground observations, pressure measurements on underground boreholes and water-level measurements in surface boreholes. In some cases aquifers may be completely drained but in many instances they are recharged, either by surface water or by lateral flow from associated hydrological basins. In any event the water must be removed by pumping, which adds significantly to the cost of mining.

Mineralogical problems within the copperbelt are referred to the Research and Development Laboratories at Kitwe, and are not normally the concern of the mine geologist. Only one example of the type of applied mineralogical research carried out will be given here. Recognition in recent years of the economic significance of cobalt has led to a changing emphasis towards mining those ore bodies where cobaltiferous minerals are significant. At Baluba it was found that a significant proportion of the cobalt which was reported from chemical assays was not recovered during the processing of the ore. Systematic studies of mill feed material from Baluba, using varying styles of sample attack, supported by a detailed mineralogical investigation, revealed that up to 30% of the cobalt was held in silicates such as biotite, talc and tremolite (N. M. Kostic, personal communication).

6.3 SYNGENETIC AND DIAGENETIC LEAD–ZINC DEPOSITS IN SHALES AND CARBONATES (SEDIMENTARY-EXHALATIVE DEPOSITS)

6.3.1 General characteristics

(a) Distribution in space and time

Sedimentary-exhalative lead–zinc deposits are of restricted geographical and temporal distribution. Their scarcity presents the geologist with a low probability of finding new deposits, but their large size and associated silver content have encouraged companies to commit funds for their exploration.

Most sedimentary-exhalative deposits are hosted by shales of Middle Proterozoic age. Hutchison (1980) has drawn attention to the fact that, whilst the Middle Proterozoic is characterized by large sediment-hosted massive sulphide deposits, there is an absence of mineralization of volcanic affiliation. However, this does not mean the products of volcanicity are absent from the stratigraphic record of sedimentary-exhalative deposits.

Several major ore deposits of this type occur within rocks of Phanerozoic age, and are similar to their Proterozoic equivalents. The Irish lead–zinc deposits share many of the characteristics of sedimentary-exhalative deposits but are associated with carbonates. Table 6.4 summarizes the principal characteristics of some Proterozoic and Phanerozoic sedimentary-exhalative deposits.

(b) Mineralogy

Galena and sphalerite are ubiquitous ore minerals in this class of deposit. Ores appear to have either similar concentrations of lead and zinc (e.g. Sullivan, Mount Isa) or else a dominant zinc component (e.g. Howards Pass, Lady Loretta). Most mineralized provinces are consistently argentiferous although some contain no silver at all. For example the Irish deposits all contain silver whereas the deposits in the Aggneys area of Namaqualand, South Africa, are devoid of silver (Rozendaal, 1978). Pyrite and pyrrhotite are common ore minerals and may locally be abundant but can also be absent (e.g. Gamsberg, S. Africa). Similarly barite may be an important gangue mineral particularly in the upper part of an ore body but is not always present.

Mineral zoning is a characteristic feature of sedimentary-exhalative deposits. In general there is a tendency for the relative proportions of lead and zinc to change laterally within the deposit. In some cases, for example McArthur River, Queensland (Williams, 1978), and Silvermines, Ireland (Taylor and Andrew, 1978), the zoning can be related to faults from which the ore fluid is believed to have issued. However, in most cases the controlling factors of mineral zoning are a matter for speculation.

(c) Tectonic setting and lithology

Table 6.4 shows that the type of tectonic setting is a consistent feature of this class of deposit. When the deposits are unmetamorphosed and undeformed they are readily interpreted as having formed within graben-type structures, normally in intracratonic basins. In the case of Broken Hill, deformation and metamorphism obscure the original tectonic framework. However, W. Laing (personal communication) considers that the evidence of two main directions of thrusting suggests incipient rifting with associated gravity sliding into the rift.

Badham (1981) suggested that ore deposits of this class can be divided into those associated with deep water sediments (Sullivan, Howards Pass) and those in relatively shallow water sediments with associated carbonates, evaporites and cherts. Subdivision may also be based on the relative importance of volcanic rocks. Tuffaceous material is closely related to the ore-bearing horizons at Mount Isa and Rammelsberg and may have been

Table 6.4 Main features of selected sedimentary-exhalative deposits

Ore deposit	Size and grade	Associated metals and minerals	Age	Metamorphic grade	Tectonic setting	Host lithology	Volcanics	Source
Howards Pass, Canada	100×10^6 t, Pb 1.5%, Zn 6.0%		Early Silurian	Negligible	Intracratonic basin	Graptolitic black shale and siltstone with chert and carbonates	Contemporaneous basalt 16 km away	1
Rammelsberg, Germany	30×10^6 t, Cu 1.0%, Pb 9%, Zn 19%	Barite, Pyrite, Ag 103 g/t	Middle Devonian	Greenschist	Intracratonic basin	Black shale with interbedded sandstone and tuff. Discordant 'Kniest' facies	Minor thin slaty tuff interbedded in ore	1
Meggen, Germany	60×10^6 t, Cu 0.2%, Pb 1.3%, Zn 10%	Barite, Pyrite	Middle Devonian	Low to middle greenschist	Intracratonic basin	Carbonaceous slates. Ore body in minor sediment-starved depression	100 m keratophyre in Early Devonian below ore body. Major mafic volcanics contemporaneous with ore in region	1
Broken Hill, Australia	180×10^6 t, Cu 0.2%, Pb 11.3%, Zn 9.8%	Ag 175 g/t	Lower to Middle Proterozoic	Granulite	Mobile trough, obscured by metamorphism and extreme deformation	Sphalerite-galena-pyrrhotite ore with quartz-calcite, Ca-Mn, Fe silicate gangue associated with alkaline rocks, carbonate, iron-formation, acid volcanics	The Potosi gneiss is considered to be volcanic	1
Sullivan, Canada	155×10^6 t, Pb 6.6%, Zn 5.7%	Ag 68 g/t, Sn, Cd, Cu, Au	Middle Proterozoic 1430 million years	Lower amphibolite	Early stage of intracratonic basin near major faults with syn-depositional movement	Footwall: argillite, siltstone and conglomerate. Siltstone interbedded with laminated sulphide above and to cast of massive sulphide ore body	Only much younger volcanics	1
Gamsberg, Namaqualand, S. Africa	93.5×10^6 t, Pb 0.55%, Zn 7.41%	Barite, Magnetite, Haematite	Middle Proterozoic Namaqualand Metamorphic Complex 900–1200 million years	Upper amphibolite facies	Restricted basin indicated by sedimentology	Basal granite gneiss, pelitic schist, metaquartzite, iron-formation, psammitic schist, amphibolite. Host rock: quartzitic, pelitic and calcareous metasediments	Cordierite-bearing units interpreted as volcanic tuff	2

Source: 1, Gustafson and Williams (1981); 2, Rozendaal (1978).

significant at Gamsberg, Namaqualand, but is not present within the stratigraphy of many sediments which host syngenetic and diagenetic base-metal sulphides.

6.3.2 Geology of the northern Australian shale-hosted stratiform lead–zinc deposits

(a) Introduction and regional geological setting

Mount Isa Mine in north Queensland has been a major producer of base metals since the 1930s. Exploration during the 1950s and 1960s in northern Queensland and adjoining parts of the Northern Territory has resulted in the discovery of a number of additional large lead–zinc deposits of similar geological style. As a result the Batten and Leichhardt River Fault Troughs are now recognized as one of the World's major base-metal provinces (Fig. 6.9).

In this account we restrict our discussion to the major ore deposits: Mount Isa, Hilton, McArthur River (HYC) and Lady Loretta. All the deposits have a similar age and tectonic setting. Their ores are contained within sediments of Middle Proterozoic (Carpenterian) age, which rest unconformably on a Lower Proterozoic basement consisting mainly of strongly metamorphosed granites and volcanics (Table 6.5). The sedimentary sequences which host the major base-metal deposits are confined to graben-type structures within which a thick succession of shallow water sediments has accumulated.

The McArthur River deposits are unmetamorphosed and are separated from the more southerly deposits by the Murphy Tectonic Ridge (Fig. 6.9), to the south of which the Middle Proterozoic sediments have been subjected to metamorphism of varying grade.

(b) Depositional environment

The correct interpretation of the depositional environment of the Middle Proterozoic sediments within the Batten and Leichhardt River Fault Troughs is of importance for the construction of a

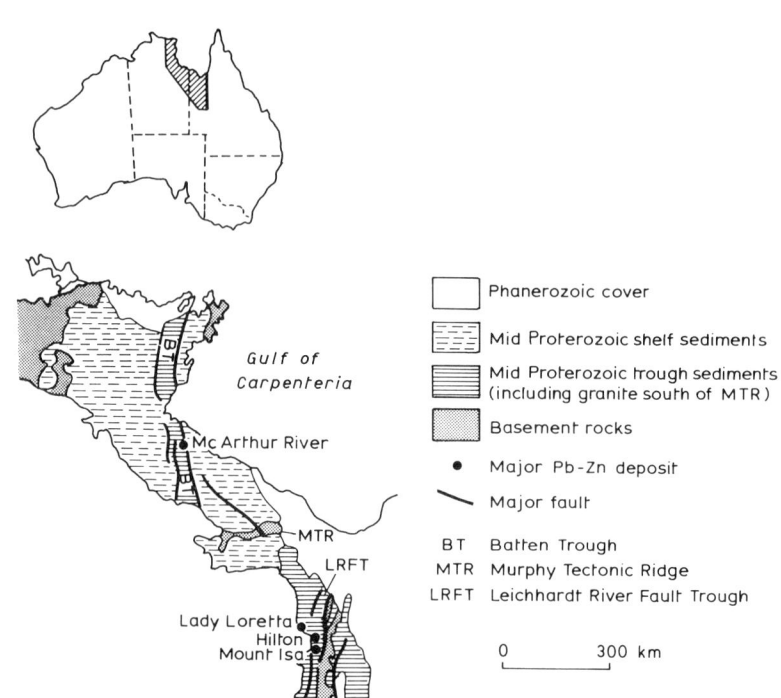

Fig. 6.9 Location of major ore deposits in the Batten and Leichhardt River Fault Troughs (modified from Lambert, 1976).

Table 6.5 Stratigraphy of the McArthur River and Mount Isa areas

	Mt Isa/Hilton	Lady Loretta	McArthur River
Middle Proterozoic	Mt Isa Group*	Paradise Creek Formation*	Reward Dolomite
			Barney Creek Formation*
	Judenan Beds	Gunpowder Creek Formation	Teena Dolomite
	Myally Beds	Myally Beds	Emerugga Dolomite
	Eastern Creek volcanics		Tooganinie Formation
	Mount Guide quartzite		Tatoola Sandstone
			Amelia Dolomite
			Mallapunya Formation
Lower Proterozoic	Leichhardt metamorphics and Argylla Formation		

*Pb–Zn ore bodies.

model of ore genesis. For example, evidence of evaporite sediments within the sedimentary sequence tends to favour a brine expulsion model. In contrast Russell's (1983) convective model requires a rapid deepening of the sedimentary basin. Most research has pointed towards shallow water conditions of sedimentation.

There are four major cycles of sedimentation in the McArthur Basin. The cycle of economic significance is the Barney Creek Formation, which comprises water-lain detrital sediment. Muir (1983) described sedimentary structures within the Barney Creek Formation which indicate that the sediments were formed in shallow water or emergent conditions. Muir considered that carbonate pseudomorphs after anhydrite within the Barney Creek Formation indicate a sabkha-type environment, in which groundwater brines circulated during diagenesis. However, Oehler and Logan (1977) have described bacteria preserved in black cherts within the mineralized sequence which, they suggested, lived below the photic zone (approximately 80 m).

The lead–zinc ore bodies at Mount Isa and Hilton Mines are hosted by the Urquhart Shale which occurs within the Mount Isa Group (Fig. 6.10). The horizons above and below the Urquhart Shale contain algal stromatolites which indicate emergent conditions whereas the Urquhart Shale itself contains casts after gypsum and anhydrite. These features are described by Neudert and Russell (1981) who envisaged a non-marine depositional model for the Mount Isa Group with intermittent streams, playa-lake stages, flash-flood deposits, stromatolitic structures and subsurface halite growth.

(c) Size and grade of deposits

The size and grade of the principal deposits are tabulated in Table 6.6. The McArthur River (HYC) deposit is clearly the largest ore body but due to mineral processing problems is not worked at present.

Mount Isa is the only operating mine (1983) and although we are considering it as an example of a shale-hosted lead–zinc deposit it is worth

Ore deposit geology and its influence on mineral exploration

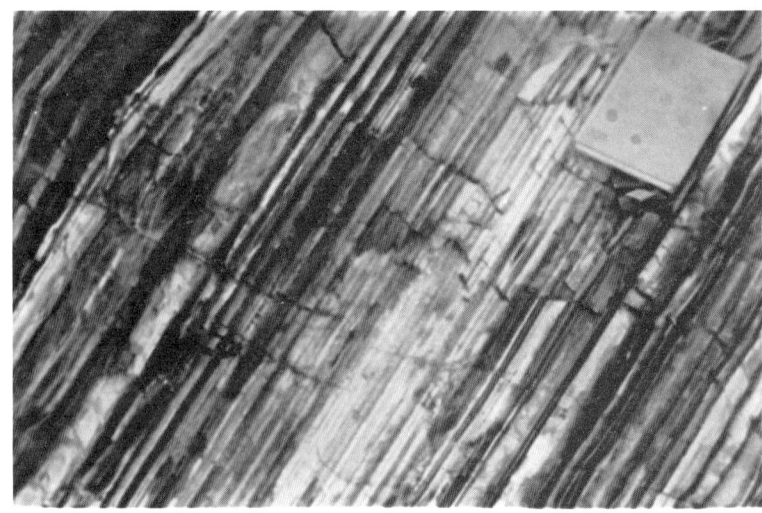

Fig. 6.10 An outcrop of barren Urquhart Shale, Mount Isa.

reflecting that copper was the main revenue earner in the early 1980s. Copper is mined from the non-stratiform silica–dolomite rock which will be discussed in a later section. At this point it should also be mentioned that the grades and thicknesses indicated in Table 6.6 are a function of assay cut-off, which is in turn determined by metal prices. Thus in principle a resurgence in the price of lead or zinc could lead to a redefinition of the assay cut-off, with resulting ore bodies which would be wider but of lower average grade. In practice this rarely happens. This is mainly due to the lead time from ore reserve calculations to mining, which makes it difficult to change the cut-off in time to reflect the change in metal prices.

(d) Mineralogy

The deposits under consideration contain lead and zinc ores of similar geological setting and presumed genesis, but there are subtle variations in mineralogy which have quite profound

Table 6.6 The size, grade and metamorphic facies of major Northern Australian lead–zinc deposits (from MIM Holdings Annual Report, 1983).

Deposit	Size ($t \times 10^6$)	Grade				Thickness (m)	Metamorphic facies
		Pb (%)	Zn (%)	Cu (%)	Ag (g/t)		
HYC (mineralization)	227	4.1	9.2	0.2	41	55	Zeolite
Mount Isa (proved and probable)	53 143★	6.1	6.6	3.1★	149	1–50	Lower greenschist
Hilton and Hilton North (proved and probable)	45	6.6	9.6	—	150	1–50	Lower greenschist
Lady Loretta (probable)	9	6.5	14.8	—	95	24	Lower greenschist

★Copper ore body.

consequences for the economic viability of the individual deposits.

The sulphide ores at Mount Isa occur as distinct bands concordant with the enclosing Urquhart Shale. The sulphide bands vary from 1 mm to 1 m in thickness, and when grouped together in sufficient density constitute ore bodies. The ore bodies tend to be regular and continuous on a large scale, but irregularities are evident on a small scale (C. Hartley, personal communication). The main ore bodies are confined to the upper 650 m of the Urquhart Shale (Mathias and Clark, 1975). The economic sulphides at Mount Isa are sphalerite, galena and freibergite. Other common sulphides include pyrite, pyrrhotite, arsenopyrite, marcasite and minor chalcopyrite (Mathias and Clark, 1975). Pyrite is the most abundant sulphide and the texture is dominantly framboidal but is locally recrystallized. Pyrrhotite is locally common and where the ore passes along strike into the silica–dolomite rock, pyrrhotite may become dominant (C. Hartley, personal communication).

Galena is highly variable in form and ranges from coarse-grained monomineralic crystals to well-bedded material associated with framboidal pyrite and carbon. Galena encloses the main silver-bearing mineral, freibergite, and the processing of the ore is therefore uncomplicated (J. Knights, personal communication). In contrast, the mineralogy of silver-bearing minerals at Hilton Mine is highly complex. Here there are ten silver-bearing minerals of which the most important are native silver, stephanite, pyragyrite, freibergite, argentopyrite and silver-bearing chalcopyrite. The silver-bearing minerals are associated with both sulphides and gangue minerals and therefore represent a formidable metallurgical problem (J. Knights, personal communication). Sphalerite at Mount Isa Mine occurs in association with galena, is normally pale brown in colour, and has an average iron content of 6–7 wt. %. It frequently occurs as conformable monomineralic layers.

The ore bodies at McArthur River differ in two important respects from the Mount Isa deposit. Firstly there is very little silver present with the highest reported value being 120 ppm (Lambert, 1976). Ridge (personal communication) considers that the presence of silver in galena is related to temperature of formation, with those deposits which contain silver having formed at temperatures above 225°C. Secondly due to the absence of metamorphism the sulphides are extremely fine-grained ($< 4\mu m$), and as a result only low recoveries are possible by conventional metallurgy during the processing of the ore.

(e) Non-stratiform ores

Non-stratiform base-metal deposits occur both at Mount Isa and McArthur River. At Mount Isa the copper ore associated with the silica–dolomite rock is of considerable economic importance and merits description for this reason alone. At both locations the discordant ores are closely associated spatially with the stratiform ores, and discussion of their genesis has been inevitably linked together. The problem to be resolved is whether the stratiform and discordant deposits are part of a single ore-forming process, or whether their association is merely spatial.

(i) The silica–dolomite rock at Mount Isa

The silica–dolomite rock is a calcareous and siliceous zone of irregular morphology within the Urquhart Shales. It has a strike length of 2600 m and a maximum width of 530 m (Mathias and Clark, 1975). The silica–dolomite includes four separate facies, of which one, the crystalline dolomite, is of restricted distribution (Fig. 6.11). A carbonate-rich facies predominates in the upper part of the silica–dolomite, where the contact with the Urquhart Shale and stratiform ores is gradational and interdigitating. The lower part of the silica–dolomite unit is highly siliceous, and consists of a poorly bedded carbonaceous chert with chalcopyrite called the brecciated siliceous shale. The lower siliceous facies of the silica–dolomite contains the highest grades of copper, and has a faulted contact with the underlying Eastern Creek Volcanics.

Syngenetic models for the copper ore have been proposed which suggest a common origin with

Ore deposit geology and its influence on mineral exploration

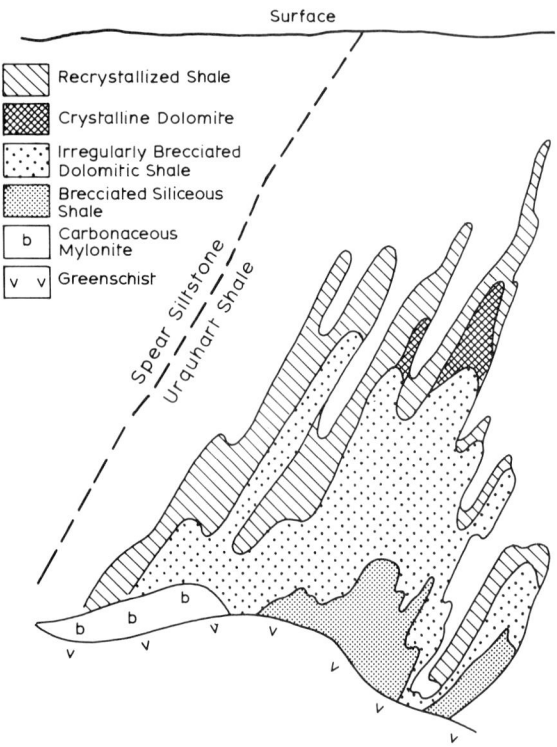

Fig. 6.11 Facies distribution within the silica–dolomite rock of Mt Isa Mine (from Mathias and Clark 1975).

They concluded that different mechanisms must be postulated for the copper mineralization. Detailed structural, microstructural and petrological studies of the silica–dolomite suggest that it has formed by replacement of the Urquhart Shales (Perkins, 1981; Swager, 1983). Three structural events are now recognized within the Urquhart Shale (D_1 to D_3). These structures have also been identified in the silica–dolomite and permit the sequence of metasomatic events to be determined. Three separate metasomatic processes are recognized and are interpreted as being syn-D_3 in age. Siliceous solutions are considered to have invaded and silicified the dolomitic Urquhart Shale resulting in an outer and expanding metasomatic front of dolomite. Chalcopyrite replaces the crystalline dolomite and is therefore believed to have been introduced at a late stage. The implications of the structural analysis are that the stratiform ores are separated from the copper ores by at least two stages of deformation. Perkins (1984) estimates, from the interpretation of radiometric age determination, that there is a 190-million-year difference between the deposition of the stratiform Pb–Zn ores and the emplacement of the copper ore body.

the stratiform lead–zinc deposits. The main basis for this model is the close spatial association between the two ore types. Stanton (1963) utilized geochemical evidence to support a syngenetic model for the silica–dolomite, and interpreted it as a facies variation of the shales which host the lead and zinc deposits. Finlow-Bates and Stumpfl (1979) proposed that copper was deposited in a subsurface breccia from a boiling hydrothermal solution, which subsequently reached the seafloor and deposited lead, zinc and iron.

Sulphur isotope and structural studies indicate that the copper ores have a separate genesis from the stratiform ores. Smith et al. (1978) have demonstrated that $\delta^{34}S$ has a more restricted range in sulphides from the silica–dolomite compared with $\delta^{34}S$ values from stratiform ores (Fig. 6.12).

(ii) Discordant mineralization at McArthur River

Small discordant lead–zinc deposits have been described from the McArthur River area by Williams (1978), and he refers to them as Cooley I, Cooley II, Cooley III and Ridge I.

The deposits are discordant veins or disseminations in which pyrite, sphalerite and galena are the main ore minerals. Chalcopyrite, bornite, marcasite and tetrahedrite also occur in order of decreasing abundance. The principal gangue minerals are dolomite, barite and quartz. The discordant mineralization occurs in breccias which are in turn hosted by the Emerugga and Cooley Dolomites (Fig. 6.13).

These discordant deposits are at present mainly of genetic interest. Williams (1978) has shown from the ratio Cu:Pb–Zn that there is a well-defined pattern of metal zoning within the deposits

Sediment-hosted copper–lead–zinc deposits

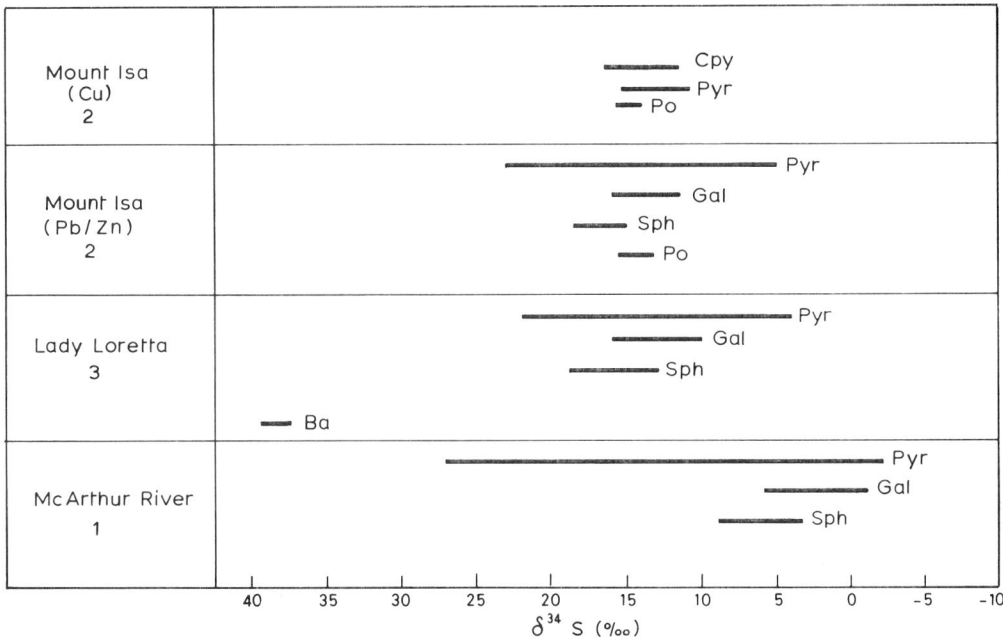

Fig. 6.12 Range of $\delta^{34}S$ from sediment-hosted lead–zinc and copper ore deposits, Northern Australia. Sources: 1, Smith and Croxford (1975); 2, Smith, Burns and Croxford (1978); 3, Carr and Smith (1977). Abbreviations: Pyr, pyrite; Gal, galena; Sph, sphalerite; Po, pyrrhotite; Ba, barite.

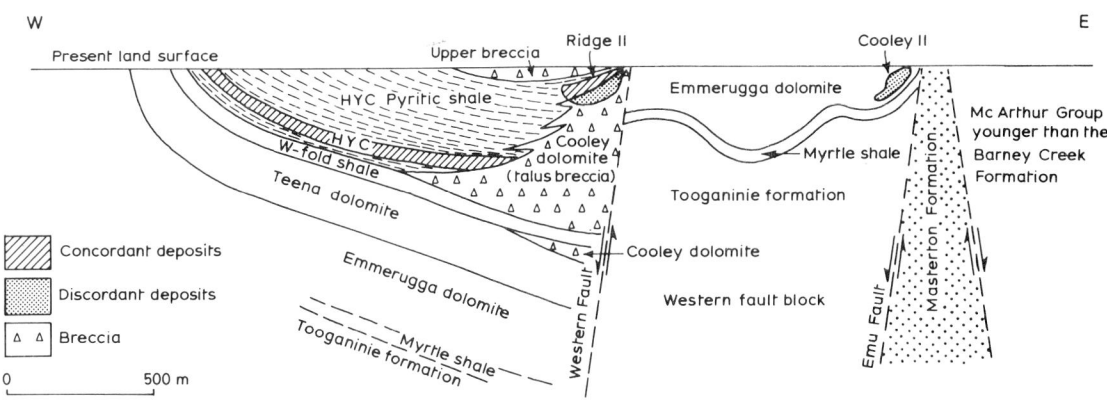

Fig. 6.13 Schematic cross-section through the area of most intense mineralization at McArthur River, looking north (from Williams 1978, after Walker et al. 1977).

which he relates to mineralizing fluids emanating from the Emu and Western Faults. The epigenetic discordant deposits have many affinities with the concordant Ridge and HYC deposits.

(f) Structure

The Northern Australian sediment-hosted base-metal deposits have similar tectonic settings, but differences in their structural evolution are

237

reflected in contrasting lithologies, in the differing proportions of extrusive volcanics and in dissimilar styles of deformation.

The McArthur River–HYC deposits are located on the eastern side of a northerly trending rift valley, the Batten trough, which is at least 500 km in length and 100 km in width. Subsidence of the rift floor allowed the accumulation of shallow water sediments which attain a thickness of more than 5000 m but which thin both towards the margin of the rift and towards a median ridge (Murray, 1975). During the later stages of sediment accumulation (Barney Creek Formation) localized subsidence on the eastern side of the rift created a depression extending over a surface of 400 km^2. Uneven subsidence resulted in the formation of subsidiary basins within which pyritic bituminous shales accumulated. Base-metal deposits are associated with these sub-basins, the most important of which hosts the HYC deposits (Murray, 1975). The mineralized beds exhibit many features of soft sediment deformation, particularly near the rift margins, but tectonic folding of the lithified sediments is characterized by open basin and domal folding disrupted by normal faulting (Murray, 1975).

The Mount Isa and Hilton deposits are located within the Leichhardt River Fault Trough, which is a northerly trending rift valley approximately 600 km in length and 50–60 km wide. The trough appears to have been initiated on a basement about 1800–1865 million years old, and continued subsidence resulted in the accumulation of 16 km of sediments and volcanics (Derrick, 1982). Derrick (1982) provided detailed stratigraphic and sedimentological evidence for the development of a positive structure during subsidence of the rift which he termed the Mount Gordon Arch. This feature subdivides the Leichhardt River Fault Trough into two second-order basins. The rift valley is filled predominantly with shallow water sediments but there is also a significant component of extrusive volcanics of which basalts assigned to the Eastern Creek volcanics are the most extensive.

The base-metal deposits within the Leichhardt River Fault Trough have been intensively deformed by both folding and faulting (Fig. 6.14). Folding at Mount Isa Mine has been described by McClay (1979, 1983) who considers that the deformation is predominantly due to tectonism rather than being the product of penecontemporaneous slumping. Four stages of folding are recognized by McClay, who analyses the structural style of folding in terms of the differing competencies of the ores and their enclosing sediments. Rapid variation of sediment thickness within the Leichhardt River Fault Trough is interpreted by Derrick (1982) as being controlled by west and west–north–west trending growth faults which he suggests may have acted as channelways for the ore-bearing fluids. Bell (1983) believes that the rapid change in sediment thickness can be better explained by thrusting. Bell's structural investigation is based partly on an area to the north of Mount Isa which he considers can be interpreted as a thrust duplex structure. Furthermore Bell (1983) concludes from his structural studies that the Mount Isa group rocks have been displaced by thrusting for more than 234 km from north to south.

6.3.3 Genesis of the Northern Australian lead–zinc deposits

The northern Australian sediment-hosted lead–zinc deposits have been the subject of differing interpretations by successive generations of geologists. Genetic theories range from synsedimentary (Stanton, 1963) to hydrothermal replacement (Smith and Walker, 1972). In recent years sedimentological, structural and sulphur isotope studies have added a substantial data base to genetic models. In particular, the investigation of the unmetamorphosed McArthur River deposits has provided an insight into the mode of formation of the more highly metamorphosed deposits south of the Murphy Tectonic Ridge.

(a) Source of sulphur
Fig. 6.12 shows a comparison of sulphur isotope values for pyrite, galena, sphalerite and barite

Sediment-hosted copper–lead–zinc deposits

Fig. 6.14 Intensely contorted shale bands from one of the Pb–Zn ore bodies within the Urquhart Shale, Mount Isa Mine.

from the deposits under consideration. It can be seen that in each deposit galena and sphalerite appear to be in isotopic equilibrium. Pyrite has a wide range of $\delta^{34}S$ values which Solomon and Jensen (1965) attributed to oxidation–reduction reactions of possibly biogenic origin. Smith and Croxford (1973) investigated the sulphur isotope systematics at McArthur River. They concluded that there were dual sources of sulphur, with that in pyrite being generated by bacterial reduction of seawater sulphate. The sulphur in sphalerite and galena was considered to be brought in by the same fluids which transported the lead and zinc. Smith and Croxford (1973) also noted that the $\delta^{34}S$ for pyrite became progressively heavier towards the top of the HYC deposit, which they attributed to bacterial reduction of a limited supply of sulphate. In a subsequent paper Smith and Croxford (1975) concluded that isotopic variation in pyrite at McArthur River reflected water depth, availability of sulphur and the degree of reduction of sulphate reservoirs.

Stable isotope studies on the Lady Loretta deposit were carried out by Carr and Smith (1977). They concluded that sphalerite and galena were formed under equilibrium conditions from the same sulphur source. However, neither mineral formed in isotopic equilibrium with pyrite. They also concluded that lead and zinc were transported as sulphides and injected into a restricted reservoir of seawater sulphate.

(b) Source of metals

The source of metals for the Northern Australian stratiform lead–zinc deposits is not only an interesting genetic problem, but is of considerable significance for future exploration. Several alternative sources have been proposed. These include subaqueous metal-rich fluids released during volcanic activity (Murray, 1975), metal-rich solutions expelled from the sedimentary pile (Williams, 1978) and metals leached from sediments and volcanics by downward-penetrating fluids (Russell, 1983).

The main supporting evidence for a volcanic origin for the metal comes from the abundance of tuffaceous material within the rocks which host the ores. Tuffs characterize the stratigraphy of both the McArthur Basin and the Leichhardt River Fault Trough and they become particularly conspicuous in those sediments which enclose the ore bodies. For example, at Mount Isa the 14/30 composite tuff marker band is a major horizon about 1 m wide. Above this horizon the main ore bodies occur, but below it the mineralization is weak (C. Hartley, personal communication). Is this volcanic detritus of any genetic significance?

Lambert (1976) considered that the tuffs in the McArthur Group were erupted on land or in shallow water, which is in accord with more recent sedimentological studies (Muir, 1983). In these circumstances it is difficult to envisage how the metals associated with volcanism would be trapped and preserved. Furthermore, recent volcanic activity provides evidence that in some cases tuffs are deposited several hundreds of kilometres from the initial eruption.

An alternative mechanism is the ascent of metal-rich fluids, expelled along major faults as a result of compaction. Williams (1978) and Rye and Williams (1981) have investigated this possibility at McArthur River with the aid of oxygen, sulphur and carbon isotopes. In addition they studied the pattern of metal zoning associated with the Emu fault. They concluded that both the discordant vein-type deposits and the concordant stratiform deposits were formed from the same fault-derived fluid.

At Mount Isa major structures such as the Mount Isa Fault show little evidence of hydrothermal alteration, or of a control on mineral deposition. However, Smith (1969) cited field evidence in support of penecontemporaneous faulting, and suggested that these faults may have been the loci of fumarolic activity.

Russell (1983) has suggested that the thick sedimentary pile at Mount Isa (and other sedimentary-exhalative deposits) is the source of metals. He considers that the metals were leached by descending seawater acidified by early reaction with clays and feldspars.

(c) Models of ore genesis

In constructing a genetic model for the northern Australian lead–zinc deposits it must be clearly recognized that some evidence is based upon carefully researched and unambiguous data, but that several critical elements of the model are largely speculative. For example, little evidence is available about the nature of the ore fluid, and temperatures of deposition based upon isotope studies have a wide range of values. Thus Rye and Williams (1981) proposed temperatures ranging from 120 to 240°C for the McArthur River deposit. Similarly caution must be exercised in using the McArthur River deposit as a model, because the data are obtained from boreholes. It would be surprising if development of the deposit did not produce major changes in ideas concerning its genesis.

The following aspects appear to be well established:

(1) The environment of deposition of the host sediments was dominantly shallow water, or emergent, with evidence of sabkha-type evaporite minerals having formed.

(2) The tectonic environment of sedimentation and ore deposition was a graben structure.

(3) The stratiform Pb–Zn ores are syndiagenetic, but post-date pyrite formation.

(4) Separate sulphur sources contributed to the formation of pyrite (biogenic reduction of sea-water sulphate) and galena and sphalerite.

Within this framework of well-researched data we can assess some of the genetic models which have been proposed for the formation of the Pb–Zn ore bodies in northern Australia. We intend to review two alternative models: the basinal compaction model (Gustafson and Williamson, 1981) and the formation of base-metal deposits by hydrothermal convection (Russell, 1983).

(i) Basinal compaction model

The essential elements of this model are that lead and zinc are extracted by the interaction between highly saline formation water and rock forming silicates. The metal-bearing brines are then expelled towards the surface along fault zones as a result of compaction. This model draws on the well-documented migration of hydrocarbons in sedimentary basins. In the context of Northern Australia the model is supported by the evidence of evaporite sediments within the stratigraphy of the ore deposits. This means that highly saline fluids were almost certainly present within the sediment during the early stages of diagenesis. Further support for the model comes from the evidence that the metals in the HYC deposit emanated from fault zones (Williams, 1978).

If we pursue the analogy with petroleum deposits, we find that most of them are formed between depths of 700 and 3000 m. This is in close accord with calculations which suggest that about 75% of the pore fluid is expelled from clays at depths of 700–2000 m (Chapman, 1973). The temperature at which the pore fluids are expelled can be inferred by considering the type of geothermal gradient which might have prevailed in a graben in Proterozoic times. This information cannot be known for certain, but if we assume that, under extensional tectonic conditions, heat flow will be at the high end of the spectrum for a continental regime it is reasonable to suggest a gradient in the order of 40°C/km. Under these circumstances the bulk of the ore fluid would be expelled at temperatures in the order of 70–100°C. Russell (1983) comments that additional water would be driven off during the montmorillonite–illite transition at 90–120°C. The limited information on the temperature of the ore fluid in the McArthur River district indicates temperatures in the range of 120–240°C (Rye and Williams, 1981). This poses a problem for the supporters of the basinal compaction theory which has not yet been resolved.

(ii) Hydrothermal convection

Studies of porphyry copper systems have led geologists to recognize the importance of hydrothermal convection as an ore-forming process. However, in this context one can identify the pluton as the principal energy source which drives the convective system. In the Mount Isa area it is possible that the Sybella granite provided an appropriate heat source. There are two problems here. Firstly, the absence of comparable igneous activity in the other mining districts. Secondly Russell (1983) has drawn attention to the fact that the Mount Isa ore bodies have grown larger with time. If the intrusion of a granitoid pluton is the driving mechanism of convection, one would anticipate that the efficacy of the ore-forming process would decline with time.

We are therefore led to consider Russell's (1983) model of a downward-penetrating convection cell. Russell considers that all sedimentary-exhalative deposits are characterized by an 'extraordinary foundering' of the seafloor at the site of deposition. He suggests that this would result in extensional strain in the upper part of the crust. Furthermore, he proposes that the condition of extensional strain would result in extensive microfracturing of the crust, producing a sufficiently enhanced permeability to allow convective circulation at relatively low temperatures. We are not convinced of the evidence for foundering of the seafloor in the Mount Isa area. On the contrary, the Urquhart Shale suggests monotonously uniform conditions of sedimentation (Fig. 6.10), similar to those in which the underlying shales and siltstones of the Mount Isa Group were deposited.

We are also sceptical about some other aspects of Russell's model. In particular, we find it difficult to visualize the ingress of fluid into the upper part of the crust. Recent experimental work at the Camborne School of Mines Geothermal Project (Pine and Batchelor, 1984) reveals that the downward growth of microfracturing associated with hydraulic injection is dependent, not only on an anisotropic stress field, but also requires the presence of strongly jointed rock. We doubt whether the partly lithified sediment of the Leichhardt River and Batten Troughs would have had sufficient rigidity for joint formation. In spite of these misgivings we must accept that Russell's model does explain the high temperature of the ore fluid and accounts for the upward increase in the size and grade of the stratiform deposits.

Perhaps a solution to this problem may lie in a genetic model which utilizes elements from both basinal compaction and downward convection theory. There may be a combination of suitable conditions during the late stages of compaction, when the rock has become sufficiently lithified to allow joint formation. Under these circumstances fluid expelled during the montmorillonite–illite transition might begin to migrate downwards rather than towards the surface.

6.3.4 Exploration for shale-hosted lead–zinc deposits in Northern Australia

(a) Historical perspective

The Mount Isa ore body was discovered in 1923 by J. C. Miles, a prospector, who recognized the significance of the gossanous outcrops. Mining began in 1927 but regional exploration did not take place until 1947. By this stage both the lead–zinc and copper ore bodies at Mount Isa were developed and concepts of ore genesis had been formulated (Doust, 1979). The principal guides to mineral exploration at that time were believed to be as follows (Clark, 1975):

(1) The strike extension of the Urquhart Shale was considered to be the main prospective area.
(2) Structural control of the ore body was considered to be important, particularly bedding plane slip associated with folding.
(3) The presence of gossanous outcrops, derived from the halo of pyritization around the ore body, was recognized as the surface expression of mineralization.

Exploration beyond the limits of the mine property was initiated in the late 1940s by a rival company. The geologists who designed the programme believed the ore body was structurally controlled, and considered that the chances of finding ore decreased with increasing distance from the Mount Isa fold (Doust, 1979). The exploration effort was therefore concentrated within an area extending for 15 km north from the mine property. Although unsuccessful, this exploration activity stimulated the prospecting activities of Mount Isa geologists, who concentrated their efforts some 20 km north of the mine property utilizing the gossanous expression of the Mount Isa ore body as their principal guide. The main exploration technique utilized was detailed geological mapping, with particular reference to the distribution of ferruginous zones (Clark, 1975). Five gossanous areas were identified but the subsequent evaluation programme was curtailed, partly due to problems of groundwater and partly because the mineralization which was intersected during drilling appeared inferior to that at Mount Isa. However, this prospect was ultimately to become the Hilton Mine (Figs. 6.15 and 6.16).

During the 1950s the concept of syngenetic ore deposition gained gradual acceptance in Australia. As a result geologists began to widen the scope of their exploration programmes beyond areas thought to be structurally favourable for ore deposition. The Urquhart Shale and its correlatives became recognized as the principal exploration target, not only near Mount Isa but potentially over a much wider area. New areas were therefore selected for exploration, on the basis of being underlain by sedimentary units of comparable age to the Urquhart Shale. Emphasis was placed on regional geological mapping and geochemical prospecting became increasingly used. In 1955, whilst soil sampling within a $60\,\text{km}^2$ target in the McArthur River area, a field assistant collected a sample of gossan. The gossan was found to contain hemimorphite, and subsequent drilling intersected Pb–Zn mineralization within the HYC pyritic shale (Murray, 1975). The target area had initially been selected because it occurred

Sediment-hosted copper-lead-zinc deposits

Fig. 6.15 The Hilton Mine (1983). In the foreground are the gossanous ridges which originally led geologists to the discovery of the deposit.

on a favourable lineament and therefore may be considered as a discovery based on geological prediction (M. J. Russell, personal communication).

A climate of exploration success and improved metal price led to a reappraisal of the Hilton deposit during the period 1966–1969. The acceptance of the syngenetic theory, combined with the recognition of tuff beds as important marker horizons, led to the reinterpretation of the Hilton geology (Clark, 1975). Using the revised interpretation, the first hole which was drilled intersected seven ore zones which were mineable by Mount Isa standards.

The grass-roots exploration of the 1950s and 1960s was accompanied by a phase of prospect evaluation, which led to the discovery of the Lady Loretta deposit. In 1968 the Lady Annie copper mine was re-evaluated which entailed the examination of 5000 m of borehole core (Doust, 1979). At the same time reconnaissance soil sampling was carried out in the area, with more detailed geochemistry and induced polarization surveys (IP) confined to the immediate vicinity of the mine.

Results of the geochemical survey revealed three areas containing abnormally high lead values, which coincided in part with anomalies detected by the IP survey. The initial hole drilled on the lead anomaly intersected 7.6 m of oxidized mineralization with an average of 21.2% Pb (Cox and Curtis, 1977). Subsequent drilling has defined the Lady Loretta ore body which has a reserve of 9×10^6 tonnes at 6.5% Pb and 14.8% Zn (*MIM Holdings Annual Report*, 1983).

Ore deposit geology and its influence on mineral exploration

Fig. 6.16 Gossan adjacent to Hilton Mine.

(b) Evaluation of exploration techniques

(i) Remote sensing

Aerial photographs have been widely used as a basis for geological mapping, but photo interpretation cannot be directly linked with successful ore discovery. This is surprising in view of the fact that many of the deposits discovered have had an expression of gossanous material, which should be evident in colour photography.

(ii) Geophysics.

The main value of geophysics has been in providing data to support geological mapping. In the Mount Isa region magnetic surveys have been used to define the contact between the Mount Isa group sediments and the underlying Eastern Creek volcanics (C. Towsey, personal communication). The main exploration benefit in this context relates to potential copper sulphides associated with possible repetitions of the silica–dolomite rock. In spite of the presence of pyrrhotite in some of the stratiform lead–zinc ores, magnetic surveys have not been successful in defining specific ore-bearing horizons within the Urquhart Shale.

Extensive leaching of rocks near the surface has severely limited the application of standard electromagnetic methods in Northern Australia. IP anomalies at Lady Loretta provided supporting data for the geochemical work, and enhanced their value as drilling targets (Doust, 1979). However, IP has been used with little success in the Mount Isa area due to a combination of deep weathering (50 m), saline groundwater and carbonaceous shales (C. Towsey, personal communication). At present down-the-hole geophysics appears to have the best potential.

(iii) Geochemistry

Geochemical prospecting has been the most successful exploration technique used in the search for base-metal deposits in Northern Australia. It has been applied both as a reconnaissance exploration tool, and for the evaluation of gossanous material.

Stream sediment sampling
There are few documented examples of stream sediments having been used successfully to identify base-metal mineralization in Northern Australia. However, orientation studies in the Dugald River area indicate that analysis of the silt-sized fraction (less than 80 mesh) for Cu, Pb, Zn and Ag may be used to define dispersion patterns which would lead to the discovery of new ore deposits (Connor et al., 1982).

Soil geochemistry
Soil geochemistry has been the most widely used geochemical technique and its success has vindicated its use in semi-arid terrain. Soil geochemistry combined with geological studies led to the discovery of the Lady Loretta lead–zinc deposit (Cox and Curtis, 1977). The reconnaissance survey involved sampling on a 400 × 100 ft (122 × 30.5 m) grid, and initially samples were analysed for copper alone. Recognition of the similarity between the Lady Annie and Mount Isa sediments led to the inclusion of lead, zinc and silver in the geochemical survey at a later stage. The contoured values of lead are shown in Fig. 6.17 and define three areas in which lead exceeds 421 ppm. Preliminary diamond drilling was carried out in area B, and intersected oxidized lead ore. In spite of the fact that the geochemical survey was carried out adjacent to the Lady Annie copper mine, no high copper values were recorded in the soils over the Lady Loretta deposit. There is an important lesson to be learnt from this case history: exploration programmes with narrowly defined objectives may miss important targets, or, alternatively, single element surveys may please accountants but are in fact a foolish economy.

The McArthur River deposit was located by the recognition of hemimorphite in gossanous material. Subsequent orientation studies have shown that distinctive Pb–Zn anomalies occur in residual soil overlying mineralization but no geochemical signature was detected in areas covered by transported alluvium (Haldane, 1980).

Rock geochemistry
The Mount Isa, Hilton and McArthur River deposits are all characterized by the development of a gossanous outcrop which proved to be the basis of their discovery. However, in other instances gossans have been drilled without identifying any associated base metals at depth. Geochemistry is probably the best discriminator between gossans overlying base-metal mineralization and barren ironstones. The most important element for identifying gossans of base-metal association is lead, which reflects the common presence of secondary lead minerals within the oxidized zones of weathered ores. Zinc

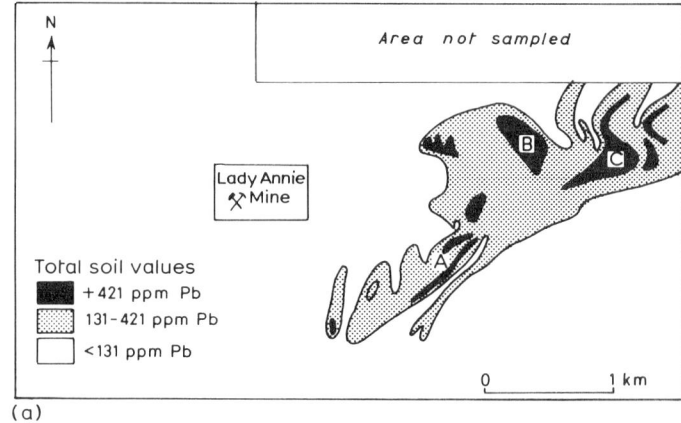

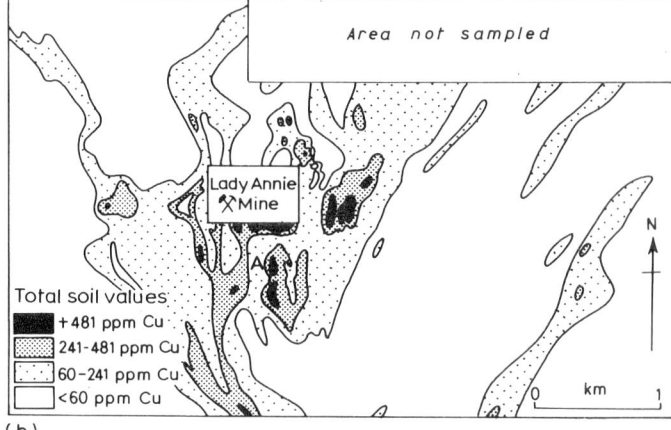

Fig. 6.17 Results of reconnaissance geochemical survey which led to the discovery of the Lady Loretta deposit. (a) lead; (b) copper. (From Cox and Curtis 1977.)

is considered to be an unreliable indicator (Taylor and Scott, 1982) in spite of the fact that hemimorphite in gossan led to the discovery of the McArthur River deposit. Arsenic, barium and antimony may be used as additional pathfinder elements but a wider suite of trace elements is preferred. Taylor and Scott (1982) recommend analysis for Pb, P, Ba, As, Zn, Mn, Co, S and Sb followed by interpretation using stepwise discriminant analysis. Using this technique Taylor and Scott were able to correctly categorize 90% of the gossanous samples they collected from the Mount Isa area.

Gulson and Mizon (1979) have shown that lead isotope measurements on gossans can distinguish between those overlying 'stratiform' Cu–Pb–Zn deposits and those developed on barren ferruginous rocks.

(c) Summary

Concepts in ore genesis have played an important role in base-metal exploration in Northern Australia. Early exploration was constrained by an incorrect emphasis on the structural control of mineralization. The application of syngenetic theory widened the area of prospective terrain and led ultimately to the discovery of new deposits.

Exploration techniques have mainly included a combination of geological mapping, soil and rock-chip geochemistry resulting in the discovery of several outcropping or near-surface deposits. At the present time neither exploration techniques nor

6.3.5 Geology of the Irish carbonate-hosted lead–zinc deposits

(a) Introduction

The Central Plain of Ireland is an area of significant base-metal production. The discovery of the Tynagh deposit in 1961 resulted in a greatly increased exploration programme which led eventually to the development of four additional mines: Gortdrum, Silvermines, Navan and Ballynoe (barite) (Fig. 6.18). Gortdrum, Tynagh and Silvermines are now closed but Ballynoe is still operating and Navan remains the most important lead–zinc deposit in Western Europe. New base-metal discoveries have been made during the last decade, but depressed metal prices and uncertainty concerning taxation have discouraged new mine development.

The main reason for discussing this mining district is because it has ore deposits with both syngenetic and epigenetic age relationships between sulphide minerals and carbonates. It therefore represents a transitional type between sedimentary-exhalative and Mississippi Valley-type deposits.

(b) Host lithology

The carbonate rocks which host the Irish base-metal deposits are of Lower Carboniferous age, and represent a change from predominantly continental conditions in Devonian times to a marine environment. The sea transgressed northwards in Lower Tournaisian times across a continent which had been extensively

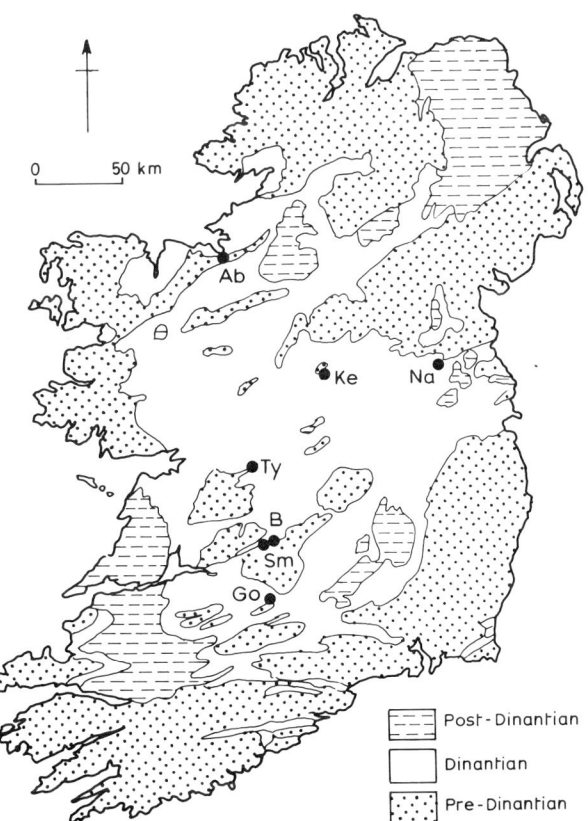

Fig. 6.18 Location of principal base-metal deposits in Ireland. Ty, Tynagh; Ab, Abbeytown; Go, Gortdrum; Ke, Keel; Na, Navan; Sm, Silvermines; B, Ballynoe (Magcobar). (From Boast *et al.* 1981a, after Morrisey *et al.* 1971.)

peneplained, although at Navan there is evidence of topographic variation in the palaeo-relief.

The lowermost part of the Carboniferous succession reflects the transition from continental to marine conditions. At Tynagh and Silvermines the Lower Tournaisian sediments are initially calcareous shales, whilst in the Navan area littoral and fluviatile siliciclastics fill erosional hollows around palaeohighs (Andrew and Ashton, 1982). For the main part Tournaisian sedimentation in Ireland is represented by variable proportions of carbonate and shale which reflects a shallow water environment. At Navan, facies variation was controlled laterally by the influence of palaeohighs, and the sedimentary sequence changed in time from a tidal flat complex, through quiescent lagoonal conditions to a high-energy environment (Andrew and Ashton, 1982). At Tynagh and Silvermines a more consistent pattern of sedimentation prevailed, although at Silvermines variation in sediment thickness can be attributed to the early influence of faulting (Boyce et al., 1983a).

In Upper Tournaisian times the sediment of the principal mining areas is characterized by Waulsortian reefs, banks of micrite and biomicrite which attain 1 km in thickness in the west of Ireland. The reefs reflect subsidence of the sedimentary environment to depths of seawater which may have attained 200 m (Boyce et al., 1983a).

The transition from Tournaisian to Visean times marks a fundamental change in the environment of sedimentation. At Navan this is represented by an erosional surface of considerable magnitude, which Andrew and Ashton (1982) interpret as representing a major phase of tectonic uplift. However, Boyce et al. (1983a) suggested that the unconformity is a preserved submarine slump scar, formed by the downslope movement of sediment on a relatively steep slope. At Silvermines the increase in tectonic activity is reflected in extensive brecciation of both the carbonate sediments and the associated syngenetic sulphides.

The upper part of the Lower Carboniferous stratigraphy is characterized by a gradual transition into barren argillaceous limestones, which are locally referred to as calp.

(c) Mineralogy, size and grade

The Irish lead–zinc deposits usually have a geometry which is either lensoid or wedge-shaped and are associated with major normal faults. In size and grade they are highly variable but usually contain less than 15×10^6 tonnes (Morrisey et al., 1971) with the notable exception of Navan which had an original size of 69.9×10^6 tonnes with grades of 10.1% Zn and 2.6% Pb.

The mineralogy of the Irish base-metal deposits is relatively simple in terms of mineral species but the textural relationships of the ore minerals are complex. Sphalerite and galena are the metallic minerals of economic significance. Sphalerite is low in iron and is commonly fine-grained although pale yellow recrystallized sphalerite may also occur. Galena varies from fine to coarse-grained and is argentiferous. The silver is normally associated with sulpho-salts which occur as intergrowths within the galena. Friebergite (argentiferous tetrahedrite) and pyrargyrite are the main silver minerals at Navan (Andrew and Ashton, 1982) whereas argentian boulangerite and jordanite were the main silver-bearing phase at Silvermines (Taylor and Andrew, 1978). Pyrite is a common ore mineral and varies in concentration and texture within individual deposits. Numerous sulpho-salts occur in trace amounts but only those containing silver are of economic significance; in some cases they are an undesirable constituent of the ore, and at Tynagh high concentrations of tetrahedrite in the concentrate resulted in payment of a penalty to the smelter (J. Hutchings, personal communication).

Gangue minerals usually include varying proportions of calcite, dolomite, quartz and clay minerals. Barite may occur as a gangue mineral but in some cases is mined, either as a separate ore body (Ballynoe/Magcobar) or is extracted from the mill tailings (Tynagh).

All the lead–zinc deposits exhibit complex textural relationships between the base-metal sulphides and their associated carbonate host rocks which reflect a number of variables. These include the time relationships between sedimentation and the introduction of sulphides, the effects of tectonism in terms of both increased porosity and permeability of the carbonate host and soft sediment deformation, the pressure under which the ore fluid was introduced into the sediment, and the structural level at which the ore fluid was introduced into the environment.

At Silvermines recent exploitation of the carbonate-hosted sulphides was from two distinct styles of mineralization (Taylor and Andrew, 1978). Stratabound ore bodies (subdivided into Lower G and K zones) occur within dolomitized carbonates of Tournaisian age and exhibit a wide spectrum of textures. In the vicinity of faults the mineralization is typically coarse-grained, and occurs in intraclastic spaces in brecciated dolomite, and in thick symmetrically banded tensional veins and minor en echelon gash zones. Beyond the influence of faulting, sulphides fill diagenetic fractures or occur as irregular replacements.

The stratiform ore bodies (subdivided into Upper G, B and Magcobar) occur between Waulsortian carbonates of Late Tournaisian–Visean age, and have textures which are typical of syngenetic or syndiagenetic deposition. In the Upper G zone the sulphides are microcrystalline and display a wide range of colloform textures. Boyce *et al.* (1983b) interpret colloform pyrite from Ballynoe as hydrothermal chimneys or mounds analagous to those forming at the present day on the East Pacific Rise. Both the carbonates and sulphides exhibit a variable degree of brecciation, which reflects the influence of tectonism on the partly lithified sediment.

The ore textures within the No. 5 ore lens at Navan have been described by Andrew and Ashton (1982). Most of the textures are interpreted as the contemporaneous deposition of sulphides and carbonates, although some of the sulphides were deposited epigenetically in zones of high permeability.

The Silvermines and Navan ore bodies are characterized by zoning of their lead and zinc values. At Silvermines the zoning is developed within the B and Upper G ore bodies. Contours of metal values within these ore bodies reveal both an ENE trend, parallel to the Silvermines fault, and a NW trend which reflects a structural control by NW trending faults which channelled the ore fluid at depth and constrained its site of emergence on the seafloor (Taylor and Andrew, 1978).

At Navan the distribution of lead and zinc values have been studied in the No. 5 lens (Andrew and Ashton, 1982). The dominant trend of the high metal values is north–easterly, although a subsidiary north–westerly trend is also apparent (Fig. 6.19). The main north–easterly trending zone of high metal concentration is coincident with an area of north–easterly trending, steeply dipping veinlets containing complex bands of fine-grained sphalerite and galena. It also coincides with the projected trace of the intersection of the 'A' and 'B' faults at depth (Fig. 6.20). Andrew and Ashton (1982) believe this indicates the control of the ore fluid by a deep north-easterly trending structure, and by smaller north-easterly trending fractures near to the surface.

Contour plans of the metal values at Tynagh have not been published. However, as the ore body is traced northwards from the Tynagh fault the sulphides pass gradationally into banded iron-formation (Fig. 6.21). Russell (1975) has shown that a halo of manganese extends for 7 km from the primary ore body. These observations support the concept of an ore fluid introduced from the Tynagh fault on to the seafloor at the same time as carbonate sedimentation.

(d) Structure

The Lower Carboniferous rocks of the Irish Central Plain are deformed by both folding and faulting which have an ENE Caledonoid trend but which are attributed to Hercynian movement. Both the Silvermines and Navan ore bodies are located near to significant ENE trending folds, but these structures have minimal significance in terms of either the genesis or the deformation of base-metal sulphides.

Faulting plays a vital role in the location of the

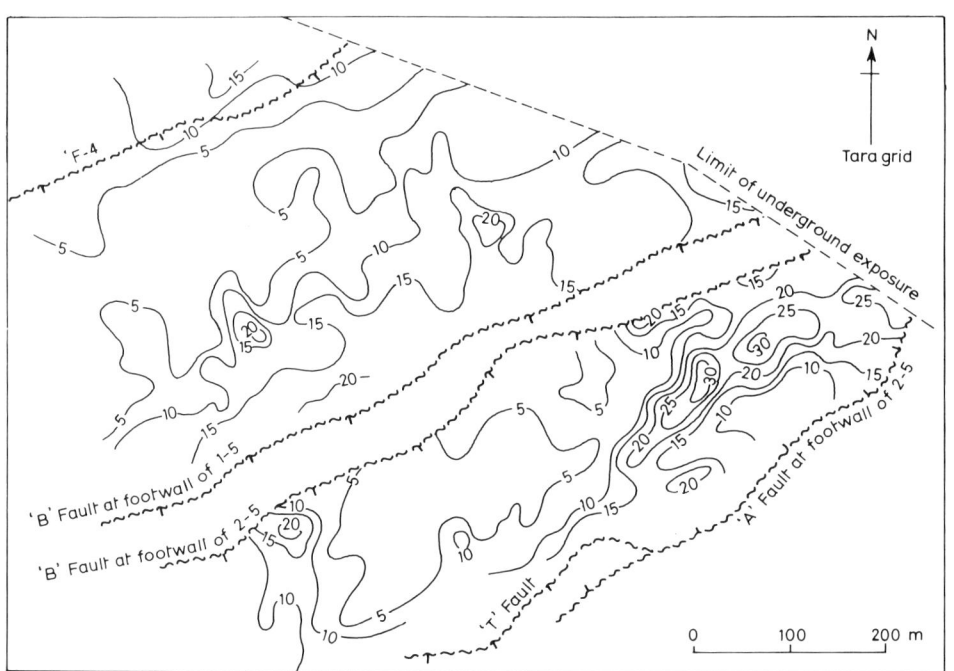

Fig. 6.19 Isograde contour plot for percentage of zinc and lead within the No. 5 Lens at Navan (from Andrew and Ashton 1982).

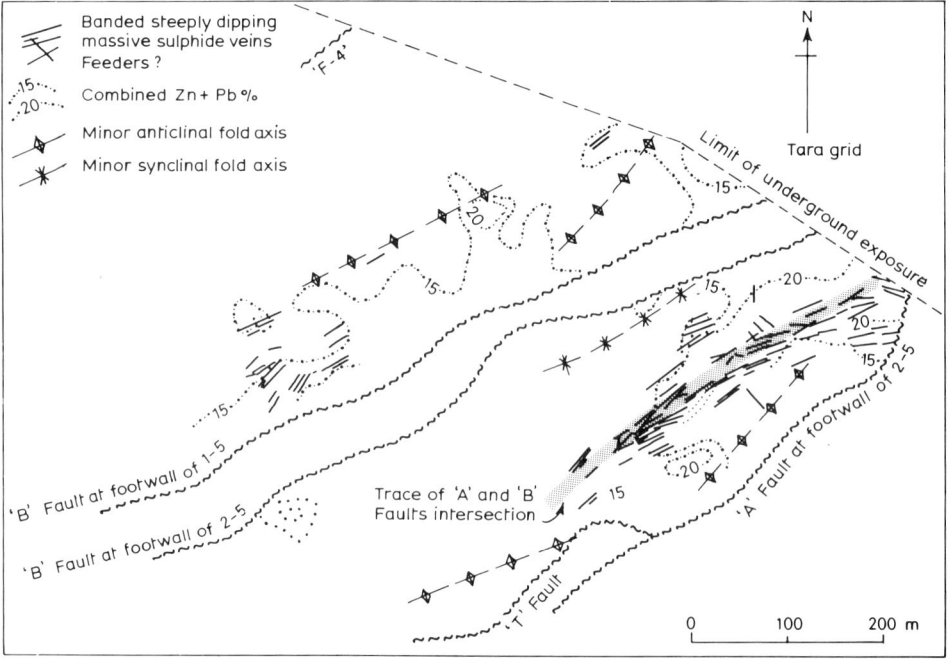

Fig. 6.20 Composite diagram showing the position of mapped postulated 'feeder' veins, major structures and percentage zinc and lead isograde contours at Navan (from Andrew and Ashton 1982).

Fig. 6.21 Banded iron formation at Tynagh mine. Banded iron formation is the lateral equivalent of the lead–zinc sulphide ore.

known ore deposits. Faults of differing age, scale and trend play a part in the genesis of the base metals and may be classified as follows.

(i) Deep-seated structures within the Lower Palaeozoic basement

Russell (1968) has drawn attention to the northerly alignment of some of the Irish base-metal deposits and suggested the presence of a deep-seated structure which acted as a channelway for ore fluids. Subsequently Phillips (reference in Halls *et al.*, 1979) proposed that the most significant deep-seated structure was the ENE trending Navan–Shannon fault which traced the position of the Iapetus suture through central Ireland. Evidence for a deep ENE trending fault at Navan is provided by Andrew and Ashton (1982).

(ii) ENE trending faults in Lower Carboniferous rocks

All the known carbonate-hosted base-metal deposits in Ireland are associated with major ENE trending normal faults (Fig. 6.22). The evidence at Silvermines and Tynagh suggests that these faults were active both during and after the deposition of carbonate sediments and the introduction of associated sulphide ores.

There is evidence that in all the major mines the ENE trending faults acted as conduits for the ore fluid. The reason why the ore bodies are restricted to particular parts of these major faults is not fully understood. It has been argued that they are developed in those sections of the fault which have the greatest throw (Morrisey, 1979) but Taylor and Andrew (1978) consider that at Silvermines the ore bodies are controlled by later NW trending faults.

Ore deposit geology and its influence on mineral exploration

Fig. 6.22 The open pit at Tynagh from which oxidized ore was exploited. Note that the Tynagh fault plane defines one side of the pit.

(iii) NW trending faults

NW trending faults have been recognized at Silvermines and appear to have acted as the main conduits for the ore fluid for the Upper G and B ore bodies (Taylor and Andrew, 1978). There is some evidence of a NW trend in the distribution of metal values recorded at Navan by Andrew and Ashton (1982), although these authors state that there is no evidence for NW trending faults within the mine.

6.3.6 Genesis of the Irish lead–zinc deposits

(a) Source of sulphur

The mineral paragenesis of the Irish sulphide ores is highly complex and sulphur isotope studies must therefore be closely allied to the interpretation of ore textures. Boast *et al.* (1981a) have carried out the most comprehensive approach to the problem in their study of the Tynagh deposit and much of this section is based upon their research. At Tynagh four stages of primary mineralization can be distinguished on the basis of mineralogy and texture. The main ore-forming phases are associated with the second and third stages.

Barite is present in stages 2 and 3 and is characterized by a narrow range of values with a mean of $\delta^{34}S + 19^0/_{00}$. Very similar values are obtained from the North Pennine ore field and from Mississippi Valley-type deposits. Most geochemists consider that this reflects the composition of seawater sulphate in Lower Carboniferous times. Pyrite is dominant in stage 1

Sediment-hosted copper–lead–zinc deposits

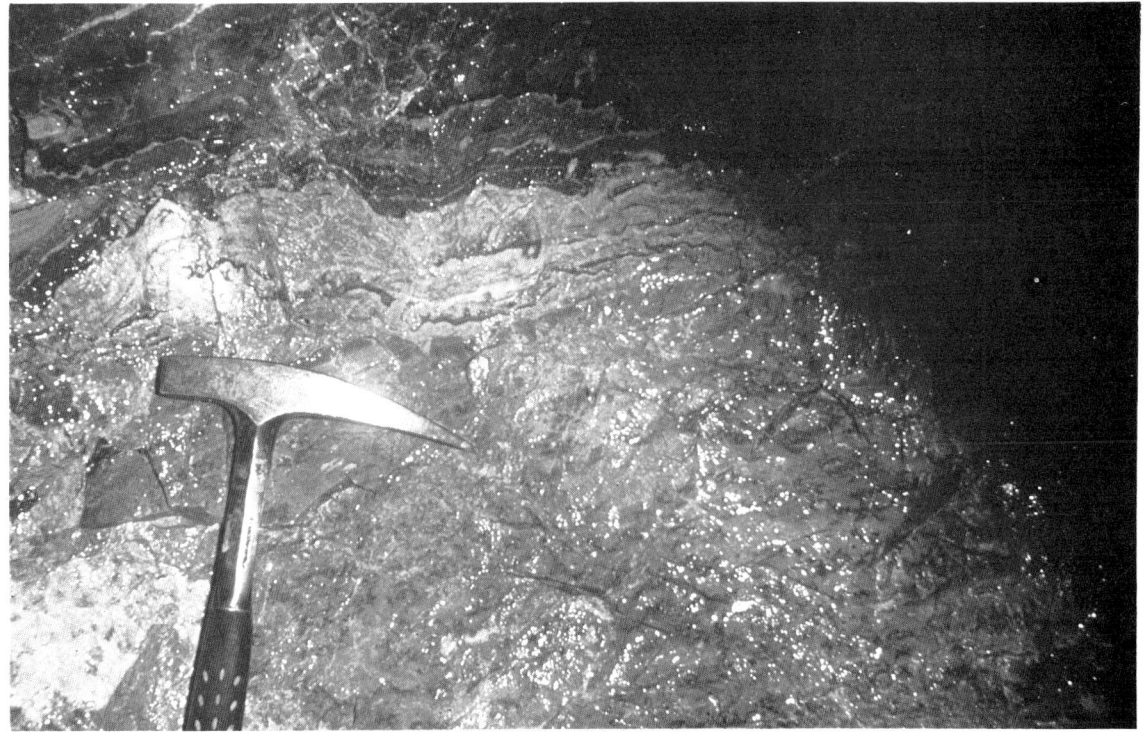

Fig. 6.23 Bedded sphalerite–galena ore, Tynagh Mine, Ireland.

and is interpreted as being of early diagenetic origin. Sulphur isotope values for pyrite range from −3.1 to −8.0‰ and are interpreted as being the product of bacterial reduction of seawater sulphate. Sphalerite is the main sulphide in stage 2 (Fig. 6.23), and sulphur isotopes for sphalerite range from −4.1 to −26.0‰ with a mean of −17.2‰. The sulphur isotopes from stage 2 show no correlation with host lithology, elevation or distance from the Tynagh fault and are interpreted as resulting from bacterial reduction of seawater sulphate.

Stage 3 mineralization is characterized by veining (Fig. 6.24) and replacement textures. Galena is the dominant sulphide. Sulphur isotope values for galena show a broad range of values from +11.1 to −23.0‰. More significantly the sulphur isotope variation can be correlated with distance from the Tynagh fault and it seems probable that an appreciable proportion of the sulphur for stage 3 mineralization was fault derived.

Sulphur isotope studies at Silvermines are described by Taylor and Andrew (1978) and Coomer and Robinson (1976). In both cases the authors conclude from the isotope data that there were two sources of sulphur. The $\delta^{34}S$ values for the Lower G ore body, which is closely related to faulting, have a narrow range suggesting the derivation of sulphur from a deep-seated source. $\delta^{34}S$ values for the stratiform ores have a wide spread of values and sulphur is considered to be derived from biogenic reduction of seawater sulphate.

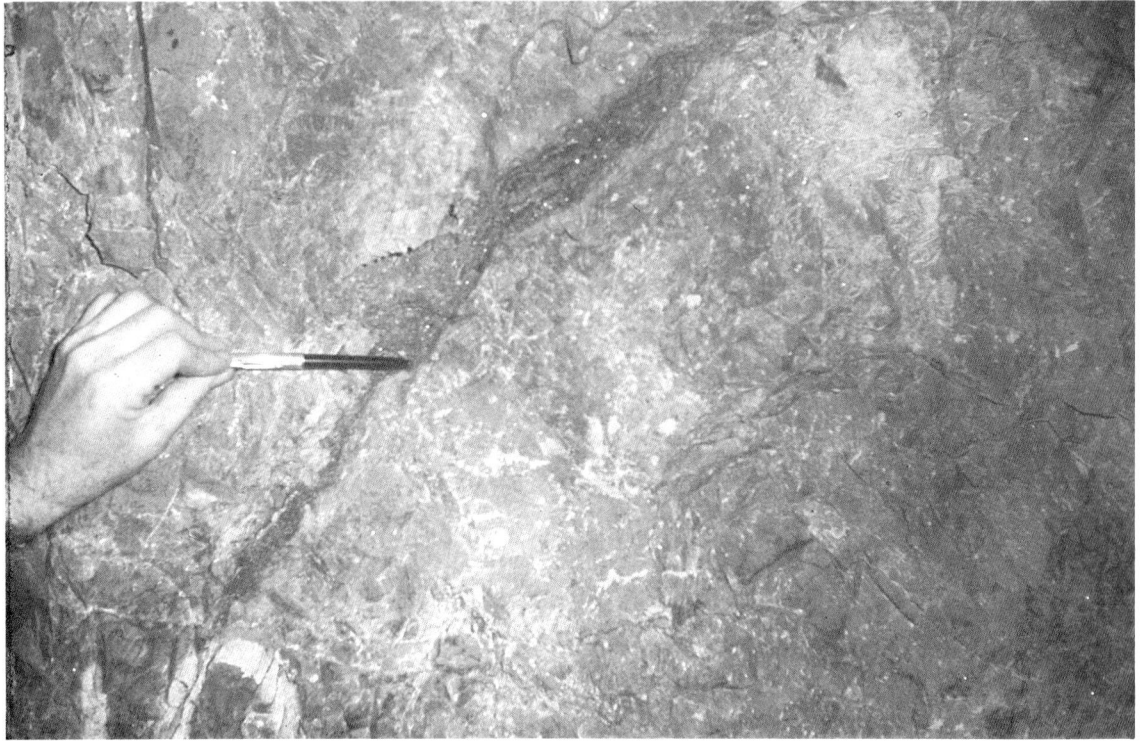

Fig. 6.24 Late stage galena veinlet at Tynagh Mine, Ireland.

(b) Source of metals

A number of possible sources of lead and zinc are available. The source favoured by most writers is a hydrothermal fluid, generated at depth, and channelled into a shallow water environment by faults. The main evidence supporting this alternative is the close spatial association of the lead and zinc ores with faults, particularly those of ENE trend. Whilst there is some unanimity on the role of faulting in controlling the movement of the ore fluid in its final stages there is less agreement on the earlier evolution of the ore fluid and the ultimate source of metal.

Russell (1968) has argued in favour of the Lower Palaeozoic geosynclinal sediments as the source of metal, and provided supporting analytical data from Lower Palaeozoic rocks from Ireland. Using the lithospheric profile of Great Britain Russell (1978) was able to show that there is a correlation between those parts of the British Isles where the Caledonian geosynclinal pile is thickest and those areas where the largest tonnages of lead and zinc have been extracted.

Boast *et al.* (1981b, 1983) have investigated the problem using lead isotopes. Their data indicate that the source of lead is similar at Silvermines and Navan but distinct from that at Tynagh (Fig. 6.25). However, at the present time the identity of the parent materials remains the subject of speculation.

(c) The ore fluid

Wall-rock alteration in the Irish base-metal deposits is limited to dolomitization of part of the succession, and information concerning the ore fluid is therefore restricted to the study of fluid inclusions. Samson and Russell (1983) have carried out a detailed investigation of fluid inclusions in

Sediment-hosted copper–lead–zinc deposits

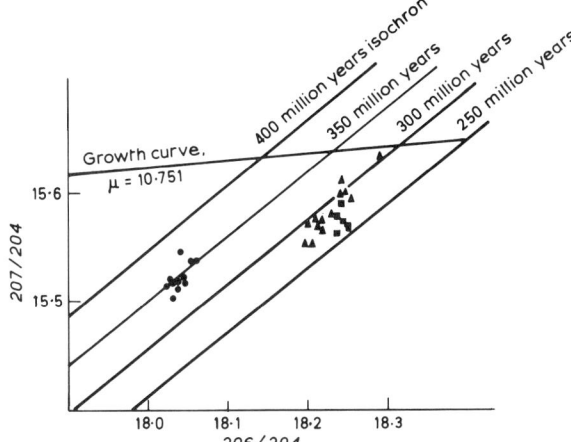

Fig. 6.25 Plot of $^{207}Pb/^{204}Pb$ versus $^{206}Pb/^{204}Pb$ ratios in galenas from Tynagh (●), Silvermines (■), and Navan (▲). (From Boast *et al.* 1983.)

the Silvermines deposit. They report that there is a large range of both temperature (50–260°C, mode at 200°C) and salinity (0–25 wt % equiv. NaCl, mostly in the range 12–22 wt% equiv. NaCl). There is a negative correlation between homogenization temperature and salinity which Samson and Russell (1983) interpret as the mixing of an ascending high temperature, low-salinity hydrothermal fluid and a near-surface, cool, highly saline brine. A few fluid inclusions are vapour-rich, which suggests that on occasion the hydrothermal solutions boiled during their ascent, but precipitation of sulphides was due to cooling by mixing with seawater and the addition of H_2S and SO_4^{2-}.

(d) Model of ore genesis

Recent detailed studies have resolved many of the problems of the genesis of Irish deposits. Ironically the elucidation of the genesis of Tynagh and Silvermines coincided with the closure of these mines. Nonetheless, Navan remains a base-metal deposit of considerable importance, and the understanding obtained from Tynagh and Silvermines may help in the discovery of new deposits.

Most geologists now accept that the Irish lead–zinc deposits are predominantly of syndiagenetic origin, and the evidence for this is presented in earlier sections. The role of faulting is seen as being a fundamental element in the control of both palaeo-environment and as a conduit for the ore fluid. The confusing array of ore textures, which led initially to a polarization between syngenetic and epigenetic schools of thought, has now led to the recognition of syndiagenesis as an important style of sediment-hosted mineralization.

Two aspects of the ore genetic model remain unresolved: the source of metals and evolution of the hydrothermal fluid and the factors which constrain the deposition of ore at particular sites on major ENE trending faults.

The first aspect has been addressed by Russell (1978) who considered that the ore fluid was initially seawater which penetrated downwards and eventually leached metal from Lower Palaeozoic metasediments. Russell considered that increased permeability of the Lower Palaeozoic sequence led to the development of a convective system, which migrated gradually downwards until an impermeable crystalline basement prevented further fluid movement (Fig. 6.26). The upward cycle of the convective system is believed to have utilized permeable zones and in particular fracture intersections.

The factors which have led to the concentration of ore deposits at particular sites on major ENE trending faults has not been fully explained. Unfortunately in exploration terms this is probably the most vital aspect of the genetic model. At Silvermines there is strong evidence (Taylor and Andrew, 1978) that NW trending faults have localized ore deposition. At Navan the evidence is less compelling but a NW trending element is apparent from the metal contour maps (Fig. 6.19). At Tynagh there is no evidence for the existence of NW trending structures and the localization of ore deposition is attributed to the zone of maximum throw on the Tynagh fault. In view of these differing mechanisms it is interesting to note that the lead isotope data (Boast *et al.*, 1983) indicate a differing source of lead for Tynagh compared with the other mines.

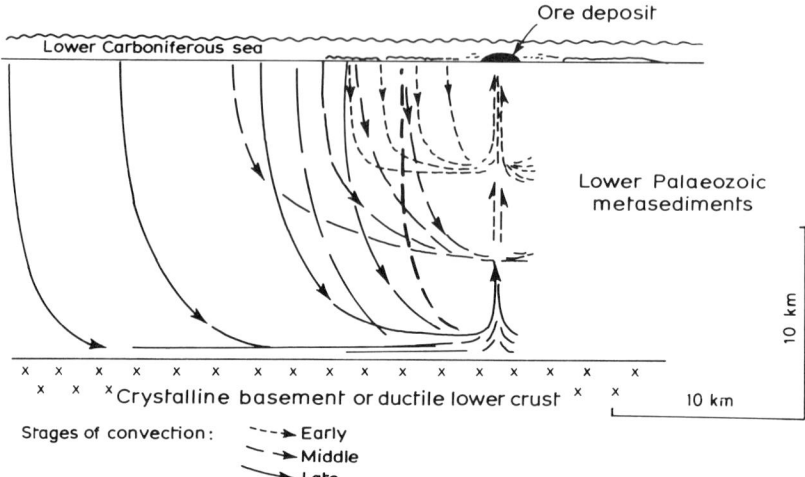

Fig. 6.26 Concept of a downward-excavating convective system (from Russell 1983).

6.3.7 Exploration in the Central Plain of Ireland

The Central Plain of Ireland is covered by an extensive mantle of glacial drift. The glacial deposits are the results of three successive advances of major ice sheets and include ground moraine or till, sands and gravels, ablation moraine, lake clays and glaciomarine sediments. Whilst the general movement of glaciation was north-west and south-eastwards there are areas where the pattern of movement is more complex (Synge, 1979). The thickness of the glacial deposits is highly variable and locally may exceed 50 m.

Although most base-metal deposits discovered in recent years have occurred near to the surface there is little outcrop in the Central Plain to aid in either geological mapping or conventional prospecting. Nonetheless mineralized outcrop, or mineralized glacial float material, have played an important part in the discovery of most of the significant base-metal deposits in Ireland. Schultz (1971) recorded that one of the first clues leading to the discovery of the Tynagh deposit was the description by the Irish Geological Survey of copper-bearing sandstone boulders, whose distribution indicated that they had been derived from the Tynagh fault. Gortdrum and Ballinalack are some of the important discoveries which can be attributed in part to the identification of mineralized boulders or outcrop.

(a) Remote sensing

Remote sensing is little used as a reconnaissance technique in glaciated terrain. McM. Moore (1980) noted that even where the cover of boulder clay in Ireland is relatively thin as at Tynagh there is little evidence of faulting on aerial photographs; he explains that this is because the faults are generally gouge-filled and have little topographic expression. Nonetheless C. J. Andrew (personal communication) states that the current trend of geologically directed exploration is based in part on computer analysis of combined remote sensing and geophysical data.

(b) Geochemistry

During the exploration boom of the 1960s stream sediment and soil surveys were widely used for reconnaissance work, but soil surveys have proved to be of most value. The main limitation on the use of stream sediments in the Central Plain arose from the high pH of the streams and groundwaters which inhibited the mobility of the base metals. Nonetheless it was a series of geochemical maps, based on stream sediment sampling, prepared by the Irish Agricultural Institute, which led to the more detailed exploration of the Navan area (O'Brien and Romer, 1971). Initially soil sampling was carried out at shallow depth using a conventional hand-held auger. Reconnaissance survey grids utilized a sample spacing of 500 ft (152.4 m) and for more detailed follow-up surveys the sample spacing was reduced to 50 or 100 ft (15.2 or 30.4 m). This type of survey defined the surface expression of mineralization at Tynagh and Navan.

During the last decade increasing use has been made of deep overburden sampling as it became apparent that the main prospective areas had been covered by conventional soil sampling. The methods used for overburden sampling in Ireland are described by Steiger and Proustie (1979) who prefer a percussion drill with a flush 'Holman-type' through-flow sampler. Woodham (1982) states that hand-held percussion drills are rarely successful when the overburden depth exceeds 30 ft (9.1 m) and he describes the application of reverse circulation drilling which has greater depth penetration. The trend from hand-held augers to more sophisticated drilling has been accompanied by a rise in costs which is justified by the increased effectiveness of the newer techniques.

Case histories of geochemical surveys which have used overburden sampling are described by Steiger and Proustie (1979) and Cazalet (1982). Cazalet (1982) describes an interesting example from the Harberton Bridge area in County Kildare (Figs 6.27 and 6.28). This prospect was first detected by surface soil sampling, which defined a 0.2 km^2 anomaly containing values up to 630 ppm Zn. Resampling the area on a 61 m grid revealed a stronger, and slightly more extensive, cluster of anomalous zinc values. Overburden sampling using a 75 m square grid, and collection of samples every 2 m to a maximum depth outlined a subsurface anomaly of 1.5 km^2 in which zinc values reached 10.8%. Subsequent diamond drilling detected a limonitic sand, which contained combined lead and zinc concentrations up to 18%. The sand was interpreted as a preglacial, or interglacial, gossan incorporated in and redistributed to varying degrees by the overlying glacial material. Ironically bedrock mineralization

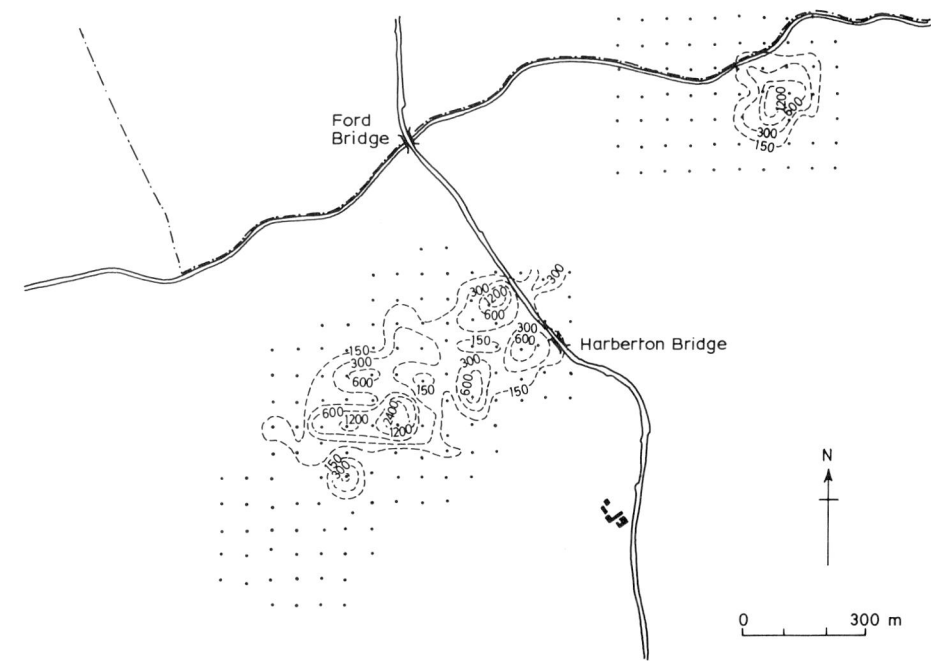

Fig. 6.27 Zinc follow-up surface soil samples, Harberton Bridge prospect. Contours at 150, 300, 600, 1200 and 2400 ppm (from Cazalet 1982).

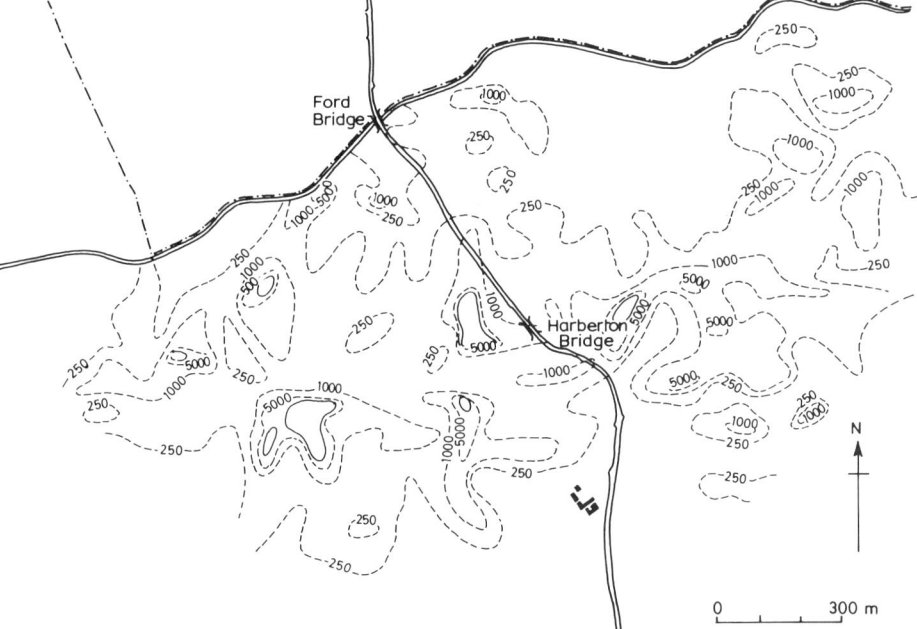

Fig. 6.28 Zinc values, maximum depth, in overburden at Harberton Bridge prospect. Contours at 250, 1000, 5000 and 10 000 ppm (from Cazalet 1982).

was detected in a part of the area without significant geochemical expression.

Up to the present nearly all the geochemical work in Ireland has been based upon secondary dispersion. However, it is apparent from research at Navan (Finlay *et al.*, 1984) that lithogeochemical studies have good potential for locating sub-outcropping mineralization.

(c) Geophysics

Airborne geophysical surveys have proved of little value in Ireland, which contrasts with the success of airborne techniques in the search for base-metal deposits in glaciated terrain in other areas. Schulz (1971) suggested four reasons why airborne surveys have been less effective in Ireland:

(i) Commercially significant base-metal sulphide deposits may not respond to electromagnetic methods (e.g. Gortdrum).

(ii) Man-made conductors (e.g. power lines) cause considerable interference and can mask important geological conductors.

(iii) Irregular distribution of prospecting licences involves the inefficient use of aircraft.

(iv) Poor weather conditions can cause considerable down time.

As a consequence the main application of geophysical methods has been in investigating the extension of known ore bodies and in evaluating geochemical anomalies. In this context the most useful techniques have been induced polarization, resistivity and gravity methods.

(i) Induced polarization–resistivity

The induced polarization method was first applied in Ireland in 1960 and continues to be the most useful geophysical technique available. Both the variable-frequency (frequency-domain) and pulse-transient (time-domain) methods have been used. The standard electrode configurations used are three electrode arrays for the pulse-type and dipole–dipole arrays for the variable-frequency methods. Gradient-array layouts have been used in some pulse-type surveys. For primary coverage electrode spacings of 200 or 300 ft (60.9 or 91.4 m) and line intervals of 500 ft (152.4 m) are most commonly used. For detailed follow-up surveys closer line intervals are used and some or all lines are surveyed at two or more different electrode spacings ranging from 50 to 800 ft (15.2 to 243.8 m). Direct-current resistivity readings are taken at all points in addition to the IP measurements (Schulz, 1971).

Williams (1982) describes the recent use of phase spectral IP (complex resistivity) in Ireland. The principal benefit of the system is its ability to indicate any inductive coupling effects and to allow these to be removed by calculation. In addition the system may be able to distinguish graphite from sulphides and thus eliminate many spurious anomalies.

IP surveys proved effective in evaluating geochemical anomalies and improving target definition at Gortdrum, Keel, Navan and Moate. They also led to the discovery of extensions to sulphide ore bodies at Silvermines and Ballyvergin (Schulz, 1971).

(ii) Gravimetric methods

Reconnaissance gravity surveys using widely spaced station intervals (2.5 km approx.) have been completed over the whole of Ireland. Deeny (1982) has utilized the regional gravity data, in conjunction with airborne magnetics, as the basis for constructing a model of the regional structure and sedimentation in Devonian and Carboniferous times. The main element of his model is a major north-easterly trending graben.

Detailed gravimetric studies at Navan, and over a blind sulphide deposit in the Tynagh area, gave encouraging results (Schulz, 1971). As a consequence closely spaced gravity surveys have been used in several integrated mineral exploration programmes in Ireland, although the results have been disappointing. One of the major interpretational problems is the frequent presence of a karstified limestone surface beneath the glacial overburden, which produces anomalies of similar shape and size to that anticipated for base-metal mineralization (Williams, 1982).

6.4 EPIGENETIC CARBONATE-HOSTED LEAD–ZINC DEPOSITS (MISSISSIPPI VALLEY-TYPE)

Mississippi Valley-type deposits contain a substantial proportion of the World's reserves of lead and zinc, and this is particularly the case in North America and Europe. The term Mississippi Valley type derives from the early development of mining in the states adjoining the upper and middle reaches of the Mississippi River. At the present time the most important mining district in the USA is the Viburnum Trend in SE Missouri. Deposits within the Viburnum Trend contain the largest lead and zinc reserves of the USA and contribute 85% of current US lead production (Vineyard, 1977).

6.4.1 General characteristics

(a) Distribution in space and time

Mississippi Valley-type deposits are characteristically of Phanerozoic age, and Stanton (1972) has drawn attention to the paucity of carbonate-hosted base-metal sulphides in rocks of Precambrian age. Badham (1981) suggested that the explanation may lie in the porosity of the lithified carbonate sediment. Phanerozoic reef carbonates consist mainly of coral debris and have a high initial porosity and permeability which is enhanced by later dolomitization. In contrast, Precambrian carbonates are commonly stromatolitic and often primarily dolomitic; as a consequence they tend to have much lower porosities and are less receptive to the ingress of an ore fluid. If the uneven temporal distribution of carbonate-hosted lead–zinc deposits is seen within the wider perspective of sediment-hosted base-metal deposits the problem is diminished. Examples of epigenetic carbonate-hosted lead–zinc deposits of Proterozoic age include the Black Angel Mine,

Greenland (Fish, 1974) and the Coxco deposit, McArthur River, Australia (Walker et al., 1983).

Examples of Phanerozoic deposits are as follows: Cambrian, SE Missouri, USA; Ordovician, Eastern Tennessee, USA; Devonian, Pine Point, NWT, Canada; Carboniferous, Tri-State field, USA; Permian, Trento Valley, Italy; Triassic, Eastern Alps; Jurassic, Silesian and Cracovian provinces (Devonian to Jurassic); Cretaceous, Northern Algeria and Tunisia.

Mississippi Valley-type deposits are of restricted geographical distribution although carbonates are common sediments within the Phanerozoic sequence. Dunsmore and Shearman (1977) have demonstrated that the host limestones for Mississippi Valley-type deposits in the Phanerozoic were deposited within the spread of latitudes 30° north and south of the palaeo-equator, which would have included zones with an arid climate. Under these climatic conditions sabkha-type evaporitic sediments would have been a common associate of carbonate rocks. Dunsmore and Shearman (1977) suggested that the proximity of evaporitic sediments to reef carbonates is a prerequisite for the formation of lead–zinc mineralization, and they indicated that the main role of evaporites is to provide a source of sulphur.

(b) Size and grade

Table 6.7 shows the size and grade of some important mining districts where lead and zinc are produced from Mississippi Valley-type deposits. It is evident that, with the exception of the Viburnum Trend, one of the main characteristics is the dominance of zinc over lead. Table 6.7 also shows that the grade of Canadian deposits is higher than those being exploited in the USA or Europe. This is because higher grades of ore are required to justify the cost of developing mine infra-structure in the remoter areas of Canada.

The mining districts listed in Table 6.7 often extend over wide areas and contain many separate ore bodies of varying size. For example the Illinois–Wisconsin district extends over an area of 10 000 km^2 and includes ore bodies which range from small gash-veins containing a few thousand tonnes of ore to replacement deposits which contain $>5 \times 10^6$ tonnes of ore (Heyl, 1983).

High concentrations of cadmium, indium and gallium are reported in galena and sphalerite from some mining districts but it is sometimes difficult to determine from the literature whether they are commercial by-products. Galena from Mississippi Valley-type deposits is not normally argentiferous.

(c) Mineralogy

Mississippi Valley-type deposits are mined principally for sphalerite and galena, but they have a variable mineralogy and there are considerable contrasts between mining districts. Essentially three subtypes can be distinguished on the basis of mineralogy.

Table 6.7 Size and grade of selected Mississippi Valley-type deposits

Source	Mining district	Size (10^6 tonnes)	Pb (%)	Zn (%)
1	Pine Point, Canada	94.5	2.5	6.2
1	Cornwallis, Canada	24.1	4.2	13.8
2	Illinois–Wisconsin, USA	100	0.5	5.0
2	Tri-State, USA	500	0.6	2.3
2	Viburnum Trend, Missouri, USA	420★	6.0	1.0
3	Alpine district: Austria, Italy and Yugoslavia	86	1.0	5.0

★Estimated.
1, Gibbins (1983); 2, Heyl (1983); 3, Klau and Mostler (1983).

(i) Zn-dominant subtype

Most Mississippi Valley-type ore bodies belong to this category. Sphalerite is the principal ore mineral with lesser quantities of galena which generally has a low silver content. Copper is subordinate to zinc and lead and is most commonly present as chalcopyrite, but other copper-bearing species may occur. There is some regional variation. For example nickel and cobalt species occur in the Upper Mississippi Valley district (Heyl, 1983) but are not reported from Pine Point. The mineralogy of the Pine Point district is discussed more fully in Section 6.4.2.

(ii) Pb-dominant subtype

The SE Missouri district, including the Viburnum Trend, is the main example of this category. Galena is the dominant ore mineral and has a low content of silver. Sphalerite is next in importance and contains cadmium, indium and gallium. Nickel–cobalt minerals are also present and, if metallurgical research is successful, will be produced as by-products (Heyl, 1983). The ore mineralogy and paragenesis of the SE Missouri district is highly complex and is described by Horrall *et al.* (1983).

(iii) F-dominant subtype

The Illinois–Kentucky district is the prime example of this category and provides about 75% of US fluorspar production (Heyl, 1983). The Pennine ore field (UK) may also be grouped in this category (Dunham, 1983).

Fluorite is the mineral of principal economic importance in the Illinois–Kentucky district. Sphalerite is the dominant sulphide and contains germanium and cadmium which are sold as by-products (Heyl, 1983). Galena is subordinate to sphalerite and is unusual for Mississippi Valley-type deposits in being argentiferous.

(d) Host rock lithology

One of the most consistent features of Mississippi Valley-type deposits is the ubiquity of limestones as hosts for the sulphide ore bodies. However, there are regional variations in the nature of the carbonate host rocks which require further explanation:

(i) Relative importance of dolomite

In most Mississippi Valley-type deposits dolomites are of regional importance within the mining district. However, in a few cases (Tri-State district, E. Tennessee) the dolomites are restricted to the vicinity of the ore body (Sangster, 1983).

(ii) The presence of erosional surfaces

Prominent erosional surfaces, which represent extensive karstification, characterize all the major Mississippi Valley-type deposits. The probable significance of this feature is that they have provided sites of high permeability for the passage of metal-bearing brines. Sangster (1983) notes that, with the exception of the SE Missouri district, the erosional surface usually occurs less than a few hundred metres above the ore body.

(iii) The significance of lateral facies change

Sangster (1983) reviewed the geological characteristics of nine major Mississippi Valley-type mining districts and concluded that in most cases there is no marked lateral facies change in the ore-hosting sedimentary units. In this respect the Pine Point district which we have selected for detailed discussion in Section 6.4.2 is atypical.

(e) Alteration

All Mississippi Valley-type deposits have low-temperature alteration halos. The most common alteration types are dolomitization and solution of the beds which host mineralization. The subject of dolomitization is discussed more fully in the context of the Pine Point mining district (Section 6.4.2). Where leaching of the carbonate beds, which host sulphide mineralization, has occurred it appears to have taken place prior to or during ore deposition (Heyl, 1983). In some cases this has resulted in sagging and collapse of the overlying beds, thereby increasing permeability and providing favourable sites for ore deposition. Silicification and pyritization are also common alteration types. In addition there may be localized patterns of metasomatism which may include K_2O, MgO, Al_2O_3, TiO_2, MnO (Heyl, 1983).

(f) Tectonic setting

Mississippi Valley-type deposits are normally located in an intracratonic setting. Laznicka (1976) suggested that the deposits of Central USA are associated with incipient rifting, and cited as evidence their association with major lineaments and the proximity of alkaline intrusions to some deposits. Mississippi Valley-type deposits are also associated with continental separation by rifting. One example is the Alpine mining district where ore deposits were formed on an unstable highly mobile shelf area at the boundary of the spreading Tethys Ocean (Bechstadt et al., 1978).

Most Mississippi Valley-type deposits are only disturbed by broad arching and faulting. However, in some mining districts, such as eastern Tennessee and Metaline, Washington, USA the ore bodies have been intensely deformed (Sangster, 1983).

6.4.2 Geology of the Pine Point District, NWT, Canada

(a) Introduction

The Pine Point district is located on the southern side of the Great Slave Lake in the North-West Territories, Canada. The area comprises about forty known Pb–Zn sulphide ore bodies within an area of 7000 km^2.

(b) Size, grade and mineralogy of deposits

The ore bodies of the Pine Point district vary in size, geometry, metal content, textures and host rock relationships. In size they range from 100 000 to 15 million tonnes. The ores have been classified on the basis of geometry into tabular and prismatic types. The prismatic ores are small discrete pods of high grade ore whereas the tabular ore comprises narrow branching mineralization (E. Lantos, personal communication).

The metal content of the ore bodies ranges from 3 to 11.5% Zn and 0.8 to 9% Pb. The district average is about 5.8% Zn and 2.2% Pb. Silver is absent from all the ore bodies.

The most distinctive feature of ore body distribution is their close relationship with the Presqu'ile dolomite and regions of karstification. On a more detailed scale it can be demonstrated that the ore bodies are associated with collapse structures, or with zones of brecciation and enhanced permeability within the limestone.

The Pine Point sulphide ore bodies consist of sphalerite, galena, pyrite and marcasite with dolomite and calcite being the main gangue minerals. Minor amounts of pyrrhotite, celestite, barite, gypsum, anhydrite, fluorite, sulphur and bitumen have been recorded from the host rocks.

Sphalerite is the most common sulphide in most of the deposits. It has a wide range of textures but occurs commonly as colloform crusts (Kyle, 1981). The iron content of sphalerite varies from 0.15% to 10.3%. Galena occurs in varying amounts in all the ore bodies and has a wide variety of forms.

The Pine Point district has a relatively simple paragenesis. Dolomite and calcite deposition precedes the introduction of sulphides. Marcasite is the earliest sulphide to form in most deposits followed by, and often intergrown with, pyrite. Sphalerite overlaps the iron sulphide stage and is covered with galena. Celestite, fluorite, sulphur and bitumen occur late in the paragenetic sequence (Kyle, 1981).

(c) Host rock lithology

The Pine Point ores are located in a carbonate sequence of Devonian (Givetian) age which has been extensively dolomitized (Fig. 6.29). Skall (1975) considered that the regional pattern of sedimentation was structurally controlled by a series of easterly trending 'hinges'. However, detailed investigations by Rhodes et al. (1984) on the stratigraphy of the Pine Point mining district have not identified any faulting which might have

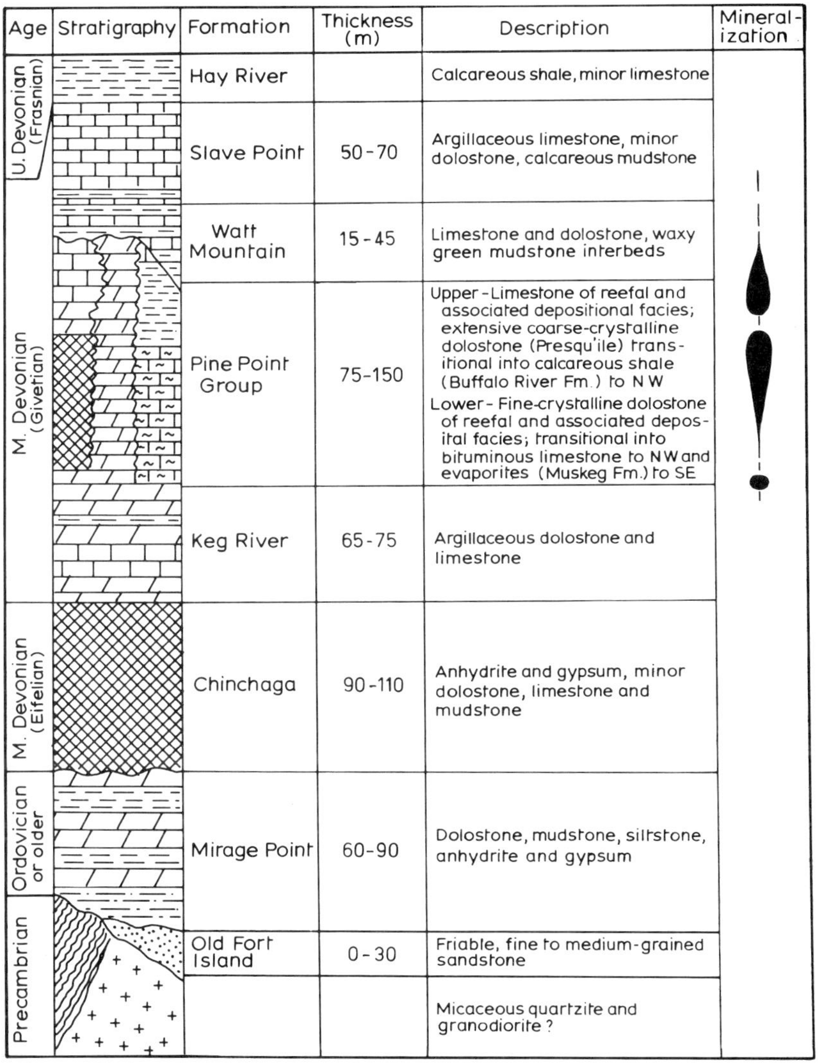

Age	Stratigraphy	Formation	Thickness (m)	Description	Mineralization
U. Devonian (Frasnian)		Hay River		Calcareous shale, minor limestone	
M. Devonian (Givetian)		Slave Point	50-70	Argillaceous limestone, minor dolostone, calcareous mudstone	
		Watt Mountain	15-45	Limestone and dolostone, waxy green mudstone interbeds	
		Pine Point Group	75-150	Upper - Limestone of reefal and associated depositional facies; extensive coarse-crystalline dolostone (Presqu'ile) transitional into calcareous shale (Buffalo River Fm.) to NW. Lower - Fine-crystalline dolostone of reefal and associated depositial facies; transitional into bituminous limestone to NW and evaporites (Muskeg Fm.) to SE	
		Keg River	65-75	Argillaceous dolostone and limestone	
M. Devonian (Eifelian)		Chinchaga	90-110	Anhydrite and gypsum, minor dolostone, limestone and mudstone	
Ordovician or older		Mirage Point	60-90	Dolostone, mudstone, siltstone, anhydrite and gypsum	
Precambrian		Old Fort Island	0-30	Friable, fine to medium-grained sandstone	
				Micaceous quartzite and granodiorite?	

Fig. 6.29 Stratigraphic column of Palaeozoic formations, southern Great Slave Lake area (from Kyle 1981, after Skall 1975).

been expected in the 'hinge zones'. During middle Givetian times the carbonates formed a barrier complex which controlled the regional facies distribution. Marine shales are predominant to the north of the barrier, whilst an evaporite facies is important to the south.

Initiation of the barrier complex occurred at the end of early Givetian times, due to the gentle arching of the marine sediments of the Keg River Formation. This led to the establishment of a bank of carbonate sands, with associated facies, which have been described in detail by Skall (1975) and Rhodes et al. (1984).

During its infancy the barrier was incomplete, and marine conditions prevailed on both northern and southern sides. However, the early development of extensive tidal flat areas established a more effective barrier, which isolated the area to the south and led to the development of an evaporite facies. At the same time subsidence in the southern part of the barrier resulted in an accumulation of bedded anhydrite which interfingers with dolomite near the barrier complex.

At a late stage a higher rate of sedimentation in the southern part of the Pine Point area reduced the effectiveness of the barrier. As a consequence the evaporitic sediments of the Muskeg Formation pass upwards into a varied range of carbonitic facies which are mainly dolomites.

Sedimentation was terminated by a marine regression during which Skall (1975) estimated the barrier was raised at least 100 ft (30.4 m) above sea level. This period of emergence was characterized by a phase of karstification before the resumption of marine conditions and deposition of the Watts Mountain Formation.

(d) Alteration

The carbonates of the Pine Point Group have been extensively dolomitized and two varieties are recognized on textural grounds. Fine-grained dolomite is the result of pervasive alteration of the lower part of the barrier complex. Skall (1975) attributed the fine-grained dolomite to the results of refluxing brines from evaporitic sediments of the Muskeg Formation. However, Kyle (1981) stated that there are hydrological objections to the reflux model and suggested that evaporative pumping of seawater through carbonate sediments or the influx of meteoric water are preferred alternatives.

The coarse-grained dolomite (Presqu'ile dolomite) occurs between the lower barrier units and the Watts Mountain Formation. Isolated remnants of the original limestone lithologies are distributed within the coarse dolomite, and it can therefore be demonstrated that this unit is the product of diagenesis superimposed on a variety of carbonate facies within the barrier complex.

Rhodes et al. (1984) question whether the two types of dolomitization are separate and distinct phenomena. They prefer the inter-mixing of saline Mg-rich waters with meteoric waters at a sub-surface interface as the most likely mechanism for both types of dolomitization.

(e) Structure

The carbonates which host lead–zinc ore deposits at Pine Point are remarkably little deformed. Rhodes et al. (1984) report that the main structural disturbances are very gentle undulations of low amplitude which are partly contemporaneous with barrier sedimentation. Rhodes et al. (1984) suggest that these structures may reflect differential compaction of contrasting sediment types and thicknesses. Some ore zones are affected by high angle faults which have displacements of 10–20 m. The relationship between ore deposits and major faults in the underlying basement is not fully understood. The tracing of major south-west trending faults towards Pine Point was instrumental in the discovery of the mineralization (Fig. 6.30). However, there is no evidence that these structures were active during carbonate sedimentation or ore deposition (Rhodes et al., 1984).

6.4.3 Genesis of the Pine Point lead–zinc deposits

(a) Source of sulphur

Kyle (1981) estimated that 15 million tonnes of sulphur are present in sulphide form in the Pine Point ore bodies. The role of sulphur is critical in controlling the deposition of lead and zinc and its origin is vital for the understanding of ore genesis. Sulphur isotopes from the Pine Point ores have been studied by Sasaki and Krouse (1969) and their results indicate remarkable uniformity. The means of $\delta^{34}S$ for galena, marcasite, pyrite and sphalerite are $+18.4$, $+19.3$, $+19.7$ and $21.6^0/_{00}$ respectively. Sasaki and Krouse (1969) suggested that the sulphur was transported in connate brines derived from the Middle Devonian evaporites which have a similar isotope ratio ($+19$ to $20^0/_{00}$) to that of the sulphide ores.

(b) Source of metal

All the lithologies in the Pine Point area have been proposed at some stage as potential source rocks by different workers. Beales and Jackson (1966) envisaged a metal supply from compacting shales

Sediment-hosted copper–lead–zinc deposits

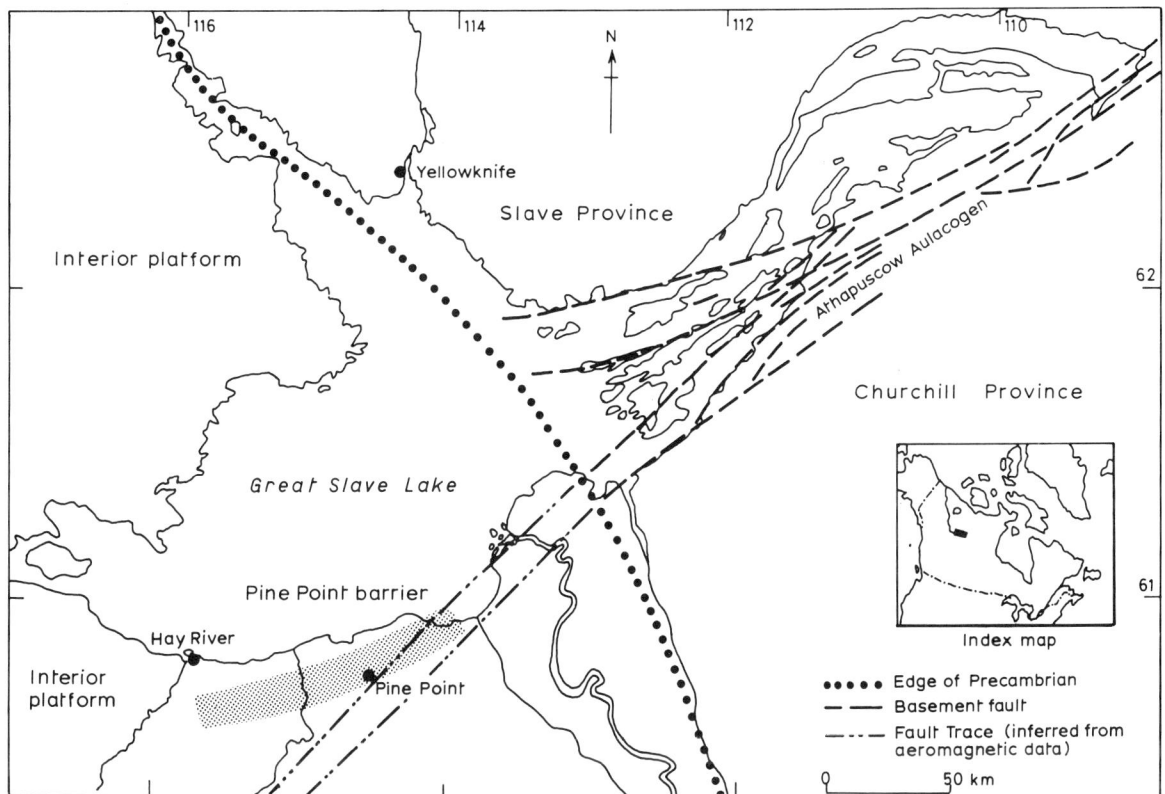

Fig. 6.30 Location and geological setting of Great Slave Lake area illustrating the relationships between basement faults and the trend of the Pine Point barrier complex (from Rhodes *et al.* 1984).

in the Mackenzie Basin in their basinal evolution model for Pine Point. Support for this theory is provided by the known expulsion of formation water during the compaction of argillaceous sediments and by the relatively high contents of lead and zinc in shales, particularly when they are carbonaceous. Thiede and Cameron (1978) have demonstrated that lead and zinc were concentrated in the Elk Point evaporites which are adjacent to Pine Point. A source in the evaporite sequence has the attraction of deriving both metal and sulphur from the same rocks. Kyle (1981) drew attention to the fact that galena, sphalerite, fluorite and barite are common associates of Phanerozoic carbonates outside known mining districts. Bjorlykke and Sangster (1981) consider that sandstone-hosted lead, red-bed copper and carbonate-hosted lead–zinc deposits have a common genetic heritage and they suggest that base metals are derived from immature, feldspathic, red-bed deposits. The separation of copper, lead and zinc are related to the severity of weathering of the host rock and the physicochemical environment of migrating groundwater. When this model is applied to the Pine Point area some supporting evidence is gained from the proximity of the host dolomites to the underlying Precambrian basement and basal sandstones of the Palaeozoic sequence (Fig. 6.29).

(c) The ore fluids

The main evidence concerning the nature of the ore fluid at Pine Point is based upon fluid inclusions. Homogenization temperatures for

265

primary fluid inclusions from sphalerite indicate that the temperature of the depositing fluid ranged from 51 to 99°C. Inclusions in white 'hydrothermal' dolomite closely associated with sphalerite range from 90 to 100°C (Roedder, 1968). Macqueen and Powell (1983) conclude from the immature nature of the organic matter at Pine Point that the overall temperature of the Barrier Complex did not exceed 60°C. They suggest that the lack of thermal equilibrium between the ore fluid and the associated host rocks points to a limited distance of transport for the metal-bearing fluid. Macqueen and Powell also suggested that the heat source for the mineralizing fluid is the underlying Precambrian rocks although this seems improbable unless the deep circulation of fluids along fault zones is envisaged.

Freezing temperatures on fluid inclusions from sphalerite indicate that the solutions were highly saline ranging from 15 to 23 wt % equiv. NaCl. Lower salinities of 15–20 wt % equiv. NaCl were obtained from inclusions within white dolomite (Roedder, 1968). These data are in general accord with other Mississippi Valley-type deposits.

Billings *et al.* (1969) have compared the homogenization temperature and salinites of fluid inclusion from Pine Point with brines collected from the Middle Devonian Presqu'ile Reef 350 km south–west of Pine Point. Bottom-hole temperatures from 36 of the sampling sites average 76°C and range up to 119°C. Zn, Mn, Fe and Cu were determined in the brines; the Zn content is of most significance and has an average content of 18.9 μg/1. Billings *et al.* (1969) calculated that if water moving out of the basin flushed the Presqu'ile conduit twice, the volume of ore at Pine Point could result.

In our opinion the balance of evidence points to the shales of the Mackenzie Basin as the source of metals at Pine Point and to metal-bearing brines as the ore fluid.

(d) Genetic model

Our understanding of the Pine Point lead–zinc ores is based on more substantial data than many other classes of deposit. There is unanimity about the sedimentary environment in which the host rocks have been deposited, and most workers agree that the deposition of ores is controlled by the filling of voids which resulted from karstification of the emergent carbonate barrier. However, Hopwood (1977) suggested the breccias may be due (in part) to hydraulic fracturing associated with local faulting within fluid-saturated aquifers. The sulphur isotope results point unequivocally towards a source of sulphur in the Devonian evaporites. The fluid inclusion data suggest that the ore fluid was a metalliferous brine depositing its load in the temperature range 51–99°C. Metal-bearing brines have been recorded 350 km from Pine Point which closely resemble the composition and temperature of the fluid inclusions from the Pine Point sphalerite.

In our opinion this evidence can best be synthesized in terms of the basinal evolution model of Beales and Jackson (1966). Their model is consistent both with the data summarized above and with geological processes which are known to operate in sedimentary basins.

6.5 EXPLORATION FOR MISSISSIPPI VALLEY-TYPE DEPOSITS

Exploration for Mississippi Valley-type deposits has resulted in the discovery of new mining districts using both a geological approach and indirect techniques.

(a) Geology

The most successful geological approach has been the recognition of similarities in stratigraphy and structure between adjacent areas. For example, the Viburnum Trend in SE Missouri was discovered on the basis of its similar geology to that of the Old Leadbelt (J. Ridge, personal communication). The projection of major lineaments from areas of known mineralization into virgin terrain has also been remarkably successful, and led to the discovery of the Pine Point and Robb Lake areas in Canada.

(b) Geochemistry

Stream sediment and soil geochemistry have played an important part in integrated mineral exploration programmes for Mississippi Valley-type deposits. For example, high lead and zinc values in stream sediments led to the discovery of the Bonnet Plume district in the Eastern Yukon, Canada (Gibbins, 1983). Soil geochemistry was used in conjunction with geophysical surveys in the discovery of the Cornwallis lead–zinc district in the North West Territories of Canada (Gibbins, 1983).

Lithogeochemistry has good potential for discovering blind Pb–Zn mineralization. Erickson *et al.* (1983) have demonstrated that rock chippings obtained as a result of drilling for water may be utilized for exploration purposes. They carried out multi-element analysis on the insoluble residue of carbonates obtained from drill holes. Their study extended from the SE Missouri mining district westwards into areas without known Pb–Zn mineralization, and revealed the potential of certain stratigraphic units and trends of anomalous metal which had not been previously recognized.

(c) Geophysics

Geophysical methods may be used both to aid the geological interpretation of a prospective area and to locate concealed sulphide ore deposits. Spector and Pichette (1983) used airborne magnetic methods in SE Missouri to define faulting and variation in the topography of the Precambrian basement below the carbonate sequence. Both of these factors are considered to exert a control on Pb–Zn mineralization in SE Missouri.

Induced polarization (IP) and gravity surveys have proved effective in detecting concealed Pb–Zn deposits. Siegel (1968) described a case history from the Pyramid ore bodies at Pine Point. Initial discovery of the Pyramid deposit was based on a time-domain induced polarization survey. As the ore body was expected to be flat-lying and tabular a three-electrode array was employed. For reconnaissance work a line spacing of 750 ft (228.6 m) was employed using a 300 ft (91.4 m) electrode spacing and a 200 ft (60.9 m) station interval. This preliminary study indicated two anomalous areas which were investigated by more detailed induced polarization surveys. Two drill holes were sited on the basis of the detailed IP survey and intersected base-metal sulphides. The extent of the ore body was then defined using a systematic gravity survey. The gravity survey was used as a follow-up technique because the IP method is not normally affected by sphalerite. Comparative results for both induced polarization and gravity over the Pyramid no. 1 ore body are shown in Fig. 6.31.

The geophysical surveys were followed by a systematic drilling programme which defined an ore body with a reserve of 3 175 000 tonnes averaging 2.9% Pb and 9.1% Zn.

6.6 CONCLUDING STATEMENT

In this chapter we have selected a number of mining districts as examples of the principal classes of sediment-hosted base-metal deposits. The geological setting of each mining district has been outlined, and the concepts of ore genesis and principal methods of mineral exploration have been described. We must now attempt to answer the question implied in the title of our book: what influence has our greatly increased understanding of this major class of ore deposit had on the planning and success of exploration programmes? For obvious reasons few mining companies are prepared to divulge their current exploration philosophies but it is possible to evaluate the differing approaches that have been made in the recent past.

Within the mining districts which we have selected as examples the major new discoveries have come mainly from the successful application of indirect methods of exploration. Geochemical exploration has played a major role in the discovery of new deposits which is perhaps surprising when we consider that the mining districts described occur within widely different climatic regimes and are in some cases characterized by transported glacial drift. In all cases the use of soils, and to a lesser extent stream sediments have been the most reliable sampling media for

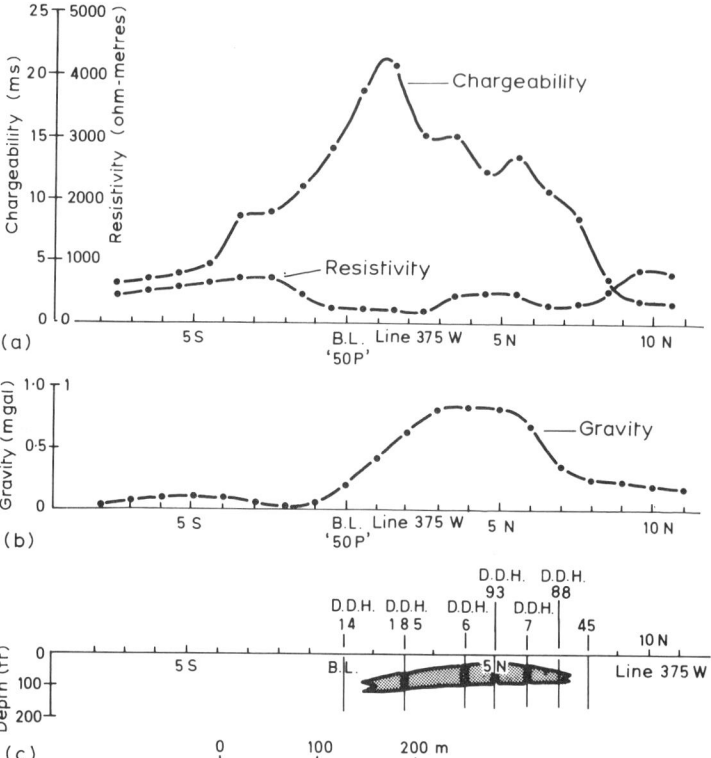

Fig. 6.31 Induced polarization and gravity profiles over the pyramid no. 1 ore body, Pine Point (a) induced polarization; (b) gravity profile; (c) cross-section of ore body from drilling. B.L., base line; D.D.H., diamond drill hole. (From Siegel 1968.)

the geochemist. It is also evident from our study that virtually all the newly discovered mineral deposits have either had some surface expression, or have been covered by relatively shallow overburden.

In many cases the geochemical and geophysical surveys have been focused on particular areas, because the geological setting of the terrain is known to be favourable from a comparison with adjacent mines. However, nearly all successful discoveries seem to have involved an element of good luck.

What has been the role of ore genesis theory in the discovery of new deposits? As far as we are able to determine there are few published examples where a particular theory of ore genesis has led to the successful discovery of a sediment-hosted base-metal deposit which is economically viable. This is a disturbing conclusion and warrants further discussion. In our opinion one reason for the lack of success lies in the separate roles of the researcher and industrial geologist. The research geologist nearly always has an academic base and little industrial experience. His material for research is often obtained after a relatively brief foray undergound. His objectives are the publication of papers in the scientific literature. In contrast the mine geologist and explorationist have the discovery of new deposits as their objective but are constrained by limited budgets and a shortage of time from considering the wider implications of much of their work. Research facilities are not usually available for the industrial geologist. We feel it may be the structure of research into the genesis of ore deposit which may lie at the root of its relative lack of success. One solution to this problem would be a much greater flow of personnel between industry and research organizations.

Failure to resolve the problem within the next

decade may lead to a shortage of supply of some metals in the early part of the 21st century. We cannot assume that near-surface deposits will continue to be found. Both new indirect methods of exploration with greatly increased depth penetration and a more sustained and industrially based programme of research into the controls of ore formation are required to provide the mineral wealth for succeeding generations.

REFERENCES

Andrew, C. J. and Ashton, J. H. (1982) Mineral textures, metal zoning and ore environment of the Navan ore body, Co. Meath, Ireland. In *Mineral Exploration in Ireland: Progress and Developments 1971–1981 (Wexford conference 1981)* (ed. A. G. Brown), Irish Association for Economic Geology, Dublin, pp. 35–46.

Annels, A. E. (1979) Mufulira greywackes and their associated sulphides. *Trans. Instn. Min. Metall. (Sect. B: Appl. Earth Sci.)*, **88**, B 15–23.

Badham, J. P. N. (1981) The origins of ore deposits in sedimentary rocks. In *Economic Geology and Geotectonics* (ed. D. H. Tarling), Blackwell Scientific Publications, Oxford, pp. 149–191.

Bartholome, P. (1974) On the diagenetic formation of ores in sedimentary beds with special reference to the Kamoto copper deposit, Shaba, Zaire. In *Gisements Stratiformes et Provinces Cuprifères* (ed. P. Bartholome), Liège Société Géologique de Belgique, pp. 203–214.

Bartholome, P., Evrard, P., Katekesha, F., Lopez-Ruiz, J. and Ngongo, M. (1973) Diagenetic ore forming processes at Kamoto, Katanga, Republic of the Congo. In *Ores in Sediments* (eds G. C. Amstutz and A. J. Bernard), Springer-Verlag, Berlin, pp. 21–41.

Beales, F. W. and Jackson, S. A. (1966) Precipitation of lead–zinc ores in carbonate reservoirs as illustrated by Pine Point ore field, Canada. *Trans. Instn. Min. Metall. (Sect. B: Appl. Earth Sci.)* **75**, B 278–285.

Bechstadt, T., Bradner, R., Mostler, H. and Schmidt, K. (1978) Aborted rifting in the Triassic of the eastern and southern Alps. *News Jahrbuch Geol. Pal.*, **156(1)**, 157–178.

Bell, T. H. (1983) Thrusting and duplex formation at Mt. Isa: a radical re-interpretation of the regional structure of the host rocks to the Mount Isa orebodies. *Nature (London)*, **304(5926)**, 493–500.

Billings, G. K., Kesler, S. E. and Jackson, S. A. (1969) Relation of zinc-rich formation waters, northern Alberta, to the Pine Point ore deposits. *Econ. Geol.*, **64**, 385–391.

Bjorlykke, A. and Sangster, D. F. (1981) An overview of sandstone-lead deposits and their relation to red-bed copper and carbonate-hosted lead–zinc deposits. *Econ. Geol. 75th Anniv. Vol.*, 179–213.

Boast, A. M., Coleman, M. L. and Halls, C. (1981a) Textural and stable isotopic evidence for the genesis of the Tynagh base metal deposit, Ireland. *Econ. Geol.*, **76**, 27–55.

Boast, A. M., Swainbank, I. G., Coleman, M. L. and Halls, C. (1981b) Lead isotope variation in the Tynagh, Silvermines, and Navan base-metal deposits, Ireland. *Trans. Instn. Min. Metall. (Sect. B: Appl. Earth Sci.)*, **90**, B 115–119.

Boast, A. M., Swainbank, I. G., Coleman, M. L. and Halls, C. (1983) Lead isotope variation in the Tynagh, Silvermines and Navan base-metal deposits, Ireland. Report of discussion, contributed remarks and authors reply. *Trans. Instn. Min. Metall. (Sect. B: Appl. Earth Sci.)*, **92**, B 101–105.

Boyce, A. J., Anderton, R. and Russell, M. J., (1983a) Rapid subsidence and early Carboniferous base-metal mineralization in Ireland. *Trans. Instn. Min. Metall. (Sect. B: Appl. Earth Sci.)*, **92**, B 55–66.

Boyce, A. J., Coleman, M. L. and Russell, M. J. (1983b) Formation of fossil hydrothermal chimneys and mounds from Silvermines, Ireland. *Nature (London)*, **306(5943)**, 545–550.

Brandt, R. T. (1961) Relationship of mineralization to sedimentation at Mufulira, Northern Rhodesia. *Trans. Instn. Min. Metall.*, **71**, 459–479.

Brock, B. B. (1961) The structural setting of the Copperbelt. In *The Geology of the Northern Rhodesian Copperbelt* (ed. F. Mendelssohn), Macdonald, London, pp. 81–89.

Carr, G. R. and Smith, J. W. (1977) A comparative isotopic study of the Lady Loretta zinc–lead–silver deposit. *Miner. Deposita*, **12**, 105–110.

Cazalet, P. C. D. (1982) A review of geochemical exploration techniques 1971–1981. In *Mineral Exploration in Ireland: Progress and Developments 1971–1981.* (ed. A. G. Brown), Irish Association for Economic Geology, Dublin, pp. 148–156.

Chapman, R. E. (1973) *Petroleum Geology*, Elsevier, Amsterdam, 304 pp.

Clark, G. J. (1975) A review of exploration and discovery of the Hilton Mine, Mt. Isa. Unpublished seminar, James Cook University, N. Queensland.

Clemmey, H. (1978) A Proterozoic lacustrine interlude from the Zambian Copperbelt. *Int. Assoc. Sedimentol. Spec. Publ.*, **2**, 257–278.

COMRATE, (1975) American National Academy of Sciences Committee on Mineral Resources and the Environment, Washington, 326 pp.

Connor, A. G., Johnson, I. R. and Muir, M. D. (1982) The Dugald River Zinc–Lead Deposit, Northwest Queensland, Australia. *Proc. Australas. Instn. Min. Metall.*, **283**, 1–28.

Coomer, P. G. and Robinson, B. W. (1976) Sulphur and sulphate–oxygen isotopes and the origin of the Silvermines Deposits, Ireland. *Miner. Deposita*, **11(2)**, 155–169.

Cox, R. and Curtis, R. (1977) The discovery of the Lady Loretta zinc–lead–silver deposit, N.W. Queensland, Australia–a geochemical case history. *J. Geochem. Explor.*, **8**, 189–202.

Dechow, E. and Jensen, M. L. (1965) Sulphur isotopes of some Central African sulphide deposits. *Econ. Geol.*, **60**, 894–941.

Deeny, D. E. (1982) A postulated Lower Carboniferous Irish rift and possible metallogenic consequences. In *Mineral Exploration in Ireland. Progress and Developments, 1971–1981* (ed. A. G. Brown), Irish Association for Economic Geology, Dublin, pp. 157–161.

Derrick, G. M. (1982) A Proterozoic rift zone at Mt. Isa, Queensland. *BMR J. Austr. Geol. Geophys.* **7**, 81–92.

Diederix, D. (1977) *The Geology of the Nchanga Mining Licence Area*. Internal Report ZCCM Nchanga Division.

Doust, G. (1979) Regional exploration for Isa-type ore, Mt. Isa western basin, N. W. Queensland. Unpublished seminar, James Cook University, N. Queensland.

Dunham, K. C. (1983) Ore genesis in the English Pennines: a fluoritic sub-type. In *International Conference on Mississippi Valley Type Lead–Zinc Deposits* (eds. G. Kisvarsanyi *et al.*), University of Missouri, Rolla, pp. 77–85.

Dunsmore, H. E. and Shearman, D. J. (1977) Mississippi Valley-type lead–zinc orebodies: a sedimentary and diagenetic origin. In *Forum on Oil and Ore in Sediments* (ed. P. Garrard), Imperial College, London, pp. 189–202.

Eden, J. G. Van (1974) Depositional and diagenetic environment related to sulfide mineralization, Mufulira, Zambia. *Econ. Geol.*, **69**, 59–79.

Ellis, M. W. and McGregor, J. A. (1967) The Kalengwa copper deposit in North-Western Zambia. *Econ. Geol.*, **62**, 781–797.

Erickson, R. L., Mosier, E. L., Viets, J. G., Odland, S. K. and Erickson, M. S. (1983) Subsurface geochemical exploration in carbonate terrain, mid-continent USA. In *International Conference on Mississippi Valley Type Lead–Zinc Deposits* (ed. G. Kisvarsanyi *et al.*), University of Missouri, Rolla, pp. 575–583.

Finlay, S., Romer, D. M. and Cazalet, P. C. D. (1984) Lithogeochemical studies around the Navan Zn–Pb orebody, Ireland. In *Prospecting in Areas of Glaciated Terrain* (ed. M. J. Jones), Institution of Mining and Metallurgy, London, pp. 35–56.

Finlow-Bates, T. and Stumpfl, E. F. (1979) The copper and lead–zinc–silver ore bodies of Mt. Isa mine, Queensland: products of one hydrothermal system. *Soc. Geol. Belg. Annal.*, **102**, 497–517.

Fish, R. (1974) Mining in Arctic lands: The Black Angel experience. *Can. Min. J.*, **8**, 24–36.

Fleischer, V. D., Garlick, W. G. and Haldane, R. (1976) Geology of the Zambian Copperbelt. In *Handbook of Strata-bound and Stratiform Ore Deposits,* (ed. K. H. Wolf), Elsevier, New York Vol. 6 pp. 223–350.

Garlick, W. G. (1961) The syngenetic theory. In *The Geology of the Northern Rhodesian Copper Belt* (ed. F. Mendelsohn), McDonald, London. pp. 146–165.

Garlick, W. G. and Fleischer, V. D. (1972) Sedimentary environment of Zambian copper deposition. *Geol. Mijnbouw,* **51(3)**, 277–298.

Gibbins, W. A. (1983) Mississippi Valley lead–zinc districts of Northern Canada. In *International Conference on Mississippi Valley Type Lead–Zinc Deposits* (eds G. Kisvarsanyi *et al.*), Proceedings Volume, University of Missouri, Rolla, pp. 403–414.

Guilloux, L. and Pelissonier, H. (1974) Les gisements de schists, marnes et grès cupriferes. In *Gisements Stratiformes et Provinces Cupriferes* (ed. P. Bartholome), Liège Société Géologique de Belgique, pp. 235–254.

Gulson, B. L. and Mizon, K. J. (1979) Lead isotopes as a tool for gossan assessment in base metal exploration *J. Geochem. Explor.* **11** 299–320.

Gustafson, L. B. and Williams, N. (1981) Sediment-hosted stratiform deposits of copper, lead and zinc. *Econ. Geol. 75th Anniv. Vol.*, pp. 139–178.

Haldane, A. D. (1980) McArthur Basin Pb–Zn deposit, McArthur Basin N. T. *J. Geochem. Explor.*, **12**, 243–247.

Halls, C., Boast, A., Coleman, M. L. and Swainbank, I. G. (1979) Discussion of S. Taylor and C. J. Andrew,

Silvermines ore bodies, Co. Tipperary, Ireland. *Trans. Instn. Min. Metall. (Sect. B: Appl. Earth Sci.)*, **88**, B 129–131.

Haranczyk, C. (1961) Correlation between organic carbon, copper and silver content in Zechstein copper-bearing shales from the Lubin–Sieroszowice Region (Lower Silesia). *Bull. Acad. Pol. Sci. Ser. Sci. Geol. Geogr.*, **IX(4)**, 183–190.

Heyl, A. V. (1983) Geologic characteristics of three major Mississippi Valley districts. In *International Conference on Mississippi Valley Type Lead–Zinc Deposits* (eds A. G. Kisvarsanyi et al.), Proceedings Volume, University of Missouri, Rolla, pp. 27–60.

Hoek, E. and Bray, J. (1981) *Rock Slope Engineering*, 3rd edn, Institution of Mining and Metallurgy, London, 358 pp.

Hoek, E. and Brown, E. T. (1980) *Underground Excavations in Rock*, Institution of Mining and Metallurgy, London, 527 pp.

Holman, R. M. (1982) Sedimentary copper deposits in the Western Anti-Atlas Mountains, S. Morocco; results of a reconnaissance exploration programme. Unpublished M.Sc. dissertation, Camborne School of Mines.

Hopwood, T. (1977) *Geological Environments of Ore Deposition*, Australian Mineral Foundation, Adelaide, Nov/Dec 1977.

Horrall, K. B., Hagni, R. D. and Kisvarsanyi, G. (1983) Mineralogical, textural and paragenetic studies of selected ore deposits of the southeast Missouri lead–zinc–copper district and their genetic implications. In *International Conference on Mississippi Valley Type Lead–Zinc Deposits* (eds. G. Kisvarsanyi et al.), Proceedings Volume, University of Missouri, Rolla, pp. 289–316.

Hunt, T. S. (1873) The geognostical history of the metals. *Trans. Am. Inst. Min. Eng.*, **1**, 331–342.

Hutchison, R. W. (1980) Massive base metal sulphide deposits as guides to tectonic evolution. In *The Continental Crust and its Mineral Deposits*, (ed. D. W. Strangway), The Geological Assoc. of Canada, special paper 20, pp. 659–684.

Klau, W. and Mostler, H. (1983) Alpine Middle and Upper Triassic Pb–Zn deposits. In *International Conference on Mississippi Valley Type Lead–Zinc Deposits* (eds G. Kisvarsanyi et al.), Proceedings Volume, University of Missouri, Rolla, pp. 113–128.

Konstantynowiz, E. (1972) Mineral distribution in the Permian formations of Western Poland. *Int. Geol. Cong. 24th Montreal 1972 Programme Sec.*, **4**, 373–380.

Kyle, J. R. (1981) Geology of the Pine Point lead-zinc district. In *Handbook of Strata-bound and Stratiform Ore Deposits* Vol. 9 (ed. K. H. Wolf), Elsevier, New York, pp. 643–741.

Lambert, I. B. (1976) The McArthur zinc–lead–silver deposit: features, metallogenesis and comparisons with other stratiform ores. In *Handbook of Strata-bound and Stratiform Ore Deposits,* Vol. 6 (ed. K. H. Wolf), Elsevier, New York, pp. 535–585.

Laznicka, P. (1976) Lead deposits in the global plate tectonic model. In *Metallogeny and Plate Tectonics* (ed. D. F. Strong), Geol. Assoc. Can. Spec. Pap., **14**, 243–270.

McClay, K. R. (1979) Folding in silver–lead–zinc ore bodies, Mount Isa, Australia. *Trans. Instn. Min. Metall. (Sect. B: Apl. Earth Sci.)*, **88**, B 5–14.

McClay, K. R. (1983) Deformation of stratiform lead–zinc deposits. In *Short Course in Sediment-Hosted Stratiform Lead–Zinc Deposits* (ed. D. F. Sangster), Mineralogical Association of Canada, Victoria B.C. pp. 283–309.

McKinstry, H. E. (1948) *Mining Geology*, Prentice Hall, Englewood Cliffe, NJ, 680 pp.

McM.Moore, J. (1980) Joint pattern interpretation: a possible aid to exploration in Ireland. *Trans. Instn. Min. Metall. (Sect. B: Appl. Earth Sci.)*, **89**, B 42–43.

Macqueen, R. W. and Powell, T. G. (1983) Organic geochemistry of the Pine Point lead–zinc ore field and region, North West Territories, Canada. *Econ. Geol.*, **78**, 1–25.

Mathias, B. V. and Clark, G. J. (1975) Mount Isa copper and silver–lead–zinc ore bodies – Isa and Hilton Mines. In *Economic Geology of Australia and Papua New Guinea* (ed. C. L. Knight), Australasian Institute of Mining and Metallurgy, pp. 351–372.

Mendelsohn, F. (1961) *The Geology of the Northern Rhodesian Copperbelt*. Macdonald, London, 523 pp.

Morrisey, C. J. (1979) Report on discussion on paper by S. Taylor and C. J. Andrew (1978) Silvermines ore bodies, County Tipperary, Ireland. *Trans. Inst. Min. Metall. (Sect. B: Appl. Earth Sci.)*, **88**, B 126–127.

Morrisey, C. J., Davis, G. R. and Steed, G. M. (1971) Mineralization in the Lower Carboniferous of Central Ireland. *Trans. Inst. Min.. Metall. (Sect. B: Appl. Earth Sci.)*, **80**, B 174–185.

Muir, M. D. (1983) Depositional environments of host rocks to Northern Australian lead–zinc deposits, with special reference to McArthur River. In *Short Course in Sediment-Hosted Stratiform Lead–Zinc Deposits* (ed. D. F. Sangster), Mineralogical Association of Canada,

Victoria, B.C., pp. 141–169.
Murray, W. J. (1975) McArthur River H.Y.C. lead–zinc and related deposits N.T. In *Economic Geology of Australia and Papua New Guinea* (ed. C. L. Knight), Australasian Institute of Mining and Metallurgy, Parkville, Victoria, pp. 329–338.
Neudert, M. K. and Russell, R. E. (1981) Shallow water and hypersaline features from the Middle Proterozoic Mt. Isa Sequence. *Nature (London)*, **293 (5830)**, 284–286.
Notebaart, C. W. and Vink, B. W. (1972) Ore minerals of the Zambian copperbelt. *Geol. Mijnbouw*, **51**, 337–345.
O'Brien, M. V. and Romer, D. M. (1971) Tara geologists describe Navan discovery. *World Min.* **24**, 38–39.
Oehler, J. H. and Logan, R. G. (1977) Microfossils, cherts and associated mineralization in the Proterozoic McArthur (H.Y.C.) lead–zinc–silver deposit. *Econ. Geol.*, **72**, 1393–1409.
Perkins, W. G. (1981) Mount Isa copper ore bodies: evidence against a sedimentary origin. Baas Becking Geological Laboratory Symposium, 2–4 March 1981.
Perkins, W. G. (1984) Mount Isa silica–dolomite and copper ore bodies – the result of a syntectonic hydrothermal alteration scheme *Econ. Geol.*, **79**, 601–637.
Peters, W. C. (1978) *Exploration and Mining Geology* John Wiley & Sons, New York, 696 pp.
Pine, R. J. and Batchelor, A. S. (1984) Downward migration of shearing in jointed rock during hydraulic injections. *Int. J. Rock Mech. Min. Sci.*, **21(5)**, 249–263.
Rentzsch, J. (1974) The "Kupferschiefer" in comparison with the deposits of the Zambian copperbelt. In *Gisements Stratiformes et Provinces Cuprifères* (ed. P. Bartholome), Liège Société Géologique de Belgique, pp. 395–418.
Rhodes, D., Lantos, E. A., Lantos, J. A., Webb, R. J. and Owens, D. C, (1984) Pine Point ore bodies and their relationship to the stratigraphy, structure, dolomitisation and karstification of the Middle Devonian Barrier Complex. *Econ. Geol.*, **79**, 991–1055.
Roedder, E. (1968) Temperature, salinity, and origin of the ore forming fluids at Pine Point, North West Territories, Canada, from fluid inclusion studies. *Econ Geol.* **63**, 439–450.
Rose, A. W. (1976) The effect of cuprous chloride complexes in the origin of red-bed copper and related deposits. *Econ. Geol.*, **71**, 1036–1048.
Rose, A. W., Hawkes, H. E. and Webb, J. S. (1979) *Geochemistry in Mineral Exploration*, 2nd ed, Academic Press, London, 657 pp.
Routhier, P. (1963) *Les Gisements Métallifères, Géologie et Principe de Recherche*. Masson et Cie, Paris, 2 vols., 1273 pp.
Rozendaal, A. (1978) The Gamsberg zinc deposit, Namaqualand. In *Mineralisation in Metamorphic Terranes 16th Congress of Geology Society of S. Africa*, (ed. V. J. Verwoerd), pp. 235–265.
Russell, M. J. (1968) Structural controls of base metal mineralization in Ireland in relation to continental drift. *Trans. Instn. Min. Metall. (Sect. B: Appl. Earth Sci,)* **77**, B 117–128.
Russell, M. J. (1975) Lithogeochemical environment of the Tynagh base-metal deposit, Ireland, and its bearing on ore deposition. *Trans. Instn. Min. Metall. (Sect. B: Appl. Earth Sci.)*, **84**, B 128–133.
Russell, M. J. (1978) Downward-excavating hydrothermal cells and Irish ore deposits: importance of an underlying thick Caledonian prism. *Trans. Instn. Min. Metall. (Sect. B: Appl. Earth Sci.)*, **87**, B 168–171.
Russell, M. J. (1983) Major sediment-hosted exhalative zinc–lead deposits: formation from hydrothermal convection cells that deepen during crustal extension. In *Sediment-Hosted Stratiform Lead–Zinc Deposits, Short Course Handbook*, vol. 8 (ed. D. F. Sangster), Mineralogical Association of Canada, Victoria, BC., pp. 251–282.
Rye, D. M. and Williams, N. (1981) Studies of the base metal sulphide deposits at McArthur River, Northern Territory, Australia. III. The stable isotope geochemistry of the H.Y.C., Ridge and Cooley deposits. *Econ. Geol.*, **76**, 1–26.
Samson, I. M. and Russell, M. J. (1983) Fluid inclusion data from Silvermines base metal-baryte deposits, Ireland. *Trans. Instn. Min. Metall. (Sect. B: Appl. Earth Sci.)*, **92**, B 66–71.
Sangster, D. F. (1983) Mississippi Valley-type deposits: a geological melange. In *International Conference on Mississippi Valley Type Lead–Zinc Deposits* (eds. G. Kisvarsanyi et al.), Proceedings Volume, University of Missouri, Rolla pp. 7–19.
Sasaki, A. and Krouse, H. R. (1969) Sulphur isotopes and the Pine Point lead–zinc mineralisation. *Econ. Geol.*, **64**, 718–730.
Schneiderhöhn, H. (1932) The geology of the Copperbelt, Northern Rhodesia. Translation in abstract. *Min. Mag. London*, **46**, 241–245.
Schulz, R. W. (1971) Mineral exploration practice in Ireland. *Trans. Instn. Min. Metall. (Sect. B: Appl. Earth Sci.)*, **80**, B 238–258.
Siegel, H. O. (1968) Discovery case history of the pyramid ore bodies, Pine Point, N.W.T., Canada. *Geophysics*, **33**, 645–656.

Skall, H. (1975) The palaeo-environment of the Pine Point lead–zinc district. *Econ. Geol.*, **70**, 22–47.

Smith, J. W. and Croxford, N. J. W. (1973) Sulphur isotope ratios in the McArthur lead–zinc–silver deposit. *Nature (London) Phys. Sci.*, **245**, 10–12.

Smith, J. W. and Croxford, N. J. W. (1975) An isotopic investigation of the environment of deposition of the McArthur mineralisation. *Miner. Deposita*, **10(4)**, 269–276.

Smith, J. W., Burns, M. S. and Croxford, N. J. W. (1978) Stable isotope studies of the origins of mineralization at Mount Isa. *Miner. Deposita*, **13**, 369–381.

Smith, S. E. and Walker, K. R. (1972) Primary element dispersions associated with mineralisation at Mount Isa, Queensland. *Bull. Bur. Miner. Resour. Geol. Geophys. Austr.*, 131 pp.

Smith, W. D. (1969) Penecontemporaneous faulting and its likely significance in relation to Mt. Isa ore deposition. *Spec. Publ. Geol. Soc. Austr.*, **2**, 225–235.

Solomon, P. J. and Jensen, M. L. (1965) Sulphur isotopic fractionation in nature with particular reference to Mount Isa, Queensland. *Proc. 8th Min. Metall. Cong.* pp. 12–21.

Spector, A. and Pichette, R. J. (1983) Applications of the aeromagnetic method to lead exploration in S. E. Missouri. In *International Conference on Mississippi Valley Type Lead–Zinc Deposits* (eds G. Kisvarsanyi *et al.*), Proceedings Volume, University of Missouri, Rolla, pp. 596–603.

Stanton, R. L. (1955a) The genetic relationship between limestone, volcanic rocks and certain ore deposits. *Austr. J. Sci.*, **17**, 173–175.

Stanton, R. L. (1955b) Lower Palaeozoic mineralization near Bathurst, New South Wales. *Econ. Geol.*, **50**, 681–714.

Stanton, R. L. (1963) Constitutional features of the Mount Isa sulphide ores and their interpretation. *Proc. Australas. Inst. Min. Metall.*, **205**, 131–153.

Stanton, R. L. (1955b) Lower Palaeozoic mineralisation York, 713 pp.

Steiger, R. and Proustie, A. (1979) Deep overburden sampling techniques in Ireland. In *Prospecting in Areas of Glaciated Terrain*, Institution of Mining & Metallurgy, London, pp. 22–30.

Swager, C. P. (1983) Microstructural development of the silica–dolomite and copper mineralisation at Mt. Isa, N.W. Queensland with special emphasis on the mechanism and timing of mineralisation. Unpublished Ph.D. thesis, James Cook University, N. Queensland.

Synge, F. M. (1979) Quaternary glaciation in Ireland. In *Prospecting in Areas of Glaciated Terrain*, Institution of Mining & Metallurgy, London, pp. 1–7.

Taylor, G. F. and Scott, K. M. (1982) Evaluation of gossans in relation to lead–zinc mineralisation in the Mount Isa inlier, Queensland. *BMR J. Austr. Geol. Geophys.*, **7**, 159–180.

Taylor, S. and Andrew, C. J. (1978) Silvermines ore bodies, County Tipperary, Ireland. *Trans. Instn. Min. Metall. (Sect. B. Appl. Earth Sci.)*, **87**, B 111–124.

Thiede, D. S. and Cameron, E. N. (1978) Concentration of heavy metals in the Elk Point evaporite sequence, Saskatchewan. *Econ. Geol.*, **73**, 405–415.

Vineyard, J. D. (1977) In preface to the issue devoted to the Viburnum Trend, S.E. Missouri. *Econ. Geol.*, **72(3)**, 337–338.

Walker, R. N., Gulson, B. and Smith, J. (1983) The Coxco deposit – a Proterozoic Mississippi Valley-Type Deposit in the McArthur River District, Northern Territory, Australia. *Econ. Geol.*, **78**, 214–249.

Walker, R. N., Logan, R. G. and Binnekamp, J. G. (1977a) Recent geological advances concerning the H.Y.C. and associated deposits, McArthur River, N.T. *J. Geol. Soc. Austr.*, **24(8)**, 365–380.

Walker, R. N., Muir, M. D., Diver, W. L., Williams, N. and Wilkins, N. (1977b) Evidence of major sulphate evaporite deposits in the Proterozoic McArthur Group, Northern Territory, Australia. *Nature (London)*, **265(5594)**, 526–529.

Webb, J. S. and Tooms, J. S. (1959) Geochemical drainage reconnaissance for copper in Northern Rhodesia. *Trans. Instn. Min. Metall. (Sect. B: Appl. Earth Sci.)*, **68**, B 125–144.

Williams, B. S. (1982) Review of current geophysical techniques for base metals in Ireland. In *Mineral Exploration in Ireland: Progress and Developments 1971–1981* (ed. A. G. Brown), Irish Association for Economic Geology, Dublin, pp. 135–147.

Williams, N. (1978) Studies of the base metal sulfide deposits at McArthur River, Northern Territory, Australia 1. The Cooley and Ridge Deposits. II. The sulphide S and organic C relationships of the concordant deposits and their significance. *Econ. Geol.*, **73**, 1005–1035; 1036–1056.

Woodham, C. R. (1982) The role of reverse circulation drilling in mineral exploration in Ireland. In *Mineral Exploration in Ireland: Progress and Development 1971–1981* (ed. A. G. Brown) Irish Association for Economic Geology, Dublin, pp. 123–134.

7

Ore deposits formed by weathering

7.1 INTRODUCTION

Many types of mineral deposit form by weathering. For discussion in this chapter we have selected bauxite, nickel laterite and kaolin; we also describe supergene manganese and the supergene enrichment of sulphides. These mineral deposits are among the most important ores formed by weathering and supergene enrichment is one of the best understood aspects of weathering. The selections were made to illustrate the several different chemical processes taking place during weathering.

Weathering is the breakdown and alteration of rocks by physical and chemical processes, both of which may be aided by organic activity. The fractionation of the rock occurs in response to changes in environmental conditions since the rock was formed. The products of weathering are materials more nearly in equilibrium with their environment than those from which they are derived. For example, the minerals of igneous rocks, formed at high temperatures in the absence of abundant water, are unstable in the cooler, wet conditions at the earth's surface; the minerals of these igneous rocks are altered to low-temperature water-bearing phases or dissolved and removed. Physical weathering disaggregates the rock creating large surface areas and greater access by fluids. This increases the susceptibility to chemical weathering. Chemical weathering usually is in two stages, alkaline then acid. The behaviour of the different elements within the silicate minerals is differentially affected by changes in pH. During the alkaline stage K, Na, Ca may be removed and the remaining rock material may be relatively enriched in Fe, Si and Al. When conditions become acid the hydroxides of aluminium and iron can migrate to a limited extent.

Rose *et al.* (1979) list the major processes of chemical weathering as hydration, hydrolysis, oxidation and solution. The earliest evidence of rock weathering is oxidation of Fe^{2+} to Fe^{3+} and removal of Na, Ca and Mg in solution (Dennen and Anderson, 1962). Na is particularly mobile during weathering but release of K into solution is restricted by its adsorption on to kaolinite.

As might be expected, the relative stabilities of the common rock-forming minerals in igneous rocks bear an antithetic relationship to the order in which they crystallize from a melt as a result of falling temperature. Thus the arrangement of

Ore deposits formed by weathering

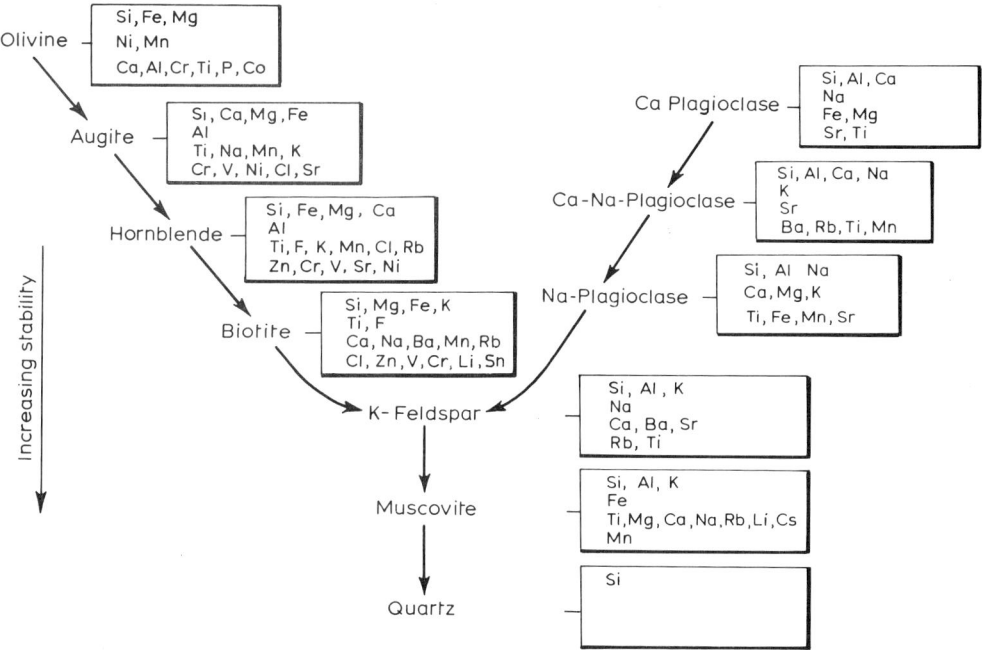

Fig. 7.1 Sequence of stability of common minerals during weathering processes (after Goldich 1938).

minerals in order of increasing stability during weathering (Goldich, 1938) corresponds to Bowen's (1922) reaction series in order of decreasing temperature of crystallization (Fig. 7.1). The relationships are not so clear when the wide-ranging compositions of metamorphic rocks are considered. It might be anticipated that those minerals which formed at the surface would be the most resistant to weathering but chemical precipitates in sedimentary rocks suffer severe weathering under changing climatic conditions. Factors which influence the nature and rate of chemical weathering include permeability, climate, relief and drainage (Ollier, 1969). Weathering is the initial stage in the cycle of events leading to the formation of sediments. Without some form of weathering rock material cannot be removed by transport to be deposited elsewhere to form alluvial deposits. Also the products of weathering may give rise to mineral deposits formed *in situ*; these are called residual deposits.

The production of an *in situ* residual deposit requires deep and long-continued weathering followed by lack of erosion. Residual deposits are therefore usually absent from glaciated regions and mountain belts where erosion is active. Two distinct types of residual deposits may be recognized. The first results from the accumulation of pre-existing mineral species in the source rock; concentration takes place by the removal of many other components from the source rock. The eluvial accumulation of cassiterite (page 178) is an example of this type. The economic concentrations of gold which occur in the weathering mantle over some porphyry copper deposits, as at Ok Tedi in Papua New Guinea, occur in this way. The second type, and more widespread economically, are the residual deposits where the valuable mineral comes into existence as a result of weathering and the economic mineral continues to accumulate. Bauxite, some of the World's china clay deposits and nickeliferous laterites are examples of this second type.

7.2 BAUXITE DEPOSITS

Bauxite is the main ore of aluminium and is produced by the weathering of a wide variety of rock types. It is a composite substance, primarily of minute crystals of one or more of three aluminium hydroxides, gibbsite, boehmite and diaspore. An amorphous form of hydrated aluminium oxide, cliachite, may be present. Impurities in bauxites include kaolinite, quartz, leucoxene and goethite. Metamorphism of bauxite leads to corundum (emery deposits).

Over 90% of bauxite output is used in the production of aluminium metal for which purified anhydrous alumina is produced from bauxite by the Bayer chemical conversion process. The process involves digestion of bauxite in a hot caustic soda solution. Before digestion the ore is blended, crushed and milled. The conditions in the pressure-leaching stage of the process are as follows:

(i) To recover Al_2O_3 from gibbsitic (trihydrate) bauxite: 33% NaOH at 140°C at 5 bars for 1–2 hours, (the American Bayer process).

(ii) To recover Al_2O_3 from boehmitic (monohydrate) bauxite: 50 % NaOH at 180–250°C at 13–25 bars for 2–8 hours, (the more costly high-temperature or European Bayer process).

If the boehmite content is low, the low-temperature process is used and the Al_2O_3 in boehmite is lost. If the boehmite content is high then the more costly high-temperature process is used and Al_2O_3 is recovered from gibbsite and boehmite. The low-temperature process does not react with quartz but does with kaolinite which uses up soda ash. In the high-temperature process both quartz and kaolinite react. Silica content is also important; if it is greater than 5% some alumina remains insoluble and this can be lost in the 'red mud' which is produced during the process. Therefore, alumina plants prefer bauxites with low silica, preferably less than 1% SiO_2. Diaspore is not generally regarded as a commercial source of alumina although bauxites containing more diaspore than boehmite and little gibbsite are processed in China and in one plant in Greece.

7.2.1 Classification of bauxite deposits

Various classes of bauxite have been proposed (Table 7.1). Harder and Greig (1960) originally sought to classify bauxites almost entirely on the composition of their bedrocks. Also in 1960 Hose suggested a scheme based purely on field occurrence. In common with earlier suggestions the Hose scheme implied that karst bauxites comprise a distinct genetic group. Valeton (1972), in an extensive survey of bauxites, returned to a classification based on bedrock lithology (Table 7.1).

Grubb (1973) has suggested that the elevation at which bauxitization took place is a suitable criterion. He divided bauxites into high-level, or upland bauxites and low-level or peneplain-type bauxites. In devising this classification Grubb dismissed the karst bauxites as a separate, genetically distinct group. He reinterpreted those karst bauxites with an earthy gibbsitic nature, such as the Jamaican bauxites, as having affinities with his upland division. He considered the pisolitic monohydrate-rich European karst bauxites as being similar to his peneplain deposits. This view is not supported by Bardossy (1982) who maintains that the Jamaican bauxites have more in common with the European karst bauxites than with the high-level lateritic bauxites.

Hutchison (1983), while recognizing there may be significant differences between bauxites formed in upland areas and those formed in lowland peneplains, combines them under the title of 'lateritic crusts' and produces a much simplified classification scheme. We consider that none of these classifications has much practical application for the exploration geologist, and since Hutchison's is easily applied and interpreted we shall adopt it. As we are concerned with finding mineable deposits we shall mention those features of Grubb's high- and low-level subdivisions which would influence economic exploitation (Table 7.2).

Table 7.1 Classifications of bauxite deposits

Harder and Greig (1960)	Hose (1960)	Valeton (1972)	Grubb (1973)	Hutchison (1983)
Surface blanket deposits	Bauxites formed on peneplains	Bauxites overlying igneous and metamorphic rocks: (i) slope type	High-level or upland bauxites	
Interlayered beds or lenses in stratigraphic sequences	Bauxites formed on volcanic domes or plateaux	(ii) plateau type on basic igneous rocks (iii) plateau type on variable rock types		Lateritic crusts
Pocket deposits in limestone, clays or igneous rocks	Bauxites formed on limestones or karstic plateaux	Bauxites on sedimentary rocks:	Low-level peneplain-type bauxites	Karst bauxites
Detrital bauxites	Sedimentary reworked bauxites	(i) on clastic sediments (ii) on carbonate rocks (iii) on phosphate rocks		Sedimentary bauxites

Table 7.2 Comparison of main bauxite types

	Lateritic crusts		
	High level	Low level	Karstic
Occurrence	Capping plateaux in tropical/subtropical regions, e.g. India	Low-relief planation surfaces, e.g. W. Australia, Guyana	Main type in countries bordering Mediterranean (40% of karst deposits) and Jamaica
Composition	Mainly gibbsitic, minor boehmite, high-iron, significant vertical variation	Gibbsite, minor boehmite, low-iron often little compositional variation	European – iron oxides < 10%, gibbsite + boehmite (50%) to boehmite + gibbsite or diaspore (may be secondary) + boehmite. Jamaica – gibbsitic, minor boehmite (<20%)
Bedrock interface with bauxite	Bauxite rests on saprolite formed on source rocks	Separated by kaolin or partly weathered rock enriched in kaolinite	Highly irregular, pillars of carbonates protrude
Texture	Relatively coarse, compact. Sparsely pisolitic with parent rock textures common	Relatively coarse-grained, compact, pisolitic, massive	European – massive, sparsely pisolitic – coarser-grained. Jamaican – earthy, finer-grained
Shape	Blankets	Blankets	Stratiform, blankets, lenses, etc.
Thickness	Up to 25 m	Less than 9 m (commonly)	Highly variable but commonly 4 to 10 m
Age	Mainly Tertiary and Quaternary	Mainly Tertiary and Quaternary	Mediterranean – Permian to Miocene. Caribbean–Miocene to Holocene
Controls	Jointing in bedrock	Progressive lowering of water table?	Distribution of carbonates – karstic surface. May be transported
Mining parameters	Very irregular footwall. Chimneys and walls extend deep into footwall. Unaffected boulders higher in profile. Significant compositional variations in vertical profile	Relatively even footwall, little compositional variation	Very irregular shapes, lenses, pockets, etc.

(a) Lateritic crusts

(i) High-level or upland bauxites

These are generally developed on volcanic or intrusive igneous source rocks and occur as blankets up to 25 m thick capping plateaux in tropical or subtropical areas. In primitive form these bauxites are earthy and sparsely concretionary with the mineralogy of the 'clay' fraction predominantly gibbsitic. Maturity is shown by the tendency to consolidate to form a compact horizon which usually maintains the parent rock texture. Eventually this becomes more massive with loss of residual textures until finally a hard, sparsely pisolitic surface horizon is formed. The upland bauxites rest upon their source rocks with little or no intervening underclay, and chimneys and walls of bauxite often extend deep into the footwall. Higher up in the deposit profile, large unaltered boulders of parent rock may be completely enclosed by bauxite. According to Grubb (1973) high-level lateritic bauxites show significant compositional variations in the vertical profile because of the control of the parent rock structures upon bauxite formation. The compositional variation, highly irregular footwall and presence of large unaltered boulders all create mining difficulties in these deposits.

(ii) Low-level peneplain-type bauxites

Bauxites of this type are commonly associated with low-relief planation surfaces such as those along the northern coast of South America and in northern Australia. Peneplain deposits are generally less than 9 m thick. They have a fairly even footwall contact with the underlying parent rock, they are separated from this rock by a conspicuous, usually kaolinitic, clay horizon. According to Grubb (1973) there is comparatively little compositional variation in these low-level bauxites. Peneplain-type deposits are frequently associated with detrital bauxite horizons of fluviatile or marine origin. It is suggested by Grubb that a significant proportion of some peneplain bauxites were originally products of

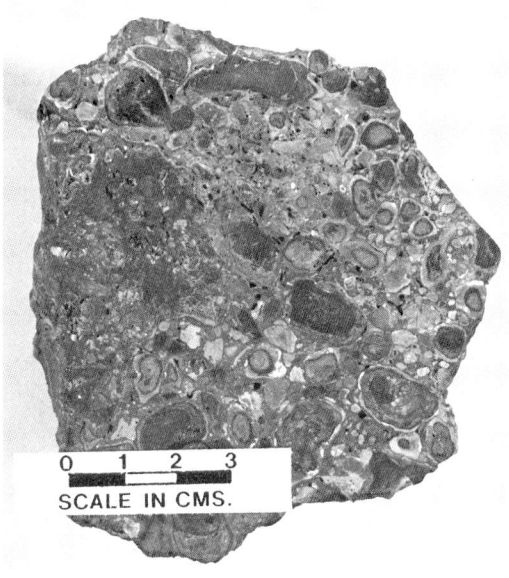

Fig. 7.2 Specimen of pisolitic bauxite, Ghana.

downwasting from regions of active lateritization.

Although predominantly pisolitic (Fig. 7.2), three distinct subhorizons can usually be recognized in a complete profile. Nearest the surface in most mature profiles is the 'A' horizon which is uniformly pisolitic and is often lithified. Below this the 'B' horizon is more coarsely pisolitic and is a semi-concretionary ore. The basal 'C' horizon is concretionary, often highly ferruginous with relatively high clay content. In more primitive deposits the entire profile may consist of the 'C' horizon ore type.

(b) Karst bauxites

This group overlies highly irregular karstified limestone and dolomite surfaces. Despite their occurrence on carbonate rocks, the majority of karst bauxites may not be genetically linked to the rocks forming their footwalls. The use of the trace-element content of these karst bauxites to establish

Fig. 7.3 Irregular nature of karst bauxite, S.E. France.

their likely parent rocks has been described by Ozlu (1983).

Bardossy recognized six different types of karst bauxites, based on depositional characterisics. We shall confine our attention to the Mediterranean type as it is the largest and the most economically important of the karstic group.

The Mediterranean-type deposits are lithologically fairly homogeneous, the high-grade bauxite core passing into clayey bauxite or bauxitic clay toward the margins of the deposit. They may occur as stratiform deposits with areas often exceeding 0.5 km^2, with an irregular bedrock interface forming the footwall and a rather smooth top. Solution cavities in the footwall may reach 10 m in depth while steep prominences from the footwall may result in the complete absence of bauxite (Fig. 7.3).

Mediterranean-type deposits also occur in the form of blankets which wrap over irregularities in the bedrock profile giving the deposit top an irregularity reflecting the bedrock contours. Strip-like deposits are also recognized within this type in southern France near the town of Baux de Provence where Berthier identified aluminium in bauxite in 1821.

The most widespread group among Mediterranean-type deposits is the lenticular deposit. Although of relatively small area, less than 0.4 km^2, the bauxite may reach thicknesses of 30 m. As the name suggests, some of these bauxite deposits are convex upwards as well as downwards, a shape difficult to explain if there has been no post-bauxite erosion as stated by Bardossy (1982). Bardossy suggests that this peculiar shape results from *in situ* weathering of windblown tuffs, an origin for these deposits previously suggested by Comer (1974). Other Mediterranean-type deposits are grouped under graben deposits, canyon-like deposits and sink-hole deposits. A final grouping for karst bauxites, 'nests' and 'bags', is also given by Bardossy to include irregular, small infillings in bedrock which rarely contain bauxite 5 m thick, 2 m being the usual maximum. Such irregular deposits are difficult to evaluate and mine but the Nagyharsany deposit in Hungary is of this type and has been mined as an underground operation.

(c) Transported or sedimentary bauxites

This is a minor grouping in which the bauxites have been deposited as a sedimentary rock. They overlie rocks to which they have no direct connection. These bauxite strata may be fossiliferous. The most extensive development of sedimentary bauxites is in the USSR and China.

7.2.2 General characteristics of bauxite deposits

(a) Distribution in space and time

The bauxite deposits of the World may be grouped into provinces (Table 7.3),

(a) The Guiana Shield Province of South America. This includes Venezuela, Guyana, Surinam, Guiana and parts of Brazil and Colombia.

(b) The Northern Brazilian Shield Province.

(c) The Caribbean Province including Costa Rica, Jamaica, the Dominican Republic, Haiti and Puerto Rico.

(d) The Guinea Shield Province extending from Guinea-Bissau to Togo with extensions into Upper Volta and the central Sahara.

(e) The Cameroon Province including the Cameroons, Zaire and adjoining countries.

(f) The Australian Province.

(g) The European Province including France, Greece, Hungary and Yugoslavia.

(h) Others, here we include the USA, China, USSR, India and Malaysia.

Accurate dating of bauxite deposits is not always possible because of unconformities between them and overlying rocks. The period of formation for a particular deposit may also span a considerable period of geological time. However, there have been well-recognized periods in the earth's history when conditions for bauxite formation were optimum. The most intense period of bauxite formation was from Cretaceous to Recent (Table 7.3). This timespan can be further subdivided into earlier (Cretaceous to Eocene) and later (Miocene to Recent) periods of bauxite formation. The Devonian to Late Carboniferous was another time of bauxite formation as was the period from Upper Proterozoic to Cambrian. The Cenozoic bauxites have formed only within the latitudes 30°S to 30°N (Fig. 7.4). For earlier deposits similar palaeogeographic constraints would have applied as strict climatic controls are necessary for bauxite formation. Tertiary bauxites tend to be dominantly gibbsitic with minor boehmite. Mesozoic deposits are invariably derived from limestones and are dominated by boehmite with local diaspore and minor gibbsite. Palaeozoic deposits mainly contain diaspore and boehmite.

The main contributors to World bauxite production are shown in Fig. 7.5. This illustrates

Table 7.3 Age of deposits in the World's bauxite provinces (data from Harben and Dickson, 1983; Shaffer, 1983).

Age	Guiana Shield	North Brazil Shield	Caribbean	Guinea Shield	Cameroon	Australia	Europe	Others
Quaternary	Colombia		Costa Rica Panama					Hawaii, Fiji, Solomon Islands
Tertiary								
Undifferentiated	Venezuela	Brazil		Guinea	Ghana, Ivory Coast	W. Australia		
Miocene			Jamaica, Haiti, Dominican Republic			Victoria		USA (Oregon, Washington)
Oligocene							Northern Ireland	
Eocene	Surinam, Guyana,					Queensland, Tasmania, Northern Territory	Italy, Yugoslavia	India, USA (Arkansas, Alabama, Georgia, Mississippi)
Palaeocene							Yugoslavia	
Mesozoic								
Cretaceous							Turkey, France, Italy, Hungary, Romania, Greece, Yugoslavia	Saudi Arabia
Jurassic							Yugoslavia	Asia
Upper Triassic							Yugoslavia	
Lower Triassic								China
Palaeozoic								
Permian								Turkey
Upper Carboniferous (Pennsylvanian)								USA (Missouri, Pennsylvania)
Lower Carboniferous (Mississippian)								China, USSR
Devonian							Spain	USSR
Cambrian								USSR
Precambrian								USSR

the impressive increase in the contribution made by Australia over the period 1962–1983. At the present time Australia produces about 40% of the World's bauxite (excluding the USSR and China). Dominating the remaining 60% are Jamaica and Guinea. Traditionally Guyana and Surinam have been important suppliers of bauxite and the former is still the World's largest supplier of refractory-grade material. Another South American country, Brazil, is becoming increasingly important and the large Trombetas deposit in the Amazon region is the first of many deposits likely to be developed in the north of the country. The USA which is the World's largest producer of aluminium was the fifth largest producer of bauxite in 1962, but by 1983 it had moved to twelfth in the list of producers.

(b) Size and grade of deposits

The trade use of bauxite classifies the material according to its end use into six grades (Table 7.4). Metallurgical grade or alumina grade (for use in the production of aluminium metal) accounts for about 90% of bauxite production. Ideally an alumina grade bauxite should contain the following (but see Table 7.4 for typical analyses):

$>50\% \ Al_2O_3$; $<6\% \ SiO_2$; $<10\% \ Fe_2O_3$; $<4\% \ TiO_2$

At the present time two of the World's leading bauxite producers, Jamaica and the Darling Range in Australia, produce lower-grade Al_2O_3 and higher Fe_2O_3 than that quoted above. The Darling Range material contains 18% Fe_2O_3 and has only 37% total Al_2O_3. Evaluation of bauxite deposits, however, is based upon the 'available alumina', recoverable by the Bayer process (Dreyer, 1978). The Darling Range material contains only 30% available Al_2O_3.

In Australia, as well as the Darling Range the important producing areas are Weipa, Queensland (proven reserves about 500 million tonnes) and Gove, Northern Territories (proven reserves 282 million tonnes). At Gove ore grade was defined as greater than 40% available alumina, less than 8% total silica and the deposit must be at least 2 m thick, unless the overburden ratio was less than 1.5:1 (Lillehagen, 1979). In Europe, which contains about 8% of the total World reserves, the bauxite occurs predominantly in Greece and Yugoslavia (Table 7.5). France has approximately 30 million tonnes of exploitable reserves. The largest stratiform karstic deposit is the Halimba Basin in Hungary which covers an area of about 7 km^2 with an average thickness between 5 and 10 m. Guinea, which has 28.6% of known World reserves (Table 7.5), contains the World's largest single bauxite mine – the Boké Mine – which produces about 9 million tonnes per year from an essentially gibbsitic ore body which contains up to 25% boehmite.

The World's major supplier of refractory-grade bauxite is Guyana which has reserves of approximately 520 million tonnes of which about 70% is classified as low-iron (less than 1.5% Fe_2O_3). China is rapidly expanding its share of this market and is the country mainly responsible for reducing Guyana's dominance from 80% to about 60%. The Trombetas area which lies to the south of the Guiana Shield region in Brazil has estimated reserves of 1000 million tonnes of 43–45% Al_2O_3, 6–15% SiO_2, 6–20% Fe_2O_3 (Aleva, 1981). Brazil currently ranks second in the list of major reserve areas of the World, containing 17.4% of the World's reserves (Table 7.5).

(c) Mineralogy

In excess of one hundred minerals have been recognised in bauxites but only ten occur in rock-forming amounts. Of these the three hydrated alumina oxides, gibbsite, boehmite, diaspore, are the sources of aluminium. The other seven minerals are corundum, goethite, haematite, kaolinite, halloysite, anatase and rutile (Bardossy, 1979). Cliachite may also be present. Gibbsite and boehmite are the main ore minerals.

Impurities commonly present in bauxite include kaolinite, quartz and goethite. True bauxite grades into an aluminous clay when the content of hydrated aluminium oxides decreases relative to the aluminosilicates such as kaolinite. An intermediate category, bauxitic clay, which may be regarded as a refractory clay is not normally commercially acceptable and contains a slight

Ore deposit geology and its influence on mineral exploration

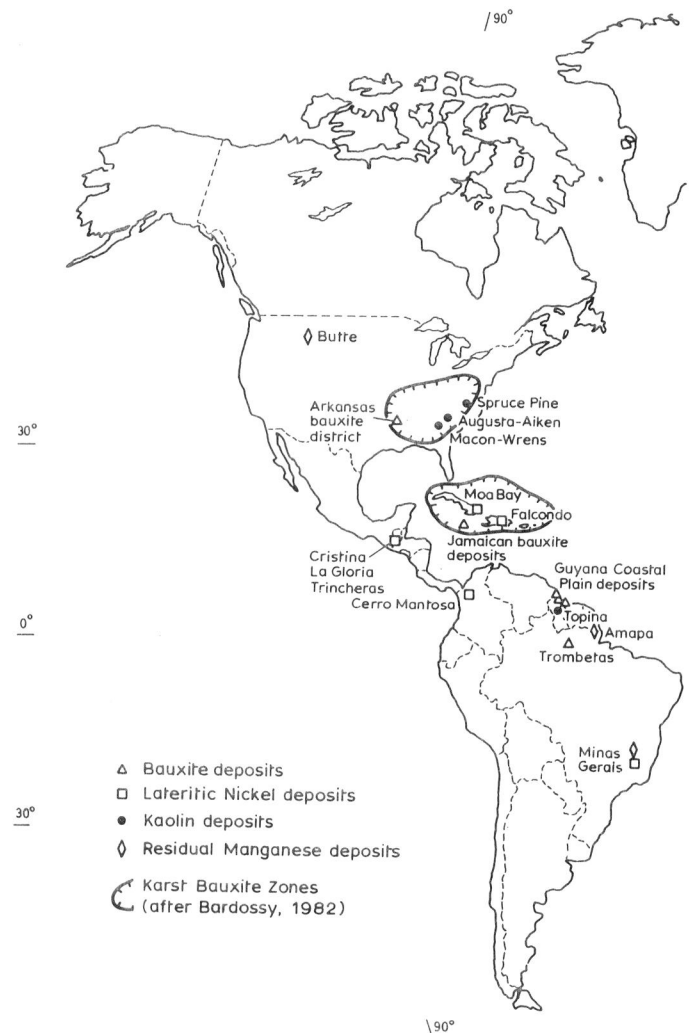

Fig. 7.4 Location map

excess of alumina over silica. With increasing proportion of hydrated oxides of iron, bauxite grades into laterite. Too high a proportion of iron may preclude the commercial exploitation of the bauxite.

(d) Host rock lithology

Most crustal rocks contain a high proportion of aluminium-bearing minerals, and bauxites may be formed from almost any rock type. Bauxites have been derived from nepheline syenite in Arkansas, USA, from limestones in many parts of the World, including Jamaica, from clays in Guyana, from shales at Gove, Australia, and from kaolinitic sands and clays in Brazil and at Weipa in Australia.

(e) Alteration

Metamorphism changes the mineralogy from gibbsite to boehmite to diaspore. Early formed Palaeozoic bauxites are dominantly diaspore, Mesozoic bauxites are predominantly boehmite and Cenozoic deposits are gibbsite. Continued metamorphism at higher grades results in the formation of corundum (emery deposits). Some

bauxite deposits contain microcrystalline corundum that has formed diagenetically (Bardossy, 1979).

(f) Tectonic setting

The large economic lateritic crust bauxites which account for about 83% of the World's supply of bauxite are found on stable continental platforms in Australia, India, Guyana and West Africa. Other lateritic bauxites, amounting to about 2% of the World's supply, are sited on younger cratons such as Malaysia and Indonesia. Karstic bauxites yield about 14% of the World's supply and are associated with younger orogenic belts. The sedimentary or transported bauxites are largely confined to stable continental platforms in China and the USSR.

7.2.3 Genesis of bauxite deposits

(a) Process of 'bauxitization'

Most crustal rocks contain a high proportion of aluminium-bearing minerals. The process by which materials of these rocks are converted into

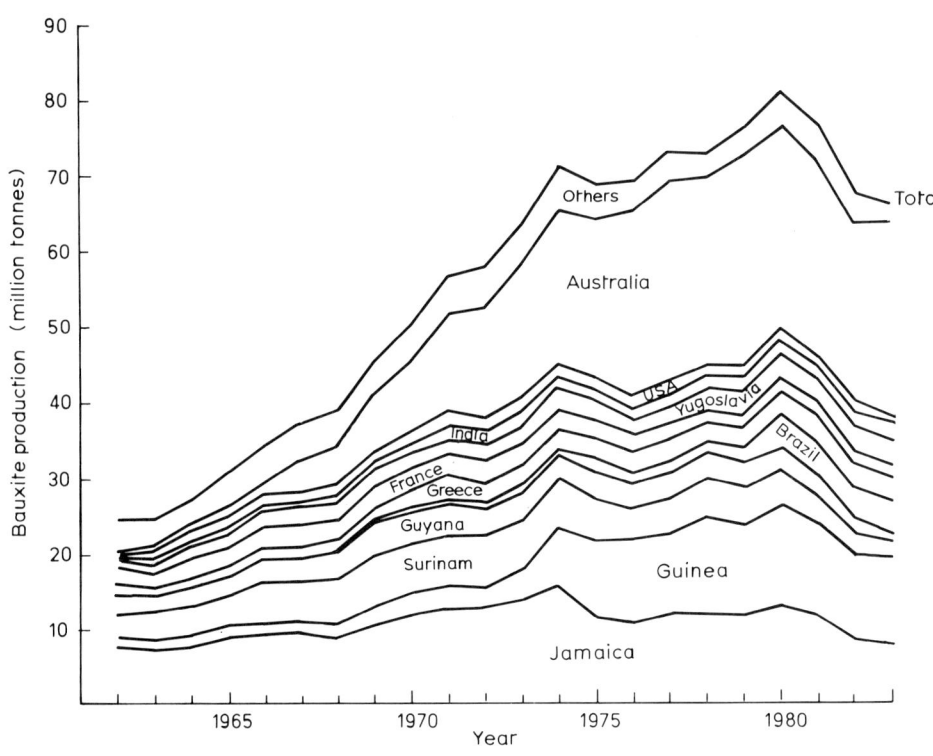

Fig. 7.5 World (excluding USSR and China) bauxite production. Source: *Mining Annual Review* for relevant years.

Table 7.4 Bauxite grades by end use. (Analyses from Harben and Dickson, 1983)

		Metallurgical	Chemical	Cement	Refractory (calcined)	Abrasive (calcined)	Proppants
Requirements		High-alumina, low-iron, low-silica, low-titania	Must be gibbsitic, very low iron, silica low but not critical	Moderately high-alumina low-silica, iron may be high, diaspore preferred	High-alumina low-iron, low-silica, very-low-alkalies	High-alumina moderately low-iron, low-silica, low-titania	High-alumina low-silica, low-clays, iron content not critical
Typical analyses	Al_2O_3	50–55%	>55%	45–55%	>84.5%	80–88%	None available
	SiO_2	0–15%	5–18%	<6%	<7.5%	4–8%	
	Fe_2O_3	5–30%	<2%	20–30%	<2.5%	2–5%	
	TiO_2	0–6%	0–6%	3%	<4%	2–5%	

Table 7.5. Major World reserves of bauxite. (Govett and Larsen, 1981)

	Bauxite* (million tonnes)	Percentage of World total
Guinea	8200	28.6
Brazil	5000	17.4
Australia	4400	15.5
Jamaica	2032	7.1
India	1600	5.6
Guyana	1016	3.5
Cameroon	1016	3.5
Greece	900	3.1
Ghana	780	2.7
Indonesia	710	2.5
Surinam	498	1.7
Yugoslavia	400	1.4
Total	26 592	92.6

*Estimate as of 1981.

bauxite is referred to as 'bauxitization'. It requires the strong aluminosilicate lattices to be broken down by the removal of silicon and the alkaline elements – potassium, sodium, calcium and magnesium – leaving aluminium in chemical compounds which are ores of aluminium. Bauxitization is confined to surface or shallow near-surface, conditions. Bauxitization (or lateritization) may take place *in situ*, or partially weathered minerals may be transported to sites where they undergo subsequent bauxitization; alternatively the bauxite itself may be eroded, transported and redeposited. On a world-wide scale truly transported bauxite is not important.

(b) Climate

The climatic conditions are of primary importance in the formation of bauxite deposits. A humid tropical or subtropical climate is essential; temperatures above 20°C favour the passage of SiO_2 into solution allowing its removal as the silicates break down. This removal of SiO_2 is the key issue in the formation of economic residual accumulations of alumina. Furthermore the intensity of bauxitization varies with the annual rainfall. If the rainfall exceeds 1.5 m per annum more or less uniformly distributed throughout the year then a clay-type (kaolinitic) laterite is formed instead of a lateritic bauxite. Consequently a climate with defined wet and dry seasons is necessary for thorough bauxitization. The leaching of silica proceeds during the dry season while during the wet season the Al_2O_3 and Fe_2O_3 are formed. The alternation of wet and dry seasons leads to the concentration of these compounds.

(c) Parent rock

The presence of a suitable parent rock for the bauxite is not as critical as the correct prevailing climatic conditions. Bauxite may be derived from almost any rock, although a well-jointed severely fractured or highly porous rock will undergo chemical weathering more rapidly than a massive or impermeable one. Pedro (1970) described experiments in which a marl was bauxitized so that the gibbsite content of the rock increased from zero to 15% in 3 years.

(d) Chemical and organic activity

Although the chemical processes involved in the formation of bauxite are not fully understood it appears that water, carbon dioxide and humic acids are some of the reagents required. In the experiments on marls mentioned above (Pedro, 1970), the samples were percolated with water at 65°C saturated with carbon dioxide. In the field there is little doubt that the process of bauxitization is speeded up if the water present is alkaline tepid rainwater, associated with tropical and subtropical climates. The activity of plants in bauxitization must also be considered, since Lovering (1959) has stated that some tropical forests can extract about 2000 tonnes of silica from an area of less than 0.5 hectares in 5000 years. Although this is a rapid rate of desilication the role of vegetation does not appear to be a dominant one since bauxite deposits occur in the Proterozoic and Cambrian. Micro-organisms probably play an important role in decomposing minerals in the parent rock.

(e) Topography and hydrology

The topography of a region is important in

laterite formation. Surfaces that permit slow downward infiltration of meteoric waters also allow chemical weathering, whereas fast run-off prevents this type of weathering. Consequently bauxite deposits lie on gently undulating erosion surfaces or peneplains. It is important that the accumulating residuum is retained and not washed away. Despite these restrictions there must be an adequate subsurface drainage to allow the withdrawal of rainwater. This rainwater will be charged with SiO_2 and other substances removed during bauxitization. The formation of bauxite deposits will not advance in areas of stagnant groundwaters. A deposit must lie above the water table for the leaching and drainage processes to proceed.

(f) Time and preservation

All the conditions described as contributing to the formation of bauxites – climate, reagents, surface and subsurface conditions – must persist for significant lengths of time to permit the accumulation of a thick weathering mantle. Bauxitization is usually regarded as a slow process but under suitable conditions it can be quite rapid. Valeton (1972) quotes the existence of a bauxitic profile on lavas aged about 10 000 years on Hawaii. The rate at which bauxitization develops is controlled by the nature of the parent rock and its permeability combined with the interaction of the various conditions mentioned above. Even if all the conditions necessary for its formation are satisfied, a bauxite will only persist for discovery and evaluation if it is preserved from erosion. Lack of protection from erosion probably accounts for the paucity of economic bauxite deposits in pre-Tertiary rocks.

7.2.4 Exploration and evaluation of bauxite deposits

Exploration for lateritic ores relies on a thorough understanding of the factors which control the occurrence of economic deposits of this type. A description of the use of such factors as guides to bauxite ore occurrences in the Mount Saddleback and associated deposits in Western Australia is given by Ward (1978). The recognition of the Rennell Island bauxite deposit from airborne gamma ray surveys and from aerial photography is outlined by Allum (1982).

According to Dreyer (1978) the evaluation of lateritic deposits is one of the most difficult problems likely to be encountered by the economic geologist. Lateral and vertical variations in laterites are very common, with high-grade ore passing into valueless rock within a few tens of millimetres. Consequently sampling must be very carefully organized; a few drill holes or grab samples are of little value except to give gross characteristics of the deposit. An evaluation programme should be based upon careful statistical studies which rely on thousands of intelligently collected samples per deposit. The most reliable method of collecting samples is from large vertical channels. If drilling is undertaken the results will not reproduce the exact information from pits but may indicate a correction factor to apply to further drill hole data. Consideration must be given to which drilling technique – auger, churn drill, rotary or core – will give results most similar to channel samples in such a heterogeneous sequence.

A regular sampling grid is preferred and a common spacing for reconnaissance is a 300 m square grid, but variations to this will depend on local conditions. To ascertain the correct spacing for calculation of grade and tonnage it is customary to take samples on 25 m centres in a 300 m square. Panagopoloulos (1983) has shown the application of geostatistical techniques to the Parnassos-Giona bauxite deposits in Greece. Here a 50×50 m drill hole grid was used and considering the silica grade as the limiting requirement, room sizes for a room and pillar mine were designed.

Bauxite analyses require special analytical facilities. Automated X-ray spectrometric (XRS) and X-ray diffraction (XRD) analytical systems are used. Large numbers of samples are processed and the results stored by computer. XRS analysis

Table 7.6 Data provided in a bauxite analysis (Strahl, 1982)

	XRS chemical								
Sample no.	Fe_2O_3	TiO_2	CaO	SiO_2	Al_2O_3	P_2O_5	MnO	ZnO	LOI
Sw 1	14.14	2.37	0.32	6.19	51.34	0.56	0.56	0.032	24.49

	Alumina distribution						
Sample no.	TA	TAA Al_2O_3	ReSil	THA Al_2O_3	ReSil	UA	Boehmite Al_2O_3
Sw 1	51.34	43.93	6.06	39.45	4.65	1.35	5.89

	Iron distribution				
Sample no.	Haematite (%)	Goethite (%)	(Mole diasp.)	H/G ratio	Settling rate (ft/h)
Sw1	6.97	7.17	23.05	0.7	8

Key

LOI	=	Loss on ignition (calculated by difference)
TA	=	Total Al_2O_3
TAA	=	Total available Al_2O_3 (boehmite digestion conditions)★
THA	=	Trihydrate available Al_2O_3 (gibbsite digestion conditions)★
UA	=	Unextractable Al_2O_3
ReSil	=	Reactive silica
H/G ratio	=	Intensity of haematite peak × (10/7)/intensity of goethite peak

★See text for details of these conditions.

is carried out for eight elements, Al, Si, Ti, Fe, Ca, P, Mn and Zn. Using XRD each sample is analysed for the boehmite, hematite, goethite and alumina in the lattice of goethite. Table 7.6 gives a typical method of reporting bauxite analyses. This method of analysis is required as the Bayer process for extracting alumina from bauxite involves the selective dissolution and reaction of specific minerals from the bauxite. A direct chemical analysis for elemental composition yields only a portion of the data required to effectively describe the potential value of a bauxitic material. Furthermore, although the important character is the 'available alumina', to fully characterize the economic value of a bauxite requires digestion, mud separation and mud-settling tests under conditions approximating to a commercial Bayer plant (Strahl, 1982). A close approximation of the process behaviour of a bauxite can also be calculated from a combination of the chemical and mineralogical analyses.

Other factors which are crucial in the evaluation stage are the anticipated metallurgical and mining problems. While the economic geologist may not be able to solve these problems it is vitally important that their presence is recognized. The grade and tonnage of reserves are affected by large irregularities in the overburden which must be smoothed out during mining. Since bauxites and other lateritic ores are mined by large-scale bulk-mining methods it is not often practical to include small or isolated pods of ore within calculated reserves. Cut-offs for bauxite are directly dependent on available alumina and inversely dependent on reactive silica. Pilot mining has to be carried out to assess the most suitable equipment and the minimum pocket size that can be economically and technically mined. When these factors are considered alongside such features as transport and the costs of establishing an infrastructure, the true ore reserve and grade are determined. An example of the interdependence of these factors is given for the low-iron bauxite deposits in the Escape River district of Queensland, Australia by White (1976).

7.2.5 Examples of bauxite deposit evaluation

To illustrate these constraints upon evaluation we shall consider three specific examples, Gove, in Australia (Lillehagen, 1979), Wagina Island, Papua New Guinea (Chapman and Evans, 1978) and Rennell Island, Papua New Guinea (Matsunaga et al., 1978). The Gove deposit is an extensive blanket with a fairly even footwall, the thickness varying between zero and 10 m. On the other hand the Rennell Island and Wagina deposits are karstic pocket types. The Rennell deposit has an average thickness of 2.4 m (maximum 9 m) and the Wagina deposit varies between 0.3 and 12 m in thickness.

The Gove deposit was evaluated from 15 cm diameter auger and core drilling on a fairly wide drilling pattern. The drilling was undertaken on a 200 m grid in favourable areas and on a 400 m grid in perimeter areas. In the latter case where the 400 m drilling encountered bauxite the grid was closed to 200 m. In this way 5 400 holes were drilled with a total length of 26 000 m. From this, ore reserves of 282 million tonnes were delineated (Lillehagen, 1979). In contrast the Wagina deposit was drilled on a 150 m grid and the holes were closed up to 75 m in certain areas. In total, 3627 holes were drilled with a total meterage of 10 127 m, and the ore reserves quoted were of the order of 30 million tonnes. In the Rennell Island deposit each bauxite pocket was examined on lines at 10 m intervals to determine grade and depth.

Sampling the Gove deposit was based upon the drill material. As bauxites often vary in the state of cementation, more than one type of drilling technique has to be used. Consequently in the Gove deposit, auger and core drilling were used. In this deposit the composition varied greatly with depth (Fig. 7.6), and therefore sampling had to determine the thickness of mineable bauxite. To achieve this, sampling was carried out at half-metre intervals. In the Wagina Island deposit drilling was used as the primary sampling method but pits were sunk to obtain bulk-density data, moisture content and to check assay data. The pits were sited to provide a density of four pits per square mile and in total 23 pits were sunk. From the orientation sampling in the Rennell Island deposit a very uniform deposit was determined and usually only one sample per pocket was taken at a regular depth of 0.75 m, more samples being taken from the larger pockets.

In the calculation of ore reserves at the Gove deposit each drill hole with an ore grade intersection was assigned an area of influence halfway to the next drill hole (40 000 m^2/drillhole). The in situ density was obtained as 1.6 g/cm^3. The cut-off grade was set at total silica less than 5% and available alumina more than 44%. Ore grade was defined as not less than 40% available alumina, not more than 8% total silica and the thickness of ore was to be at least 2 m (or 1.5 m where the overburden ratio did not exceed 1.5:1). The evaluation data were processed by computer to produce isoquality and isopach maps (Fig. 7.7a,b). The estimated reserves as mentioned above were 282 million tonnes at 3.7% total silica and 51.1% total alumina. A similar procedure was adopted in the Wagina deposit while the Rennell deposit was estimated from sections through each pocket (proven reserves contained in some 3700 pockets).

Because of processing problems the Wagina deposit did not come into production, while the small pilot plant at the Rennell deposit encountered the following problems:

(1) low settling of red mud due to the fine particle size,

(2) a high organic content which lowered the precipitation rate of aluminium hydroxide,

(3) a high phosphate content which affected the hydrate quality,

(4) extremely low silica (0.2%) which caused insufficient desilicification.

Ore deposits formed by weathering

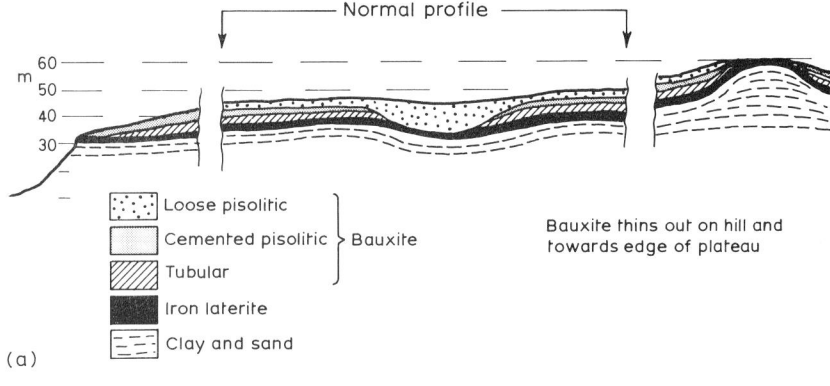

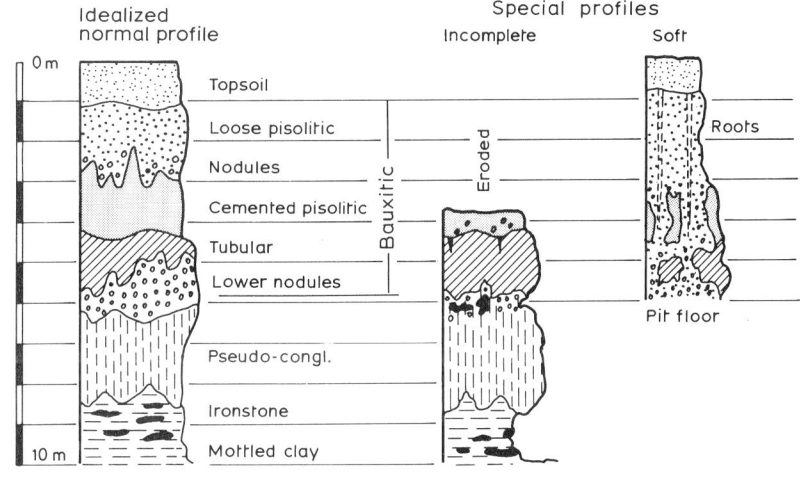

Fig. 7.6 (a) Schematic profile of laterite capping on the plateau. (b) Sections through laterite and bauxite, Gove deposit, Australia (from Lillehagen 1979).

7.3 LATERITIC NICKEL DEPOSITS

Two types of mineable nickel exist. The nickel deposits discussed in this chapter are the lateritic type. Hypogene-type nickel ores have been described in Chapter 2. A general description of nickeliferous laterites is given by Golightly (1979) in the *International Laterite Symposium*, and Golightly (1981) presents a comprehensive review of nickeliferous laterite deposits. Papers in the *International Laterite Symposium* cover many aspects of exploration, evaluation and exploitation of lateritic nickel.

The average concentration of nickel in igneous rocks is about 80 ppm, and it is only well represented in the ultramafic rocks. In these, Ni may occur as small crystals of nickel sulphides such as pentlandite and millerite, but it also substitutes for ferrous Fe and Mg in silicates (especially olivine) or oxides (magnetite). In the residual deposits such as those we are considering the nickel has been leached from the olivine, serpentine or nickeliferous magnetite and occurs in nickeliferous silicates, garnierite.

Previous authors (for example, Evans, 1980) have forecast an increased contribution from nickel laterites to meet the anticipated increased demand for nickel. This increase has not yet materialized and since the early 1980s there has been a cut-back in the contribution owing to the increased fuel costs for exploitation. Alternative fuel sources (to oil) are being sought, as are practical techniques for acid leaching.

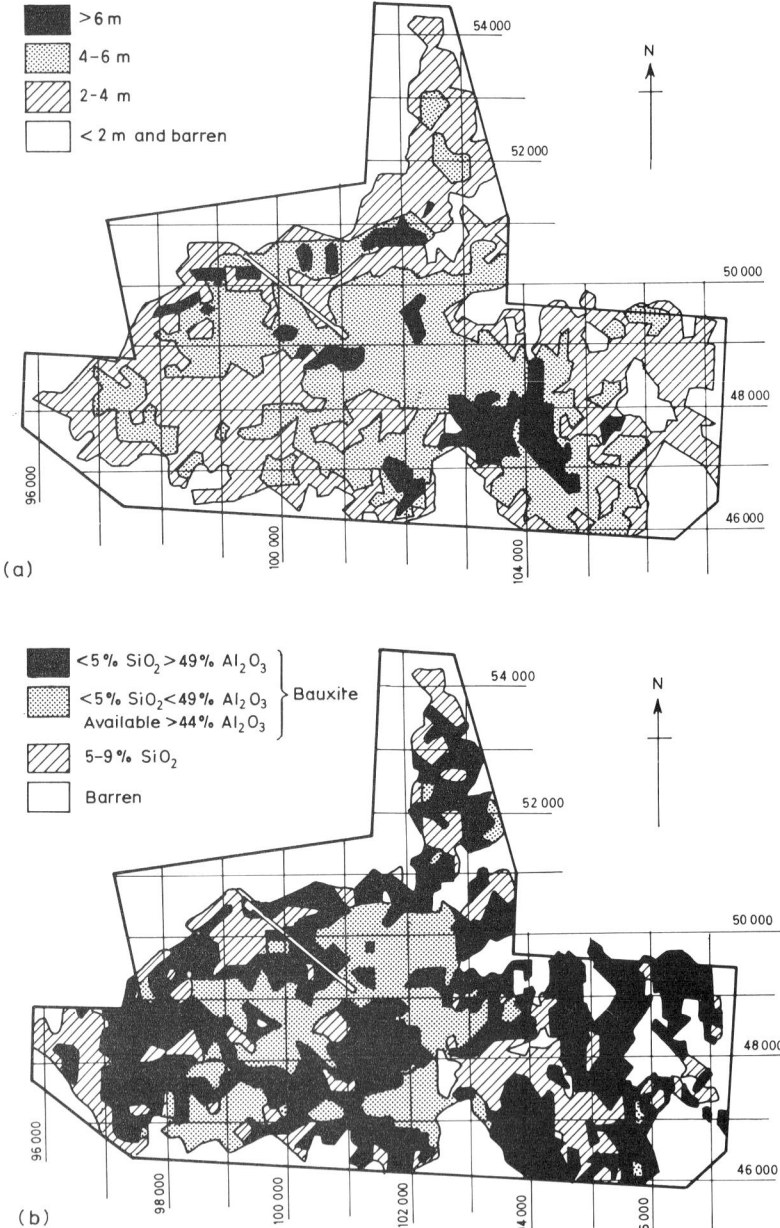

Fig. 7.7 (a) Isopachs of bauxite in the main deposit, Gove, Australia and (b) isoquality map for the main deposit (from Lillehagen 1979).

7.3.1 General characteristics of nickel laterite deposits

(a) Deposits in space and time

Lateritic nickel accounts for about 65% of the known land-based nickel reserves in Western countries, although in 1980 this ore type only accounted for 41.3% of World production, with the hypogene sulphide ores dominating the production. The most important lateritic nickel ore deposits are in New Caledonia, and these have been mined since about 1876. Other important occurrences are at Greenvale, Queensland, Australia, at Pomalaa and Soroako, Indonesia, and on Nonoc Island, Philippines. Smaller deposits are mined in Oregon, USA, Moa Bay, Cuba, and at Falcondo in the Dominican Republic. Laterites formed from ophiolites are worked by underground mines on the east coast of mainland Greece near Larymna as well as by large open pits on the island of Euboea. In Greece, nickel and cobalt are produced as by-products of bauxite operations. Some of the more promising deposits such as the Exmibal operation in Guatemala (which produced 6940 tonnes of Ni in matte in 9 months of 1980) have been forced to close owing to escalating fuel costs.

Nickeliferous laterites in South America, which could make substantial contributions to the World supply if the economics are right, are those in Minas Gerais, Brazil, and the Cerro Matoso deposit of Colombia. Within the USA, laterites form the only domestic source of Ni and the principal deposits are those on the Oregon/Californian border. These are insignificant in terms of US nickel consumption.

Geomorphological evidence and other indirect age indications suggest that the major lateritic nickel deposits have formed from about the mid-Tertiary to the present. Some deposits that occur outside the modern laterite soil belt, including those of Greece and Yugoslavia, are older and lie beneath a Cretaceous and Tertiary cover.

(b) Grade and size of deposits

Most deposits being mined range from 1.5 to 2.0% of Ni. The plant feed at Inco Metal's Soroako deposit in Indonesia averages just over 2.4% Ni (Harju, 1979b). New Caledonian ores average nearly 3% Ni, this higher grade reflecting the higher production costs. When mining began in New Caledonia the ore grade was more than 10% Ni. Because of the variation within nickel laterite deposits blending is often necessary.

The ore deposits of the Dominican Republic occur over a total area of 7 km^2 with an average thickness of 8 m and in these the cut-off is 1.2% Ni. The reserves in these deposits are about 70 million tonnes (Haldemann et al., 1979).

(c) Mineralogy

Above the parent rock is the saprolite zone, this is extremely porous and the olivine and pyroxene of the parent rock are destroyed. In the saprolite the serpentine and chlorite have reduced Mg contents and strongly increased Ni and Fe contents. Quartz and smectite minerals appear as pseudomorphous replacements of olivine and serpentine. Above the saprolite is a zone characterized by a silica boxwork in a matrix of limonite or smectite minerals. The smectite minerals, generally a nickel–chromium-rich nontronite, and goethite have replaced all parent rock minerals except talc and chromite.

The *in situ* limonite zone which lies above the nontronite zone has fine-grained nickeliferous goethite and amorphous ferric hydroxides. Talc, chlorite and chrome spinel may remain from the parent rock. Upwards this zone may grade into a ferricrete crust (the 'Canga' in Brazil) consisting of colloform goethite tubes and veinlets or pisolites of goethite and haematite.

The nickel distribution is such that the residual concentration in the limonite zone only rarely attains 2% Ni; the upper part of this zone has a much lower concentration and is often disposed of as overburden. The lower limonite zone may be mined as ore in its own right (as at Moa Bay, Cuba) or may be blended with silicate ores from the saprolite zone which usually has the richest Ni content.

(d) Host rock lithology

The nickel laterite deposits of greatest economic importance have developed on a peridotite bedrock. The primary source of nickel here is forsteritic olivine and serpentine. In principle, nickel sulphide-bearing rocks may also produce nickel laterites but these are not economically significant.

(e) Tectonic setting

The majority of commercial deposits, New Caledonia, Indonesia, Cuba and Guatemala, belong to outer island arcs and zones of complex plate tectonic boundaries where obduction or underthrusting of oceanic crust by continental crust has occurred. Shield areas in Brazil, Africa, India and Australia have serpentinized ultramafics but with outcrops generally less than 100 km^2.

7.3.2 Genesis of lateritic nickel deposits

The conditions for the formation of economic lateritic nickel deposits are similar to those described earlier for bauxites. The major exception is that weathering is only likely to produce accumulations of Ni on ultramafic rocks with very high Ni contents. Climatic conditions are again important. Under extreme tropical conditions the removal of Si is too rapid for the formation of new silicates in the weathering profile. Other features of such climatic conditions result in the Ni occurring in the oxidized ore form which tends to be of lower grade, usually less than 2% (Cuba, Guinea). If the climate is temperate or too dry the weathering is too slow to create economic deposits. Consequently the richest ore deposits are found in areas of subtropical climate.

Topography plays an important role in the accumulation and the distribution of economic deposits (Fig. 7.8). Too steep a slope will result in the removal of the weathered mantle before the nickel is released. Where such steep slopes abut plateaux as in New Caledonia, however, there may be an enrichment of Ni near the tops of the slopes with Ni moved by groundwater from elsewhere on the plateaux. As a generalization, major enrichment of saprolite occurs immediately above and below the inflection point on slopes (Fig. 7.8). This is due to a low water table at the plateau edge with meteoric circulation being

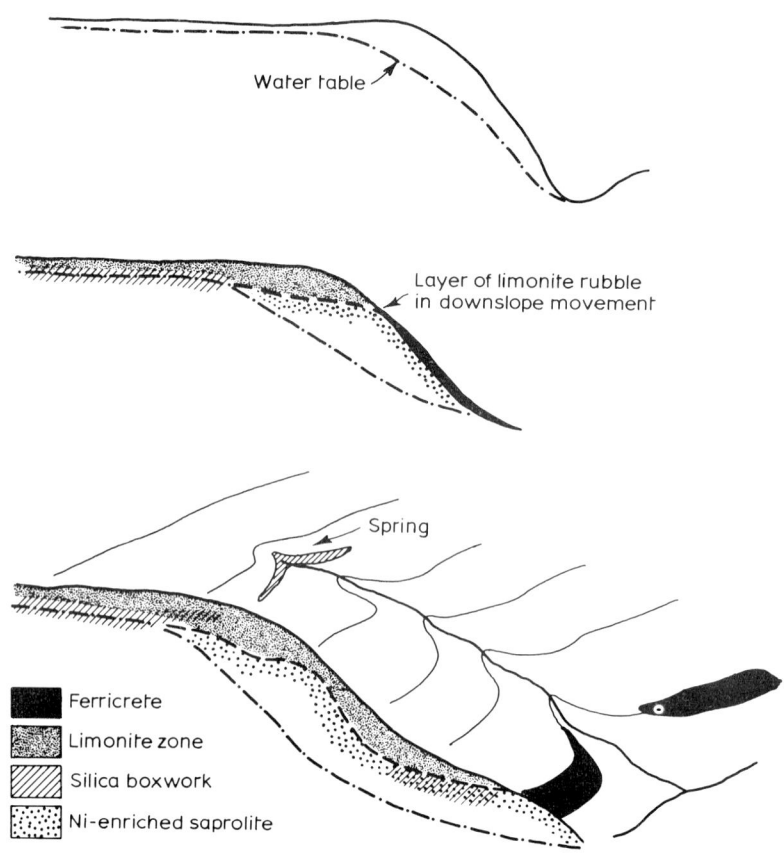

Fig. 7.8 Idealized sequence of lateritization on plateau edge (after Golightly 1979).

forced down through the saprolite. A higher water table elsewhere on the plateau, or the lowland areas, prevents nickel enrichment. Erosion removes enriched saprolite from the convex brow of the plateau edges. Any nickel in the plateau aquifers will tend to escape, via springs, at the edge of the plateau.

Tectonic activity appears to assist the formation of large economic supergene Ni accumulations although this activity must be neither too slow nor too rapid (LeLong et al., 1976). Moderate tectonic activity allows the formation of peneplains, as in New Caledonia. In this case the movement appears to be due to the collision of the Australo–Tasmanian and oceanic plates. From this it may be envisaged that peridotite massifs related to an arc environment are more likely to yield economic concentrations of nickeliferous deposits than the peridotites entombed in old basements (LeLong et al., 1976).

7.3.3 Deposits of New Caledonia

These are the most extensive economic deposits of lateritic nickel discovered to date. Approximately 5000 km^2 of the island is underlain by serpentinized peridotites. These rocks have been undergoing weathering in a humid warm climate since at least the Pliocene; also since this time the island has been suffering tectonic uplift (Dubois et al., 1973). Consequently several erosion surfaces have been formed and are covered by thick weathered blankets at different elevations on the island. Steep slopes with thin soils separate these flatter areas. A typical weathered profile on one of these plateaux is shown in Fig 7.9. The profiles are capped by a ferruginous crust overlying a reworked gravelly ferruginous horizon. Neither the gravelly horizon nor the ferruginous crust, if present, is considered economic to work. Underlying these is a fine saprolite which forms the thickest part of the profile. This horizon, from which most of the Si and Mg have been removed during weathering, is classified as a lateritic oxidized Ni ore. Until recently this zone, with average grades of 0.9 to 1.4% Ni, was considered non-economic. In addition to its low magnesia and silica it has high iron and is relatively rich in cobalt (up to 0.2 wt%). The horizon of coarse saprolite, the silicated Ni ore, is the main ore zone, but here the nickel content varies according to the site of the profile. On the plateau edges it may be as high as 5% NiO while in the high parts of the plateaux it may reduce to 2% NiO. Below the average level of weathering the bedrock is broken by faults and crush zones which often carry significant amounts of garnierite. At this level, which may extend to 10 m thickness, pseudo-lodes and boxworks occur. These form the richest ores with up to 35% NiO (LeLong et al., 1976). It was the presence of these garnieritic pseudo-lodes in the bedrock that once led to hydrothermal origins being proposed for these deposits, but it is now over thirty years since the true weathering nature of the origin has been recognized.

The significant feature about the New Caledonian nickel deposits, apart from the great lateral extent and variation due to peneplanation and uplift, is that the Ni enrichment is not seen to increase downwards to a maximum in the coarse saprolite. The Ni content remains almost constant throughout the thickness of the fine saprolite ferruginous formations (Trescases, 1973). Leaching of Ni ocurs only at the top of the profiles. Trescases proposed the following explanation:

(i) The first type of trap for the Ni released from the peridotite is the horizon of newly formed or inherited little-weathered silicates at the bottom of the profile. As the weathering front penetrates the zone of changing conditions – Mg leached, pH more acid plus the presence of some organic acids – the Ni is mobilized again from the now weathered silicates and it migrates downward to the layer where the silicates are unweathered. With prolonged weathering considerable accumulation of Ni at the bottom of the profile occurs. We would accept this as analogous to the 'normal' process of supergene enrichment.

(ii) About one-third of the New Caledonian Ni released by weathering of the silicates does not migrate and is trapped by such phases as goethite.

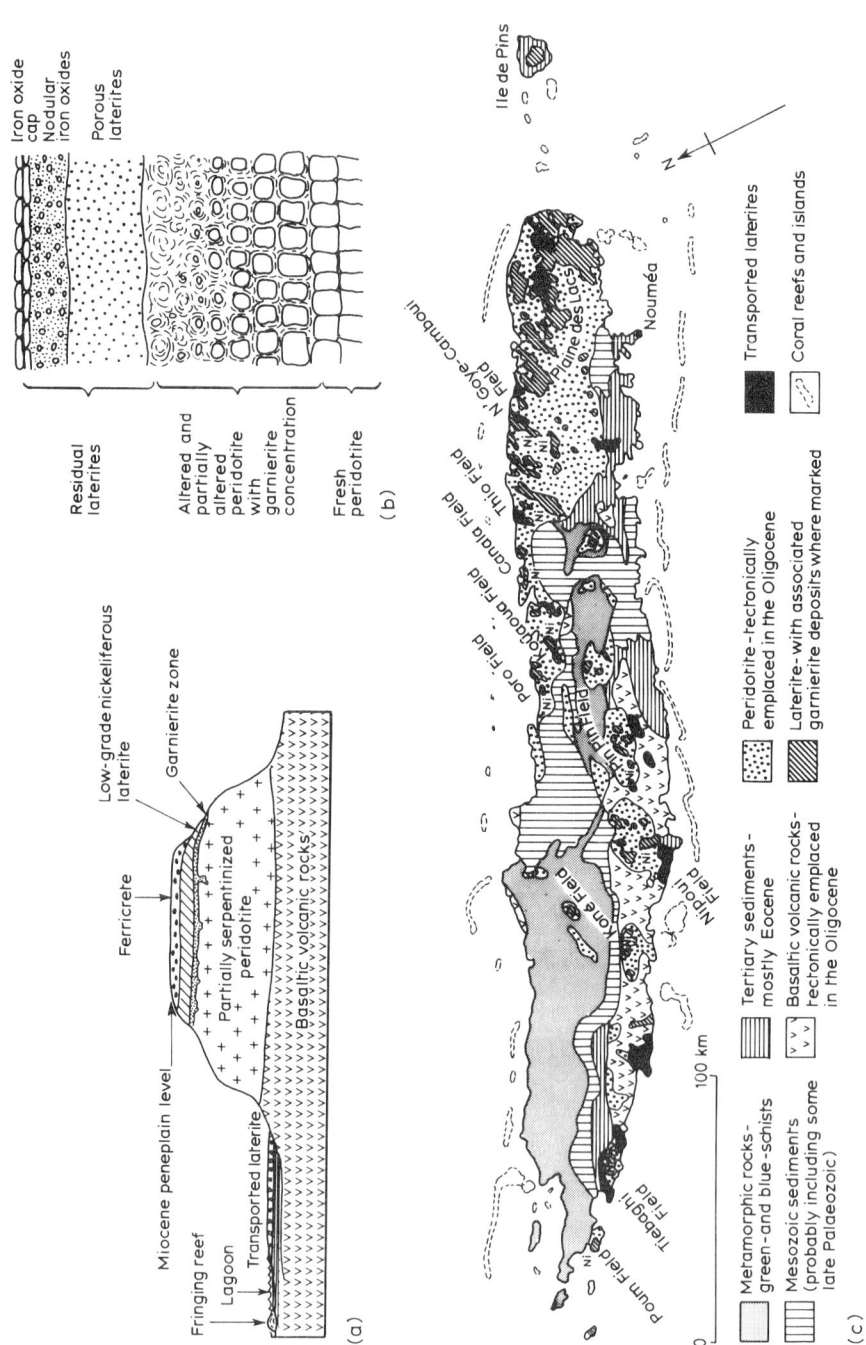

Fig. 7.9 New Caledonian nickel deposits (a) Diagrammatic profile of a peridotite occurrence showing the development of a residual nickel deposit (after Dixon 1979). (b) Section through nickeliferous laterite deposits (from Evans 1980). (c) Geology of New Caledonia (from Dixon 1979).

This entrapment is very effective and immobilizes the Ni in the major part of the oxidized zone.

(iii) Superimposed upon (i) and (ii) are localized enriched zones resulting from concentration due to circulating groundwater. Examples in New Caledonia are along the edges of the plateaux and in the boxwork structures of the bedrock.

7.3.4 Exploration and evaluation of lateritic nickel deposits

Many of the observations we have made regarding the exploration and evaluation of bauxite deposits apply equally to lateritic nickel. Consequently we will only draw attention to exploration and evaluation procedures particularly applicable to lateritic nickel. Detailed descriptions of exploration programmes for two specific deposits are given by Harju (1979a,b) (Table 7.7).

In the analysis of nickel laterite samples the critical elements generally are nickel, cobalt, iron, magnesia, silica and loss on ignition. The permissible content of various elements and ratios of elements must be determined for the particular treatment processes to be used.

In exploring for nickel laterites in tropical rainforests certain characteristics may be sought by remote sensing (Allum, 1982). These characteristics include rounded topography, general absence of jointing and uniformity of species of vegetation in comparison to the surrounding rocks. This uniformity of vegetation may be due to the paucity of phosphorus and potassium in ultramafic rocks, few species being able to thrive under such conditions. The use of aerial photography, satellite imagery and synthetic aperature radar (SAR) imagery to define nickel laterite target areas in the Amazon Basin, Brazil, is described by Allum. Ground follow-up having confirmed the presence of nickel laterites overlying ultramafics in the target area makes it possible to extrapolate the recognition of topographic, structural and vegetation features beyond the immediate area to interpret similar occurrences as possibly due to nickel laterites. The non-stereoscopic interpretation of SAR means that this method is quicker than aerial photography, but the latter provides more information on vegetation texture and is therefore more reliable in target selection.

Within the chosen target area the distribution of the nickel values will depend upon the weathering characteristics of the parent rock. This can result in very rapid changes in grade vertically and laterally (Figs. 7.10 and 7.11). Such variations must be identified and the footwall cut-off, ore zone, possible future ore zone (present stockpile) and waste must be quantified as accurately as possible (Fig. 7.10). Drilling, pitting and sampling must be undertaken in the light of the local conditions and changed if necessary during the exploration programme (Table 7.7). In lateritic nickel the drilling characteristics of the laterite and saprolitic portions are very different and these need to be appreciated by any geologist supervising a drilling programme in these deposits. Furthermore, sampling difficulties arise since in the saprolite the major nickel values are commonly in weathered zones and occur in less altered boulders.

In Indonesia Harju (1979b) states that metre by metre comparisons of drill hole data with identically situated trenches and pits have shown that the nickel content of a drill sample frequently has little relation to that of the excavated material in the saprolite zone. Drill holes are deflected by the hard rock and silica fragments, tending to oversample the soft material. Harju established a workable correlation between average nickel grade in closely spaced holes and the grade of the -5 cm screen fraction (which is equivalent to plant feed).

In Guatemala, on the other hand, Harju (1979a) suggested little difference between the average nickel values from drilling and from test pitting. Augering preferentially sampled the soft-weathered material between boulders and thus gave slightly higher iron values than test pit sampling. In these Guatemalan deposits the conventional areas-of-influence methods of grade estimation tended to give over-optimistic results.

Table 7.7 Exploration programmes for two lateritic nickel deposits (data from Harju, 1979 a,b)

	Sulawesi, Indonesia		*Cristina, La Gloria, Trincheras deposits, Guatemala*
1969–70	Aerial photographic assessment 50 km² sampled on 200 × 400 m and 200 × 200 m grid yielding 33 500 m of drilling, each metre sampled for Ni, Co, Fe (using AAS). Geological mapping recognizes two bedrock types giving different chemical and physical types of ore.	1960	Exmibal commences exploration. Hand augering on widely spaced lines – 800 holes (7000 m).
		1961–62	743 holes (11 735 m) drilled on 100 m grid in central portion of deposits. 305 test pits and 105 power auger holes (total 6025 m.)
1970–71	Bulk samples (totalling 5000 tonnes) mined for metallurgical testing.	1962–67	Test pitting (900 pits totalling 12 000 m). 1100 holes (12 630 m) on maximum 200 m grid with 50 m spacing in key areas. 50 000 samples assayed for Ni, Co, Fe by conventional titration.
1970–72	Further 39 500 m of drilling, 275 test pits (total 2450 m) plus 44 backhoe trenches.		
1973	Stage 1 of project agreed.	1966	Eight bulk samples (500 to 2500 tonnes) excavated for metallurgical testing.
1972–74	Further exploration outlines additional ore.	1967–71	Mine development drilling on 25 m grid in areas selected for initial mining. 1800 holes (totalling 30 000 m). Test patterns also drilled at 5 and 10 m spacing to study local variations. Large block in La Gloria drilled at 12.5 m spacing for overburden stripping control.
1975	Stage 2 of project agreed (annual production forecast 45 000 tonnes nickel in matte).		
1973–76	Additional 33 000 m, 800 test pits (5935 m) and 230 trenches. Also detailed mine development sampling at 25 to 12.5 m spacing 3900 holes (45 000 m), 173 large test pits (1385 m) 2000 tonnes of material from trenches and test pits processed.	1968	Trial mining exercise, 60 000 tonnes of ore mined and stockpiled.

7.4 KAOLIN DEPOSITS

Kaolinite is the chief economic mineral of the kandite group of clay minerals. Its basic structure is a two-layer lattice, a gibbsite sheet and a silica tetrahedral sheet; kaolinite does not expand with increasing water content, one of the characteristics which distinguishes it from the smectite group. Its crystal structure is generally resistant to attack by most corrosive fluids, making kaolinite inert, not readily reacting with media in which it is placed; this is an important commercial property of china clay. Commercial-quality china clay is used in the paper industry, in china manufacture, for ceramics

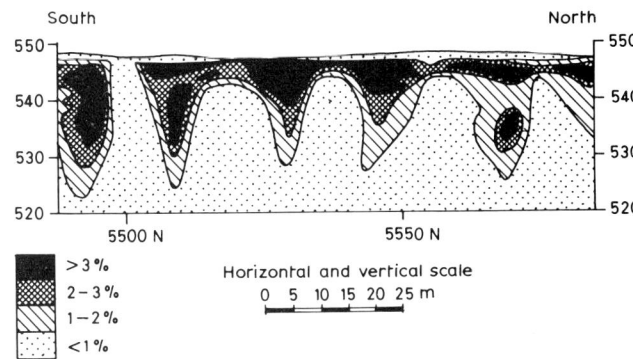

Fig. 7.10 Section showing nickel isograd patterns, Greenvale nickel laterite ore body, Australia (after Burger 1979).

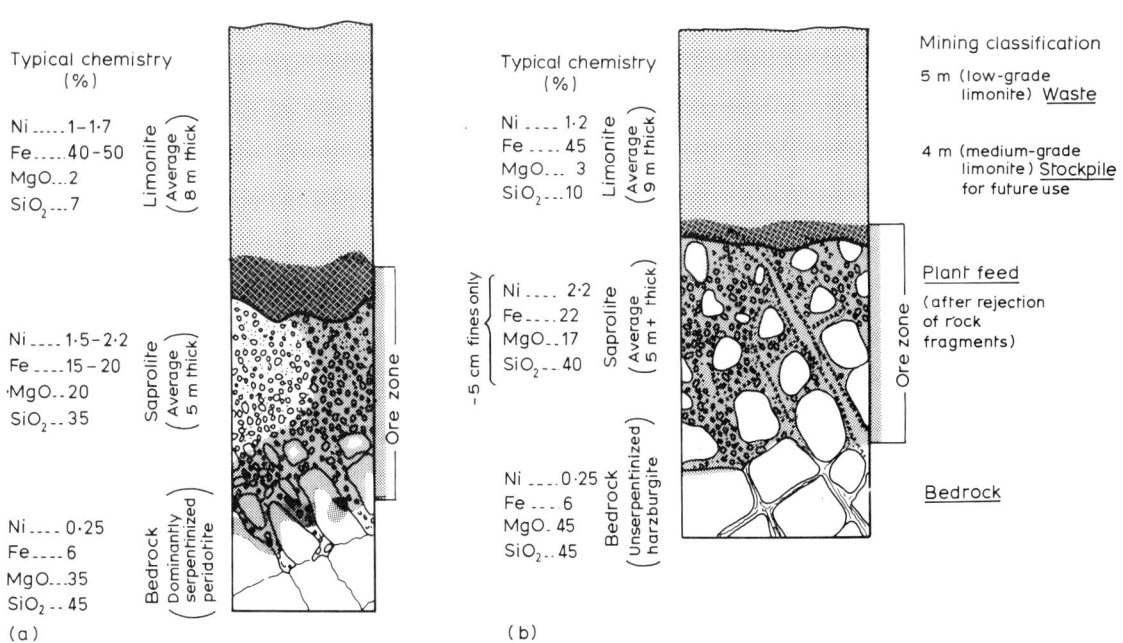

Fig. 7.11 Schematic profiles of nickel laterite deposits (a) Exmibal deposits, Guatemala; (b) Inco's Soroako deposit, Indonesia. Based on data from Harju (1979a,b).

Fig. 7.12 Sedimentary kaolin deposit ('ball clay') with interbedded lignite (black), Bovey Tracey, Devon, UK.

and refactories. Diverse other markets exist including the paint and rubber industries. In the paper industry china clay is used to fill the interstices of the pulp fibres and as a surface coating to produce a smooth, bright, often glossy finish. As this represents one of the major markets for the product the brightness of the clay is a very important criterion when a kaolin deposit is being evaluated. China clay lacks sufficient plasticity and green strength to make some ceramic products. Ball clay, a variety of kaolin having plasticity and high-strength characteristics, is commonly added to china clay to improve workability. (Fig 7.12).

7.4.1 Genesis of kaolin deposits

The hydrous aluminosilicate, kaolinite, is formed from the destruction of aluminium silicates, principally feldspars. The feldspar lattices are wrecked by ionic solution, hydration and hydrolysis. K, Na and Ca are extracted and reaction of the Al and Si with OH results in formation of kaolinite. The conversion of orthoclase to kaolinite may be illustrated as follows:

$$4KAlSi_3O_8 + 2CO_2 + 4H_2O$$
$$\rightarrow 2K_2CO_3 + Al_4Si_4O_{10}(OH)_8 + 8SiO_2$$

This can occur under late-stage hydrothermal conditions or under chemical weathering conditions in tropical or subtropical climates. A combination of these two environments has been invoked for at least one major group of deposits, those in south-west England (Bristow, 1977).

The chemical weathering conditions that produce kaolin deposits are considered broadly similar to those of bauxitization (Section 7.2.3).

Frequently lateritic bauxites have major, if impure, kaolin deposits associated with them. Kaolinization often occurs as an intermediate stage in the formation of lateritic bauxites. The presence of oxidizing sulphides hastens kaolinization. Deposits that result from supergene weathering contain kaolin stacks that are fragile and undeformed; the china clay is usually of low density and high porosity.

Kaolinization resulting from hydrothermal activity is likely to be controlled by fluid paths such as jointing and brecciation and will probably be best developed near major conduits. Deposits arising in this way will tend to be more restricted than the weathering blankets formed by supergene chemical weathering but they may extend much deeper.

Material that has suffered *in situ* kaolinization may be eroded and transported to be deposited elsewhere as any other detrital mineral. Sedimentary kaolin deposits are major sources of production in the USA and the ball clay deposits of England are thought to have originated in this way (Fig. 7.12). Sedimentary sorting processes will have been responsible for creating sufficiently concentrated deposits to be economically viable. Some material in these deposits may have undergone kaolinization within the basin of deposition.

7.4.2 Classification of kaolin deposits

We propose to divide kaolin deposits into three groups, based upon their mode of formation: (1) residual kaolin deposits; (2) sedimentary kaolin deposits; (3) hydrothermal kaolin deposits. Large concentrations of pure white kaolinite may be found within each of these groups.

7.4.3 General characteristics of kaolin deposits

(a) Distribution in space and time

Approximately half the World's production of kaolin comes from extensive china clay deposits in the USA and the UK. Major deposits occur in Czechoslovakia, Guyana and Brazil. Australia and South East Asia also have large quantities of kaolin. The kaolin deposits of south-west England are dated at 273 million years (Bray and Spooner, 1983) whereas the major kaolin-producing area of the USA, in South Carolina and Georiga, has deposits of Upper Cretaceous to Middle Eocene age. Deposits of kaolin resulting from chemical weathering have formed where climatic conditions were suitable. Those in Western Australia, which may have begun formation in the Proterozoic, have experienced many periods of kaolinization between then and the present day.

(b) Size and grade of deposits

It is not possible to give an average size for a kaolin deposit because of the different modes of formation. The typical thickness of a chemical weathering profile is between 35 and 50 m, the larger figure representing more prolonged weathering as in Western Australia. This does not represent the original thickness as erosion may have removed large quantities. The kaolin deposits of sedimentary origin form strata and lenses up to 10 or 12 m thick and extend for a few kilometres. Kaolin deposits of hydrothermal origin tend to be more restricted in area, but are much more vertically extensive.

The annual World production of kaolin is about 18 million tonnes. In 1982 the USA had about 40% of the market and the UK about 16% (Industrial Minerals, 1983). These two countries dominate the high-value paper-coating clay market. The largest single producer world-wide is English China Clays (ECC) whose British operations are centred in south-west England. Their total production amounted to 2.9 million tonnes in 1981. Some 50% of the china clay produced is used for paper filling while 30% goes for paper coating. Ceramics accounts for another 15% while the remainder is used for a wide variety of products. The trend in usage seems to indicate an increase in the paper-coating clay production compared to the filler usage.

(c) Mineralogy

Kaolin deposits that have been formed by chemical weathering may have halloysite, siderite, pyrite and limonite as accessory minerals. The kaolinite that occurs as the pallid zone in bauxite profiles may be associated with one of the bauxite minerals. Resistate minerals may also be present. Kaolinization associated with igneous rocks gives rise to deposits of kaolinite plus one or more of the essential minerals of the igneous rock. In the south west of England china clay deposits contain quartz, feldspar, mica and relict tourmaline. Such deposits do not have much limonite, siderite or pyrite.

(d) Host rock lithology

Residual deposits of kaolin result from the chemical weathering of feldspar-rich rocks, particularly granites and other aluminous rocks. At present production of good-quality china clay from the pallid zones beneath bauxite layers is not very significant.

7.4.4 Residual kaolin deposits

(a) Spruce Pine District, North Carolina, USA

These kaolin deposits were formed from the chemical weathering of small, irregular, pegmatitic, alaskite stocks (Harben, 1978). The stocks are a few hundred metres to five kilometres across and are coarse-grained alaskite with about 50% oligoclase and 25% microcline. The kaolinization took place during the early Tertiary and the deposits lie beneath a marked erosion surface which suggests very extensive weathering. The decomposition of the alaskite extends down as far as 30 m but the deposits are not worked below 15 m. The deposits are composed of kaolinite and halloysite mixed with partly decomposed feldspar, fresh quartz and muscovite. Inclusions of schists and gneiss in the parent alaskite produce waste within the deposits. The clay is mined by power shovel or hydraulic jet and the district produces about 160 000 tonnes per year of water-worked clay.

(b) Gabbin, Western Australia

This project, sited 285 km north east of Perth, is in an ancient peneplain around 400 m in elevation within the Yilgarn block of the Archaean Shield of Western Australia. The block is composed of elongate greenstone belts surrounded by large granite masses. Although in general the granite masses have ages of either 3.2 or 2.5 million years, the age of the Gabbin granite is not known. This granite is an adamellite of quartz, microcline and plagioclase. The plagioclase has weathered more rapidly than the potassium feldspar as is often the case in kaolinization. Accessory minerals are rare. Much of the Archaean of the area has been emergent since the start of the Proterozoic (Walker, 1978). Through this long period of time the area must have experienced a great variety of climates and since the Tertiary there has been a drying out of the climate accompanied by deep lateritization. Some of the lateritic profile has been removed due to post-Tertiary uplift.

The weathering front is normally 30 m below the present surface but it may be as much as 50 m. A NE–SW foliation exists in the Gabbin granite and the deepest weathering and zones of best rheology appear to follow this trend. A typical section shows the recognizably weathered granite overlain by a coloured sandy clay which underlies the ore-grade kaolin. The kaolin deposit may be 35 m thick with an average of 11 m. The thickness of kaolin seems to follow the topography but there may be some fault control (Walker, 1978). Overlying the kaolin is a varying thickness of silcrete and sandstones. The distribution of the various ore-grade clays along one section is shown in Fig. 7.13.

7.4.5 Sedimentary kaolin

Deposits in South Carolina and Georgia, USA

About 90% of the kaolin production of the USA is from sedimentary deposits in South Carolina and Georgia. The production is based on two centres, one extending from Macon to Wrens in Georgia,

Ore deposits formed by weathering

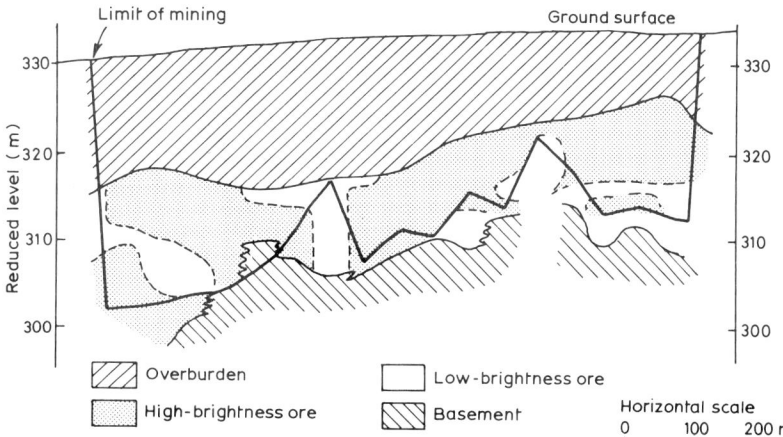

Fig. 7.13 Gabbin kaolin – geology and assay sections (Line 15800N) (from Walker 1978).

and the other near Augusta, Georgia, and Aiken, South Carolina. Here the kaolin is in lenses within the Upper Cretaceous to Middle Eocene formations. These lenses occur, apparently at random, throughout a thickness of more than 50 m of the formation and may have lengths of several kilometres. The thickness of these lenses ranges up to 12 m and all contain some detrital sand and mica. Disseminated pyrite is present in the kaolin under thick cover and as the outcrop is approached the oxidation of pyrite gives a random yellow stain to the kaolin which affects the grade of clay produced. It is generally agreed that the source of these kaolinitic sediments lay in crystalline rocks to the north-west but the presence of the lenses in otherwise coarse, cross-bedded sands has been the topic of much debate. Initially it was assumed the crystalline rocks were deeply weathered and the kaolinite was formed *in situ* and then washed into the areas where they now occur. On this hypothesis the lenses of kaolinite are difficult to explain. Kesler (1952) explained the presence of the lenses by transport of detritus from the crystalline rocks. Weathering of feldspars into kaolinite occurred within the delta sands, the kaolinite being later washed selectively to cut off stream segments or ponds (analogous to the settling ponds used in china-clay processing) containing fresh water. These kaolin-filled ponds were eventually covered by sands as delta building progressed northwards due to subsidence.

7.4.6 Hydrothermal kaolin deposits

The kaolin deposits of south-west England

Some of the finest-quality kaolin deposits in the World occur in the UK, in south-west England. These are in Devon and Cornwall with the main economic deposits in the latter county. These deposits have been described by Bristow (1969), and the kaolinites have been examined in detail by Exley (1976) and Sheppard (1977). Their genesis has been discussed by many authors, most recently Durrance *et al.* (1982), Alderton and Rankin (1983) and Bray and Spooner (1983).

The granites of south-west England are two feldspar granites and the process of kaolinization proceeded with the alteration of plagioclase to pure kaolinite, while the potash feldspar changed into a mixture of kaolinite and some fine-grained mica. Of the two other essential minerals in the granite, the quartz remained relatively unaltered while the mica either recrystallized into a finer-grained form or remained unaltered.

Although we include these deposits within the chapter dealing with weathering deposits, their place here is not fully justified as they have long been considered to originate from hydrothermal activity (Bristow, 1977). However, there have been proponents of a supergene origin, the most recent being Sheppard (1977). Bristow (1977) regarded the hydrothermal stage as a major preparation process which produced some intermediate mineral phases and which rendered the granite very susceptible to alteration when the kaolinite formation took place under supergene conditions. Following this hydrothermal 'softening up' process, and possibly when the granites were essentially consolidated and cooled, great quantities of groundwater entered the system. This probably occurred during or after the late stage of hydrothermal activity, and, according to Alderton and Rankin (1983), these low-salinity fluids, with temperatures less than 170°C, were responsible for the kaolinization. The volume of available fluids accounts for the very large size of the deposits. Highest quality, in terms of brightness and particle size, is related to the lithionite granite which was deficient in iron-rich biotite.

Bray and Spooner (1983) have examined the evidence for and against a weathering origin for these deposits. They observed that no vertical profile development is present in the south-west of England deposits and the depth of kaolinization (as much as 250 m) is significantly greater than the 35–50 m thicknesses which are typical of complete chemical weathering profiles. Furthermore the K/Ar age for the secondary muscovites in the deposits indicates that kaolinization occurred at the time of the sheeted quartz–tourmaline ± cassiterite ± wolframite mineralization in the region (267 million years). Bray and Spooner conclude that the evidence suggests a single-stage hydrothermal process.

The deposits lie on eroded cupolas of an extensive Hercynian granite batholith, and

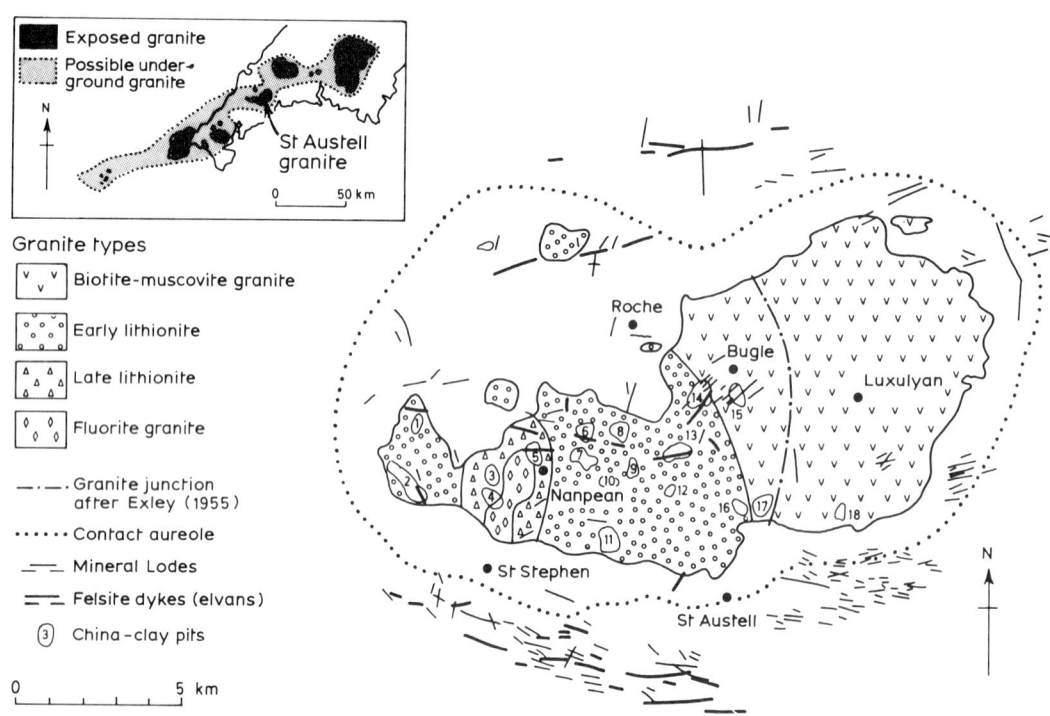

Fig. 7.14 Geological map of the composite St Austell granite showing the distributions of the different granite types, metafelsite dykes (elvans), mineral lodes and important china-clay pits (from Bray and Spooner 1983).

Ore deposits formed by weathering

7.15 Hydrothermal (?) kaolin deposit. Soft, white kaolinized granite in steeply dipping zone under massive jointed granite, St Austell, Cornwall, UK.

although five of the six main cupolas have some kaolinization which has been commercially exploited, the main mass of china clay is located in the lithionite granite of the St. Austell area (Fig. 7.14). The clay deposits often occur as funnel shapes opening upwards and extending as deep as 250 m. The funnel-shaped deposits may occur in zones, or alternatively the kaolinized zones may form elongated troughs sometimes dipping at high angles into the ground (Fig. 7.15). This dip may result in unkaolinized granite overlying kaolinized granite. Such shapes make evaluation difficult, as does the occurrence of 'core stones' where unkaolinized granite apparently overlies kaolinized rock. Laterally kaolinized material may pass into unkaolinized granite over very short distances.

7.4.7 Exploration for and evaluation of kaolin deposits

If there is sufficient contrast between the china clay and surrounding rocks, geophysical techniques may be used to delineate the deposit and reduce the amount of expensive drilling. This is true of some of the deposits of south-west England where there is sufficient resistivity contrast between the intact granite and china clay to allow the identification of hard-rock areas. Gravity has also been used in this area and from detailed gravity work certain estimates of mass of clay in the ground are possible.

The mode of formation means that it is unlikely that a very good interborehole correlation will exist in a china-clay deposit, and consequently it has to be drilled on a grid system at whatever spacing is considered optimum. In the case of the Gabbin deposit, drilling was carried out on a 200 m grid and more detailed drilling was undertaken on a 100 m grid in areas of particular interest. The Topira deposit in the Ituni area of Guyana, however, has been extensively evaluated on the basis of 42 holes with an average grid spacing of 72 m. In this case proven reserves of 3.4 million tonnes of paper-grade clay have been outlined. English China Clays use a 100 m grid and then undertake infill drilling at 60 m spacings. Elsewhere in south-west England the normal 'stope' in china-clay working is 9 or 12 m (Vincent, 1983) so evaluation borehole samples are of the same order, and minor variations in lithology are ignored.

Large-diameter cores are desirable to obtain as large a sample as possible. A major problem with drilling china clay is the flushing action of the drilling method. This tends to wash out and erode the core, limiting core recovery. Conventional and wire-line techniques suffer from this problem and the use of bentonitic drilling muds is limited by the possibility of contamination of the clays. Core recovery tends to be a function of core diameter, again favouring larger diameters, but to achieve the better recovery with larger diameters, the price has to be paid in greater weight and cost.

In evaluating a china-clay deposit the commercial parameters are of most importance and the economic geologist must be well aware of market requirements and possible future developments. The commercial parameters usually of most importance are brightness and fired colour, both having to be white, and particle size distribution. The rheological properties, linear shrinkage and other fired properties are of considerable importance. Bulk sampling from trial pits is advisable but even this may not provide material representative of the commercial-quality clay which will eventually be supplied to the customer.

Reserve evaluation is necessarily complex as the market specifications may require blending, and sample data must be collected with future markets in mind. As blending may involve numerous open pits in operation simultaneously, the geologist must ensure that none containing potentially marketable clay are allowed to be back-filled or flooded. Waste-disposal planning must be effective to avoid sterilization of future reserves and to minimize visual impact. The geologist in the clay industry may be expected to advise on plant site location and assist in site investigations for waste tips and mica dam location. As with other mining operations, mine reclamation to comply with environmental restrictions is very important.

7.5 SUPERGENE MANGANESE DEPOSITS

As well as the mobilization of Fe and Al, which may lead to the formation of iron laterites and bauxites, Mn may be mobilized during chemical weathering to form residual deposits. During the main stage of magmatic crystallization no independent manganese mineral is formed; the available Mn is fixed in the ferromagnesian minerals. These manganese-bearing minerals are subsequently weathered to give rise to economic deposits of manganese. The solubility of Mn is greater than Fe or Al, particularly in areas of high organic activity. Consequently Mn tends to travel to deeper zones and it will accumulate towards the base of the weathered zone in ultrabasic to basic rocks (where the pH is higher). This higher pH leads to the immobilization of the Mn. In acidic rocks the Mn is more mobile in the weathering profile and tends to escape from the weathered zone.

In the case of residual deposits of manganese, as with bauxites, climate and drainage control, together with the nature of the parent rock, are the main factors that determine the concentration of the deposit. The parent rock from which the residual deposit is to form should have a sufficiently high Mn content (say 10% Mn). Manganiferous carbonate and silicate–carbonate rocks are most amenable to enrichment through oxidation. The presence of pyrite in the source rock enhances the dissolution and migration of the Mn.

A comprehensive description of manganese deposits, not only those of residual origin, is given by Roy (1981). Readers interested in the wider occurrences of manganese deposits are advised to read that text.

Important deposits of residual manganese occur in West Africa, South Africa, Brazil and India. Each of these occurrences is associated with Precambrian formations which is notable when compared with the distribution of residual bauxite deposits.

7.5.1 Examples of supergene manganese ore bodies

(a) Brazilian deposits

The principal supergene manganese deposits in Brazil occur in Minas Gerais State and in the Territory of Amapa north of the Amazon. They result from weathering enrichment of metamorphosed manganese carbonate and silicate–carbonate protores. In Minas Gerais the supergene process has occurred in a weathering profile where the depth of oxidation exceeds 100 m. The manganese carbonate protores are more likely to produce oxide ores than the manganese–silicate carbonate protores; the extent of supergene enrichment is directly related to the rhodochrosite content of the protore. All large commercial deposits of supergene ore occur on hill tops or on the crests of ridges. The enriched ores of Minas Gerais contain 40–48% Mn (Roy, 1981).

The mineable deposits in the Amapa area occur as lenses which have been enriched to depths of 70–100 m. The protore was manganiferous schists, carbonates (rhodochrosite) and silicates (rhodonite and spessartite) of the Precambrian Amapa Series. There is usually a sharp change from high grade to underlying waste. With a production of about 1.2 million tonnes per year the reserves are some 30 million tonnes at 42% Mn on a washed basis (*Engineering and Mining Journal*, 1975). These deposits are mined by open pit; most of the material is amenable to ripping although up to 30% of the ore and waste must be blasted. Typical analysis of the high-grade manganese ore (dry basis) is as follows: Mn, 49%; SiO_2, 2.5%; Fe, 5.5%; Al_2O_3, 5.5%; H_2O, 5%; P, 0.05%.

(b) West African deposits

Commercial quantities of residual manganese are found at Nsuta (Ghana), Mokta (Ivory Coast) and Tambao (Upper Volta). The development of these deposits has been described by Perseil and Grandin (1978). In these three areas Precambrian manganiferous carbonate (rhodochrosite), along with garnetiferous quartzites and gondite, were the protore for the supergene oxide deposits formed under a Tertiary planation surface. Of particular interest is the Ghanaian Nsuta deposit where most of the supergene manganese oxide deposits were formed at the expense of a bedded rhodochrosite protore. In this deposit the appearance of pyrolusite and cryptomelane at an intermediate stage of oxidation rather than as an end product is considered the result of the acidic environment and the high potassium content of the host rocks.

(c) South African deposits

The best commercial-quality manganese ores in the Republic of South Africa occur in the Klipfontein Hills of the Postmasburg-Aucampsrust area, where they are associated with a siliceous breccia. The deposits form large ore bodies in hollows of a palaeokarst topography on the Transvaal dolomite. Originally it was suggested that dissolution of manganese from the underlying manganiferous dolomite had occurred during weathering. Subsequently it has been proposed that the Mn was dissolved from the dolomite in a reducing environment in an upper groundwater body rich in organic matter. The Mn was oxidized and precipitated as Mn^{4+}-bearing minerals by interaction with an oxygenated lower groundwater body in a karst hydrography. Militating against both these hypotheses is the high-temperature mineral assemblage (braunite, bixbyite, jacobsite and hausmannite). It could be argued that the ores have been subsequently modified but there is an absence of metamorphic effects in the enclosing rocks (Roy, 1981). The origin of these deposits is still in doubt.

(d) Other deposits

Extensive supergene concentrations of manganese occur in India. Generally the protores are bedded manganese oxide ore bodies of Precambrian age. At Butte, Montana, USA, hydrothermal deposits of rhodochrosite have undergone supergene oxidation.

7.6 SUPERGENE SULPHIDE ENRICHMENT

Most primary sulphide ore bodies have mineral assemblages that are unstable in near-surface conditions. Andrew (1980) lists the susceptibility of the more common sulphides to oxidation as decreasing in the order: pyrrhotite, chalcopyrite, fine pyrite, sphalerite, galena and coarse pyrite.

Above the water table, where conditions are oxidizing, the sulphides will be converted to sulphates and the metal cations are either precipitated as insoluble oxidate minerals or taken into solution in groundwaters to be precipitated elsewhere. These oxidizing ore bodies also give rise to acidic solutions. The degree of acidity depends on the metal ions present, particularly on the extent of hydrolysis or on the insolubility of hydroxides. Most acids are solutions resulting from oxidation of sulphides containing iron because the ferric ion produced is hydrolysed to an extremely insoluble oxide or hydroxide. Pyrite is a very common constituent of sulphide ore bodies and its decomposition produces the insoluble iron hydroxide (limonite) and sulphuric acid:

$$2FeS_2 + 15O + 8H_2O + CO_2 \rightarrow 2Fe(OH)_3 + 4H_2SO_4 + H_2CO_3$$

In the weathering of sulphides therefore the

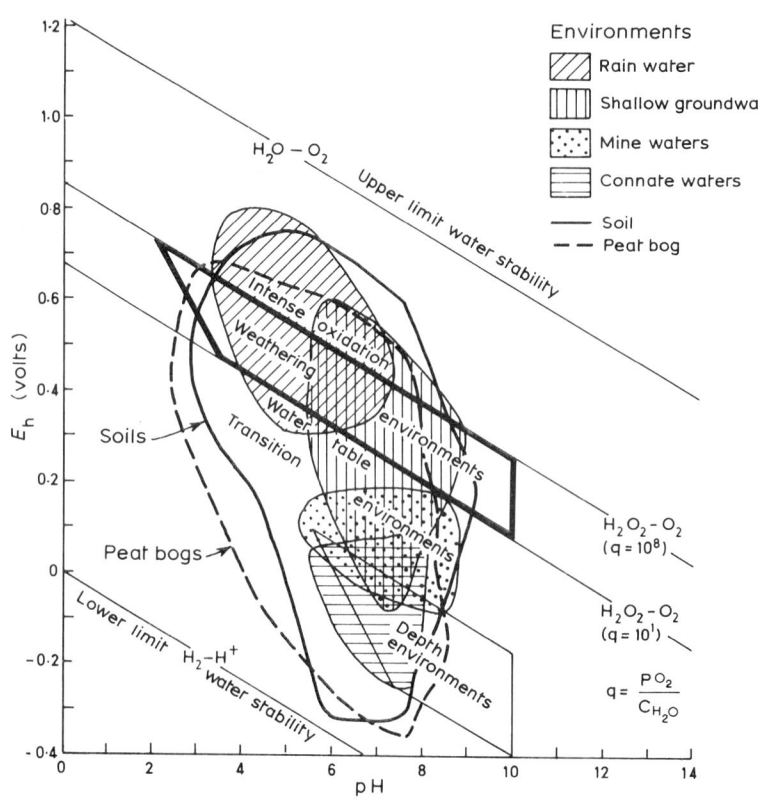

Fig. 7.16 Geochemical environments in terms of E_h and pH (after Baas Becking et al. 1960); weathering and depth environments after Sato (1960).

Ore deposits formed by weathering

significant variables are E_h and pH. The range of these two variables in the environment of oxidizing ore deposits is shown in Fig. 7.16.

The reactions outlined above result in large amounts of iron being fixed above the water table, commonly as goethite or, more rarely, haematite; these coupled with silica typify the gossans or 'iron hats' which characteristically overlie sulphide bodies (Figs 6.16 and 7.17).

The bulk of the metals dissolved from the oxidizing portion of the ore body remain in solution until they reach the water table which marks the boundary between the zones of oxidation and reduction. In this zone reactions occur which cause the precipitation of the metals and the replacement of primary sulphides by secondary sulphides resulting in a substantial increase in grade of the ore body. This is the zone of supergene enrichment. To form a thick supergene layer requires a fairly deep water table and a gradual lowering of the ground surface by erosion. Rapid movements in the water table, associated with dramatic climatic changes, may result in the preservation of a 'perched' supergene zone as that seen at Chuquicamata, Chile. The importance of the supergene zone, or blanket, in the economics of porphyry copper deposit exploitation and exploration has been covered in Chapter 3. The supergene enrichment of low-grade iron-formations is discussed in the next Chapter.

Andrew (1980) has produced an extensive work on the supergene alteration and gossan textures of base-metal ores. Sulphide weathering and the use of gossans in mineral exploration has been described by Blain and Andrew (1977). Readers requiring detailed descriptions of the processes involved and the particular textures associated with specific gossan developments are referred to these two papers.

7.6.1 Exploration and evaluation of weathered sulphide deposits

The leached nature of gossans, their texture and colour, all assist with exploration. Colour may help to locate gossans from airborne surveys as may a lack of vegetation. They may form resistant cappings which lead to positive topographic features (Fig. 6.16). Following the discovery of a gossan an attempt must be made to relate the gossan to the size and 'value' of the underlying sulphide ore body. As with the evaluation of most ore bodies only expensive drilling will provide the absolute answer to this question. To determine if drilling is warranted the evaluation begins with a specialized investigation of the mineralogical and textural features and then concentrates on the

Fig. 7.17 (a) General section through a sulphide-bearing vein showing supergene enrichment. (b) Variation of Cu and Fe with depth through this vein (modified from Evans 1980).

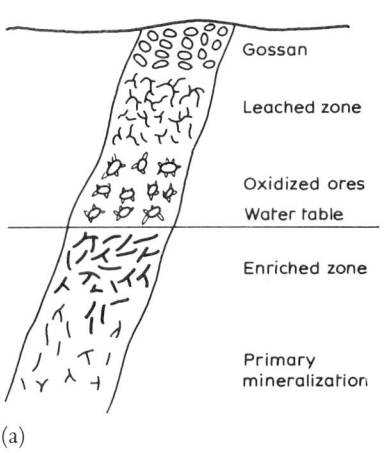
(a)

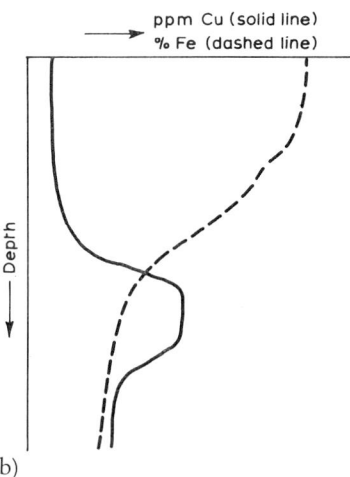
(b)

Ore deposit geology and its influence on mineral exploration

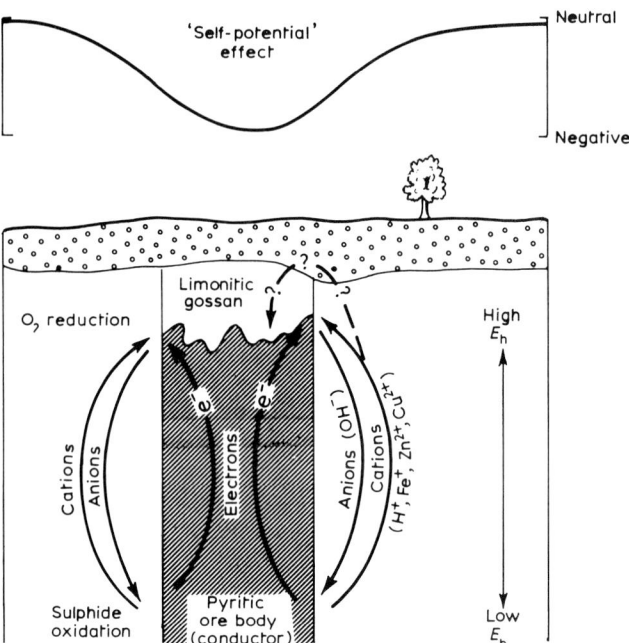

Fig. 7.18 Schematic diagram of oxidizing ore body, showing flow of electrons and ions (from Rose, Hawkes and Webb 1979).

geochemistry of the gossan (Sections 2.3.7 (b) and 6.3.4). Trace element geochemistry is probably the most reliable criterion for gossan evaluation. The iron oxide mineralogy may not be diagnostic or characteristic but metal oxidates or the residual sulphide gangue may give a guide to genesis. Andrew (1980) states that highly leached gossans commonly overlie massive sulphides with low metal/sulphur ratios. While recognizing the limitations, Andrew has found that where textural appraisal can be locally quantitative it is the most definitive of all gossan evaluation techniques. Evaluation of gossans, as guides to mineralization, may be particularly beneficial in areas where the gossan-forming process (deep-weathering) has advanced to such an extent as to make conventional geochemistry or geophysics difficult to apply. Generally, however, evaluation of the gossan will guide the choice of subsequent geophysical, geochemical and drilling techniques.

Application of geophysics
In contrast to the weathering of silicate minerals which gives rise to the economic residual deposits described in this chapter, the main types of reaction involved in the chemical weathering of the sulphide ores are electrochemical and take place between the sulphide (electronic conductor) and an aqueous medium. The supergene alteration of the sulphide ore body causes it to behave as a large electrochemical cell with electronic flow through the ore body and ionic flow through the surrounding groundwater system (Fig. 7.18). It is this phenomenon which forms the basis for exploration using self-potential or spontaneous potential (SP). Over some oxidizing sulphide bodies voltages of many hundred millivolts have been detected.

The maximum anomaly quoted by Andrew (1980) for selected deposits in Southern Africa is 0.52 volt over the semi-massive, pyrite–chalcopyrite bodies at Otjihase where the oxide zone is up to 25 m thick. The occurrence of similar SP anomalies over sulphide bodies which are submerged below the water table may indicate a relict cell, formed when the water table was lower and in which the E_h of the supergene alteration has not re-equilibrated with the E_h of the groundwaters.

Although such SP anomalies may occur over oxidizing sulphide bodies, anomalies of similar magnitude may occur as a result of completely different geological circumstances, for example the presence of anthracite. Therefore reliance on geophysical methods alone is not possible, but, combined with accurate geochemical and mineralogical interpretation and data from other exploration methods, it should reduce the amount of drilling required.

7.7 CONCLUDING STATEMENT

The mineral deposits covered in this chapter are all linked through their main agent of formation, chemical weathering. In bauxites, the genesis relies on very specific climatic conditions whereas for lateritic nickel deposits to form these climatic conditions must act upon a narrow range of rock types.

Although the genesis is well documented for most of these deposits and can be useful in their exploration, the problem faced by the geologist is more one of evaluation than exploration. Many of these deposits occur either as irregular blankets or pockets, which reflect irregularities in the bedrock. They also rapidly change in grade vertically and laterally. Drilling is undertaken earlier and probably more extensively in these deposits than in any other mineral deposit type. We have stressed the effect of the end use upon the material mined and which involves the geologist in the search for material of very specific qualities.

In dealing with supergene enrichment we have mentioned the attempts at interpreting the relict mineralogy and structure in the gossans overlying sulphide ore bodies. Earlier, when dealing with porphyry copper deposits (Section 3.3.e), we stressed the economic significance of the chalcocite blanket to these deposits. In porphyry copper deposits there have also been attempts to deduce the underlying sulphide mineralization, and possible magnetite content from the gossan mineralogy.

Part of any geological text is always devoted to the classification of mineral deposits and we explained our overall approach, in Chapter 1, as being typological. Within each mineral deposit type we have attempted a classification where this appears beneficial to the understanding of the deposit type or to the benefit of the explorationist.

In dealing with lateritic nickel no classification scheme was proposed, as at present we consider one unnecessary for the location of other similar deposits. The mode of formation and occurrence of lateritic nickel deposits does not necessitate further subdivision of the deposit type. A classification of kaolin is useful, the division into residual, sedimentary and hydrothermal providing a good indication of the environments in which further deposits may be sought. An additional classification would be applied within the deposit areas, based upon the various grades of clay available, defined by their end use. Such detailed subdivisions will change with market requirements.

The other lateritic deposit which has been described in detail is bauxite. Many attempts have been made to classify these deposits. From the classification schemes currently suggested we chose a simple division into lateritic crust and karst. As an initial, descriptive approach this is reasonably successful but for exploration and evaluation purposes it is not entirely satisfactory. We feel that a classification is needed that more adequately describes the internal composition and compositional variations of bauxites. A classification along these lines would guide geologists towards an end use for the material. In constructing Table 7.2, combining the features of bauxites, we feel that we see such a classification developing with divisions based upon iron content. We wait for other workers with more specialized knowledge of bauxites to produce a comprehensive classification of practical use to the exploration geologist.

REFERENCES

Alderton, D. H. M. and Rankin, A. (1983) The character and evolution of hydrothermal fluids associated with the kaolinized St Austell granite, S.W. England. *J. Geol. Soc. Lond.*, **140**, 297–309.

Aleva, G. J. J. (1981) Essential differences between the Bauxite deposits along the Southern and Northern edges of the Guiana Shield, South America. *Econ. Geol.*, **76**, 1142–1152.

Allum, J. A. E. (1982) Remote sensing in Mineral Exploration—case histories. *Geosci. Can.*, **8(3)**, 87–92.

Andrew, R. L. (1980) Supergene alteration and gossan textures of base-metal ores in Southern Africa. *Miner. Sci. Eng.* **12(4)**, 193–215.

Baas Becking, L. G. M., Kaplan, I. R. and Moore, D. (1960) Limits of the natural environments in terms of pH and oxidation–reduction potentials. *Jour. Geology* **68**, 243–284.

Bardossy, G. (1979) Growing significance of bauxites. *Episodes*, **2**, 22–25.

Bardossy, G. (1982) *Karst Bauxite: Bauxite Deposits on Carbonate Rocks.* Elsevier, Oxford, 441 pp.

Blain, C. F. and Andrew, R. L. (1977) Sulphide weathering and evaluation of gossans in mineral exploration. *Miner. Sci. Eng.*, **9(3)**, 119–150.

Bowen, N. L. (1922) The reaction principle in petrogenesis. *Jour. Geology*, **30**, 177–198.

Bray, C. J.. and Spooner, E. T. C. (1983) Sheeted vein Sn–W mineralization and greisenization associated with economic kaolinization, Goonbarrow china clay pit, St Austell, Cornwall, England: Geologic relationships and geochronology. *Econ. Geol.*, **78**, 1064–1089.

Bristow, C. M. (1969) Kaolin deposits of the United Kingdom of Great Britain and Northern Ireland. In *Kaolin Deposits of the World, A-Europe.* Proc. Int. Geol. Congr. 23rd, Prague, 1968, vol. 15, pp. 215–288.

Bristow, C. M. (1977) A review of the evidence for the origin of the Kaolin deposits in S.W. England. In *Proc. 8th Int. Symp. and Meeting on Alunite*, (ed. E. Galen), servicio de Ministerio de Industrie Energia, Madrid, Rome, pp. 1–19.

Burger, P. A. (1979) Greenvale Nickel Laterite Orebody. In *International Laterite Symposium*, American Institute of Mining, Metallurgy and Petroleum Engineers New York, pp. 24–37.

Chapman, J. H. and Evans, H. J. (1978) Wagina Island Bauxite: exploration and mining assessment. *Proc. 11th Commonwealth Mining & Metallurgical Congress Hong Kong*, pp. 441–447.

Comer, J. B. (1974) Genesis of Jamaican bauxite. *Econ. Geol.*, **69**, 1251–1264.

Dennen, W. H. and Anderson, P. J. (1962) Chemical changes in incipient rock weathering. *Bull. Geol. Soc. Am.*, **73**, 375–384.

Dixon, C. J. (1979) *Atlas of Economic Mineral Deposits*, Chapman and Hall, London. 143 pp.

Dreyer, R. M. (1978) Principles of evaluation of lateritic ores. *Min. Eng.* August, 1201–1202.

Dubois, J., Launay, J. and Recy, J. (1973) Les mouvements verticaux en Nouvelle-Caledonie et aux Illes Loyautes et leur interprétation dans l'optique de la tectonique des plaques. *Cah. ORSTOM, Ser. Geol. Paris*, **5**, 3–24.

Durrance, E. M., Bromley, A. V., Bristow, C. M., Heath, M. J. and Penman, J. M. (1982) Hydrothermal circulation and postmagmatic changes in the granites of South West England. *Proc. Ussher Soc.* **5**, 304–320.

Engineering and Mining Journal (1975) ICOMI: manganese technology with social progress, New York, p. 105.

Evans, A. M. (1980) *An Introduction to Ore Geology*, Blackwell, Oxford, 231 pp.

Exley, C. S. (1976) Observations on the formation of kaolinite in the St Austell granite, Cornwall. *Clay Miner.*, **11**, 51–63.

Goldich, S. S. (1938) A study in rock weathering. *J. Geol.*, **46**, 17–23.

Golightly, J. P. (1979) Nickeliferous laterites: a general description. In *International Laterite Symposium*, American Institute of Mining, Metallurgy and Petroleum Engineers, New York, pp. 1–23.

Golightly, J. P. (1981) Nickeliferous laterite deposits. *Econ. Geol. 75th Anniv. Vol.*, 710–735.

Govett, G. J. S. and Larsen, J. (1981) *The World Aluminium Industry*, Vol. 1, Australian Mineral Economics Pty. Ltd., Sydney, Australia.

Grubb, P. L. C. (1973) High-level and low-level bauxitisation: a criterion for classification. *Miner. Sci. Eng.*, **5**, 219–231.

Haldemann, E. G., Buchan, R., Blowes, J. H. and Chandler, T. (1979) Geology of lateritic nickel deposits, Dominican Republic. In *International Laterite Symposium*, American Institute of Mining Metallurgy and Petroleum Engineers, New York, pp. 57–85.

Harben, P. (1978) The Spruce Pine Mining District USA. *Ind. Miner.*, **132**, 23–37.

Harben, P. and Dickson, T. (1983) Non-metallurgical grades – bauxites creme de la creme. *Ind. Miner.* **192**, 25–43.

Harder, E. C. and Greig, E. W. (1960) Bauxite. In *Industrial Minerals and Rocks* (eds J. L. Gillson et al.) American Institute of Mining Engineering, pp. 65–85.

Harju, H. O. (1979a) Exploration of Exmibals nickel

laterite deposits in Guatemala. In *International Laterite Symposium*, American Institute of Mining, Metallurgy and Petroleum Engineers, New York, pp. 245–251.

Harju, H. O. (1979b) Exploration of P.T. Inco's nickel laterite deposits in Sulawesi, Indonesia. In *International Laterite Symposium*, American Institute of Mining, Metallurgy and Petroleum Engineers, New York, pp. 292–299.

Hose, H. R. (1960) The genesis of bauxites, the ores of aluminium. *Proc. Int. Geol. Congr. XXI Section XVI Copenhagen*, pp. 237–247.

Hutchison, C. S. (1983) *Economic Deposits and their Tectonic Setting*. Macmillan Press, London, 365 pp.

Kesler, T. L. (1952) Occurrence and exploration of Georgia's kaolin deposits In *Problems of Clay and Laterite Genesis*. Am. Inst. Mining and Metallurgical Engineers, Symposium, New York, pp. 162–177.

Lelong, F., Tardy, Y., Grandin, G., Trescases, J. J. and Boulange, B. (1976) Pedogenesis, chemical weathering and processes of formation of some supergene ore deposits. In *Handbook of Strata-bound and Stratiform Ore Deposits* (ed. K. H. Wolf), Elsevier, Amsterdam, Vol. 3, pp. 93–174.

Lillehagen, N. B. (1979) The estimation and mining of Gove bauxite reserves. In *Estimation and Statement of Mineral Reserves*, Australasian Institute of Mining and Metallurgy, Parkville, Victoria, pp. 19–30.

Lovering, T. S. (1959) Significance of accumulator plants in rock weathering. *Bull. Geol. Soc. Amer.*, **70**, 781–800.

Matsunaga, T., Akiyama, S. and Fuyie, T. (1978) Rennell Island Bauxite: exploration and mining assessment. *Proc. 11th Commonwealth Mining and Metallurgical Congress, Hong Kong*. pp. 449–453.

Ollier, C., (1969) *Weathering – Geomorphology Texts – 2*, Longman, London.

Ozlu, N. (1983) Trace element content of 'Karst Bauxites' and their parent rocks in the Mediterranean Belt. *Miner. Deposita*, **18**, 469–476.

Panagopoloulos, C. (1983) Use of exploration data for improved mineral recovery. *Trans. Inst. Min. Metall.*, **92**, A90–A92.

Pedro, G. (1970) Sur l'altération des materiaux calcaires en conditions "latéritisantes": étude experimentale de l'évolution d'une marne illitique. *Compt. Rend. Acad. Sci. Paris Ser. D*, **2**, 36–68.

Perseil, E. A. and Grandin, G. (1978) Evolution minéralogique du manganèse gisements d'Afrique de l'Ouest: Mokta, Tamboa, Nsuta. *Miner. Deposita*, **13**, 295–311.

Rose, A. W., Hawkes, H. E. and Webb, J. S. (1979) *Geochemistry in Mineral Exploration*, Academic Press, London, 657 pp.

Roy, S. (1981) *Manganese Deposits*. Academic Press, London, 458 pp.

Sato, M. (1960) Oxidation of sulphide ore bodies – I Geochemical Environments in terms of E_h and pH. *Econ. Geol.*, **55**, 928–961.

Shaffer, J. W. (1983) Bauxite Raw Materials. In *Industrial Minerals and Rocks*, 5th edn., American Institute of Mining, Metallurgy and Petroleum Engineers, Vol. 1, pp. 503–527.

Sheppard, S. M. F. (1977) The Cornubian Batholith, S.W. England: D/H and $^{18}O/^{16}O$ studies of kaolinite and other alteration minerals. *J. Geol. Soc. London*, **133**, 573–591.

Strahl, E. O. (1982) Modern analytical methods in bauxite survey programmes. *J. Geol. Soc. Jam. Proc. Baux. Symp.*, **V**, 188–203.

Trescases, J. J. (1973) Weathering and geochemical behaviour of the elements of ultramafic rocks in New Caledonia. *Bur. Miner. Resourc. Geol. Geophys., Canberra, Bull.*, **141**, 149–161.

Valeton, I. (1972) *Bauxites. Development in Soil Science*, Vol 1, Elsevier, Amsterdam, 226 pp.

Vincent, A. (1983) Exploration in the Extractive Industries. In *Prospecting and Evaluation of Non-metallic Rocks and Minerals* (eds K. Atkinson and R. Brassington). The Institution of Geologists, London, pp. 1–14.

Walker, A. L. (1978) The Gabbin kaolin deposit and its markets. *Proc. 3rd Ind. Min. Int. Congr. Paris*, pp. 87–93.

Ward, H. J. (1978) Exploration guides and methods in the discovery of Mt. Saddleback and associated bauxite deposits, Southwestern Australia. 4th Int. Congress for the Study of Bauxites, Alumina and Aluminium, Athens, Vol 2, pp. 965–995.

White, A. H. (1976) Genesis of low-iron bauxite, north eastern Cape York, Queensland, Australia. *Econ. Geol.*, **71**, 1526–1532.

8

Iron ores of sedimentary affiliation

8.1 INTRODUCTION

The bulk of the World's production of iron ore is processed into steel, and the needs of the steel industry therefore have a profound effect on the type of ore mined. An understanding of the technology of the steel industry is useful for the geologist working on iron ore deposits, and a good introduction to the subject has been provided by Strassburger (1969). Recent changes in the mining and processing of iron ore are reviewed by Peterson (1980).

One trend within the industry has been a greatly increased demand for ore. This is reflected in an increase in World steel production from 270 million tonnes in 1955 to an estimated 750 million tonnes in 1983 (*Mining Annual Review*, 1984). At the same time there has been a demand for higher grades of ore, which contrasts with commodities such as copper where the average grade has declined in the last fifty years (see Fig. 3.1). The average grade of iron ore entering international trade was 51% Fe in 1950 but was estimated to be 59% in 1975 (Canada, 1976). In addition the steel manufacturers have demanded increasingly rigorous specifications for the physical properties of the ore such as particle size distribution.

These economic forces led to world-wide exploration for high-grade large-tonnage iron ore deposits and to the evaluation of known resources during the 1950s and 1960s. At the same time metallurgical research resulted in the development of beneficiation processes which enabled some types of ore of low grade to become economically attractive. In the USA this has resulted in a major proportion of iron ore being produced by the processing of iron-formations which would have been considered 'waste' in the 1940s.

As a consequence of the dramatic success of both exploration and processing there has been a fundamental change in the types of iron ore utilized by the major steel-manufacturing countries. This is well illustrated by the changed pattern of production in Western Europe compared with Australia. Table 8.1 shows that the production from oolitic iron ores in France, UK and Germany declined rapidly from 1962 whilst those countries such as Australia which exploit ores associated with iron-formation have witnessed a spectacular surge in the production of iron ore. The advantage of proximity to the European blast furnaces has, within two decades, been outweighed by lower-priced imported ores of higher and more consistent grade.

Iron ores of sedimentary affiliation

Table 8.1 Comparative statistics of iron ore production (10^6 tonnes)

Country	Ore type	1962	1972	1982
France	Oolitic	65.2	54.8	19.7
UK	Oolitic	15.2	9.0	0.5
West Germany (FRG)	Oolitic	16.5	6.1	1.4
Australia	Supergene ore associated with banded iron-formation	4.8	62.5	89.0

Source: *Mining Annual Review* (1963, 1973, 1983).

The more recent trend in the steel industry from the blast furnace towards the use of direct reduction techniques has also changed the emphasis in demand for iron ore (Stephenson and Smailer, 1980). Ore to be utilized in direct reduction methods of iron manufacture must be of higher grade than would be acceptable for a conventional blast furnace and must also contain very low concentrations of impurities, particularly SiO_2 and Al_2O_3.

World production of iron ore is dominated by the USSR which had an estimated mine production of 245×10^6 tonnes in 1983 (*Mining Annual Review*, 1984). Brazil, Australia and China are also major producers of iron ore, and there are forty additional countries from which production was recorded in 1983.

Iron-formation and enrichment ores derived from them are the main iron ores mined at the present time but there is some geographical variation in ore type. In the USA since the 1950s there has been an increasing trend towards the exploitation of unaltered iron-formation due to the exhaustion of their high-grade ores. Conversely in Brazil it is probable that in the future banded iron-formation will play a less significant role as a source of iron due to the discovery of large reserves of high-grade enrichment ore in the Serra dos Carajas district.

In the Michipicoten district of Canada a carbonate type of iron-formation is mined and can be sold to the industry with minimal beneficiation. However, the oxide-type iron-formation mined in the USA and Brazil requires treatment before it becomes commercially viable. The main processes involved are crushing, grinding, magnetic separation and froth flotation.

Two factors arise from the beneficiation of the ores which require further comment. Firstly the crushed and up-graded iron oxide concentrate has a grain size which is too fine to be acceptable for blast furnace feed and it is normally made into pellets of three-eighths to one-half inch (9 – 13 mm) diameter. Secondly the processing of the iron ore adds to the cost of the concentrate which is sold to the steel industry, a financial burden which naturally enriched ores do not have to carry.

8.2 CLASSIFICATION OF IRON ORES

The geologist's perspective of ore deposits and therefore his attitude towards their classification is to a certain extent dependent on the country or group of countries in which he works. This is well illustrated by the case of iron ores. In the USSR, which is by far the most important World producer, iron ore is produced from a diversity of ore types and a comprehensive classification is required (Table 8.2). However, in the Western World most of the ore types shown in Table 8.2 are now largely of historical or local value. In some cases unusual types of iron ore may serve a local steel industry where low transport costs may be advantageous, and in other instances a

Table 8.2 Classification of iron ore deposits (modified from Sokolov and Grigor'ev, 1977)

Genetic group	Examples
1. Magmatic	Kachkanar (Urals), Kusinsk, USSR
	Kiruna, Sweden
2. Contact-metasomatic	Marmora, Ontario, Canada
	Magnitogorsk, USSR
3. Hydrothermal	Iron Mountain and Pea Ridge, Missouri, USA
	Korshunovsk, Eastern Siberia, USSR
4. Marine-sedimentary (weakly metamorphosed and unmetamorphosed)	Northamptonshire, UK
	North Island, New Zealand
	Nizhne-Angara, Eastern Siberia
5. Continental sedimentary	Orsk-Khalilovosk, S. Urals, USSR
6. Weathering crusts (includes enrichment ore)	Hamersley area, W. Australia
	Serra dos Carajas, Brazil
	Michailovsk, USSR
7. Metamorphic (includes iron-formation)	Lake Superior district, USA
	Itabira district, Brazil
	Krivoi Rog, Ukraine, USSR

Government subsidy may support an indigenous industry so that the price of steel is maintained at an artificial level. Having noted these exceptions it must now be emphasized that the main types of ore which dominate the non-Communist steel industry at the present time and for the foreseeable future are Precambrian iron-formations and the enrichment ores which are closely associated with them (Fig. 8.1). Iron-formation is designated as a metamorphic ore type in Table 8.2, but Western geologists prefer to emphasize their sedimentary affiliation. The classification of iron-formation has been hindered by a confusing terminology which varies with geographical location and has sometimes changed with technological development. Unfortunately this problem is still not fully resolved and the definitions presented here may well be modified in the future.

8.2.1 Iron-formation (IF)

James (1954) has defined iron-formation as a chemical sedimentary rock, typically thin-bedded and/or finely laminated, containing at least 15% iron of sedimentary origin and commonly but not necessarily containing layers of chert. James excluded Phanerozoic iron-rich sediments from this definition and selected the term 'ironstone' to distinguish them from iron-formation. Synonymous terms which are sometimes used for iron-formation include banded haematite quartzite, quartz banded ore and banded ironstone.

Trendall (1983a) has recently reviewed the nomenclature of iron-formation. He prefers to retain iron-formation as a general lithological and stratigraphic term for iron-rich sedimentary rocks and recommends that 'anomalously high content of iron' should replace the 15% limit defined by James. Trendall's definition would also apply to Phanerozoic iron-rich sediments. Whether James' or Trendall's definitions are accepted, the term 'banded iron-formation' (BIF) is reserved for those iron-rich sediments which are characterized by regular banding.

Several attempts have been made to classify iron-formation. Gross (1965) considered that iron-formation can be divided into two main types: Superior and Algoma. The Superior type is deposited in a near-shore continental-shelf environment in association with dolomites, quartzites and black shales, whereas the Algoma type is associated with volcanics. Gross's classification appears to be valid for North America, but is not so readily applicable to the major iron-formations of the Hamersley Basin in Western Australia, or the Transvaal Basin in South Africa. In spite of its limitations the distinction between Superior and Algoma types is still widely used.

Kimberley (1978) has proposed a palaeo-environmental classification of iron-formation. Six classes of ore are recognized and are described by their acronyms: for example shallow-volcanic-platform iron-formation (SVOP-IF). Trendall (1983a) considers that the environment of deposition of many iron-formations is insufficiently understood for this scheme to be widely used. At the present time no wholly satisfactory system of classification has been proposed which is generally accepted by geologists who are studying this class of deposit.

8.2.2 Itabirite

This may be defined as laminated, metamorphosed, oxide-type banded iron-formation, in which the original chert or jasper bands have been recrystallized into megascopically distinguishable grains of quartz, and in which iron is present as thin layers of haematite and magnetite (modified from Dorr and Barbosa, 1963). The term was originally applied to enrichment ores in the Itabira district of Brazil.

8.2.3 Taconite

This may be defined as iron-formation, suitable for concentration of magnetite and haematite by fine grinding and magnetic or other treatment, from which pellets containing 62–65% iron can be manufactured (modified from Bates and Jackson, 1980). The term was originally applied to bedded ferruginous chert in the Lake Superior district of Minnesota.

8.3 GENERAL CHARACTERISTICS OF IRON-FORMATION

8.3.1 Distribution in space and time

Iron-formations are largely confined to Precambrian shield areas, and the cratons of both Northern and Southern hemispheres contain deposits of economic importance.

Iron-formations occur throughout the Precambrian but there are particular periods of time when they are notably abundant. The subject has been investigated by Goldich (1973) and has been discussed more recently by James (1983). Fig. 8.2 shows the periods of most abundant deposition of iron-formation and it is apparent that the Early Proterozoic (2500–1900 million years) represents the most important time interval. Many of the World's large iron-formations were formed at this time including the Lake Superior region, USA, the Labrador Trough, Canada, the Krivoi Rog district, USSR, the Transvaal Griquatown region, S. Africa, Minas Gerais State, Brazil, and the Hamersley Iron Province, Western Australia.

James (1983) suggests that the depositional maximum of iron-formation in Lower Proterozoic times results from the combination of structural, geochemical and biological factors. The advent of the Lower Proterozoic was marked by a change in structural style from orogenesis and cratonization in the late Archaean to a period of structural stability characterized by the development of shallow intracontinental troughs and marginal basins. This provided numerous sites favourable for the deposition of iron-formation. The chief biological factor was the development of primitive organisms which were capable of producing oxygen by photosynthesis. The main consequence would have been the increased tendency for iron to be oxidized and precipitated as iron hydroxides. The prime geochemical factor was the transport of dissolved iron and silica by upwelling ocean currents to favourable depositional sites.

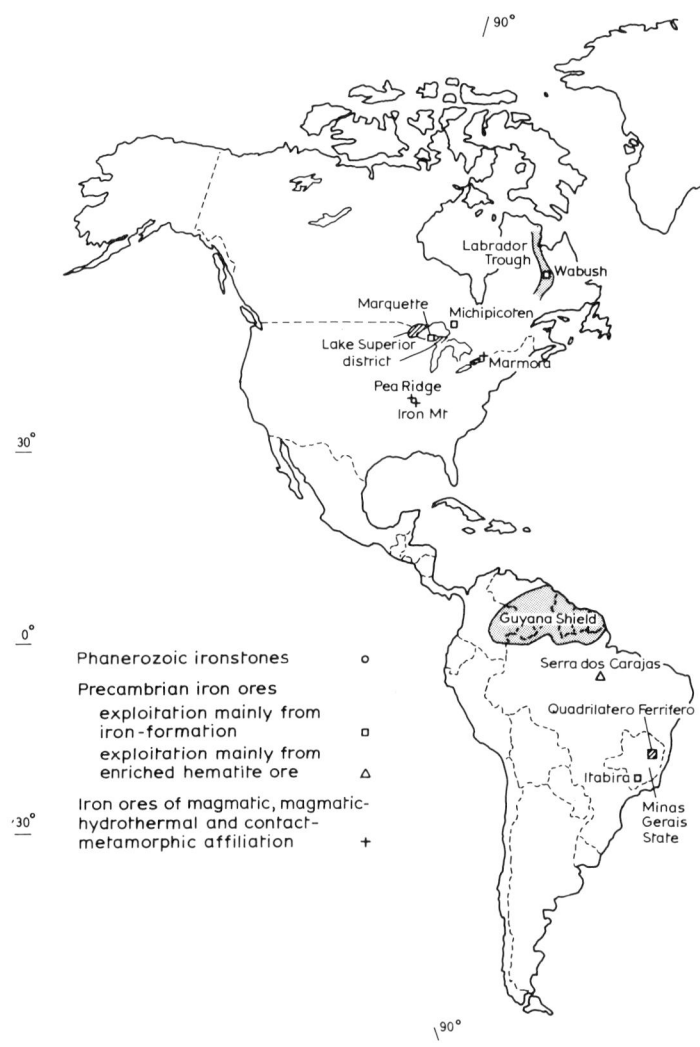

Fig. 8.1 Location map.

8.3.2 Size and grade of deposits

Iron-formation is essentially an iron-rich sedimentary rock which normally has a large area and therefore the resources of iron are enormous. The original tonnage of iron-formation estimated to contain about 30% Fe and 45% SiO_2 within selected Precambrian sedimentary basins has been computed by James (1983). He concludes that the tonnage of iron-formation ranges from 10^9 tonnes for some of the smaller deposits such as those contained in the Swaziland Supergroup, Swaziland, to 10^{14} tonnes for the larger sedimentary basins such as Krivoi Rog, USSR, and Hamersley, Western Australia. These figures are merely intended to give an idea of the scale of sediment deposited and are not meant to be of commercial significance.

The criteria used in determining the reserves of iron associated with iron-formation vary according to whether the ore is of the enrichment type or has to be beneficiated to obtain an iron-rich

concentrate. In the Hamersley Basin, Western Australia, the ores of commercial value are enrichment ores derived from banded iron-formation. For the purpose of reserve calculation a cut-off of 55% Fe is utilized and reserves are estimated to be 33×10^9 tonnes (Hamersley Iron, 1981). Within the Hamersley Basin several specific ore types with differing grades of iron and varying concentrations of contaminants are recognized and the information is summarized in Table 8.6.

In the Lake Superior district of the USA the bulk of iron-formation mined is taconite, which has to be beneficiated before it is of commercial value. The geological criteria which are applied in determining the cut-off limit for iron and the resulting size of the deposit are shown in Table 8.3. Ohle (1972) explained that whilst high grade is a definite advantage it is only one of many factors which have to be considered. The average grade of magnetic iron which could be profitably exploited in the early 1970s was above 20%. However, taconite with a magnetic iron content as low as

Ore deposit geology and its influence on mineral exploration

8.3.3 Mineralogy

The valuable minerals in iron-formation are magnetite, haematite and siderite. Within the Lake Superior district of the USA the iron silicates stilpnomelane and minnesotaite may be locally important, and in the Wawa district of Canada pyritiferous shales are closely associated spatially with the siderite ores.

The average grade of iron-formation is typically in the range 20–35% Fe (James, 1983). Normally for iron-formation to be considered as ore it must be either enriched by supergene processes or beneficiated to form a high-grade concentrate. The main economic minerals in supergene iron ore are haematite, martite and goethite. Their distribution in the Hamersley Iron Province of Western Australia is described in Section 8.6.

James (1954) studied iron-formation in the Lake Superior district and suggested that four main facies could be distinguished: oxide, carbonate, sulphide and silicate. James considered that the first three facies were controlled by the water depth and oxygen content of the marine environment. More specifically the oxide facies was interpreted as having been deposited in shallow well-oxygenated water, whereas the sulphide facies was considered to have been deposited in a deeper, oxygen-deficient environment. James illustrated this concept diagrammatically but his cross-section showing a lateral transition between facies types has been

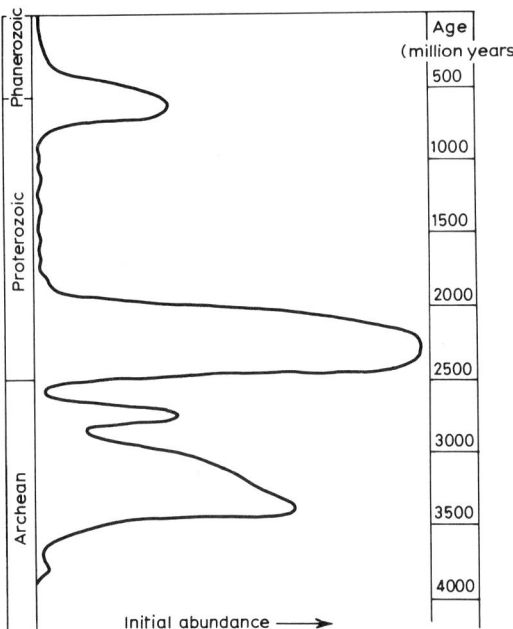

Fig. 8.2 Estimated abundance of iron-formation deposited through geological time. Horizontal scale is non-linear, approximately logarithmic; range $0–10^{15}$ tonnes (from James 1983).

12–15% may be mined if it can be profitably treated, whereas those ores which are difficult to grind may need to contain 25% magnetic iron before they are worth exploiting (Ohle, 1972). In 1979 the reserves of iron ore in the Lake Superior district were estimated to be 16×10^9 tonnes (Peterson, 1980)

Table 8.3 Geological factors in the evaluation of iron ore deposits (from Ohle, 1972)

1. Type and grade: direct shipping, wash, heavy media, taconite, etc., iron: silica ratio, impurities, and associated elements.
2. Tonnage: crude and product, effect on capital cost and recuperation schedules, weight recovery.
3. Grain-size: grind size, texture, liberation of ore minerals, elimination of impurities.
4. Grindability: kWh per tonne to reduce the ore to concentrating and agglomerating sizes.
5. Mineralogy: magnetite, haematite, goethite, silicate or carbonate. Impurity mineralogy – effect on the ability to separate the impurities in processing.
6. Distribution of ore types: grades, textures, mineralogies – can selective mining be done?
7. Depth and nature of overburden: sand and gravel, or rock. Open pit versus underground mining.
8. Shape and attitude of the ore body: tonnes per vertical metre, effect on stripping ratio.
9. Location: topographic effects, climate.

interpreted by many geologists as a true section rather than as a conceptual model. In practice, exposures which demonstrate the lateral transition between facies have yet to be described. Trendall (1983a) therefore suggests that such terms as carbonate-facies iron-formation should not be applied in the context of a particular basinal model but should be retained as a descriptive term.

8.3.4 Lithology

We explained in Section 8.2 that geologists are not in agreement about the definition of iron-formation. As a result specialists who are describing iron-formation from a particular sedimentary basin often find it necessary to define the sense in which they are using the term. For the purpose of this chapter we prefer the definition proposed by James (1954): 'a chemical sedimentary rock, typically thin-bedded and/or finely laminated, containing at least 15% iron of sedimentary origin and commonly but not necessarily containing layers of chert'. However, we agree with Trendall that 15% is an arbitrary figure which is probably best replaced by a qualitative term such as iron-rich. We also find James's use of the term 'ironstone' for oolitic iron ores of Phanerozoic age a helpful distinction.

The term banded 'iron-formation' is used when the iron-rich and chert-rich components are separated into distinct layers. In the case of the Krivoi Rog Basin, USSR, and the Hamersley Basin, Western Australia, banding is a widespread feature of iron-formation (Trendall, 1983a; Belevtsev *et al.*, 1983). In other sedimentary basins the banding may be more restricted. For example irregular to wavy, bedded chert material characterizes part of the Ironwood iron-formation in the Animikie Basin, Lake Superior Region (Morey, 1983).

The scale of banding in banded iron-formation usually varies from a few millimetres to several

Fig. 8.3 Banded iron-formation, Lucy open pit, Wawa, Ontario. The darker coloured horizons represent a veneer of oxidation on siderite. The lighter coloured layers consist of chert.

Fig. 8.4 Banded iron-formation, Mount Tom Price, Western Australia. Alternating layers of haematite/magnetite and chert.

centimetres (Figs 8.3 and 8.4). In the Hamersley district, Western Australia, three scales of banding are recognized (Trendall, 1975).

(1) Large-scale macrobands defined by alternations between thin layers of iron-formation and pyroclastics and thicker layers of banded cherty iron-formation. The change between lithologies ranges from 0.6 to 15.0 m (Fig. 8.5).

(2) Medium-scale mesobands. This is the scale of banding which gives banded iron-formation its characteristic appearance. The banding is caused by alternations between iron-rich and silica-rich layers. Mesobands vary from millimetres to centimetres in thickness (Fig. 8.4).

(3) Microbands are laminations in the chert bands caused by the distribution of iron minerals.

Other sedimentary structures and textures are also evident in iron-formation. Dimroth (1977) has described sedimentary textures from the Sokoman Formation, Labrador, which include micritic, pelleted, intraclastic, peloidal, oolitic, pisolitic and stromatolitic types. On this evidence Dimroth compared the sedimentary environment of the Labrador trough with the present-day coastline of the Persian Gulf. Similar sedimentary textures have been described from the Griqualand West and Transvaal Basins in South Africa (Beukes, 1983). Prelithification slump structures, stylolites and shrinkage cracks are also described in the literature (Eichler, 1976).

8.3.5 Tectonic setting

Precambrian iron-formations are of both Archaean and Proterozoic age and we therefore propose to discuss their tectonic settings separately.

Iron ores of sedimentary affiliation

Fig. 8.5 Large-scale macrobands in banded iron-formation Mount Tom Price, Western Australia. A scale is provided by people standing in the open cut.

(a) Archaean iron-formation

James (1983) distinguishes deposits of Middle (3500–3000 million years) and Late (2900–2600 million years) Archaean age. The largest deposits of Middle Archaean age occur within the Guyanan Shield of South America and the Liberian Shield in Africa. Prior to continental separation the iron-formations occupied an area of 250 000 km^2 (James, 1983). Gruss (1973) suggested that deposition took place on an epicontinental shelf with localized basins of thick iron-formation.

Some deposits of Late Archaean age have a cratonic setting, for example those of the Yilgarn block, Western Australia. However, they are more commonly associated with vulcanicity in greenstone belts. The Archaean iron-formations of the Canadian Shield have been described by Goodwin (1973). They formed in small sedimentary basins in which the iron-formation is characterized by distinctive patterns of facies variation.

(b) Proterozoic iron-formation

The Early Proterozoic represents the major period of deposition of iron-formation. Interpretation of the environment of deposition of iron-formation from a range of Early Proterozoic deposits indicates the importance of a broad, gently subsiding area of shallow water open or partly open to the sea (Mitchell and Garson, 1981). Iron-formations of Lower Proterozoic age are therefore typical of passive continental margins and it is probably the establishment of this tectonic setting in Lower Proterozoic times which in part at least accounts for the abundance of iron-formation during this epoch.

8.4 GENESIS OF IRON-FORMATION

The widespread distribution of iron-formation and the enormous size of the enrichment iron ore deposits associated with them have largely resolved the identification of the World's iron resources for the foreseeable future. The debate concerning the genesis of iron-formation is therefore more of scientific concern than of pressing economic importance. The problem which remains is the discovery of very high-grade ores which are in demand from the steel industry at the present time. The factors which control the enrichment of banded iron-formation will therefore be considered in greater detail in a later section (8.6.6).

The genesis of Precambrian banded iron-formation has been reviewed by Eichler (1976) and more recently discussed by Ewers (1983) and Morris and Horwitz (1983). The main problems which need to be resolved are as follows:

(1) The source of iron. The main alternatives are as follows:
 (a) The intensive weathering of a continental land mass (Lepp and Goldich, 1964).
 (b) Subaquatic exhalations from volcanic activity (Goodwin, 1956).
 (c) Subaquatic decomposition of lava (Huber, 1959).
 (d) The up-welling of sea bottom water (Holland, 1973; Morris and Horwitz, 1983).
 (e) The re-solution of iron from detritus on the seafloor (Borchert, ref. in Stanton (1972).

(2) The source of silica. The main alternatives are as follows:
 (a) Weathering of a continental land mass under climatic conditions different from those which leach iron.
 (b) Exhalations from acid volcanism (Beukes, 1973).
 (c) Intense biological activity possibly related to an influx of silica of volcanic origin.

The proposal that the weathering of a continental land mass could provide the components of iron-formation has been severely criticized by Trendall (1965). His principal criticism is the enormous areal extent and depth of bauxitization which would be required to produce the requisite amount of iron. Trendall (1965) pointed out that there is little evidence in the rocks peripheral to iron-formations of such extensive bauxitization. The transport of iron by rivers to a sedimentary basin also raises the question of why clastic sediment is rarely associated with banded iron-formation. In view of these problems the majority of geologists regard the iron as being of marine derivation, and most would consider that it has been transported in solution as Fe^{2+}.

(3) Factors which control the banded structure in banded iron-formation. Firstly it is necessary to remind the reader that banding is recognized on differing scales and it is the nature of mesobanding which has largely dominated discussion in the literature. The following mechanisms have been suggested:

 (a) Mutual precipitation if iron hydroxides and colloidal silica from seawater (Moore and Maynard, 1929).

 This proposal implies that the banding is a diagenetic feature. Some support for co-precipitation of silica with $Fe(OH)_3$ is provided by experimental work described by Ewers (1983).

 (b) Rhythmic banding by precipitation from alternating silica- and iron-containing volcanic emanations (Goodwin, 1956; Morris and Horwitz, 1983).

 (c) Seasonal delivery of silica in dry seasons and of iron in rainy periods with solutions derived from the adjacent land mass (Hough, 1958).

 In discussing the sources of material we have already noted the improbability of iron and silica being of continental derivation. This must therefore be regarded as an unlikely mechanism for banding.

 (d) Diagnetic replacement of carbonate (Dimroth, 1977).

Ewers (1983) has pointed out that this mechanism requires ferrous iron to be introduced uniformly into sediment over a vast area with subsequent intermittent introductions of an oxidizing agent. It seems more probable that iron has been precipitated as a primary sediment as a result of oxidation of Fe^{2+}. Ewers (1983) mentions oxidation by atmospheric oxygen, oxidation as a by-product of photosynthesis, and photochemical oxidation as alternative mechanisms.

The cause of banding in banded iron-formation remains a matter for speculation. It may be that we are seeing the result of a combination of mechanisms. The iron and silica seem most probably to be of volcanic derivation but this does not imply a common source. The variation between the mesobands probably reflects an original compositional layering within the sediment which has been exaggerated by the redistribution of elements during diagenesis.

8.5 ENRICHED HAEMATITE ORE DEPOSITS

Enriched haematite ores are those which have been formed by natural processes from low-grade iron-formation. They represent the principal ore type which is being mined at the present time. The USSR is the leading World producer of iron ore, and the high-grade ores of the Krivoi Rog iron-ore district represent the most important source of production (Sokolov and Grigor'ev, 1977). Table 8.4 summarizes the mineralogical and chemical characteristics of the ore.

Brazil is the second largest World producer of iron ore. Current production is mainly from the 'Iron Quadrilateral' of Minas Gerais State where the ore types include both banded iron-formation (itabirite) and ore resulting from the enrichment of itabirite.

Enormous reserves of iron were discovered in the Serra dos Carajas district, 550 km south of Belem, in 1967 (Tolbert et al., 1971). The current development of the Carajas district (*Mining Journal*, Jan. 1st, 1982) will result in a greatly increased proportion of high-grade ore being exploited in Brazil compared with itabirite. The reserves of the 'Iron Quadrilateral' were $11\,000 \times 10^6$ tonnes in 1977 and contained 40–69% Fe. The reserves of the Serra dos Carajas amount to 15.7×10^9 tonnes of which the measured reserve is 2.4×10^9 tonnes (Hughes and Hughes, 1979).

The first ore body which has been designated for mining in the Serra dos Carajas district is referred to as N4E. The average chemical composition of the ore is shown in Table 8.5.

Australia is the World's third largest producer of iron ore. Virtually all of the production is from high-grade enriched deposits from the Hamersley Iron Province in the north-west of Western Australia. These ores are discussed in further detail in Section 8.6 and will not therefore be considered at this point.

All enriched haematite ore deposits have similarities in the nature of their protore and in their morphology, mineralogy and chemistry. There are, however, important differences in the quality of the ore which is available. For example the ores from the Carajas district are designated as

Table 8.4 Mineralogy and chemistry of rich ores from the Krivoi Rog Basin (from Sokolov and Grigor'ev, 1977)

	Amount (wt%)		
Ore type	Iron	Phosphorus	Sulphur
Martite and haematite–martite	63.7	0.26	0.043
Martite–haematite–hydrohaematite	62.3	0.08	0.03
Haematite–hydrohaematite–hydrogoethite	57.5	0.088	0.01–0.001
Magnetite and magnetite–specularite	54.0	0.04	0.15

Table 8.5 Chemical composition of the iron ores of the Serra Dos Carajas district, Brazil (from Hughes and Hughes, 1979)

	Amount (wt%)				
	Fe	P	Mn	$Al_2O_3 +$ SiO_2	Loss on ignition
Normal ore	66.31	0.04	0.23	2.51	2.01
High-Mn ore	59.16	0.03	5.93	2.77	2.98

'first class' ores by the steel industry with very low concentrations of SiO_2, Al_2O_3 and P compared with many other mining districts.

8.6 THE HAMERSLEY BASIN – AN EXAMPLE OF BANDED IRON-FORMATION AND ASSOCIATED ENRICHMENT ORES

8.6.1 Introduction

The Hamersley Basin occurs in the north-west of Western Australia. Though iron ores were known in 1920 to occur in the area their full potential was not appreciated until the early 1960s. At the present time production is mainly from the Tom Price, Paraburdoo and Newman deposits which are characterized by their high-iron and low-phosphorus contents.

8.6.2 Regional geology

The rocks of the Hamersley Basin crop out over an area of 100 000 km² (Trendall, 1983b) and comprise some 15 000 m of sediment and volcanics. Trendall (1983b) interprets the results of radiometric age determinations as suggesting initiation of sedimentation within the Hamersley Basin at about 2750 million years. Age determinations and assumed rates of sedimentation indicate that deposition was completed by about 2300 million years.

There are three major stratigraphic subdivisions within the Hamersley Basin. The lowest Fortescue group rests on an Archaean basement, and comprises mainly volcanics with subsidiary clastic sediments. The thickness is variable but is known to attain 4.25 km. The overlying Hamersley Group varies in thickness from 1.86 to 2.16 km and consists of five major, banded iron-formation units, with intercalated carbonates and shales of volcanic derivation. The main units of economic significance are the Dales Gorge member of the Brockman iron-formation and the Marra Mamba iron-formation in which the banded iron-formation is dominated by an oxide mineral assemblage. A volcanic component is present within the Hamersley Group which includes thick and extensive rhyolites. The third stratigraphic subdivision is the Turee Creek Group, and the overlying Wyloo Group, which consist of 9 km of clastic sediments of greywacke facies with dolomite and basalt being locally important. These three stratigraphic units occur within the Mount Bruce Supergroup.

The chemical and pyroclastic sediments, which constitute the Hamersley Group, indicate a relatively shallow sedimentary environment of wide lateral extent, which was isolated from terriginous detritus. Morris and Horwitz (1983) believe that the sediments were deposited on a shallow submarine platform, and they suggest that the Bahamas provide the most appropriate modern analogue. The sedimentary model proposed by Morris and Horwitz involves the introduction of iron and silica by upwelling, marine-bottom currents, which derive their major constituents from a large, oceanic rift or hot spot.

The sediments of the Mount Bruce Supergroup have been little deformed. The predominant structural style is gentle, open folding with more intense deformation being confined to the elliptical bounding zone of the Hamersley Basin (Trendall, 1983b). Two sets of vertical faults commonly displace the sedimentary succession. Smith et al. (1982) recognize four zones of regional metamorphism ranging up to (prehnite)-epidote-actinolite facies. They conclude, from the relationship of metamorphic zoning to stratigraphy, that metamorphism is entirely controlled by depth of burial.

8.6.3 The protore

The Dales Gorge member of the Brockman iron-formation is the main stratigraphic unit within which the highest-grade ores are concentrated. Trendall and Blockley (1968) subdivided the Dales Gorge member into 33 macrobands. These consist of 16 oxide iron-formations, which range from 2 to 15 m in thickness. The intervening macrobands comprise shales, chert, siderite and ankerite. The oxide banded iron-formation exhibits the mesobanding which is characteristic of this lithology, and in addition microbanding has been described from the chert layers. The ore horizons are thinned with respect to the banded iron-formation. For example, at Tom Price the ore thickness is 90 m, compared with 156 m of banded iron-formation in the type-section (Gilhome, 1975).

The main ore mineral present within the 'oxide facies' is magnetite, although haematite varies from a major to an accessory component. Morris (1980) described considerable textural variation in the magnetite in the Hamersley Iron Province.

Chemical analyses of both the oxide facies macrobands and of the intervening lithologies have been published by Ewers and Morris (1981). The average composition of 17 oxide facies macrobands at Paraburdoo is Fe_2O_3 46.37%; SiO_2 43.15%; Al_2O_3 0.09% and 0.21% P_2O_5. Other major elements include MgO, CaO, Na_2O, TiO_2, MnO and S. Ewers and Morris (1981) concluded that the primary deposits were chemically similar throughout the Hamersley Basin.

8.6.4 Size and grade of ore bodies

The currently mined ore bodies in the Hamersley Iron Province are at Tom Price (Figs 8.6, 8.7 and 8.8), Paraburdoo and Newman. There is also a number of deposits which have been evaluated but are not yet exploited. Information about the size and grade of some selected deposits is summarized in Table 8.6.

The rich haematite ores are concordant with the banded iron-formation, and their configuration reflects the structural complexity of the host rock. Figures 8.9 and 8.10 demonstrate the relatively undisturbed nature of the sediments at Koodaideri, compared with the deformed structure at Mount Whaleback. All of the known ore bodies have a surface expression and in some the ore extends to more than 400 m below surface, as for example at Mount Whaleback.

8.6.5 Mineralogy

The mineralogy of the supergene iron ores of the Hamersley Iron Province has been described by Morris (1980, 1985).

Table 8.6 Size and grade of selected ore deposits in the Hamersley Iron Province (source: Hamersley Iron, 1981)

Deposit	Size (tonnes $\times 10^6$)	Fe (%)	P (%)	SiO_2 (%)	Al_2O_3 (%)
Tom Price	700	64	0.05	4	2
Koodaideri	720	62	0.12	3	2
Nammuldi/Silvergrass	300	62	0.06	3	1–2

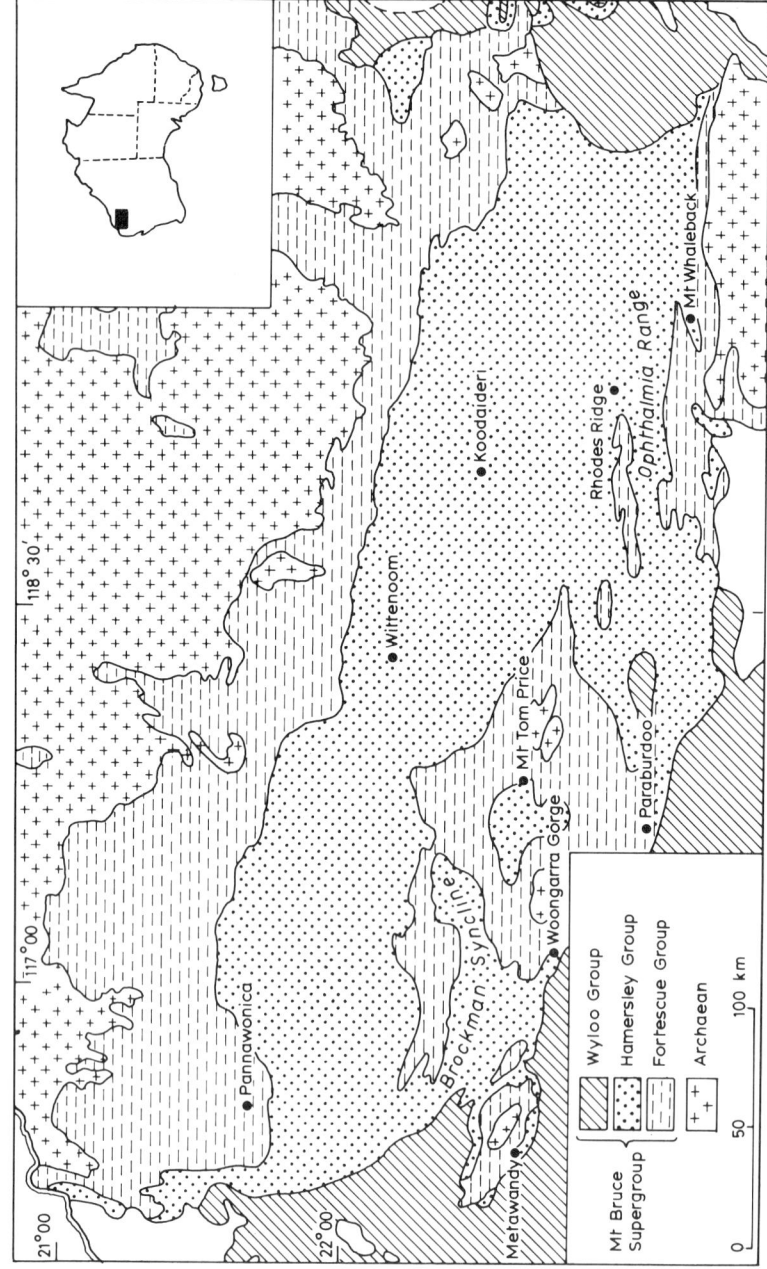

Fig. 8.6 Geological sketch map of the Hamersley Iron Province, Western Australia (from Ewers and Morris 1981).

Fig. 8.7 Mount Tom Price open pit (1977).

Fig. 8.8 Blasting at Mount Tom Price.

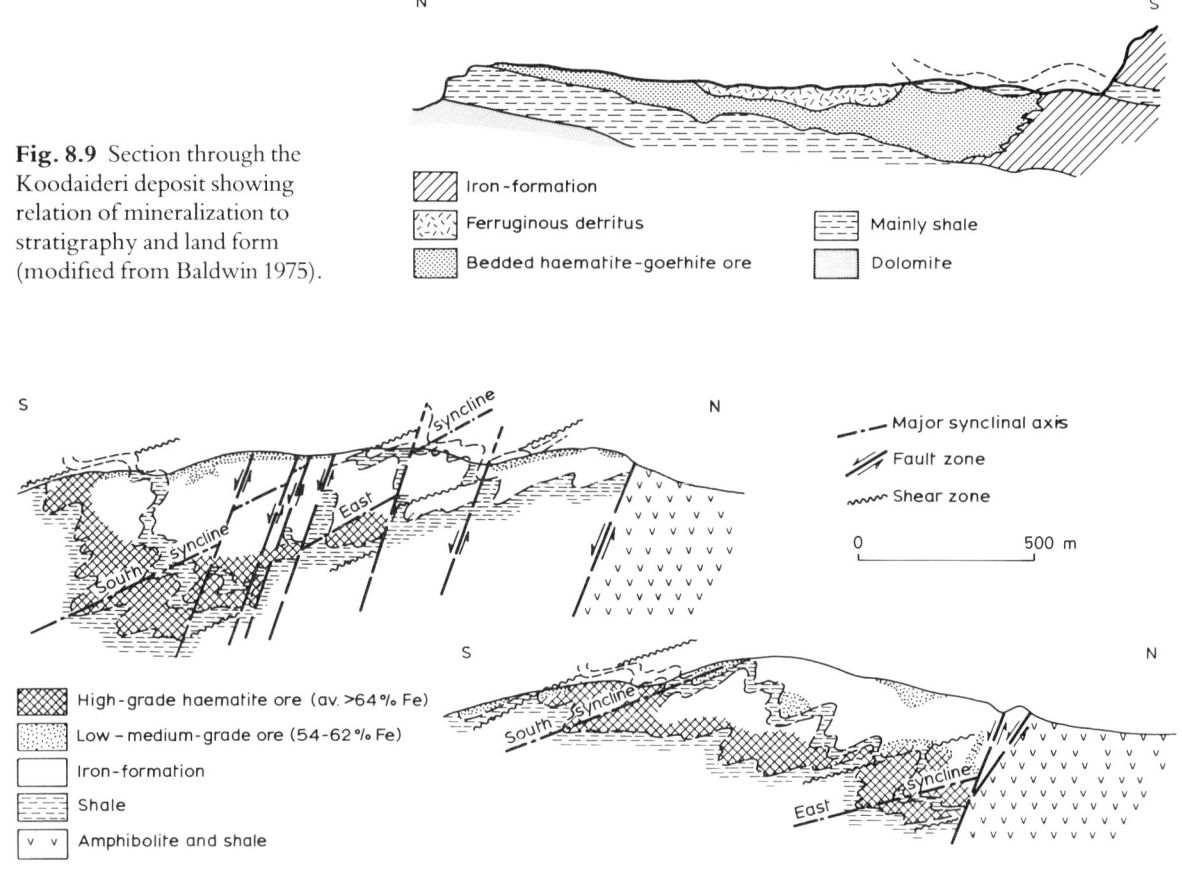

Fig. 8.9 Section through the Koodaideri deposit showing relation of mineralization to stratigraphy and land form (modified from Baldwin 1975).

Fig. 8.10 Cross-sections through the Mount Whaleback ore body showing the association of high-grade ore with synclinal structures (modified from Kneeshaw, 1975).

Morris (1985) divides the Hamersley ores into two classes: haematite–goethite ore, and haematite-rich ore characterized by the presence of microplaty haematite. The former class is dominant, but the latter is the most commercially valuable ore.

Both classes of ore are derived from banded iron-formation which is initially altered to martite–kenomagnetite–goethite ore by supergene enrichment (oxidation of magnetite, and iron-metasomatism of silicates and carbonates). Differing conditions and processes act upon this intermediate-stage ore, and lead to the development of the two main classes of high-grade ore. Most commonly, alteration of the kenomagnetite, by inversion to haematite, and hydration to goethite, leads to the formation of haematite–goethite ore. However, in some instances fine platy and irregular haematite grow within the goethite, and result in the formation of the high-grade haematite ores. Morris (1985) speculates that this class of ore may result from shallow burial, when temperatures of about 100°C are attained. However, the restricted distribution of this ore type indicates that factors additional to temperature must play an important role. Perhaps the localized channelling of deep meteoric water may have played a part in the development of the haematite-rich ores. In addition to the two main classes of iron ores, there are several subtypes which Morris (1985) believes have formed by dehydration and/or the solution, removal and reprecipitation of iron.

8.6.6 Genesis

The genetic model for newly discovered mineralization is always provisional, because observations are limited to surface outcrops, trenches and borehole cores. It is usual for ideas to evolve as more detailed information becomes available from mining and as researchers publish their results on the geochemistry and mineralogy of the ores. In the Hamersley district the close spatial and genetic association of the enriched ores with banded iron-formation has been widely recognized from the outset and has not been seriously challenged. The debate has focused on the timing and mechanism of the removal of large quantities of silica and the concurrent addition of iron to the parent iron-formation.

Campana (1966) recognized three main controls of iron enrichment: stratigraphic, structural and palaeoclimatic. The stratigraphic control was demonstrated by the concentration of major ore bodies within the Dales Gorge member of the Brockman iron-formation over a wide area. Structural control was initially inferred from the localization of ore bodies in axial flexures, and down-faulted segments of the Brockman iron-formation. Subsequently, Trendall (1975) noted that the range of associated structures was wider than at first thought, and he also stated that ore bodies are not always present in favourable structures. The influence of climate was assumed from the coincidence of the upper plane of the ore body, with a horizon which is not only the present-day topography, but one which is recognized as representing an old erosional surface of Tertiary age. This led initially to the conclusion that the ores were of supergene origin, with the leaching of silica being attributable to the copious rainfall, high temperature and variable water table in Tertiary times.

The most recent review of the evidence concerning the genesis of the supergene iron ores of the Hamersley Iron Province is provided by Morris (1980). There are two principal areas where concepts have changed since the mid-1960s. Firstly, the distinction between two classes of enrichment ore: haematite–goethite and haematite-rich ore (Morris, 1985). Secondly, the recognition that much of the enrichment ore is Precambrian in age. The best evidence for the age of the ore is based on the presence of fragments of haematite-rich ore within conglomerates at the base of the Wyloo Group which is dated at about 1800 million years.

Morris (1980) considered that the haematite–goethite ore and its derivatives can be explained by processes of supergene enrichment. The differences within these ore types largely reflect differing stages of maturity.

The genesis of the haematite-rich ore has proved a more contentious subject. Trendall (1975) in a review of the Hamersley Iron Province considered that the haematite-rich ores had formed at a temperature of 150°C and suggested burial to a depth of 4 km with a geothermal gradient of 30°C/km. Morris (1980) accepts that the depth of burial may have been as much as 4 or 5 km but does not believe it is necessary to invoke such a high temperature. He suggests, partly on thermodynamic criteria, that the temperature of ore formation has not exceeded about 100°C. The mineralogy of the haematite-rich ores may have been modified by the subsequent effects of the solution and redeposition of iron.

8.7 EXPLORATION

The discovery of the larger iron ore deposits in the 'New World' has followed a similar pattern, in spite of the widely differing conditions between the Amazon jungle of Brazil, and the sparsely vegetated terrain of the Hamersley area of Western Australia. In all cases the objectives have been to define the limits of the iron-formation and to quantify the ore reserves.

Marsden (1968) commented that in the early

years of exploration in the Lake Superior region, prior to the 1880s, 'the dip-needle for magnetic surveys and the pick and shovel were used effectively by the pioneer explorers.' Nearly a hundred years later in South America, Africa, India and Australia the essential tools for exploration have been the geological pick and geophysical equipment of more sophisticated design than the dip needle. In Brazil, Liberia and Australia geological mapping has formed the basis for exploration, owing to adequate exposure, and the recognition that the high-grade ores have well-defined stratigraphic and structural controls. In the Hamersley district of Australia the scale of preliminary geological mapping varied from 1:10 000 to 1:40 000 (W. G. Burns, personal communication).

Geological mapping is commonly supplemented by airborne and ground-based magnetic surveys. These have proved most effective in areas where magnetite is a major ore mineral. For example ground magnetic and seismic surveys were used successfully in exploration for new deposits in the Marquette Range, Lake Superior district, using a 25 × 50 ft (7.6 × 15.2 m) grid (T. Wentworth Slitor, personal communication). In the Hamersley district magnetic and gravity surveys have proved to be of limited value owing to the general absence of definitive signatures over the ore bodies (W. G. Burns, personal communication). In the evaluation of the Nimba deposit in Liberia, only limited use was made of magnetic and gravimetric techniques, because the rough terrain made interpretation difficult, and because good exposure made geophysics superfluous (P. Ericsson, personal communication).

8.8 EVALUATION

Drilling has proved to be an essential stage in the definition of iron ore reserves. The usual procedure is to define the geometry of the ore body with widely spaced holes, and then evaluate the quality of the ore with more detailed infill drilling and bulk sampling. Diamond drilling is normally used in the preliminary stages of evaluation and recourse is sometimes made to percussion drilling at a later stage in the programme. The grid size selected for drilling varies according to the ore type. For example the Empire ore body, Marquette Range, was evaluated by drilling vertical holes on a 600 ft (182.8 m) grid which was later reduced to 300 ft (91.4 m). Test work was carried out on 45 ft (13.7 m) composite samples (T. Wentworth Slitor, personal communication).

In the evaluation of enriched ores, where the specifications of chemical composition are more rigorous, a more detailed spacing of boreholes is required. For example, the drilling of the Tom Price and Paraburdoo deposits, in the Hamersley district, was initially carried out on a 400 × 160 m grid, which was subsequently narrowed to 60 × 60 m spacing, with samples collected for analysis at 2 m intervals (W. G. Burns, personal communication).

Normally the core is split, with half being sent for chemical analysis for the following constituents: Fe, FeO, Fe_2O_3, SiO_2, Al_2O_3, P, Mn and loss on ignition. The Fe content represents the grade of the deposit. The relative proportions of ferrous and ferric iron must be ascertained because they react differently in the subsequent processing of the ore. In general, the oxidized ore is preferable, because it is more readily reduced to iron in the blast furnace (R. H. Parker, personal communication).

Mn at high concentrations is a contaminant, but at low levels may be beneficial in that it provides a natural alloying element, thus reducing the amount of ferromanganese which would have to be added in the manufacture of most types of steel (R. H. Parker, personal communication). SiO_2, Al_2O_3 and P are contaminants, as are CO_2 and H_2O which are measured by the loss on ignition.

Definition of the size and quality of the ore body is usually followed by the collection of large bulk samples for metallurgical test work. Some of this material may have been obtained at an early stage from the drill core, but usually adits or tunnels are driven into the ore body at differing depths to obtain large representative samples. In the evaluation of the Carajas district, Brazil, 3400 m of tunnelling was carried out in the final stages of evaluation. In addition, a 108 m raise was driven from the tunnel to the surface along the axis of a diamond drill hole. This was partly to check the reliability of the information obtained from drilling, and partly to enable samples to be collected for physical testing (Ruff et al., 1974). Some of the problems identified during metallurgical test work on the Marra Mamba iron ore deposit Western Australia are described by Slepecki (1981).

When iron ore production commences the mine geologist is primarily concerned with grade control. In the Hamersley district careful grade control is required because the shale macrobands contain contaminants, such as Al_2O_3 and P, and must be mined separately from the ore. At Tom Price and Paraburdoo Mines open pit mapping is carried out at a scale of 1:1000 (W. G. Burns, personal communication). The Nimba deposit, Liberia, has a very inhomogeneous ore, and careful grade control is required which is based on geological mapping at a scale of 1:500 (P. Ericsson, personal communication).

In some iron ore deposits the combination of geological mapping and drilling may prove to be an inadequate basis for grade control. O'Leary (1979) has described the improved reliability of ore prediction which arose from an integrated geological and geostatistical approach at the Scully mine, near Wabush in SE Labrador. Initially the stratigraphy and structure were based on conventional mapping and petrographic studies, which were interpreted in terms of three separate ore-bearing horizons, deformed by a simple style of folding. The preliminary interpretation did not adequately explain the distribution of iron and manganese, and furthermore a large discrepancy was found to exist between the grade of ore estimated and that obtained from the mill.

For the statistical study four parameters were selected: the iron and manganese in the original sample, the concentrate weight at 6% SiO_2, and the corresponding Mn assay. The plotting of these data in the form of frequency diagrams indicated trimodality, which was found to persist even when the data were restricted to one ore-bearing horizon. As a result a detailed re-examination of borehole information was carried out, with the eventual recognition of eleven separate units within the stratigraphy of the mine. More detailed mapping also revealed two separate phases of folding and two sets of faults, which post-date the folding.

In order to investigate the chemical variation within the eleven stratigraphic units the system was unfolded to form a series of maps, which were later used as a basis for trend surface analysis. A series of polynomial trend surfaces were generated on the eleven ore units, and the first-order trend surfaces were found to be statistically acceptable as models for grade distribution.

8.9 CONCLUDING STATEMENT

In this chapter we have attempted to demonstrate the rapid change in the type of iron ore which has been mined during the last twenty-five years. During this period the World has witnessed a dramatic change towards the exploitation of iron-formation, and in particular the enrichment ores derived from it. At the same time there has been a change in the pattern of production. Production has declined from the iron ore deposits of Europe and North America, and new mining districts have been developed in Africa and Australia. One consequence for the geologist is that the ore deposits which have been most intensively studied and described are now of secondary importance. Furthermore the studies on ore deposits in Australia have called into question much of the conventional wisdom of the last thirty years. As a result many fundamental aspects such as nomenclature and classification are still unresolved.

To what extent has our knowledge of the geology of iron ore deposits aided in their exploration? In our opinion there has been a tendency for geologists to concentrate too much attention on the genesis of Precambrian banded iron-formation. This is a fascinating area of research but the various speculations which have been advanced have provided little aid to the exploration geologist. The identification of new reserves of Precambrian banded iron-formation can largely be attributed to geological mapping and geophysics.

In contrast the genesis of the enrichment ores which are now recognized as being of prime economic importance has received scant attention. It is now recognized that these high-grade ores are not the uniform products of supergene enrichment. Morris (1980, 1985) has classified the supergene ores of the Hamersley Iron Province into two classes, and it is probable that similar distinctions will become evident from other major mining districts. The classification and distribution of the main classes of supergene ore are of pressing economic importance, because haematite-rich ores are in increasing demand by the steel industry. What are the factors which control the distribution of valuable haematite-rich supergene ores? This is an area where genetic models based on theoretical studies, laboratory experiments, interpretation of mineral paragenesis and geological mapping would be of value. However, such an approach is still at an early stage.

REFERENCES

Baldwin, J. T. (1975) Paraburdoo and Koodaideri ore deposits and comparisons with Tom Price iron ore deposits, Hamersley Iron Province. In *Economic Geology of Australia and Papua New Guinea. 1. Metals. Monograph Series 5* (ed. C. L. Knight), Australasian Institute of Mining and Metallurgy, Parkville, Victoria, pp. 898–905.

Bates, R. L. and Jackson, J. A. (1980) *Glossary of Geology* 2nd edn, American Geological Institute, Falls Church, Virginia, 749 pp.

Belevtsev, Ya. N., Belevtsev, R. Ya. and Siroshtan, R. I. (1983) The Krivoi Rog Basin. In *Iron Formation: Facts and Problems* (eds A. F. Trendall and R. C. Morris), Elsevier, Amsterdam, pp. 211–249.

Beukes, N. J. (1973) Precambrian iron-formations of Southern Africa. *Econ. Geol.*, **68**, 960–1004.

Beukes, N. J. (1983) Palaeoenvironmental setting of iron-formations in the depositional basin of the Transvaal Supergroup, South Africa. In *Iron Formation: Facts and Problems* (eds A. F. Trendall and R. C. Morris), Elsevier, Amsterdam, pp. 131–198.

Campana, B. (1966) Stratigraphic, structural and palaeoclimatic controls of the newly discovered iron ore deposits of Western Australia. *Miner. Deposita*, **1**, 53–59.

Canada. Energy Mines and Resources (1976) *Iron Ore*, Mineral Policy Series MR 148, 35 pp.

Dimroth, E. (1977) Facies models – diagenetic facies of iron formation. *Geosci. Can.*, **4**, 83–88.

Dimroth, E. and Kimberley, M. J. (1976) Precambrian atmospheric oxygen: evidence in the sedimentary distribution of carbon, sulphur, uranium and iron. *Can. J. Earth Sci.* **13(9)**, 1161–1185.

Dorr, J. V. N., II and Barbosa, A. L. de Miranda (1963) Geology and ore deposits of the Itabira District, Minas Gerais, Brazil. *U.S. Geol. Surv. Prof. Pap.*, **341-C**, 110 pp.

Eichler, J. (1976) Origin of the Precambrian banded iron-formations. In *Handbook of Strata-bound and Stratiform Ore Deposits* (ed. K. H. Wolf), Elsevier, Amsterdam, Vol. 7, pp. 157–197.

Ewers, W. E. (1983) Chemical factors in the deposition and diagenesis of banded iron-formation. In *Iron Formation: Facts and Problems* (eds A. F. Trendall and R. C. Morris), Elsevier, Amsterdam, pp. 491–510.

Ewers, W. E. and Morris, R. C. (1981) Studies of the Dales Gorge Member of the Brockman Iron-Formation, Western Australia. *Econ. Geol.*, **76**, 1929–1953.

Gilhome, W. R. (1975) Mount Tom Price ore body, Hamersley Iron Province. In *Economic Geology of Australia and Papua New Guinea. 1. Metals. Monograph Series 5* (ed. C. L. Knight) Australasian Institute of Mining and Metallurgy, Parkville, Victoria, pp. 892–897.

Goldich, S. S. (1973) Age of Precambrian banded iron-formations. *Econ. Geol.*, **68**, 1126–1134.

Goodwin, A. M. (1956) Facies relations in the Gunflint iron-formation. *Econ. Geol.*, **51**, 565–595.

Goodwin, A. M. (1973) Archaean iron-formations and tectonic basins of the Canadian Shield. *Econ. Geol.*, **68**, 915–930.

Gross, G. A. (1965) Geology of iron deposits in Canada. *Geol. Surv. Can. Econ. Geol. Rep.*, **22**.

Gruss, H. (1973) Itabirite iron ores of the Liberia and Guyana shields. *UNESCO Earth Sci. Ser.*, **9**, 335–359.

Hamersley Iron: Resources, Technology, Operations (1981) Company booklet produced by Hamersley Iron (Pty) Ltd.

Holland, H. D. (1973) The oceans: a possible source of iron in iron-formations. *Econ. Geol.*, **68**, 1169–1172.

Hough, J. L. (1958) Fresh-water environment of deposition of Precambrian banded iron-formations. *J. Sediment. Petrol.*, **28**, 414–430.

Huber, N. K. (1959) Some aspects of the origin of the Ironwood iron-formations of Michigan and Wisconsin. *Econ. Geol.*, **54**, 82–118.

Hughes, G. E. F. and Hughes, P. D. (1979) Brazil iron ore. *Austr. Min.*, **71(4)**, 42–67.

James, H. L. (1954) Sedimentary facies of iron-formation. *Econ. Geol.*, **49**, 235–293.

James, H. L. (1983) Distribution of banded iron-formation in space and time. In *Iron Formation: Facts and Problems* (eds A. F. Trendall and R. C. Morris), Elsevier, Amsterdam, pp. 471–486.

Kimberley, M. K. (1978) Palaeoenvironmental classification of iron-formations. *Econ. Geol.*, **73**, 215–229.

Kneeshaw, M. (1975) Mt. Whaleback iron ore body, Hamersley Iron Province. In *Economic Geology of Australia and Papua New Guinea. 1. Metals. Monograph Series 5* (ed. C. L. Knight), Australasian Institute of Mining and Metallurgy, Parkville, Victoria, pp. 910–915.

Lepp, H. and Goldich, S. (1964) Origin of Precambrian iron-formations. *Econ. Geol.*, **59**, 1025–1060.

Marsden, R. W. (1968) Geology of the Iron ores of the Lake Superior Region in the United States. In *Ore Deposits of the United States 1933–1967. The Graton-Sales Volume* (ed. J. D. Ridge), American Institute of Mining, Metallurgical and Petroleum Engineers, New York, pp. 489–507.

Mitchell, A. H. G. and Garson, M. S. (1981) *Mineral Deposits and Global Tectonic Settings*, Academic Press, New York, 405 pp.

Moore, E. S. and Maynard, J. E. (1929) Solution, transportation and precipitation of iron and silica. *Econ. Geol.*, **24**, 272–303, 365–402, 506–527.

Morey, G. B. (1983) Animikie Basin, Lake Superior Region U.S.A. In *Iron Formation: Facts and Problems* (eds A. F. Trendall and R. C. Morris), Elsevier, Amsterdam, pp. 13–60.

Morris, R. C. (1980) A textural and mineralogical study of the relationship of iron ore to BIF in the Hamersley Iron Province of W. Australia. *Econ. Geol.*, **75**, 184–209.

Morris, R. C. (1985) Genesis of iron ore in banded iron-formation by supergene and supergene metamorphic processes – a conceptual model. In *Handbook of Stratabound and Stratiform Ore Deposits* (ed. K. H. Wolf) Vol. 12, Chapter 12, Elsevier, Amsterdam.

Morris, R. C. and Horwitz, R. C. (1983) The origin of the iron-formation rich Hamersley Group of Western Australia - deposition on a platform. *Precambr. Res.*, **21**, 273–297.

Ohle, E. L. (1972) Evaluation of iron ore deposits. *Econ. Geol.*, **67**, 953–964.

O'Leary, J. (1979) Ore reserve estimation methods and grade control at the Scully Mine, Canada. *Min. Mag.*, April, 300–314.

Peterson, E. C. (1980) Iron ore. In *Minerals Facts and Problems*, 1980 edn, United States Department of the Interior Bulletin, 671, pp. 433–454.

Ruff, A., Tremaine, J. H., Bernardelli, A. L., Beisiegel, V. and Drummond, N. F. (1974) Exploration of one of the largest iron ore districts; Brazil's Serra dos Carajas. *Min. Eng.*, January, pp. 30–33.

Slepecki, S. (1981) Marra Mamba iron ore – a case study in exploration and development of a new ore type. In *Sydney Conference 1981 Parkville, Victoria*, Australasian Institute of Mining and Metallurgy, Parkville, Victoria, pp. 195–207.

Smith, R. E., Perdrix, J. L. and Parks, T. C. (1982) Burial metamorphism in the Hamersley Basin, Western Australia. *J. Petrol.*, **23(1)**, 75–102.

Sokolov, G. A. and Grigor'ev, V. M. (1977) Deposits of iron. In *Ore Deposits of the USSR* (ed. V. I. Smirnov), Pitman Publishing Ltd, London. pp. 7–109.

Stanton, R. L. (1972) *Ore Petrology*, McGraw-Hill, London, 713 pp.

Stephenson, R. L. and Smailer, R. M. (eds) (1980) *Direct Reduced Iron – Technology and Economics of Production and Use*, Iron and Steel Society of AIME, Warrendale, Pennsylvania.

Strassburger, J. H. (ed.) (1969) *Blast Furnace – Theory and Practice*, Vols I and II, Gordon and Breach Science Publishers, New York.

Tolbert, G. E., Tremaine, J. W., Melcher, G. C. and Gomes, C. B. (1971) The recently discovered Serra dos Carajas Iron deposits, Northern Brazil. *Econ. Geol.*, **66**, 985–994.

Trendall, A. F. (1965) Origin of Precambrian iron-formations. Discussion of paper by Lepp and Goldich *Econ. Geol.*, **60**, 1065–1070.

Trendall, A. F. (1975) The Hamersley Basin – regional geology. In *Economic Geology of Australia and Papua*

New Guinea 1. Metals. Monograph Series 5 (ed. C. L. Knight) Australasian Institute of Mining and Metallurgy, Parkville, Victoria, pp. 411–412.

Trendall, A. F. (1983a) Introduction. In *Iron Formation: Facts and Problems* (eds A. F. Trendall and R. C. Morris), Elsevier, Amsterdam, pp. 1–11.

Trendall, A. F. (1983b) The Hamersley Basin. In *Iron Formation: Facts and Problems* (eds A. F. Trendall and R. C. Morris), Elsevier, Amsterdam, pp. 69–129.

Trendall, A. F. and Blockley, J. G. (1968) Stratigraphy of the Dales Gorge Member of the Brockman Iron-Formation in the Precambrian Hamersley Group of Western Australia. *W. Austr. Geol. Surv. Annu. Rep.* 1967, 48–53.

9

Uranium ores of sedimentary affiliation

9.1 INTRODUCTION

Uranium is used as a fuel mineral for nuclear power, and in the manufacture of nuclear weapons. In this chapter we begin by briefly reviewing the adequacy of uranium resources to supply an expanding nuclear industry and then consider the relative importance of the principal deposit types. The diversity of uranium occurrences precludes a detailed consideration of each deposit type but the reader is referred to Nash et al. (1981) for an excellent overview of the subject. We have selected unconformity and sandstone-type deposits for particular consideration. The reasons are partly because together they constitute more than 60% of World resources but also because they serve to illustrate contrasting approaches to exploration. Uranium contained in quartz–pebble conglomerates is discussed in Chapter 5.

In 1983 non-Communist World production of uranium metal was 37 400 tonnes (*Mining Annual Review*, 1984). The principal producing countries were the USA, Canada, South Africa, Namibia and Australia. Production is forecast to grow slowly to 42 500 tonnes in 1990, when the major producing countries will be Canada and Australia (*Mining Annual Review*, 1983, 1984).

Uranium reserves and resources are estimated using several different cost levels at which the ore might be mined. Normally the cost levels used are <US$80/kg of U and US$80–130/kg of U. In addition uranium resources are subdivided into reasonably assured resources and estimated additional resources. The term 'reasonably assured resources' is roughly equivalent to the term 'measured and indicated reserves' conventionally used by the mining industry (Fig. 1.2). Estimated additional resources are based on geological evidence, for example, deposits believed to exist along well-defined geological trends (Patterson, 1980). An authoritative estimate using these criteria is shown in Table 9.1.

Table 9.1 Uranium resources (from Uranium, 1983)

Cost of production	Reasonably assured resources (tonnes × 10^3)	Estimated additional resources (tonnes × 10^3)
<US$80/ kg of U	1468	914
>US$80< 130/ kg of U	575	308

Ore deposit geology and its influence on mineral exploration

Fig. 9.1 Location map.

To what extent are these resources adequate to supply the demand for uranium during the next 25 years? Basically the question is unanswerable, because the forecasts for energy consumption and in particular nuclear energy requirements are invariably incorrect. In addition it is impossible to forecast the size of future discoveries. This is well illustrated by the discovery of Olympic Dam (Fig. 9.1) in 1976 in an area without any previously known uranium potential (Haynes, 1979). The present indicated reserves at Olympic Dam are 1 million tonnes which make it the largest uranium resource in the Western World (*The Australian*, 7.7.83). Most forecasts indicate that uranium supplies are adequate to support the predicted growth in nuclear energy up to the year 2000 (Bowie, 1981; Uranium, 1983).

Table 9.2 shows the relative importance of differing categories of uranium deposit. However this type of information is liable to change very rapidly as existing deposits become depleted and new discoveries are made. Indeed the discovery of Olympic Dam will require the addition of a new deposit type to the inventory (Roberts and

Uranium ores of sedimentary affiliation

Hudson, 1983). It is possible to predict that during the late 1980s sandstone-type deposits will decline in importance and an increasing proportion of production will be derived from unconformity deposits in Canada and Australia.

9.2 GEOCHEMISTRY OF URANIUM IN THE SECONDARY ENVIRONMENT

Detailed theoretical studies combined with extensive field observations have enabled geochemists to explain the factors controlling uranium mobility in the secondary environment. In essence uranium geochemistry can be described in terms of the reduced form U:(IV) and the oxidized form U:(VI). Langmuir (1978) has demonstrated that the uranous ion U^{4+} can only reach significant concentrations as the fluoride complex, in reduced groundwaters, which have a pH below 3. As these conditions are uncommon in the natural environment, tetravalent uranium occurs principally as the minerals uraninite and coffinite, although the uranous titanate, brannerite, is locally important (Nash et al., 1981).

Table 9.2 Relative importance of deposit types (modified from Nash et al., 1981)

Type of deposit	Typical grade (% U_3O_8)	Percentage of total	Principal countries
Quartz-pebble conglomerate	0.1–0.2 / 0.03	20	Canada / South Africa
Unconformity	0.2–4.0	16	Canada and Australia
Ultrametamorphic	0.02–0.25	6	Namibia
Classical vein	0.1–1.0	5	France, Canada
Igneous associated	0.02–0.3	6	Brazil, Greenland
Sandstone	0.10–0.35	45	USA, Niger, Argentina
Calcrete	0.1–0.3	2	Australia, Namibia

The stability of uraninite in the secondary environment has been investigated by Hostetler and Garrels (1962) and Langmuir (1978). Figure 9.2 shows the stability field of uraninite, and clearly demonstrates that interaction between uraninite and either groundwater or rain water will result in oxidation to uranyl complexes. The precise nature of the uranyl complexes which are dominant is a matter for debate (Langmuir, 1978; Tripathi, 1979).

The solubilities of uranyl minerals under most conditions form a generalized series of decreasing solubility: carbonates > sulphates > phosphates and arsenates > silicates > vanadates (Nash et al., 1981). As a result the most common uranyl minerals are the vanadates, carnotite and tyuyamunite, the phosphates of the autunite group and the silicate uranophane.

In addition to being precipitated as discrete mineral phases, uranyl ions can also be removed from solution by adsorption. Organic matter, clays, zeolites and iron hydroxides and oxyhydroxides are the main adsorbents. Humic materials, formed by the breakdown of plant debris, are the most important type of organic adsorbent. As a result the 'A horizon' of podzolic soils, peats, lignites and organic-rich swamps and lake sediments is often found to be associated with abnormally high uranium contents. In the context of the formation of sedimentary uranium deposits it is important to understand the precise role of organic matter. The results of current investigations indicate that the main role of organic matter is in adsorbing the uranyl ion (Nash et al., 1981). Reduction to the uranous ion takes place more slowly, and may be influenced by several factors. Often organic matter is spatially associated with uranium deposits, but is not itself enriched in uranium. This suggests that other adsorbents may be effective. Giblin (1980) has demonstrated experimentally that uranium is adsorbed on kaolinite, particularly within the pH range 6.5–8.5. It is probable that such clays as montmorillonite are even more effective adsorbents. We are not aware of experimental studies on the adsorption of uranyl ions by ferric hydroxides, but the evidence of uraniferous limonite suggests it is an effective barrier to the mobility of uranium.

It is clear that uranium mobility may be restricted either by the formation of uranyl minerals or by adsorption. Most sedimentary uranium ores are characterized by uraninite and coffinite, and therefore reduction to the uranous ion is an important aspect of ore genesis. Reduced

Uranium ores of sedimentary affiliation

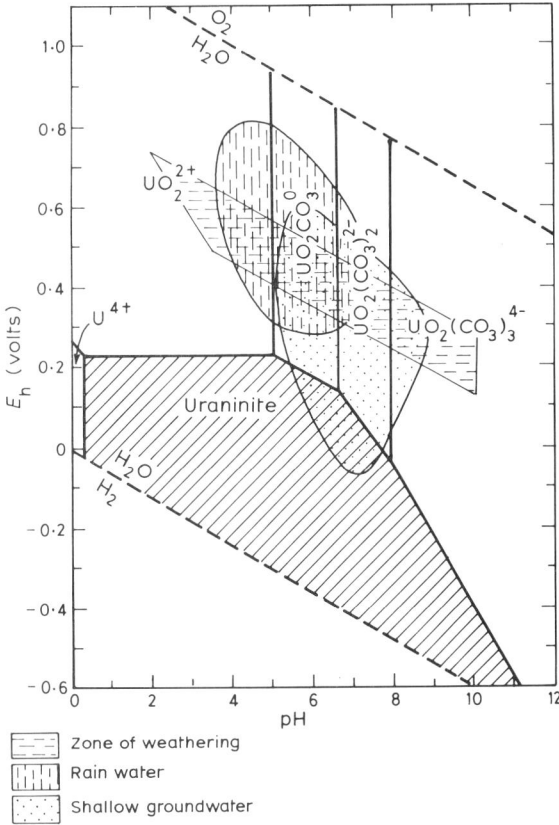

Fig. 9.2 Stability fields of uranium species in the surface environment (modified from Langmuir 1978). E_h–pH diagram in the U–O$_2$–CO$_2$–H$_2$O system at 25°C for $pCO_2 = 10^{-2}$ atm. Uraninite, UO$_2$ complexes, solution boundaries are drawn at 10^{-6} M (0.24 ppm) dissolved uranium species. E_h and pH conditions of geochemical environments from Hansuld (1967).

9.3 UNCONFORMITY-TYPE URANIUM DEPOSITS OF THE NORTHERN TERRITORY, AUSTRALIA, AND NORTHERN SASKATCHEWAN, CANADA

9.3.1 Introduction

Unconformity-type uranium deposits represent a new class of uranium ore, which was not recognized until the early 1970s. Although the Rum Jungle deposit was discovered near Darwin, in the Northern Territory of Australia in 1949, the potential of the East Alligators River area was not appreciated until the late 1960s (Needham and Roarty, 1980). At about the same time the discovery of the Rabbit Lake deposit, in Canada, initiated a surge of exploration in Saskatchewan which has resulted in the discovery of 21 new ore bodies (Saracoglu *et al.*, 1983). It is evident that two new mineralized districts have been identified, which together constitute about 16% of the World's uranium reserves (Nash *et al.*, 1981). The proximity of all the mines and prospects in both areas to a conspicuous Mid-Proterozoic unconformity has led to the designation of this style of ore deposit as being of unconformity type (Derry, 1973). Whilst the geological setting of both mining districts is very similar (Table 9.3) there are major differences in the genetic models which appear most appropriate. This has in turn resulted in differing exploration philosophies which must be put into practice in widely contrasting climatic regimes.

9.3.2 Geological setting

(a) Stratigraphy and host rock lithology

(i) Northern Saskatchewan

The Canadian Shield in northern Saskatchewan consists of Archaean granitoid basement rocks, and early Proterozoic metasediments, unconformably overlain by sandstones of the Athabasca Formation. The Lower Proterozoic sequence is divided into lithostructural domains, of which the

sulphur species are considered to be the most important reducing agents, although there is disagreement about the derivation of the sulphur. Some authorities consider that the sulphur is derived by bacteriogenic reduction of sulphate (Rackley, 1976), but the controlled oxidation of pyrite has also been suggested as a mechanism for the production of reduced sulphur species (Granger and Warren, 1969).

Table 9.3 Important characteristics of unconformity-type uranium deposits

Geological characteristics	N. Saskatchewan, Canada	Northern Territory, Australia
Host lithology	Lower Proterozoic: graphitic metapelites, calcareous metasediments and quartzites. Middle Proterozoic: quartz sandstone and conglomerate	Lower Proterozoic: pelites (sometimes graphitic) adjacent to massive carbonate rocks. Middle Proterozoic: quartz sandstone
Mineralogy	Mineralogy is complex and highly variable	Simple mineralogy
Structure	Mineralization is fault- and fracture-controlled. Reverse faults important	Mineralization associated with faulting and collapse breccias
Relationship to Archaean granite	No spatial association	Important spatial and probably genetic relationship
Wall-rock alteration	Ubiquitous chloritization	Ubiquitous chloritization
Relationship between ore bodies and unconformity	Ore bodies occur either a few hundred metres below or immediately above the unconformity	No ore bodies occur above the unconformity
Regional setting of the ore bodies	Ore bodies are always spatially associated with the Mid-Proterozoic unconformity but they occur within a range of Lower Proterozoic metasediments	Stratabound within the Cahill and Masson Formations near to Archaean basement.

Wollaston domain is most commonly associated with uranium ore deposits (Fig. 9.3). Hoeve and Sibbald (1978a) emphasized the sedimentary parentage of the Wollaston domain, which may be subdivided into meta-arkose and quartz-amphibole units. These lithologies are interpreted as having been deposited in a shallow-water marginal marine environment. The metasediments have been subjected to four phases of deformation and amphibolite-facies conditions of metamorphism and, as a consequence, many geologists would describe the rocks as migmatites (Nash *et al.*, 1981). Graphitic metasediments are the most important host rocks for uranium deposits but calcareous metasediments and quartzites may also be significant (Hoeve and Sibbald, 1978b).

The highly deformed and metamorphosed Lower Proterozoic sediments are overlain unconformably by the Athabasca Formation, which is of Middle Proterozoic age. The Athabasca Formation consists of up to 1750 m of fluviatile sandstones. Below the unconformity is an extensive regolith, which is interpreted as resulting from lateritic conditions of weathering prior to the deposition of the Athabasca Formation (Hoeve and Sibbald, 1978a).

The uranium mineralization may be confined to

Uranium ores of sedimentary affiliation

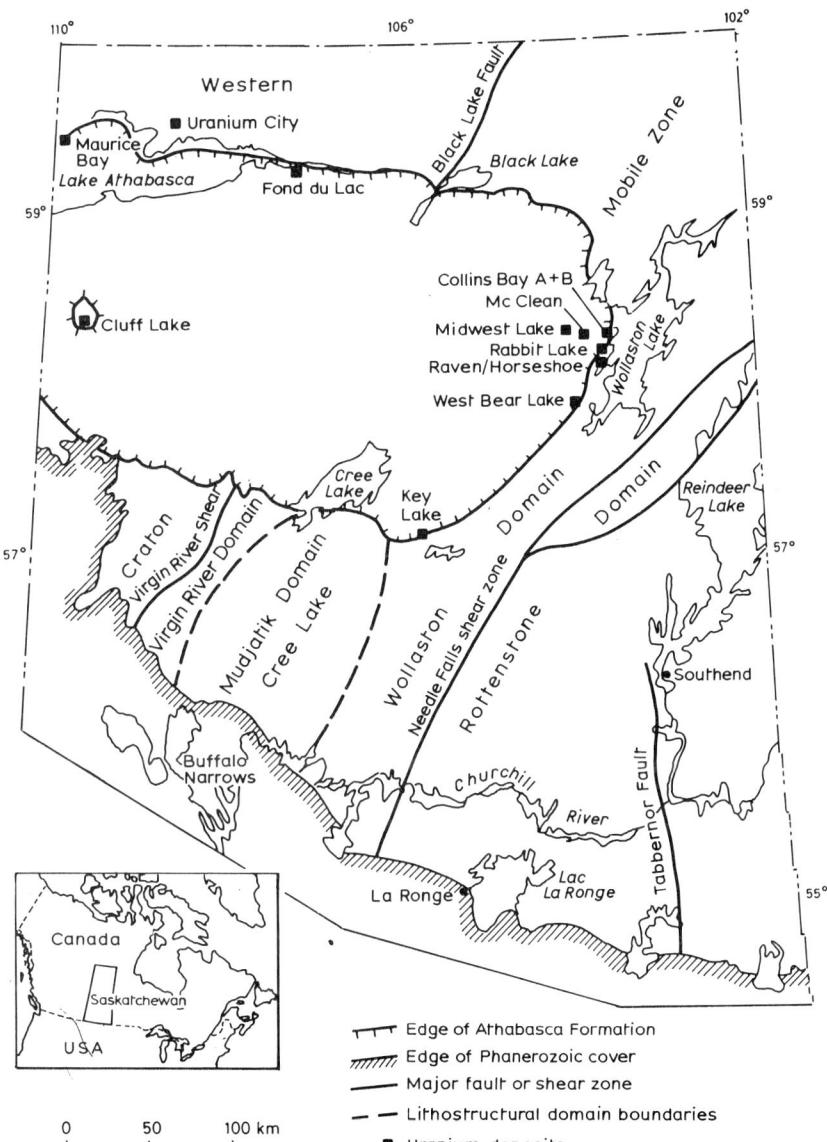

Fig. 9.3 Major geological subdivisions of the Canadian Shield in northern Saskatchewan, and location of selected pitchblende deposits (modified from Hoeve and Sibbald, 1978b).

Lower Proterozoic metasediments below the unconformity (e.g. Rabbit Lake) but may also extend into the overlying Athabasca Formation (e.g. Key Lake, Fig. 9.4). The ore deposits are fault- and fracture-controlled, and reverse faults which trend parallel to the foliation of the Lower Proterozoic metasediments are particularly significant (Hoeve and Sibbald, 1978b).

(ii) Northern Territory

In the Australian Northern Territory uranium deposits are located within the Pine Creek geosyncline (Fig. 9.5). The geosyncline had an intracratonic setting and formed by rifting of the Archaean crust. Seven structural units are identified of which the most important are the

Fig. 9.4 Generalized cross-section of the Deilmann ore body, Key Lake, Saskatchewan (modified from Dahlkamp 1978a).

South Alligator trough and adjoining shelf areas (Stuart-Smith *et al.*, 1980). The Lower Proterozoic metasediments consist principally of an alternating sequence of psammites and pelites. The uranium deposits are stratabound and are hosted by the Cahill and Masson Formations of the Namoona Group (Needham and Roarty, 1980). The Cahill Formation consists principally of quartz-schists, carbonaceous schists, dolomite–magnesite and calc-silicate gneisses. The Masson Formation comprises mainly ferruginous shale. The main lithology associated with uranium mineralization

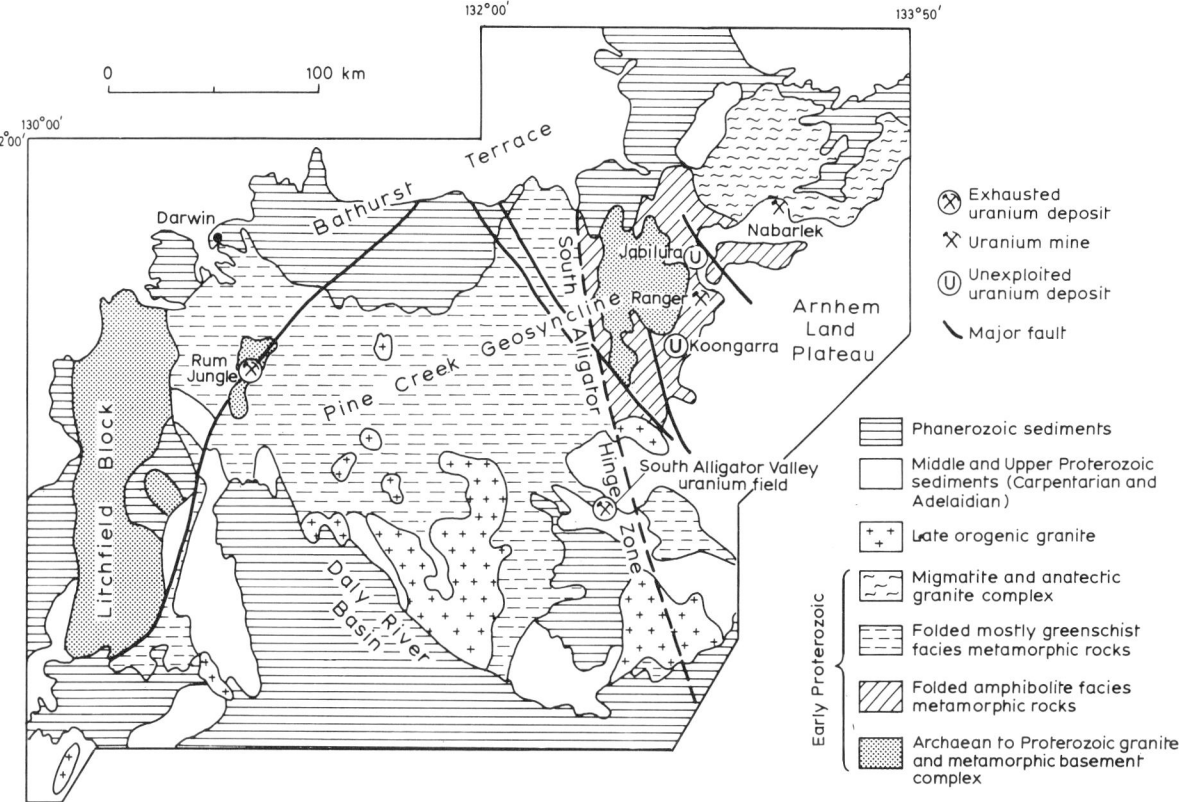

Fig. 9.5 Geological setting of uranium deposits in the Pine Creek geosyncline (modified from Ferguson and Goleby 1980, reproduced with permission from the Director, Bureau of Mineral Resources, Geology and Geophysics, Canberra, Australia).

is carbonaceous pelites, adjacent to massive carbonate rocks (Needham et al., 1980). However, there is an antipathetic relationship with carbonate in the ore zone (Hegge et al., 1980). Regional metamorphism within the Pine Creek geosyncline is divided into two provinces by the South Alligator Hinge zone (Ferguson, 1980). To the west of this divide the rocks are of low metamorphic grade, whilst to the east granulite and amphibolite facies of metamorphism prevail (Fig. 9.5). The Lower Proterozoic metasediments rest unconformably on an Archaean basement which includes two-mica granites. The granites contain accessory uraninite, and are considered to be the probable source of uranium (Needham and Roarty, 1980). The proximity of mineralization to the Archaean basement is an important aspect of the geological setting of uranium in the Pine Creek geosyncline.

The Lower Proterozoic metasediments are overlain unconformably by the Kombolgie Formation, which is of Mid-Proterozoic age (Fig. 9.6). The Kombolgie Formation consists of a sequence of well-sorted quartz sandstones. Immediately below the unconformity the Lower Proterozoic sediments have been exposed to a period of intensive chemical weathering.

The uranium deposits in the Pine Creek geosynclines are all confined to Lower Proterozoic metasediments below the unconformity and no uranium has been recorded from the Kombolgie Formation.

Ore deposit geology and its influence on mineral exploration

Fig. 9.6 The Ranger 1 Mine Northern Territory, Australia (1983). Background: sandstones of the Kombolgie Formation. Foreground: open pit temporarily filled with water.

(b) Mineralogy and wall-rock alteration

(i) Saskatchewan

The uranium deposits of northern Saskatchewan are characterized by their diversity of mineral species and complex mineral paragenesis. The grade of ore is high compared with other classes of uranium deposit and typically ranges from 0.3 to 4% U_3O_8, although localized concentrations of ore may exceed 59% U_3O_8 (Dahlkamp, 1978a). The most detailed geological studies have been carried out at Rabbit Lake, where three stages of mineralization are recognized (Hoeve and Sibbald, 1978a). Stage 1 is characterized by veins of pitchblende and coffinite, with a wide range of accessory minerals. Stage 2 comprises euhedral quartz in veins which normally contain little uranium, and usually have a simple mineralogy.

Stage 3 mineralization consists mainly of impregnations of sooty pitchblende and coffinite. The mineralization is both preceded by and coeval with extensive wall-rock alteration, which indicates changing redox conditions during mineral deposition. The wall-rock alteration associated with stage 1 mineralization is characterized by chlorite, with pervasive haematization, which indicates an oxidizing ore fluid during part of the period of ore deposition. The colour of stage 2 gangue minerals indicates that conditions were mainly reducing, and the abundance of pale green chlorite alteration associated with stage 3 indicates the persistence of reducing conditions.

The mineralogy of the Key Lake deposit has been described by Dahlkamp (1978a). The deposit differs from Rabbit Lake in that nickel-bearing

346

species are abundant in addition to pitchblende and coffinite. Gersdorffite is the dominant nickel mineral accompanied by lesser quantities of millerite, niccolite, rammelsbergite and bravoite. Kaolinite and Fe-rich chloritic alteration zones are associated with the uranium ore bodies.

(ii) Northern Territory

The ore deposits of the Pine Creek geosyncline have a simple mineralogy and a uniform style of wall-rock alteration. The average grade of the deposits varies from 0.20% at Ranger 1 to 0.39% U_3O_8 at Jabiluka 2.

The primary uranium ore mineral is uraninite. Other uranium minerals are of limited occurrence but include coffinite, brannerite, thucholite and amorphous mixtures of uranium with titania and phosphates (Ewers and Ferguson, 1980).

Chlorite is the only gangue mineral which is common to all the deposits. The degree of chloritization ranges from minor in the Rum Jungle deposits to massive and pervasive in the Alligator River deposits (Needham and Roarty, 1980). Chloritization tends to be intense within the vicinity of the deposit but is rare in rocks which are distant from the mineralization (Needham and Roarty, 1980).

Haematization is associated with uraninite at Nabarlek, but is not characteristic of the other deposits. The mineralization in the Pine Creek geosyncline therefore appears to have been largely deposited under uniformly reducing conditions.

Secondary uranium minerals are of some economic significance. At Ranger 1 the zone of weathering generally extends down to about 18 m (Eupene *et al.*, 1975), and contains principally the phosphate saleeite, with lesser quantities of sklodowskite, gummite and metatorbernite (Fig. 9.7).

(c) Structure

In both Saskatchewan and the Northern Territory, Lower Proterozoic rocks, which form the principal hosts for uranium mineralization, have been subjected to polyphase deformation. In both mining districts the ore bodies are associated with zones of structural disturbance, and it is pertinent to both ore genesis and exploration to establish the nature of the structures, and their controls on ore deposition.

Uranium deposits in the Athabasca region are commonly associated with faults, but ore deposition is not consistently associated with a particular style of structure. The Rabbit Lake deposit is spatially associated with north-easterly trending low-angle reverse faults which may have caused brecciation of the stage 1 mineralization (Hoeve and Sibbald, 1978a). At Key Lake, both the Gaertner and Deilmann ore bodies are aligned along a prominent north-east-trending shear zone. Three types of cataclasite are recognized. A kaolinitic mylonite is the principal ore host, and an Fe-rich chloritic mylonite is also uraniferous. In contrast, Mg-rich chloritic–sericitic mylonite is barren (Dahlkamp, 1978a). It therefore appears that a combination of physical and chemical factors is controlling ore deposition in the Athabasca region.

The uranium deposits in the East Alligator district of the Pine Creek geosyncline are all associated with brecciation (Hegge *et al.*, 1980). The zone of brecciation associated with the Ranger 1 ore body has been described from drill core by Eupene *et al.* (1975), who attributed the brecciation to volume reduction following the replacement of carbonate by chert. Alternative suggestions are that the breccias are collapse structures, related to near-surface karstification (Ferguson *et al.*, 1980), and that they are diapiric breccias, resulting from the density instability of evaporites (Crick and Muir, 1980). Field observations in the newly developed Ranger 1 open pit suggest that shear zones are more important than had been previously recognized.

It is clear that whilst there is no consistency of structural style associated with this class of ore deposit, there is evidence of a physical control of many of the ore deposits. The most probable explanation is that brecciation has enhanced the permeability of the host rock for migrating fluids.

Fig. 9.7 Mottled zone of laterite overlying the Ranger 1 deposit. Uranium is normally concentrated in the pisolitic zone but depleted in the mottled zone.

9.3.3 Ore genesis

(a) The source of uranium

There is little agreement amongst geologists who have studied the deposits of the Athabasca basin concerning the source of the uranium. Uraniferous pegmatites have been described from the Archaean gneisses (Beck, 1969), and the majority of geologists envisage the uranium as having been derived from Archaean basement rocks. However, there is no consensus on the mechanism by which the uranium has been mobilized. Alternatives which have been proposed include the concentration of uranium by chemical weathering, prior to the deposition of the Athabasca Formation (Robertson and Lattanzi,

1974; Dahlkamp and Tan, 1977) and remobilization of uranium by ascending hydrothermal fluids (Morton, 1977). Hoeve and Sibbald (1978a) are of the opinion that the uranium was originally present in the sandstones of the Athabasca Formation, but has been leached during intrastratal solution by descending meteoric water. The main evidence for this mechanism comes from post-diagenetic oxidation of the Athabasca sandstone.

The source of uranium for the deposits in the Pine Creek geosyncline is generally agreed to be the Archaean two-mica granite. The granite contains a range of uraniferous accessory minerals, including uraninite (McAndrew and Finlay, 1980). Further evidence is provided by the close spatial relationship between uranium deposits and the Archaean basement (Needham and Roarty, 1980). Whilst Archaean granitoids seem to be the ultimate source of uranium, many of the metasediments within the Pine Creek geosyncline also have abnormally high uranium contents. The greatest concentrations of uranium away from the ore zone are in pyritic, carbonaceous black shales of the Koolpin, Masson and Cahill Formations (Ferguson and Winer, 1980). Crick and Muir (1980) have demonstrated that many of the carbonate rocks in the Rum Jungle and Alligator River areas are replacive after evaporites. They argue that uranium leached by weathering of the Archaean rocks would have been concentrated in evaporitic sediments. Subsequently uranium-rich brines could have been released during diapirism, during the conversion of gypsum to anhydrite, and during regional metamorphism.

The uranium deposits of the Pine Creek geosyncline are located near to Mid-Proterozoic sandstones of the Kombolgie Formation but there is no evidence that these sediments have acted as source rocks for the mineralization.

(b) The ore fluid

What type of fluid was responsible for transporting uranium to its present site adjacent to the Mid-Proterozoic unconformity? The best evidence available is derived from the study of fluid inclusions. However, it must be borne in mind that several stages of uranium transport have probably taken place. Also it is important to recall that fluid inclusion studies are based on gangue minerals obtained from assemblages of complex paragenesis.

Pagel (1977) and Pagel et al. (1980) have investigated fluid inclusions obtained both from the Saskatchewan ores and from the overlying Athabasca Formation. Their studies at Rabbit Lake indicate values of $160 \mp 10°C$ for euhedral quartz, $>130°C$ for dolomite and $120°C$ for calcite. The salinities of the fluids decrease from 30 wt.% equivalent NaCl for euhedral quartz to 1 wt.% equivalent NaCl for calcite. Fluid inclusions obtained from quartz overgrowth within the Athabasca Formation have high salinities which indicate that silicification of the sandstone was from brines. From salinity measurements, Cl/Br ratios and isotope data, Pagel and his colleagues (1980) conclude that the solutions which silicified the Athabasca Formation and deposited the gangue minerals associated with uranium were formation waters similar in composition to those found in modern sedimentary basins.

Homogenization temperatures indicated that temperatures of $220°C$ were attained at the base of the Athabasca Formation. If a geothermal gradient of $35°C/km$ is assumed then it is apparent that a depth of burial approaching 6 km has been attained.

The main study of fluid inclusions from the Pine Creek geosyncline is based on material from Jabiluka and Nabarlek (Ypma and Fuzikawa, 1980). Two distinctive homogenization temperatures have been established for vein quartz: early vein quartz has a modal value of $150°C$, and later vein quartz has a modal value of $110°C$. The fluid inclusions from both generations of quartz have $CaCl_2$ as their dominant solute and exhibit widely varying salinities from low values up to 30 wt.% equivalent NaCl. Ypma and Fuzikawa (1980) suggest that the mixing of highly saline brines and meteoric water is the most reasonable mechanism to explain the variation in salinity.

(c) Factors controlling uranium deposition

We have already discussed the close spatial association between unconformity-type deposits and zones of structural disturbance and brecciation. We concluded that areas of high permeability are required for the concentration of uranium deposits. However, what are the factors which cause uranium to be immobilized within zones of high permeability? There are a number of common mineralogical and lithological associations which characterize the uranium deposits in both the Athabasca Basin and the Pine Creek geosyncline. These will now be evaluated as guides to ore deposition.

(i) Chloritic alteration

Chloritic alteration is a common feature of both the Saskatchewan and Northern Australian deposits and is described in an earlier section. Most geologists consider that at least part of the chlorite is coeval with the deposition of uranium. However, Pagel and his colleagues (1980) disagree that this is the case in Saskatchewan. On the basis of the D/H values of water equilibrated with chlorite they consider that this mineral was deposited before the gangue minerals commonly associated with uranium.

The chloritic alteration associated with the East Alligator River area has been studied by Ewers and Ferguson (1980). The chlorites have a wide range of composition but tend to be magnesium-rich. X-ray-diffraction studies indicate the presence in some samples of poorly crystalline true clay chlorite, illite-like mica and gibbsite in interlayer positions within the chlorite. Ewers and Ferguson (1980) also report that there is a tendency for the Fe^{2+}/Fe^{3+} ratio of chlorite to decrease with increasing uranium content in whole-rock samples. It is possible that the chlorite has formed by diagenetic alteration of clay minerals. If this is the case adsorption of uranyl ions by clays is an obvious control of ore deposition.

Chloritic alteration of the Kombolgie Formation overlying the Jabiluka deposit has been described by Gustafson and Curtis (1983). From detailed petrological and chemical studies, they conclude that the chlorite was deposited after lithification of the sandstone by circulating groundwater. Gustafson and Curtis (1983) attribute the virtual absence of uranium in the Kombolgie Formation to the lack of an effective reducing agent.

(ii) Graphitic metasediments

In the Athabasca basin graphitic metasediments are commonly associated with uranium deposits. Graphitic metapelites occur adjacent to the uranium ore bodies at Collins Bay and Key Lake whilst graphitic gneisses are associated with mineralization at Cluff Lake and Maurice Bay. The Rabbit Lake deposit is rooted in graphitic metasedimentary rocks (Hoeve and Sibbald, 1978b).

A more variable association is found between graphitic metasediments and uranium ore in the Pine Creek geosyncline. In the Rum Jungle area graphitic shales are a common associate of uranium mineralization (Fraser, 1980). However, in the East Alligator River district graphitic material is absent from the Ranger 1, Koongarra and Nabarlek deposits.

Hoeve and Sibbald (1978a) considered that in the Athabasca Basin graphite has been the source of a mobile reductant. Their viewpoint is shared by Pagel and his colleagues (1980), who believe that the carbon associated with pelites at the base of the Athabasca Formation has undergone a sharp decrease of the H/C ratio. They argue that CO_2, H_2O and hydrocarbons would have been released during the burial of organic material. The generation of a mobile reductant from graphitic metasediments appears to be a valid mechanism in Saskatchewan but it can only be applied to a minority of deposits in the Pine Creek geosyncline. However, Binns et al. (1980) consider that graphitic schists have contributed to the deposition of uranium at Jabiluka by reducing the ore fluid.

(d) The age of mineralization

The age of uranium mineralization in Northern Saskatchewan and the Northern Territory is summarized in Tables 9.4 and 9.5. It is evident that in Saskatchewan the uranium deposits associated with the Mid-Proterozoic unconformity have similar ages between 1100 and 1200 million years. Whereas vein-type uranium mineralization at Beaver Lodge has a Lower Proterozoic age, the unconformity-type deposits are significantly younger than the Athabasca Formation.

The age of mineralization in the Pine Creek geosyncline is more complex, and suggests remobilization of uranium over a prolonged period of time. Chloritization is a ubiquitous alteration process associated with uranium ores and probably reflects an important stage of uranium deposition. Significantly the chloritization is dated as younger than the Kombolgie Formation which is in accord with field observations.

(e) Genetic models

Many genetic models have been proposed for unconformity-type deposits. In essence most of the models may be grouped into two types. Firstly there are those in which the ore-forming processes predate the deposition of the overlying Mid-Proterozoic sandstone, which merely provides a protective cover for the mineralization. In the second type of genetic model the overlying sandstone plays a dynamic role in ore formation. We review some examples of both types with two questions in mind. Do the Saskatchewan and Australian deposits have a common origin or merely similar geological settings? Does the genetic model affect the practicality of mineral exploration?

(i) Models of ore genesis in which the Mid-Proterozoic sandstones have a passive role

Dahlkamp (1978a) has proposed a genetic model

Table 9.4 Ages of major geological events in the Athabasca Basin

Major geological events	Uranium mineralization (million years)			
	Beaverlodge	Key Lake	Rabbit Lake	Cluff Lake
Hudsonian metamorphism of Lower Proterozoic metasediments 1570–1880 million years	Initial mineralization 1780 ± 20			
Deposition of Athabasca Formation 1350 ± 50 million years				
Diabase intrusion 938–1230 million years	Reworking of uranium 1110 ± 50	Initial mineralization 1160–1228	Initial mineralization 1075	Initial mineralization 1050
	Reworking of uranium 270 ± 20			

Sources: In Hoeve and Sibbald (1978a).

Table 9.5 Ages of major geological events in the Pine Creek geosyncline

Major geological events	Uranium mineralization (million years)	
	Alligator River Deposits	South Alligator Valley
Metamorphism of Lower Proterozoic sediments 1782–1825 million years[1]		
Deposition of Kombolgie Formation 1648 million years[1]	900–1700[2]	
Chloritization[1] 1610 million years		800[3]
		500[4]

Sources: 1, Page *et al.* (1980); 2, Hills and Richards (1976); 3, Hills and Richards (1972); 4, Cooper (1973).

for the Key Lake deposit, in which the Athabasca Formation has an essentially protective role. Dahlkamp envisaged an initial stage of metamorphic mobilization, and concentration of uranium within the Lower Proterozoic metasediments. The main stage of ore concentration is related to a period of prolonged chemical weathering during which the pre-Athabasca Formation regolith developed. The overlying sandstones of the Athabasca Formation are considered to have acted as a protective cover, although some remobilization of uranium into the sandstone is related to epeirogenic movements. Critics of Dahlkamp's model point to the post-Athabasca Formation age of the uranium ore and the relatively high temperatures of formation, indicated by fluid inclusion studies.

A similar model is proposed for the Jabiluka, Ranger and Koongarra deposits by Ferguson *et al.* (1980). Their model differs in that the structural sites for uranium concentration are interpreted to be collapse structures resulting from the development of a karst topography on emergent carbonate rocks. The dolines were subsequently filled with Mg-rich clays or chlorite and organic matter. The uranium is considered to have been transported at low temperature as uranyl carbonate complexes within migrating groundwaters. Adsorption on clays and organic matter is thought to be the main mechanism of uranium concentration. The subsequent deposition of sandstones belonging to the Kombolgie Formation is regarded as providing a seal for the uranium deposits. One can understand geologists disregarding the possible genetic significance of the Kombolgie Formation as it is virtually devoid of uranium.

(ii) Models of ore genesis in which the Mid-Proterozoic sandstones have a dynamic role

Genetic models have been proposed for both the Athabasca Basin and the Pine Creek geosyncline in which descending fluids are responsible for the main stage of uranium concentration.

Hoeve and Sibbald (1978a) have presented a genetic model for the Rabbit Lake deposit, which involves the flushing out of uranium from the Athabasca Formation by descending groundwater. Petrological studies of the sandstones within the Athabasca Formation show that they have been subjected to extensive intrastratal solution by oxidizing groundwater. Owing to the lower porosity of the metamorphosed Lower Proterozoic metasediments, the descending fluid would have been channelled into zones of structural disturbance where mobile reductants such as methane have reduced the E_h of the ore fluid and precipitated out uraninite. Pagel and co-workers (1980) suggest a modification of Hoeve and Sibbald's model. They characterize the ore fluid as descending saline formation water derived from the Athabasca sandstone. However, they identify carbonaceous pelites of Lower Proterozoic age as the probable source of uranium which is remobilized by the hot, saline-oxidizing ore fluid. Mobile reductants derived from the alteration of carbonaceous material during burial are suggested as appropriate chemical barriers to uranium mobility.

Several genetic models of uranium formation *par descensum* have been proposed for the deposits of the Alligator Rivers area. Ypma and Fuzikawa (1980) suggest that the sandstones of the Kombolgie Formation have provided a thermal insulating blanket, which enabled radiogenic heat to build up in the underlying uraniferous granitoids. They also consider that fractures within the Kombolgie Formation have acted as channelways for descending oxidizing groundwater which has leached uranium from granitic basement rocks. Adsorption of uranyl ions on to clay minerals is regarded as the main mechanism of uranium concentration. A similar model is proposed by Gustafson and Curtis (1983) who draw attention to the magnesian metasomatism of the lower part of the Kombolgie Formation at Jabiluka. Gustafson and Curtis emphasize the role of collapse structures in carbonates of the Lower Cahill Formation as channelways for ore fluids and regard carbonaceous material as the most probable reductant.

Are the Saskatchewan and Northern Territory deposits formed by the same processes? In both cases there is good evidence that the uranium was derived from Archaean rocks and may have been upgraded by sedimentary, diagenetic and metamorphic processes within Lower Proterozoic metasediments. Fluid inclusion data suggest that whilst the temperatures of ore deposition are similar between the two mining districts differing types of fluid may have been involved. For example Pagel *et al.* (1980) interpret the fluid trapped within gangue as formation water identical to that associated with the Athabasca Formation. However, Ypma and Fuzikawa (1980) envisage the mixing of descending groundwater and highly saline brines in their genetic model of the Nabarlek and Jabiluka deposits. Structural and chemical controls of ore deposition are recognized as being critical for ore concentration in all unconformity-type deposits. In Canada the structures are mainly fault-controlled whereas many geologists interpret brecciation in the Pine Creek geosyncline as collapse structures resulting from the solution of carbonates. There is considerable difference in emphasis on the factors which immobilize uranium to form high-grade ore deposits. Canadian geologists agree that a mobile reductant derived from hydrocarbons is the most appropriate mechanism for reducing and precipitating uranium. This is not accepted by all geologists who have studied the Australian deposits, because carbon is a variable constituent of the host rocks. In contrast, chlorite is a ubiquitous gangue mineral, and either chlorite or precursor clay minerals may have acted as adsorbents for uranyl ions. In conclusion, it appears that the main contrasts in ore genesis between the two mining districts are the physical and chemical controls of ore deposition, although the role of the Kombolgie Formation is not fully resolved. We might expect these differences in the controls of mineralization to be reflected in the approaches which have been made to exploration.

9.3.4 Exploration

There are important parallels in the geological setting and in the history of exploration between the mining districts of North Saskatchewan and the Northern Territory. However, it must be appreciated that in Canada there has been a sustained exploration effort since the discovery of the Rabbit Lake deposit in 1968. In contrast exploration in Australia has been paralysed by the uncertainty of Government policy towards uranium mining, and by the designation of part of the East Alligator River district as an Aboriginal Reserve. As a result there has been a more intense investigation of the Athabasca Basin with greater emphasis in recent years on the search for deposits concealed at depth below the Athabasca Formation.

The first major uranium discovery in Northern Saskatchewan was in the Beaverlodge area in 1946 (Tremblay, 1978). In the 1950s and 1960s airborne radiometrics played a dominant role in the search for uranium in Canada. During this period attention was focused on the search for vein- or conglomerate-hosted deposits within Archaean or Proterozoic rocks. The Athabasca Formation was generally considered to represent a low-priority exploration target, although Blake (1956) expressed a dissenting viewpoint.

During the 1960s geologists began to apply the supergene model of ore genesis, developed for uraniferous Proterozoic sandstones in West Africa to the Athabasca Formation (Strnad, 1980). The concept of uranium deposits developed within the Athabasca Formation was reinforced by the occasional discovery of mineralized sandstone boulders. Although the supergene model directed geologists' attention to the Athabasca Formation, it limited their vision of ore concentration to the sandstone above the Mid-Proterozoic unconformity.

Johns (1970) was one of the first geologists to recognize the potential of the regolith below the Athabasca Formation. He suggested that labile minerals including uraniferous species were leached by descending meteoric water and concentrated in the regolith due to changes in permeability. On this basis Johns recommended a number of areas for exploration which included the Carswell structure – now the site of the Cluff Lake deposit. Johns' concept of uranium formation was later presented in more sophisticated form by Hoeve and Sibbald (1978a) in their model of the Rabbit Lake deposit.

The first major discoveries of unconformity-type deposits were at Rabbit Lake (1968), Fond du Lac (1969) and Cluff Lake (1969) at the margin of the Athabasca Formation. All the early discoveries were found from airborne γ-ray spectrometer surveys. The radiometric anomalies were found to be associated with mineralized boulder trains which led explorationists to the concealed ore deposits (McH. Clark et al., 1982). The success of γ-ray spectrometry in this terrain is remarkable because in principle the technique has the ability to penetrate to a depth of only 0.5 m (Hambleton-Jones, 1978). In practice the ore deposits are covered by up to 100 m of glacial till. Armstrong and Brewster (1980) suggest that a combination of glacial plucking and frost heaving has produced a surface expression of the sub-outcropping mineralization.

The exploration concepts and methodology used in the discovery of unconformity-type deposits in Saskatchewan are explained in case histories by Scott (1981), Gatzweiler et al. (1981) and Saracoglu et al. (1983).

Saracoglu and his colleagues (1983) describe the discovery of the McClean deposit beneath 168 m of Athabasca sandstone. Their initial geological model, based on the Rabbit Lake deposit, comprised 'a structurally controlled deposit with a well developed alteration halo occurring near the eroded surface of the underlying basement'. An appraisal of the basement geology below the Athabasca sandstone using airborne magnetics combined with a reconnaissance radon survey led to a preliminary drilling programme. These authors describe how the description by Dahlkamp and Tan (1977) of the Key Lake deposit led them to modify their geological model, placing much greater emphasis on the presence of graphitic metapelites associated with the uranium ore body. As a consequence of the association with graphitic material, airborne and ground-based electromagnetic methods became of major importance. In the event the McClean deposit was discovered by deep drilling of a target which had both an electromagnetic signature and a radon anomaly.

Uranium was first discovered in the Northern Territory at Rum Jungle in 1949 by prospectors using hand-held Geiger-Muller counters. Subsequently the small South Alligator Valley uranium field was discovered using relatively unsophisticated airborne radiometrics (Armstrong and Brewster, 1980). The exciting discovery of the potential of the East Alligator River district was based upon the reinterpretation of the geology of the Pine Creek geosyncline which revealed an analogous geological setting to the Rum Jungle area (Ryan, 1972). It was therefore the unconformity between the Lower Proterozoic and Archaean rocks which directed geologists to search for uranium in the eastern part of the Pine Creek geosyncline. The Mid-Proterozoic unconformity was not seen at this stage as having any exploration significance.

The discovery of the Ranger, Koongarra and Nabarlek deposits can be attributed to the use of radiometric surveys from fixed-wing aircraft (Armstrong and Brewster, 1980). However, the Jabiluka 1 deposit was found using a ground-based radiometric survey. In describing the discovery of Jabiluka, Rowntree and Mosher (1981) state that the radiometric anomaly was a third-order priority at the end of the initial phase of exploration, and they emphasize the importance of flexibility in assessing the relative importance of radiometric anomalies. Rowntree and Mosher (1981) explain that in the initial stages of exploration in the East Alligator River area it was important to determine the lithologies associated with radiometric anomalies. If this could not be achieved by geological mapping, additional information was obtained from trenching or auger drilling. The discovery of the Jabiluka 2 deposit was based upon diamond drilling, which was carried out to test the hypothesis that there was a periodicity in the distribution of uranium deposits in the East Alligator River area (Rowntree and Mosher, 1981).

In conclusion, it is evident that the discovery of the uranium deposits of Saskatchewan and the Northern Territory must rank amongst the major exploration successes of the 1960s and 1970s. In both mining districts geologists were guided to the areas of interest by simple genetic models. In Saskatchewan, the Athabasca Formation and its underlying regolith have been the main focus of exploration whereas in the Northern Territory the unconformity between Archaean basement rocks and Lower Proterozoic metasediments has been the principal exploration guide.

9.4 SANDSTONE-HOSTED URANIUM DEPOSITS OF THE WESTERN USA

9.4.1 Introduction

Sandstone-type deposits account for more than 95% of the uranium resources of the United States (Nash *et al.*, 1981). In terms of World uranium production the United States was pre-eminent until the early 1980s. In 1980 US production of uranium metal was 16 156 tonnes, but it is predicted that this will decline to 8000 tonnes in 1985 when Canada will assume the leading role in uranium production (*Mining Annual Review*, 1984). The main reason for the decline stems from a combination of low uranium prices during the early 1980s and the high cost of exploration and mining in the western United States, compared with near-surface deposits in other parts of the World.

Uranium is mined from sandstones in three provinces within the United States: the Colorado Plateau, the sedimentary basins of Wyoming and South Dakota and the south Texas coastal plain region. The Colorado Plateau has been the source of most production and contains more than half of the uranium resources of the United States (World Uranium, 1980). Within the Colorado Plateau there are four principal mining districts: the Grants district, Monument Valley–White Canyon district, the Uravan district and the Lisbon Valley district (Fig. 9.8). The deposits within the Grants

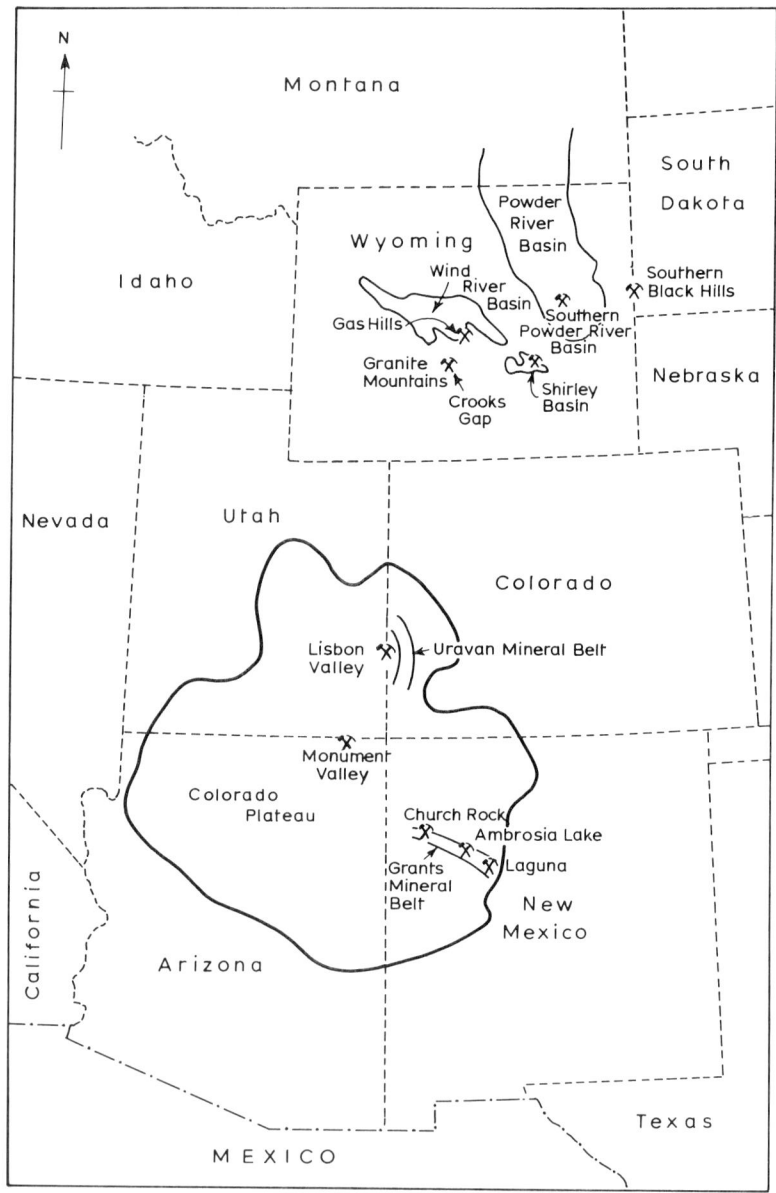

Fig. 9.8 Location of major uranium mining districts in the Western USA (modified from Rackley 1976).

district contain the most important reserves (Nash et al., 1981). The deposits of the Colorado Plateau are similar in many respects to those of Wyoming and S. Dakota. Rackley (1976) groups the occurrences together using the term 'Western-States-type mineralization', a nomenclature which we have adopted for this chapter. Uranium deposits also occur in tuffaceous sandy sediments, of Late Eocene to Pliocene age, in south Texas, but they amount to only a small proportion of the United States resources (World Uranium, 1980). The Texas deposits are not described here, and the reader is referred to papers by Eargle and Weeks (1973), Reynolds *et al.* (1982) and Goldhaber *et al.* (1983) for information concerning this area.

9.4.2 Classification of deposits

Sandstone-hosted uranium deposits are recognized as a distinct ore type although there are similarities in some respects with conglomerate-hosted and unconformity-type deposits. Dahlkamp (1978b) described a threefold subdivision of sandstone-type deposits based upon the geometry of the ore deposits (Fig. 9.9). Peneconcordant deposits occur parallel to the bedding of the enclosing fluviatile sandstones and are the dominant type within the Colorado Plateau. Roll-type deposits are stratabound, but discordant, and roughly resemble a crescent in cross-section. They are the predominant type within Wyoming and the Texas Gulf coast. Stack deposits consist of tongue-shaped impregnations of sandstones, adjacent to permeable fault zones. Their main occurrence is in the Grants district, New Mexico.

Adler (1974) suggested that peneconcordant deposits can be further subdivided on the nature of the reductant which has immobilized uranium. Adler distinguished those uraniferous sandstones which contain an indigenous reductant, normally plant debris, from those in which a mobile reductant, such as H_2S, has been introduced epigenetically. The uranium deposits of the Colorado Plateau are mainly of the former type, whilst the reductant in the Grants district is considered to have been mobile.

Many geologists prefer to use the term 'geochemical-cell' to describe the discordant type of mineralization (Rackley, 1976). This term has a genetic connotation and emphasizes the control of changing redox conditions on uranium deposition. Nash *et al.* (1981) subdivide the roll or geochemical-cell deposits into primary and secondary types. They also separate the uranium–vanadium deposits of the Uravan mineral belt, on the grounds that redox changes result from the mixing of two solutions.

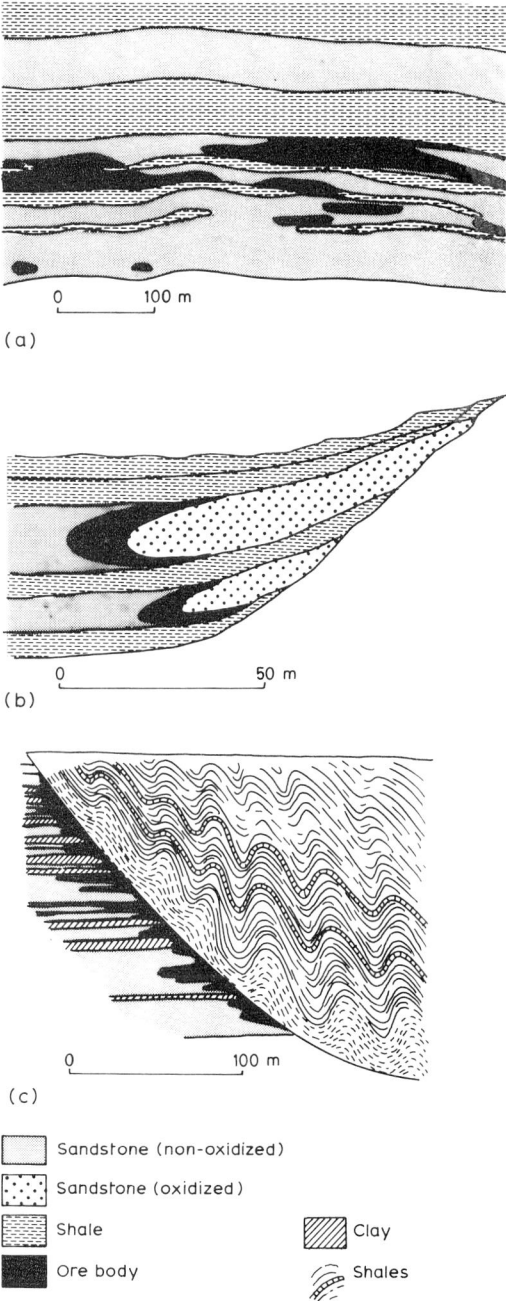

Fig. 9.9 Classification of sandstone-hosted uranium deposits (modified from Dahlkamp, 1978b). (a) Sandstone-type/penecordant; (b) sandstone-type/roll-front; (c) sandstone-type/tectolithological.

9.4.3 Geological setting

(a) Tectonic setting
The sediments which host the Western States uranium deposits formed in a continental environment, and the sedimentary forms and facies suggest that epeirogenic movement was dominant. The tectonic setting of the Colorado Plateau has been described as a miogeosyncline filled with molasse-type sediment (Mitchell and Reading, 1969). The Wyoming uranium deposits are deposited in sandstones formed in a back-arc continental setting (Mitchell and Garson, 1981).

The intimate relationship between tectonic evolution and the formation of uranium deposits in the Western States has been described by Rackley (1976). Some of the factors of importance include rates of uplift in the source area, sediment supply, rainfall, stream gradient, vegetation growth, and rates of sediment accumulation. During the final stages of tectonic evolution the erosion of the basin margin provided the conditions for the ingress of well-oxygenated water which initiated the process of ore formation.

(b) Stratigraphy and host rock lithology
The sediments which host uranium deposits in the Western States range from Lower Triassic to Eocene in age. In the Colorado Plateau area, the main stratigraphic horizons of economic significance are the Chinle Formation of Lower Triassic age and the Morrison Formation of Upper Jurassic age. The Morrison Formation contains the bulk of the uranium reserves in the United States (Fischer, 1968). In Wyoming the principal uranium deposits occur in the Inyan Kara Group of Lower Cretaceous age and in the Fort Union, Wind River and Wasatch Formations of Palaeocene and Eocene age (Harshman, 1968a).

The sediments which host the uranium deposits are mainly fluviatile sandstones. The dominant sedimentary forms are alluvial fans within which braided streams are the main channels for sediment transport and deposition. A number of important uranium mines are located within this sedimentary environment, for example the Gas Hills district, Wyoming. Here the ore bodies are located on the margins of the fan (Rackley, 1976).

Streams emerging from the distal sector of the fan assume a meandering style. The coarse sandstones, which are deposited in point bars, chute bars and channel-lag deposits, host the most important uranium deposits in the Western States. For example, in the Lisbon Valley district the long, linear uranium ore bodies are generally confined to channel floor and interchannel facies. However, in the Rio Algom Mine the host rock for the deposits is believed to be of crevasse splay origin (Huber, 1980).

With increasing distance from the fan the sediment deposited within the meander system becomes fine-grained. In this environment fine-grained point bars may host uranium as in the case of the Shirley Basin and Powder River districts, Wyoming (Rackley, 1976).

It is difficult to generalize about the petrology of the sediments which host the uranium deposits other than to emphasize their variability. Normally sandstones are predominant, with variable proportions of conglomerate, mudstones and siltstones. The sandstones are commonly arkosic or feldspathic and usually contain a significant content of clay minerals. The clays can often be attributed to the alteration of volcanic clasts by post-depositional processes (Adams et al., 1978). Some of the clay content of the sandstones may have been incorporated as clasts resulting from bank collapse. Fossilized plant material is commonly present within the sandstone and is considered to have been incorporated into the sediment at the time of deposition. However, in the Grant district the organic matter is authigenic in nature and is thought to be derived from humic materials (Adams et al., 1978).

Rackley (1976) argued that the main processes which convert sediment into an appropriate host rock for uranium occur either during or soon after deposition. Post-depositional changes are mainly the result of compaction and dewatering.

(c) Mineralogy

The mining districts of the Western States share a consistent assemblage of primary uranium minerals, but exhibit distinctive regional variations. Uraninite and coffinite are the dominant ore minerals; the main variables are the presence of vanadium-bearing species, the nature of carbonaceous material and the proportion of secondary uranium minerals.

Vanadium minerals are significant by-products of uranium mining in the states of Utah, Colorado and Arizona. Montroseite and vanadium-bearing mica, chlorite and clay are of most importance. The presence of abundant vanadium in the primary ore is reflected in the common occurrence of stable secondary uranium minerals such as carnotite and tyuyamunite in the oxidized zone above the water table. The primary uraniferous minerals are found predominantly replacing carbonized fossil wood (Fischer, 1968).

Coffinite, uraninite and uraniferous carbonaceous material are the principal ore minerals in the Grants district, New Mexico. The carbonaceous material is considered to be authigenic in nature, having been introduced epigenetically into the host sandstones (Kelley et al., 1968). Although vanadium species are not reported from the primary mineral assemblage the most common secondary minerals are vanadates.

The Wyoming deposits are also characterized by the presence of uraninite and coffinite. However, they differ from the Colorado Plateau deposits in having no consistent association with organic carbon (Harshman, 1968a). There is a variable distribution of secondary uranium species, and where vanadium is deficient, uranium is normally leached from the supergene zone.

The average grade of uranium mined within the Western States varies geographically and has declined dramatically in recent years. In 1969 the average grade of ore mined from the Colorado Plateau was 0.25% U_3O_8 whilst in Wyoming the grade was more variable, ranging up to 1% U_3O_8 (Fischer, 1968; Harshman, 1968b). However, during the ensuing decade the average grade of US uranium production declined to 0.10% U_3O_8 (Darmayan, 1981).

(d) Structure

The Colorado Plateau became a structural entity during the Late Cretaceous Laramide orogeny. It is characterized by major zones of uplift (Fig. 9.10). Fischer (1968) described the larger structural blocks as 'being tilted, like partly raised trap doors, with thousands of feet of structural displacement'. The principal deformational style of the Mesozoic sediments associated with the uplift is the development of prominent monoclinal flexures. Mineralization is often associated with the zones of major uplift. For example, the Uravan mineral belt occurs within the Uncompahgre uplift. However, uranium deposits may also be associated with faulting and folding. The Lisbon Valley deposits occur on the western limit of a faulted asymmetric anticline formed by the flowage of evaporites (Wood, 1968). The uranium mineralization in the Grants district is associated with the Zuni uplift, but the deposits appear to be controlled in part by north to north-east trending faults (Kelley et al., 1968). Major faulting tends to be confined to the margins of the plateau. The western margin is defined by normal faults of large displacement, whilst the eastern margin is characterized by high-angle reverse faults (Fischer, 1968).

In Wyoming the principal structures are north-westerly trending asymmetric anticlines. The uranium deposits are located in the axial portions, or on the flanks of intermontane basins which are simple downwarps, in some cases modified by folding or faulting (Harshman, 1968a).

(e) Age of mineralization

The age of mineralization in the Western States is debatable although most geologists accept that the uranium is epigenetic in origin. The problem of dating the mineralization accurately is due to uncertainty concerning the reliability of uranium–lead ages in sandstones (Rackley, 1976). In the secondary environment, uranium is readily

Ore deposit geology and its influence on mineral exploration

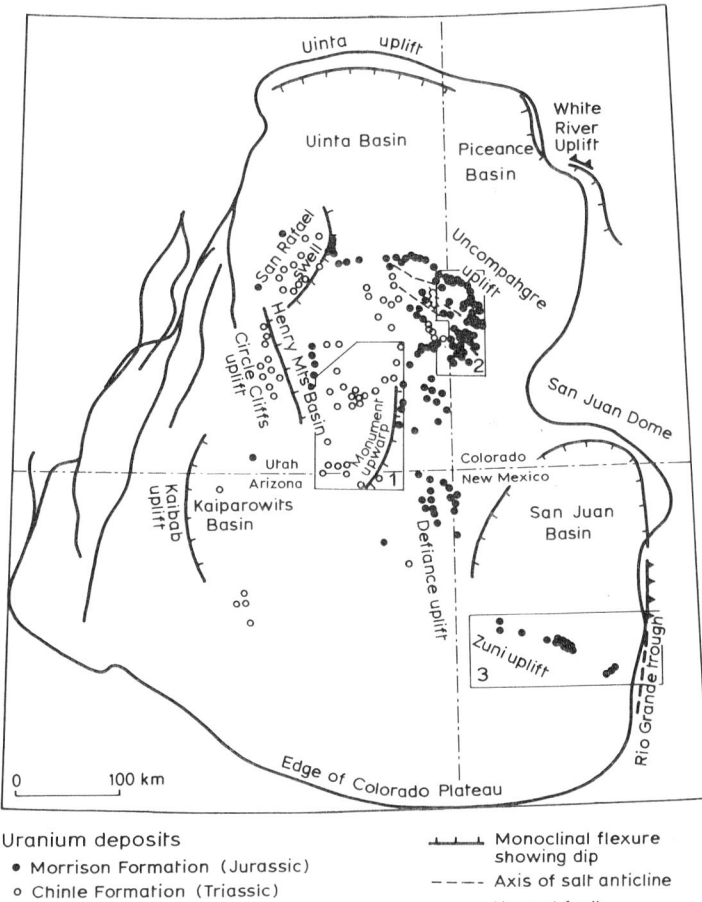

Fig. 9.10 Principal uranium deposits and major structural features in the Colorado Plateau region. Blocks show approximate areas covered by the following districts: 1, Monument Valley – White Canyon; 2, Uravan mineral belt; 3, Grants mineral belt. Modified by Nash *et al* (1981) from an original map in Fischer (1970).

remobilized due to changes in oxidation conditions, which cause it to become separated from its daughter minerals. As a result geologists tend to rely mainly on field relationships to determine the approximate age of ore deposition. Radiometric age determinations tend to be used as supporting evidence where appropriate.

The field evidence suggests that the uranium mineralization in the Colorado Plateau is diagenetic (Fischer, 1968), whereas the roll-type deposits are thought to have formed some time after deposition of the sandstones which now host the uranium mineralization. However, radio-metric age determinations (Dooley *et al.*, 1974; Miller and Kulp, 1963) suggest that in both the peneconcordant and roll-type deposits there is a time interval of between 25 and 30 million years between sediment deposition and the introduction of uranium.

9.4.4 Ore genesis

(a) Source of uranium

The provenance of uranium is essentially a geochemical problem, and the relative merits of the source rocks must be measured mainly by geochemical criteria. The most comprehensive geochemical data are based on rocks associated with the Shirley Basin, Wyoming (Fig. 9.8). Arkosic sandstones within the Shirley Basin were derived from a granite within the Granite Mountains, Central Wyoming. Analysis of the granite has shown that the uranium concentration is about four times the World average (Rackley, 1976). Harshman (1968b) stated that groundwater issuing from tuffaceous beds of the White River Formation contains 10–20 ppb U. This compares with values of 0.4–4 ppb in groundwater from Colorado (Wedepohl, 1972). It therefore seems probable that there are multiple sources for uranium in the Shirley Basin.

Waters and Granger (1953) considered that the uranium and vanadium in the Colorado Plateau may have been derived from the leaching of devitrified volcanic ash within the Morrison and Chinle Formations. However, whereas uranium is typically associated with felsic lithologies, vanadium is normally enriched in mafic igneous rocks (Wedepohl, 1972). More than one source rock is therefore required for minerals within the Colorado Plateau.

Gabelman (1977) has suggested that uranium, vanadium and other elements may have been introduced into the sedimentary sequences of the Western States by mantle-derived gases. Gabelman points to the occurrence of diatremes within the Colorado Plateau which he believes could have acted as the conduits for metals derived from the mantle.

(b) The ore fluid

The absence of gangue minerals which are coeval with uraninite and coffinite precludes the use of fluid inclusions as indicators of the nature of the ore fluid in sandstone-hosted uranium deposits. However, it is known that oxidizing groundwater can leach uraniferous source rocks, such as granite, and can transport the uranium as soluble uranyl species (Section 9.2). The main uncertainty is the direction of groundwater movement. Sandford (1982) has attempted to resolve this problem by modelling groundwater movement during Cretaceous times in the Colorado Plateau. Sandford considered that the main controls of groundwater movement are aquifer permeability (including a decrease in permeability with lateral facies change to finer-grained sediments), thickness variation within the aquifer and the damming effects on groundwater movement, resulting from horsts of crystalline basement, such as the Uncompahgre uplift. Sandford (1982) concluded that the main concentrations of uranium are related to zones of upwelling saline groundwater.

Groundwaters in the Colorado Plateau at the present day have uranium concentrations of 0.4–4.0 ppb (Wedepohl, 1972), and it is reasonable to assume that groundwaters in Mesozoic times had similar values. It is therefore necessary to conclude that prolonged periods of groundwater flow are necessary to produce a uranium deposit of economic value.

(c) Factors controlling the deposition of uranium

Most geologists consider that uranium has been introduced into the host fluviatile sandstones shortly after deposition by migrating, well-oxygenated meteoric water. We now consider the factors which have immobilized uranium to form the variety of ore deposits that are evident in the Western States. In an earlier discussion of the classification of sandstone-type deposits the distinction was drawn between peneconcordant and roll-type or geochemical-cell deposits. As this separation partly reflects differing controls on ore deposition, the two types will be discussed separately:

(i) Peneconcordant deposits

In peneconcordant deposits the principal uraniferous minerals, uraninite and coffinite, are closely associated with carbonaceous material in fluviatile sandstone. Throughout much of the Colorado Plateau the carbonaceous material is

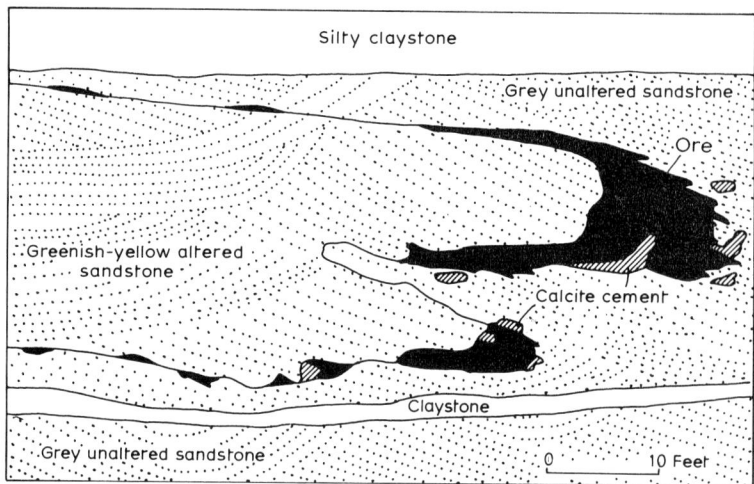

Fig. 9.11 Cross-section through a roll-type deposit from the Shirley Basin, Wyoming (from Harshman 1968b).

coalified wood, which is considered to have been rafted into position at the same time as sedimentation. In the Grants district, the uranium is associated with tabular layers of organic-rich material (Adams et al., 1978). The organic material is a humate and is regarded as being epigenetic. In all the peneconcordant deposits where organic material is conspicuous it is probable that uranium is immobilized by a combination of adsorption on carbonaceous material and reduction by both organic material and biogenically derived H_2S.

(ii) Roll-type or geochemical-cell deposits

The term 'roll-type deposit' refers to the cross-cutting relationship between uranium mineralization and its host rock (Fig. 9.11). It was first applied to deposits in the Colorado Plateau in 1956 but is now more widely used to describe the deposits of Wyoming and Texas. The concept of a geochemical-cell was first developed by Shockley et al. (1968). This summary of the conditions prevailing within a geochemical-cell is taken from Rackley (1976).

The crux of the geochemical-cell is the complementary roles played by separate populations of bacteria on either side of the 'roll front' (Fig. 9.12). Within the reduced zone, anaerobic bacteria break down carbonaceous material to form simple organic molecules, CO_2 and H_2, which produce low E_h conditions. A separate species of bacteria (*Desulfovibrio*) utilizes CO_2 and inorganic sulphate to create methane and hydrogen sulphide. In effect the organic material acts as the fuel for bacterial reactions, the by-products of which act as reductants for uranium.

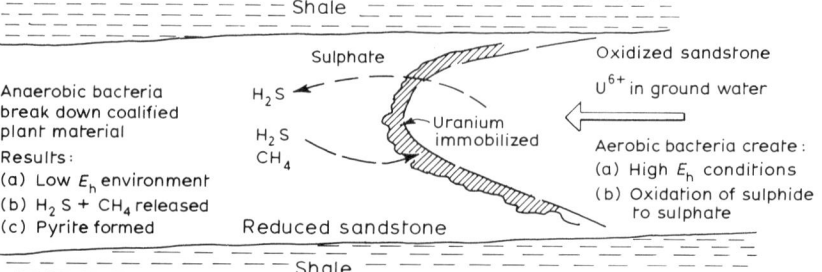

Fig. 9.12 Mechanism of formation of roll-type uranium deposit (based on Rackley, 1976).

Within the oxidized part of the roll front aerobic bacteria create high E_h conditions within which uranyl species are the stable form of uranium, and sulphide minerals are oxidized to sulphates. Soluble sulphate migrates into the reduced zone thus contributing to continuing cycles of the geochemical-cell. Uranium also migrates for a short distance into the reduced zone, but is quickly immobilized by reduction of soluble uranyl species to insoluble uranous oxide (uraninite).

The essence of the geochemical-cell is the transport of uranium in oxidizing groundwater through highly permeable channelways, which pass through zones containing reduced facies of the sandstone (controlled by organic matter or sulphide species).

(d) Genetic model

The generally accepted model of ore genesis in the Western States is that uranium is leached from either granite or tuffs and transported along permeable palaeochannelways by oxygenated meteoric water. Uranium concentration is related either to adsorption by carbonaceous material or rapid changes in redox conditions. An important aspect of the genetic model which needs to be explained is the restriction of uranium to particular stratigraphic horizons. This may be explained by recalling Rackley's (1976) thesis that the occurrence of uranium deposits is closely associated with the tectonic evolution of a sedimentary basin. Thus the development of a uranium deposit depends upon a combination of circumstances which have an underlying tectonic control. The main factors include uplift of the source area with eventual exposure of uraniferous rocks, a climate which favours the luxuriant growth of vegetation, and the preservation of permeable sediments with an appropriate content of organic matter, pyrite and clay minerals. The deformation and erosion of the sedimentary basin is also critical in determining the availability of sites for the ingress of oxygenated water into the palaeochannel system.

Gabelman (1977) has criticized the evidence for the 'consensus' genetic model, and believes that it has resulted in an unnecessarily restricted attitude to exploration. He argues that uranium may also have been introduced into the sandstone environment by hypogene processes, and suggests that diatremes and tectonic intersections would provide favourable channelways for mantle-derived metal-bearing volatiles.

9.4.5 Exploration

Large-scale uranium mining and exploration in the Western States has developed since 1948 when the US Atomic Energy Commission commenced a major ore-buying programme. Prior to this date there had been some intermittent extraction of uranium from the Colorado Plateau but the tonnages involved were diminutive compared with the production of vanadium (Fischer, 1968).

All of the uranium-mining districts were originally characterized by a surface manifestation of mineralization, which was identified by prospectors or detected by radiometric surveys. However, many of the larger uranium deposits have no surface expression, and therefore the bulk of exploration has been based on drilling. The scale of activity may be gauged from the records: between 1948 and 1978 some 117 000 km of exploration and development drilling was carried out in the search for uranium in the United States (World Uranium, 1980). During this period there has been a trend towards deeper holes and in 1979 the average depth drilled was 136 m (World Uranium, 1980). In some cases, for example the Ambrosia Lake District in New Mexico, important discoveries have been made by wild-cat drilling (Kelley *et al.*, 1968). However, the bulk of ore deposits have been found either by drilling near known deposits or have been located by the systematic search for favourable guides in areas without a previous history of mining. Porter (1981) describes an interesting case history where a knowledge of the guides to uranium deposits in the Grants district was applied successfully in the previously unexplored region of Bernabe Montano, New Mexico.

Within the Colorado Plateau the following

guides are recognized as being useful for uranium exploration (Malan, 1968; Motica, 1968; Wood, 1968; Porter, 1981).

Favourable host lithology
In most of the mining districts of the Colorado Plateau the favoured host lithology for uranium mineralization is a poorly sorted arkosic sandstone. The key factor appears to be the presence of zones of high permeability within the sandstone. However, Motica (1968) commented that in the Uravan Belt the permeability of the host sandstone does not appear to be critical, but the presence of numerous clay seams appears to be a more important lithological association.

Favourable sedimentary forms
In many areas the sandstones which host uranium have been deposited within river channels. Malan (1968) observed that the uranium deposits in the Monument Valley–White Canyon districts are restricted to the more deeply scoured portions of the channelways. In the Uravan belt, the thickness of the host sandstone is an important control on uranium deposition (Motica, 1968). The channel-ways which host uranium often have a persistent trend and this factor is of critical importance in defining the drilling pattern for exploration. For example, in the Uravan belt initial exploration is normally based on reconnaissance drilling in which the holes are sited at 1000 ft (304.8 m) centres at right-angles to the channel trend (Motica, 1968).

The presence of carbonaceous material
Within the states of Colorado, Utah and Arizona uranium deposits have a close spatial association with coalified wood. Typically the uranium occurs replacing carbonaceous material, but in some cases the coalified wood is merely adjacent to the mineralization. Motica (1968) stated that in the Uravan Belt 'the size and grade of an ore deposit is not always governed by the amount of carbonaceous material present'. In the Grants district the organic material is a humate which is believed to have been introduced epigenetically into the host sandstone (Adams *et al.*, 1978).

Alteration
In many of the mining districts of the Colorado Plateau conspicuous alteration patterns are associated with the ore bodies. In the Uravan Belt alteration is evident as colour changes in the host sandstone (Motica, 1968). In the Lisbon Valley district the zone of alteration is defined by bleaching and is also reflected in a halo of anomalous molybdenum values around the ore deposit (Wood, 1968). Riese and his colleagues (1978) have demonstrated that in the Grants district there is a distinctive mineralogical zoning. Montmorillonite is dominant in the reduced zone, chlorite is enriched in the ore, and kaolinite is the main clay mineral in the oxidized facies.

In addition to the guides which are of regional application there also appear to be controls which operate on a district basis. For example, in the Lisbon Valley area most of the uranium ore bodies are restricted to within 500 ft (152.4 m) of a particular structure contour interval. The reason for this is obscure but it may reflect a control by a previous water table or a water/oil/gas interface (Wood, 1968).

Exploration guides have also been described from Wyoming and South Dakota. In some cases the guides are similar to those of the Colorado Plateau but their relative importance is not the same. The presence of a permeable host lithology containing coalified woody material is significant, but the role of stream channelways in controlling ore formation is less easily demonstrated (Harshman, 1968a). In the Powder River district uranium ore is spatially associated with the contact between red and drab-coloured sandstones and this acts as an important exploration guide (Mrak, 1968).

Drilling for favourable guides to mineralization is the principal exploration technique used in the Western States even at the reconnaissance stage. The approach therefore differs from most other mining districts where drilling is normally used at a relatively late stage in the exploration sequence.

Normally non-core drilling is used, especially in the early stages of exploration, but cores of favourable stratigraphic horizons are obtained for detailed lithological study and trace element analysis. It is normal procedure for the holes to be logged with γ-ray and resistivity probes. The equipment and procedure used in down-the-hole logging are described by Hawkins and Gearhart (1969) and Dodd *et al.* (1970).

9.5 CONCLUDING STATEMENT

Uranium can occur in a diversity of geological settings, and occurrences of uranium are documented in many parts of the World. However, the bulk of the uranium deposits outside the Communist bloc occur in a limited number of countries, principally Canada, Australia, USA, South Africa, Niger and Namibia. Therefore, although there is sufficient uranium to satisfy World demand for the next 15 years the supply is as susceptible to disruption as supplies of oil. Western European countries which have opted for a rapidly expanding nuclear power programme should be concerned about this state of affairs. One solution to the problem is to find additional sources of supply, which brings us to the subject of exploration. In our opinion there are four aspects of uranium exploration which need particular attention.

9.5.1 Concepts of area selection

Concepts of area selection appear to be poorly developed for uranium deposits. In the examples we have chosen for discussion in this chapter the mining districts were originally located by radiometric surveys and subsequent exploration spread out from near-surface manifestations of uranium mineralization. The newly discovered Olympic Dam deposit in Australia appears to be an exception as it is completely blind. However, its discovery was based on an exploration model for copper, and the uranium content may be regarded as an unexpected bonus. There should by now be sufficient information for reliable criteria of area selection to be developed. However, this is not evident in the published literature.

9.5.2 Genetic and exploration models

Two points need to be mentioned here:

(a) Genetic models for sandstone-hosted uranium in the USA have been developed at a relatively late stage in mining development. They have been of less value as a predictive tool than an empirical approach. In Canada and Australia simple genetic models have aided exploration. In spite of similarities between the deposits differing genetic models have been used.

(b) Three of the deposit types listed in Table 9.2 were unknown prior to 1965. It is probable that further styles of uranium remain undiscovered. It is therefore important that exploration models should remain very flexible.

9.5.3 Factors controlling uranium concentration

More than 80% of the World's uranium reserves are contained in sedimentary or metasedimentary rocks (Table 9.2). It is therefore of paramount importance for the success of future exploration efforts that the source rocks, ore fluids and factors controlling uranium deposition in the secondary environment are better understood.

9.5.4 Discovery of blind uranium deposits

The bulk of the World's uranium mining districts have been discovered by the use of radiometric surveys. However, it is unlikely that many more major uranium deposits will be located using this technique due to its limited depth penetration. There is an urgent need for new concepts and new techniques which will provide increased depth of penetration in uranium exploration.

REFERENCES

Adams, S. S., Starr Curtis, H., Hafen, P. L. and Salek-Nejad, H. (1978) Interpretation of post-depositional processes related to the formation and destruction of the Jackpile-Paguate uranium deposit, N.W. New Mexico. *Econ. Geol.*, **73**, 1635–1654.

Adler, H. H. (1974) Concepts of uranium ore formation in reducing environments in sandstones and other sediments. In *Formation of Uranium Ore Deposits*, Proceedings of a symposium organized by the Intl Atomic Energy Agency, Vienna, 1974, pp. 141–166.

Armstrong, C. W. and Brewster, N. E. (1980) A comparison of exploration techniques in Saskatchewan and the Northern Territory. In *Uranium in the Pine Creek Geosyncline* (eds J. Ferguson and A. B. Goleby), I.A.E.A., Vienna, pp. 733–741.

Beck, L. S. (1969) Uranium deposits of the Athabasca region, Saskatchewan. *Sask. Dept. Min. Res. Rep.* 126.

Binns, R. A., McAndrew, J. and Sun, S.-S. (1980) Origin of uranium mineralisation at Jabiluka. In *Uranium in the Pine Creek Geosyncline* (eds J. Ferguson and A. B. Goleby), I.A.E.A., Vienna, pp. 543–562.

Blake, D. A. W. (1956) Geological notes on the region south of Lake Athabasca and Black Lake, Saskatchewan and Alberta. *Geol. Surv. Pap. Can.*, **55(33)**, 12.

Bowie, S. H. U. (1981) World and U.K. nuclear fuels. Assessment of energy resources. *The Watt Committee on Energy*. Report No. **9**, 51–55.

Cooper, J. A. (1973) On the age of uranium mineralisation at Nabarlek N.T. Australia. *J. Geol. Soc. Austr.*, **19**, 483–486.

Crick, I. H. and Muir, M. D. (1980) Evaporites and uranium mineralisation in the Pine Creek geosyncline. In *Uranium in the Pine Creek Geosyncline* (eds J. Ferguson and A. B. Goleby), I.A.E.A., Vienna, pp. 531–542.

Dahlkamp, F. J. (1978a) Geologic appraisal of the Key Lake U–Ni deposits, N. Saskatchewan, Canada. *Econ. Geol.*, **73**, 1430–1449.

Dahlkamp, F. J. (1978b) Classification of uranium deposits. *Miner. Deposita*, **13(1)**, 83–104.

Dahlkamp, F. J. and Tan, B. (1977) Geology and mineralogy of the Key Lake U–Ni deposits, Northern Saskatchewan, Canada. In *Geology, Mining and Extractive Processing of Uranium* (ed. M. J. Jones), Institution of Mining and Metallurgy, London, pp. 145–157.

Darmayan, P. (1981) The balance of supply and demand: a reassessment. In *Uranium and Nuclear Energy: 1980*, Proceedings of the Fifth International Symposium held by the Uranium Institute, London, pp. 13–22.

Derry, D. R. (1973) Ore deposition and contemporaneous surfaces. *Econ. Geol.*, **68**, 1374–1380.

Dodd, P. H., Drouillard, R. F. and Lathan, C. P. (1970) Borehole logging methods for exploration and evaluation of uranium deposits. *Geol. Surv. Can. Econ. Geol. Rep.*, 26.

Dooley, Jr., J. R., Harshman, E. N. and Rosholt, J. N. (1974) Uranium–lead ages of the uranium deposits of the Gas Hills and Shirley Basin, Wyoming. *Econ. Geol.*, **69**, 527–531.

Eargle, D. H. and Weeks, A. M. D. (1973) Geologic relations among uranium deposits, south Texas coastal plain region U.S.A. In *Ores in Sediments* (eds G. C. Amstutz and A. J. Bernard), Springer-Verlag, Berlin, pp. 101–113.

Eupene, G. S., Fee, P. H. and Colville, R. G. (1975) Ranger One Uranium Deposits. In *Economic Geology of Australia and Papua New Guinea* (ed. C. L. Knight), Australasian Institute of Mining and Metallurgy, Parkville, Victoria, pp. 308–316.

Ewers, G. R. and Ferguson, J. (1980) Mineralogy of the Jabiluka, Ranger, Koongarra and Nabarlek uranium deposits. In *Uranium in the Pine Creek Geosyncline* (eds J. Ferguson and A. B. Goleby), Intl. Atomic Energy Agency, Vienna, pp. 363–374.

Ferguson, J. (1980) Metamorphism in the Pine Creek geosyncline and its bearing on stratigraphic correlations. In *Uranium in the Pine Creek Geosyncline* (eds J. Ferguson and A. B. Goleby), Intl. Atomic Energy Agency, Vienna, pp. 91–100.

Ferguson, J., Ewers, G. R. and Donnelly, T. H. (1980) Model for the development of economic uranium mineralisation in the Alligator Rivers uranium field. In *Uranium in the Pine Creek Geosyncline* (eds J. Ferguson and A. B. Goleby), Intl. Atomic Energy Agency, Vienna, pp. 563–574.

Ferguson, J. and Goleby, A. B (eds) (1980) *Uranium in the Pine Creek Geosyncline*, I.A.E.A., Vienna, 758 pp.

Ferguson, J. and Winer, P. (1980) Pine Creek geosyncline: statistical treatment of whole rock chemical data. In *Uranium in the Pine Creek Geosyncline* (eds J. Ferguson and A. B. Goleby), Intl. Atomic Energy Agency, Vienna, pp. 191–208.

Fischer, R. P. (1968) The uranium and vanadium deposits of the Colorado Plateau region. In *Ore Deposits of the United States 1933–1967. The Graton-*

Sales Volume (ed. J. D. Ridge), Am. Inst. Mining, Metallurgical and Petroleum Engineers, pp. 735–746.

Fraser, W. J. (1980) Geology and exploration of the Rum Jungle Uranium Field. In *Uranium in the Pine Creek Geosyncline* (eds J. Ferguson and A. B. Goleby), Intl. Atomic Energy Agency, Vienna, pp. 287–298.

Gabelman, J. W. (1977) Migration of uranium and thorium – exploration significance. *Studies in Geology* No. 3 American Association of Petroleum Geologists, 149 pp.

Gatzweiler, R., Schmeling, B. and Tan, B. (1981) Exploration of the Key Lake uranium deposits, Saskatchewan, Canada. In *Uranium Exploration Case Histories*, Intl. Atomic Energy Agency, Vienna, pp. 195–220.

Giblin, A. M. (1980) The role of clay adsorption in genesis of uranium ores. In *Uranium in the Pine Creek Geosyncline* (eds J. Ferguson and A. B. Goleby), Intl. Atomic Energy Agency, Vienna, pp. 521–530.

Goldhaber, M. B., Reynold, R. L. and Rye, R. O. (1983) Role of fluid mixing and fault related sulphide in the origin of the Ray Point Uranium District, South Texas. *Econ. Geol.*, **78**, 1043–1063.

Granger, H. C. and Warren, C. G. (1969) Unstable sulfur compounds and the origin of roll-type uranium deposits. *Econ. Geol.*, **64**, 160–171.

Gustafson, L. B. and Curtis, L. W. (1983) Post-Kombolgie metasomatism at Jabiluka, Northern Territory, Australia and its significance in the formation of high grade uranium mineralisation in Lower Proterozoic rocks. *Econ. Geol.*, **78**, 1–25.

Hambleton-Jones, B. B. (1978) Theory and practice of geochemical prospecting for uranium. *Min. Sci. Eng.*, **10(3)**, 182–197.

Hansuld, J. A. (1967) E_h and pH in geochemical prospecting. *Geol. Surv. Can. Pap.* **66(54)**, 172–187.

Harshman, E. N. (1968a) Uranium Deposits of Wyoming and South Dakota. In *Ore Deposits of the United States 1933–1967, The Graton-Sales Volume* (ed. J. D. Ridge) Am. Inst. Mining, Metallurgical and Petroleum Engineers, New York, pp. 815–831.

Harshman, E. N. (1968b) Uranium deposits in the Eocene sandstones of the Powder River Basin, Wyoming. In *Ore Deposits of the United States 1933–1967, The Graton–Sales* Volume (ed. J. D. Ridge), Am. Inst. Mining, Metallurgical and Petroleum Engineers, New York, pp. 838–848.

Hawkins, W. K. and Gearhart, M. (1969) Gamma-ray logging in uranium prospecting. In *Nuclear Techniques and Mineral Resources*, Int. Atomic Energy Agency, Vienna, pp. 213–222.

Haynes, D. W. (1979) Geological technology in mineral resource exploration. In *Mineral Resources of Australia* (eds D. F. Kelsall and J. T. Woodcock), Australian Academy of Technical Sciences, Parkville, Victoria, pp. 73–96.

Hegge, M. R., Mosher, D. V., Eupene, G. S. and Anthony, P. J. (1980) Geological setting of the East Alligator uranium deposits and prospects. In *Uranium in the Pine Creek Geosyncline* (eds J. Ferguson and A. B. Goleby), I.A.E.A., Vienna, pp. 259–272.

Hills, J. H. and Richards, J. R. (1972) The age of uranium mineralization in northern Australia. *Search*, **3**, 382–385.

Hills, J. H. and Richards, J. R. (1976) Pitchblende and galena ages in the Alligator Rivers region N.T. Australia. *Miner. Deposita*, **11**, 133–154.

Hoeve, J. and Sibbald, T. I. I. (1978a) On the genesis of Rabbit Lake and other unconformity-type uranium deposits in Northern Saskatchewan, Canada. *Econ. Geol.*, **73**, 1450–1474.

Hoeve, J. and Sibbald, T. I. I. (1978b) Mineralogy and geological settings of unconformity-type uranium deposits in Northern Saskatchewan. In *Uranium Deposits: Their Mineralogy and Origin* (ed. M. M. Kimberley), Mineralogical Association of Canada, Canada Books International, London, pp. 457–472.

Hostetler, P. B. and Garrels, R. M. (1962) Transportation and precipitation of uranium and vanadium at low temperatures, with special reference to sandstone-type uranium deposits. *Econ. Geol.*, **57**, 137–167.

Huber, G. C. (1980) Stratigraphy and uranium deposits, Lisbon Valley District, San Juan County, Utah. *Color. Sch. Mines Q.*, **75(2)**, 1–45.

Johns, R. W. (1970) The Athabasca sandstone and uranium deposits. *Western Miner*, Oct., 42–52.

Kelley, V. C., Kittel, D. F. and Melancon, P. E. (1968) Uranium deposits of the Grants Region. In *Ore Deposits of the United States, 1933–1967. The Graton-Sales Volume* (ed. J. D. Ridge), Am. Inst. Mining, Metallurgical and Petroleum Engineers, pp. 747–770.

Langmuir, D. (1978) Uranium solution-mineral equilibria at low temperatures with applications to sedimentary ore deposits. *Geochim. Cosmochim. Acta*, **42**, 547–569.

McAndrew, J. and Finlay, C. J. (1980) The nature and significance of the occurrence of uranium in the Nanambu complex of the Pine Creek geosyncline. In

Uranium in the Pine Creek Geosyncline (eds J. Ferguson and A. B. Goleby), Intl. Atomic Energy Agency, Vienna, pp. 357–362.

McH. Clark. R. J., Homeniuk, L. A. and Bonnar, R. (1982) Uranium geology in the Athabasca and a comparison with other Canadian Proterozoic basins. *CIM Bull.*, **75(840)**, 91–98.

Malan, R. C. (1968) The uranium mining industry and geology of the Monument Valley and White Canyon Districts, Arizona and Utah. In *Ore Deposits of the United States 1933–1967. The Graton-Sales Volume* (ed. J. D. Ridge), Am. Inst. Mining, Metallurgical and Petroleum Engineers, pp. 790–804.

Miller, D. S. and Kulp, J. L. (1963) Isotope evidence on the origin of the Colorado Plateau uranium ores. *Geol. Soc. Am. Bull.*, **74**, 609–630.

Mitchell, A. H. and Reading, H. G. (1969) Continental margins, geosynclines, and ocean floor spreading. *J. Geol.*, **77**, 629 pp.

Mitchell, A. H. G. and Garson, M. S. (1981) *Mineral Deposits and Global Tectonic Settings*, Academic Press Geology Series, London, 405 pp.

Morton, R. D. (1977) The Western and Northern Australian uranium deposits – exploration guides or exploration deterrents for Saskatchewan? In *Uranium in Saskatchewan* (ed. C. E. Dunn), *Sask. Geol. Soc. Spec. Publ.*, **3**, 211–254.

Motica, J. E. (1968) Geology and uranium–vanadium deposits in the Uravan mineral belt, S.W. Colorado. In *Ore Deposits of the United States 1933–1967, The Graton-Sales Volume* (ed. J. D. Ridge), Am. Inst. Mining, Metallurgical and Petroleum Engineers, pp. 805–814.

Mrak, V. A. (1968) Uranium deposits in the Eocene sandstones of the Powder River Basin, Wyoming. In *Ore Deposits of the United States 1933–1967, The Graton-Sales Volume* (ed. J. D. Ridge), Am. Inst. Mining, Metallurgical and Petroleum Engineers, pp. 839–848.

Nash, J. T., Granger, H. C. and Adams, S. S. (1981) Geology and concepts of genesis of important types of uranium deposits. *Econ. Geol. 75th Anniv. Vol.*, 63–116.

Needham, R. S., Crick, I. H. and Stuart-Smith, P. G. (1980) Regional geology of the Pine Creek geosyncline. In *Uranium in the Pine Creek Geosyncline* (eds J. Ferguson and A. B. Goleby), Intl. Atomic Energy Agency, Vienna, pp. 1–22.

Needham, R. S. and Roarty, M. J. (1980) An overview of metallic mineralisation in the Pine Creek geosyncline. In *Uranium in the Pine Creek Geosyncline* (eds J. Ferguson and A. B. Goleby), Intl. Atomic Energy Agency, Vienna, pp. 157–174.

Page, R. W., Compston, W. and Needham, R. S. (1980) Geochronology and evolution of the late Archaean basement and Proterozoic rocks in the Alligator Rivers Uranium Field, Northern Territory, Australia. In *Uranium in the Pine Creek Geosyncline* (eds J. Ferguson and A. B. Goleby), Intl. Atomic Energy Agency, Vienna, pp. 39–68.

Pagel, M. (1977) Microthermometry and chemical analysis of fluid inclusions from the Rabbit Lake uranium deposit, Saskatchewan. *Can. Inst. Min. Metall. Trans*, **86**, 157.

Pagel, M., Poty, B. and Sheppard, S. M. F. (1980) Contribution to some Saskatchewan uranium deposits mainly from fluid inclusion and isotopic data. In *Uranium in the Pine Creek Geosyncline* (eds J. Ferguson and A. B. Goleby), Intl. Atomic Energy Agency, Vienna, pp. 639–654.

Patterson, J. A. (1980) Long term uranium supply outlook. In *Uranium and Nuclear Energy: 1980*, Proceedings of 5th International Symposium held by the Uranium Institute, London, pp. 64–85.

Porter, D. A. (1981) Exploration of Bernabe Montano Complex of uranium deposits, New Mexico, U.S.A. In *Uranium Exploration Case Histories*, Intl. Atomic Energy Agency, Vienna, pp. 279–291.

Rackley, R. I. (1976) Origin of Western-states type uranium mineralization. In *Handbook of Stratabound and Stratiform Ore Deposits* (ed. K. H. Wolf), Elsevier, Amsterdam, Vol. 7, pp. 89–152.

Reynolds, R. L., Goldhaber, M. B. and Carpenter, D. J. (1982) Biogenic and non-biogenic ore-forming mechanisms in the south Texas uranium district: evidence from the Panna Maria deposit. *Econ. Geol.*, **77**, 541–556.

Riese, W. C., Lee, M. J., Brookings, D. G. and Della Valle, R. (1978) Application of trace element geochemistry to prospecting for sandstone-type uranium deposits. In *Geochemical Exploration 1978* (eds J. R. Watterson and P. K. Theobald), *Proc. 7th Int. Geochem. Expl. Symp.*, pp. 47–64.

Roberts, D. E. and Hudson, G. R. T. (1983) The Olympic Dam copper–uranium–gold deposit, Roxby Downs, South Australia. *Econ. Geol.*, **78**, 799–822.

Robertson, D. S. and Lattanzi, C. R. (1974) Uranium deposits of Canada. *Geosci. Can.*, **1**, 8–19.

Rowntree, J. C. and Mosher, D. V. (1981) Discovery of the Jabiluka uranium deposits, East Alligator River

Region, Northern Territory of Australia. In *Uranium Exploration Case Histories*, Intl. Atomic Energy Agency, Vienna, pp. 171–194.

Ryan, G. R. (1972) Ranger 1 : a case history. In *Uranium Prospecting Handbook*, Institution of Mining and Metallurgy, London, pp. 296–300.

Sandford, R. F. (1982) Preliminary model of regional Mesozoic groundwater flow and uranium deposition in the Colorado Plateau. *Geology*, **10**, 348–352.

Saracoglu, N., Wallis, R. H. and Brummer, J. J. (1983) The McClean uranium deposits, northern Saskatchewan – discovery. *CIM Bull.*, **76(852)**, 63–79.

Scott, F. (1981) MidWest Lake uranium discovery, Saskatchewan, Canada. In *Uranium Exploration Case Histories*, Intl. Atomic Energy Agency, Vienna, pp. 221–242.

Shockley, P. N., Rackley, R. I. and Dahill, M. P. (1968) Source beds and solution fronts. *Remarks Wyo. Met. Sect.* Am. Inst. Mining, Metallurgical and Petroleum Engineers, Feb. 27, 7 pp.

Strnad, J. G. (1980) Genetic models and their impact on uranium exploration in the Athabasca sandstone basin, Saskatchewan, Canada. In *Uranium in the Pine Creek Geosyncline* (eds J. Ferguson and A. B. Goleby), Intl. Atomic Energy Agency, Vienna, pp. 631–638.

Stuart-Smith, P. G., Wills, K., Crick, I. H. and Needham, R. S. (1980) Evolution of the Pine Creek geosyncline. In *Uranium in the Pine Creek Geosyncline* (eds J. Ferguson and A. B. Goleby), Intl. Atomic Energy Agency, Vienna, pp. 23–38.

Tremblay, L. P. (1978) Geological setting of the Beaverlodge type of vein-uranium deposit and its comparison to that of the unconformity type. In *Uranium Deposits: Their Mineralogy and Origin* (ed. M. Kimberley), Mineralogical Association of Canada, Canada Books International, London, pp. 431–455.

Tripathi, V. S. (1979) Comments on 'Uranium solution-mineral equilibria at low temperatures with applications to sedimentary ore deposits'. *Geochim. Cosmochim. Acta*, **43**, 1989–1990.

Uranium – Resources, Production and Demand (1983) A joint report by the Organisation for Economic Cooperation and Development, the Nuclear Energy Agency and the Intl. Atomic Energy Agency, December 1983, 348 pp.

Waters, A. C. and Granger, H. C. (1953) Volcanic debris in uraniferous sandstones and its possible bearing on the origin of precipitation of uranium. *U.S. Geol. Surv. Circ.*, **224**, 26 pp.

Wedepohl, K. H. (1972) *Hand-book of Geochemistry*, Springer-Verlag, Berlin.

Wood, H. B. (1968) Geology and exploitation of uranium deposits in the Lisbon Valley area, Utah. In *Ore Deposits of the United States 1933–1967, The Graton-Sales Volume* (ed. J. D. Ridge), Am. Inst. Mining, Metallurgical and Petroleum Engineers, pp. 771–789.

World Uranium (1980) World Uranium: Geology and Resource Potential, International Uranium Resources Evaluation Project, OECD Nuclear Energy Agency and International Atomic Energy Agency, 524 pp.

Ypma, P. J. M. and Fuzikawa, K. (1980) Fluid inclusion and oxygen isotope studies of the Nabarlek and Jabiluka uranium deposits, Northern Territory, Australia. In *Uranium in the Pine Creek Geosyncline* (eds J. Ferguson and A. B. Goleby), Intl. Atomic Energy Agency, Vienna, pp. 375–396.

10

Ores formed by metamorphism

10.1 INTRODUCTION

The processes of metamorphism give rise to some of the World's major deposits of tungsten and some important copper occurrences. These processes may also result in important high-grade sources of molybdenum, lead, tin, iron and zinc as well as by-product base and precious metals. Industrial minerals are also worked from deposits formed by metamorphism. Among the industrial minerals are asbestos, wollastonite, magnesite, talc and graphite.

The mineralogy of the deposits is often very complex and although some of these deposits have been extensively studied, metamorphism as an ore-forming process does not receive the coverage in some textbooks that it deserves. As some of the deposits are high grade, they can be worked economically even at a time of high costs and difficult marketing. Consequently there has been a renewed interest in exploration for these deposits in the late 1970s and early 1980s. This has been accompanied by more work on their genesis, characteristics and field relations (Meinert, 1983).

Metamorphism is usually regarded as a relatively high-temperature, high-pressure phenomenon, and most of the economic ore deposits covered in this chapter belong to such a regime. However, the effects of alteration and replacement in the formation of mineral deposits may be observed at much lower temperatures and pressures than those normally considered by metamorphic petrologists. In this context may be considered diagenesis and cell-by-cell replacement. These may be relatively easily simulated and subjected to experiment and they will provide evidence of the genesis of certain ore types, not only those of metamorphic origin. Consequently we will mention two examples cited recently before considering the major ore deposit type formed by metamorphic processes – skarns.

Alteration of recent sediments begins relatively soon after burial commences and these changes, otherwise referred to as diagenesis, may have direct effects on the formation of mineral deposits. During diagenesis trace elements may be released from detrital minerals. These trace elements can be retained in saline interstitial fluids which then migrate to suitable sites where they precipitate their metal contents by reaction with reduced sulphur or hydrocarbons (Holmes *et al.*, 1983). The accumulation of metals may continue at these

Ores formed by metamorphism

sites to form potentially economic deposits or the deposit in turn may suffer further metamorphism, alteration and, eventually, destruction. The release of trace elements may also occur during the stoping of enormous volumes of metamorphosed sediments resulting from magmatic activity and these could provide the source, in part at least, of the large tonnages of metals contained in some of the other mineral deposits considered in this textbook, such as those in Chapters 3 and 4. Replacement deposits of various types occur but possibly one of the most intriguing is the formation of a Holocene zinc ore body at Howard's Pass, Yukon, which seems to have formed by cell-by-cell replacement of moss which formed interstitially in talus (Jonasson *et al.*, 1983).

In what follows within this chapter we shall be concentrating upon the alteration and replacement of rocks around plutonic intrusions, or supposedly caused by unseen plutonic intrusions. Therefore emphasis will be placed on the ore deposits which form by metamorphic changes related to hydrothermal and magmatic activity. Attention will be concentrated upon metallic ores associated with thermally altered country rocks.

Although regional metamorphism will be largely ignored, some mention should be made here of the regional metamorphic effects on sedimentary ores. These effects may be summarized as: (i) effect on mineralogy; (ii) effect on grain size; and (iii) effect on lithology. These effects can greatly influence the viability of a mining prospect. For example the HYC deposit (see Chapter 6) has not been developed mainly because the deposit is unmetamorphosed and the grain size is therefore very fine. This poses serious problems for the processing of the ore.

Where the mineralogical and chemical changes occur in rocks as a result of interaction with fluids from an external source, the process is called metasomatism. Such metasomatic activity is one of the mechanisms which leads to the formation of the ore deposit type known as 'skarns', to which the remainder of this chapter is devoted. Skarns may also form by metamorphic recrystallization of impure carbonate rocks and basic volcanics.

10.2 SKARNS

10.2.1 Introduction

'Skarn' was originally a term applied to coarse-grained calc-silicate gangue associated with the iron ore deposits of Sweden. Subsequently it has been used to include a variety of calc-silicate rocks rich in calcium, iron, magnesium, aluminium and manganese that formed by replacement of originally carbonate-rich rocks. The term is used irrespective of the presence or absence of potentially economic mineralization.

Skarns may form by (i) metamorphic recrystallization of impure carbonate rocks, (ii) metasomatic reaction between unlike lithologies, or (iii) infiltration metasomatism caused by hydrothermal fluids.

10.2.2 Classification of skarns

Skarns may be classified according to the rock type they replace and the terms 'exoskarn' and 'endoskarn' have been applied to replacements of carbonates and intrusive rocks respectively in contact zones where the intrusive rock was assumed to be genetically related to the skarn-forming fluids (Einaudi and Burt, 1982). Endoskarn is favoured in those areas where fluid flow is presumed to have been dominantly into the pluton or upward along its contacts with carbonates rather than where metasomatic fluid flow is largely out of the pluton. Endoskarns show mineral zoning reflecting addition of calcium. Under reducing conditions the sequence towards limestone is biotite–amphibole–pyroxene–(garnet). Any potassium feldspar disappears whereas biotite and plagioclase remain important except in the rare cases where garnet becomes dominant. Therefore the diagnostic assemblage is pyroxene–plagioclase. This sequence is typical of most tungsten skarns and some copper skarns. Under relatively oxidizing conditions, epidote–quartz is favoured over pyroxene–plagioclase, and garnet tends to be more abundant. Such cases are common in most copper and lead–zinc skarns.

Exoskarns may be classified on the dominant

Table 10.1 Characteristics of major skarn types (from Einaudi et al., 1981 and Einaudi and Burt, 1982)

Type	Tungsten (calcic)	Tin (calcic)	Tin (magnesian)
Metal association (minor metals)	W,Mo,Cu (Zn, Bi)	Sn,F (Be,W)	Sn,F (Be,B)
Tectonic setting	Continental margin Syn-/late orogenic	Continental margin Late to post-orogenic	
Associated intrusives	Quartz diorite/ quartz monzonite	Granite	Granite
Intrusive morphology	Large batholithic plutons	Stocks batholiths	Stocks batholiths
Intrusive alteration	Local mica, calcite, pyrite; endoskarn	Greisen	Greisen
Ore	Scheelite, molybdenite, chalcopyrite pyrrhotite, pyrite	Cassiterite, arsenopyrite, pyrrhotite, stannite	Cassiterite, minor arsenopyrite, pyrrhotite, stannite, sphalerite
Typical size (million tonnes)	0.1–2	0.1–3	1
Typical grade	0.7% WO_3	0.1–0.7%Sn	?

calc-silicate mineral assemblage, which in turn reflects the composition of the carbonate rock replaced, into (a) magnesian skarn – skarn that replaces dolomite and consists largely of magnesian silicates (for example forsterite and serpentine) and (b) calcic skarn – skarn that replaces limestone and consists largely of Fe–Ca silicates (for example andradite and hedenbergite). The bulk of the World's economic skarn deposits occurs in these calcic exoskarns (Table 10.1). In some cases calcic skarn may replace magnesian skarn, for example Mason Valley Mine (Einaudi, 1977) and at Costabonne (Guy, 1980). Skarns resulting from the replacement of basalts will also have a distinctive mineralogy. For instance, the tin-bearing skarns in the Land's End peninsula of the UK are associated with a cordierite–anthophylite–biotite hornfels, and in this area a garnet–copper ore has been worked in the past (K. F. G. Hosking, personal communication).

10.3 SKARN DEPOSITS

Ore deposits associated with skarn were referred to as 'hydrothermal metamorphic' by Lindgren (1905), 'tactite' by Hess (1919) and 'pyrometasomatic' by Lindgren (1922). More recently the terms 'igneous metamorphic' (Park and MacDiarmid, 1975) or 'contact metasomatic' (Jensen and Bateman, 1979) have been adopted, although Evans (1980) has revived the use of 'pyrometasomatic'. As ore deposits of this 'type' can occur in environments suggesting either

Ores formed by metamorphism

Copper (calcic)	Zinc-lead (calcic)	Molybdenum (calcic)	Iron (magnesian)	Iron (calcic)
Cu (Mo,W,Zn)	Zn,Pb,Ag (Cu,W)	Mo,W (Cu,Bi,Zn)	Fe (Cu, Zn)	Fe (Cu, Co, Au)
Continental margin		Continental margin		Oceanic island arc; rifted continental margins
Syn-orogenic to late orogenic		Late orogenic	Syn-orogenic	
Granodiorite to quartz monzonite	Granodiorite to granite, diorite to syenite; plutons commonly absent	Quartz monzonite to granite	Granodiorite to granite	Gabbro to syenite mostly diorite
Stocks, dykes breccia pipes	Large stocks to dykes – if present	Stocks	Small stocks, dykes sills	Large to small stocks, dykes
Local endoskarn, K – silicate, sericitic	Extensive endoskarn?	Quartz veins K – silicate	Minor endoskarn, propylitic	Extensive endoskarn Na – silicates
Chalcopyrite, bornite, pyrite, haematite, magnetite	Sphalerite, galena, chalcopyrite arsenopyrite	Molybdenite, scheelite, bismuthinite pyrite, chalcopyrite	Magnetite, pyrite, chalcopyrite, sphalerite, pyrrhotite,	Magnetite, chalcopyrite, cobaltite, pyrrhotite,
1–100	0.2–3	0.1–2	5–100	5–200
1–2% Cu	9% Zn, 6%Pb, 5 oz/ton Ag	?	40% Fe	40% Fe

wholly metamorphic or entirely metasomatic processes, or virtually anywhere between the two end members, and as igneous contacts are not always necessarily present, none of these terms is satisfactory. We prefer to follow Einaudi *et al.* (1981) in using the term 'skarn deposits', a term less restrictive in its genetic connotations and one recognizable by most economic geologists.

The metamorphic origin of these deposits is recognized by the vast majority of authorities but some authors have attempted to assign a syngenetic origin to certain of the deposits, for example Hutchinson (1979) for the Tasmanian tin deposits, and Stumpfl (1977) suggested that many tungsten deposits may be syngenetic exhalative. Hutchinson's interpretation of much of the evidence has been questioned by Solomon (1980).

Hutchinson (1980) re-emphasizes that collectively the factors he has considered present a strong case for an exhalative origin. Hutchinson's views find very little support from other workers on the Renison and Cleveland deposits (K. F. G. Hosking, personal communication). However Stumpfl's suggestion of a syngenetic exhalative origin appears to be correct for the Austrian tungsten deposits.

The paper by Einaudi *et al.* (1981) is an excellent review of skarn deposits and provided much of the inspiration for this chapter.

10.4 CLASSIFICATION OF SKARN DEPOSITS

The subdivision of skarn deposits has been based on the protolith rock type or composition, the

Ore deposit geology and its influence on mineral exploration

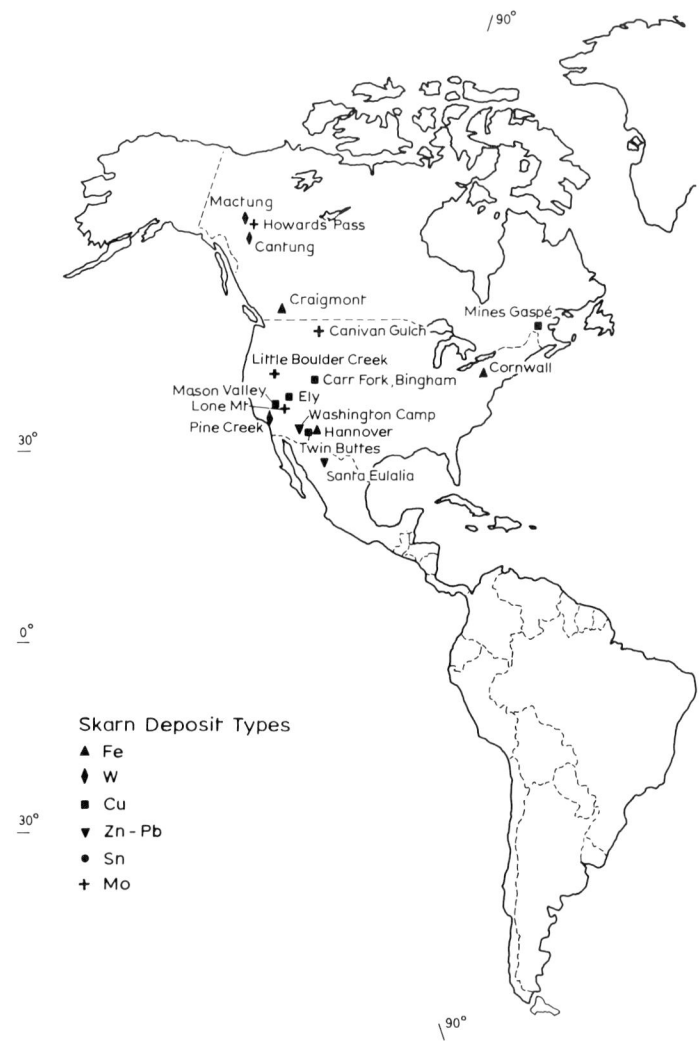

Fig. 10.1 Location map of the skarn deposits mentioned in text.

mechanism of fluid movement or the temperature and extent of magmatic involvement. None of these provides a classification that assists exploration and since some of these characteristics are difficult to determine without very extensive research we feel skarn deposits are best classified by the dominant economic metal Fe, W, Cu, Zn–Pb, Mo and Sn (Einaudi *et al.*, 1981; Meinert, 1983). All of these can occur in each skarn type but W, Cu and Zn–Pb deposits are notably rare in magnesian skarn. Variations and gradations within the skarn deposits are a result of magma type, depth of formation, reducing capacity and composition of host rocks, distance of carbonate horizon from the magmatic source and degree of meteoric water involvement.

10.4.1 General characteristics

(a) Distribution of deposits in space and time
Most skarn deposits are Mesozoic or younger, but there are a few important Palaeozoic tungsten and tin skarns such as the King Island deposits. Copper and lead–zinc skarn deposits are dominantly of

Ores formed by metamorphism

Tertiary age. Skarn deposits are distributed worldwide. They were major contributors of copper and iron at the beginning of the twentieth century. Their importance for these metals decreased during this century with geological attention for sources of copper being focused on porphyry deposits, stratiform and volcanogenic sulphides and for iron on iron-formations. They still maintain an important position for the supply of tungsten and recently interest has been reawakened in their importance in porphyry copper districts such as the south-west USA (Atkinson and Einaudi, 1978). The USSR, Japan and China have major skarn deposits as do the USA, Canada and Tasmania (Fig. 10.1).

(b) Size and grade of deposits

Although in general skarn deposits are low in tonnage by comparison with volcanogenic sulphide or sediment-hosted sulphide deposits, they are important suppliers of certain metals (Table 10.1). Approximately 58% of the Western World's tungsten production is from scheelite from skarns, and this proportion may well

Fig. 10.2 Malayaite from the type area of Chenderiang, Perak, Malaysia. Specimen photographed in (a) ordinary light and (b) ultraviolet light.

SCALE IN CMS.

increase (Bender, 1979). In tungsten-bearing skarns the WO_3 content usually varies from 0.4% to over 1%. Two of the most important skarn tungsten deposits are Canada Tungsten in the North American Cordilleran belt, with reserves over 30 million tonnes of 0.9% WO_3, and the King Island deposit, Tasmania, which has a yearly output in excess of 400 000 tonnes at 0.69% WO_3 (Kwak et al., 1982).

Tin skarns contain grades of 0.1 to 0.7% with tonnages up to 30 million tonnes but not all this tin is economically recoverable, largely because an appreciable amount may occur as an essential element in malayaite or as a non-essential component particularly in andradite garnet (Fig. 10.2a,b).

Iron skarns range in size from 2 million to 300 million tonnes; zinc in skarns may grade up to 12%

(c) Mineralogy

The main ore minerals for each skarn type are summarized in Table 10.1; it can be seen that magnetite is the main source of iron, scheelite of tungsten, chalcopyrite of copper, cassiterite of tin, and sphalerite and galena of zinc and lead respectively. The main gangue minerals in the exoskarns are garnets, pyroxene and amphibole (Figs. 10.3 and 10.4). These minerals are the most useful in classification and exploration since they are present in all skarn types and show marked compositional variations. Figures 10.3 and 10.4 show the different compositions of the gangue in the various skarn deposit types (as defined by the main metal present). Garnets and pyroxene have been widely used for skarn comparisons because their compositional variations can be shown relatively simply in such ternary plots as Figs. 10.3 and 10.4.

The relative abundance and compositions of the main gangue minerals will be controlled by the original rock type, the environment and fluid involved in the skarn formation. Detailed investigation is necessary to determine whether rock type or fluid was the major source of components for the skarn. A summary of the exoskarn compositions for the main skarn types, as defined by the principal metal present, is given in Table 10.2. Detailed descriptions of the mineralogy of each of the main skarn types are given below.

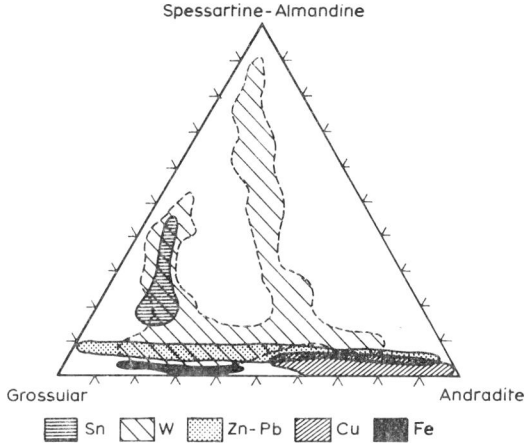

Fig. 10.3 Garnet species from calcic skarn deposits (after Einaudi and Burt 1982).

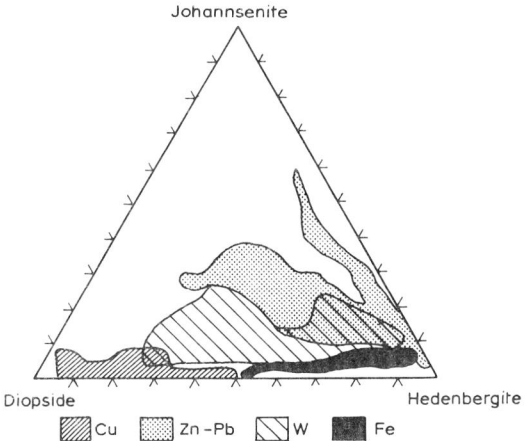

Fig. 10.4 Pyroxene species from calcic skarn deposits (after Einaudi and Burt, 1982).

(d) Host rock lithology and alteration

The associated intrusives for each of the main skarn types are given in Table 10.1. The host rock lithology is fundamental to the formation of skarn deposits and it has long been recognized that skarns are usually associated with calcareous rocks. Thermal (contact) metamorphism results in calcium–aluminium silicates in calcareous shales, in calcium–magnesium silicates in silty dolomites and wollastonite in cherty limestones. These may be grouped under the terms iron-poor calc-silicate marbles and hornfelses. If impure carbonates and calcareous shales and sandstones are abundant then metamorphic calc-silicates may comprise a large

and these skarns may have lower grades of lead. Although these zinc–lead skarns contain negligible copper, they may contain up to 255 g/t silver.

Some of the major porphyry copper systems of the World have copper skarn deposits associated with them (for example Bingham and Twin Buttes). The average grade of these is 1% copper and the tonnage ranges from 50 to 500 million tonnes, for instance Ok Tedi has about 50 million tonnes at 1.46% Cu (Pintz, 1984).

Table 10.2 Exoskarn composition and main gangue mineralogy of calcic skarn deposit types. (Source: data from Einaudi *et al.*, 1981 and Einaudi and Burt, 1982)

Type	Iron	Tungsten	Copper	Zinc–lead	Tin–tungsten
Exoskarn composition Early minerals	High Fe Low S, Mn Ferrosalite (Hd20–80), grandite (Ad20–95), epidote, magnetite	High Al, Fe Low S Ferrosalite–hedenbergite (Hd60–90, Jo5–20), grandite (Ad10–50), idocrase, wollastonite	High Fe, S Low Al, Mn Andradite (Ad60–100), diopside (Hd5–50), wollastonite	High Fe, Mn, S Low Al Manganoan hedenbergite (Hd30–90, Jo10–40), andraditic garnet (Ad20–100, Spess. 2–10) bustamite rhodonite	High Al, F Low Fe, S Idocrase spess.-rich grandite, Sn andrad., malaya., danbur. datolite
Late minerals	Amphibole, chlorite, ilvaite.	Spessartine (5–35) almandine (5–40)– grandite, biotite, hornblende, plagioclase.	Actinolite (chlorite, montmorill.)	Mn actin. ilvaite chlorite dannemor. rhodochr.	Amphi. mica chlor. tourm. fluo.

Abbreviations: actin, actinolite; ad, andradite; amphi, amphiboles; andrad, andradite; chlor, chlorite; danbur, danburite; dannemor, dannemorite; fluo, fluorite; Hd, hedenbergite; Jo, johannsite; malaya, malayite; montmorill, montmorillonite; rhodochr, rhodochrosite; spess, spessartine; tourm, tourmaline.

portion of a skarn deposit. These metamorphic calc-silicates will reflect the composition of the original sedimentary rocks ('sedimentary protolith'). Most commonly the dominant impurities are magnesium and aluminium, therefore the resulting calc-silicate mineralogy is grossularitic garnet and diopsidic pyroxene. If iron-rich sedimentary protoliths occur, as at King Island, Tasmania (Edwards *et al.*, 1956), iron-rich metamorphic calc-silicates are formed.

These early-formed metamorphic aureole phenomena are often overprinted by the metasomatic processes of skarn growth. Such metasomatic processes are necessary to the formation of economic deposits since the early metamorphic event is not associated with ore formation, but may be accompanied by the creation of increased permeability. Two varieties of metasomatic skarns have been recognized – reaction skarn and igneous metasomatic skarn. The former is generally early in skarn formation and is usually confined to moderate- to high-grade metamorphic terrains. This type generally has a bulk composition reflecting the host rocks and has a restricted association of major minerals (grossularite–diopside; feldspar, clinozoisite). Ore formation is not commonly associated with this type of metasomatic skarn. Igneous metasomatic skarns, on the other hand, show much more varied major mineral associations and have bulk compositions bearing little obvious relation to the host rocks. They are caused by magmatic-hydrothermal systems and are late in the skarn-forming episode. Ore deposition is generally associated with this type of metasomatic skarn.

Following the end of the episode of skarn

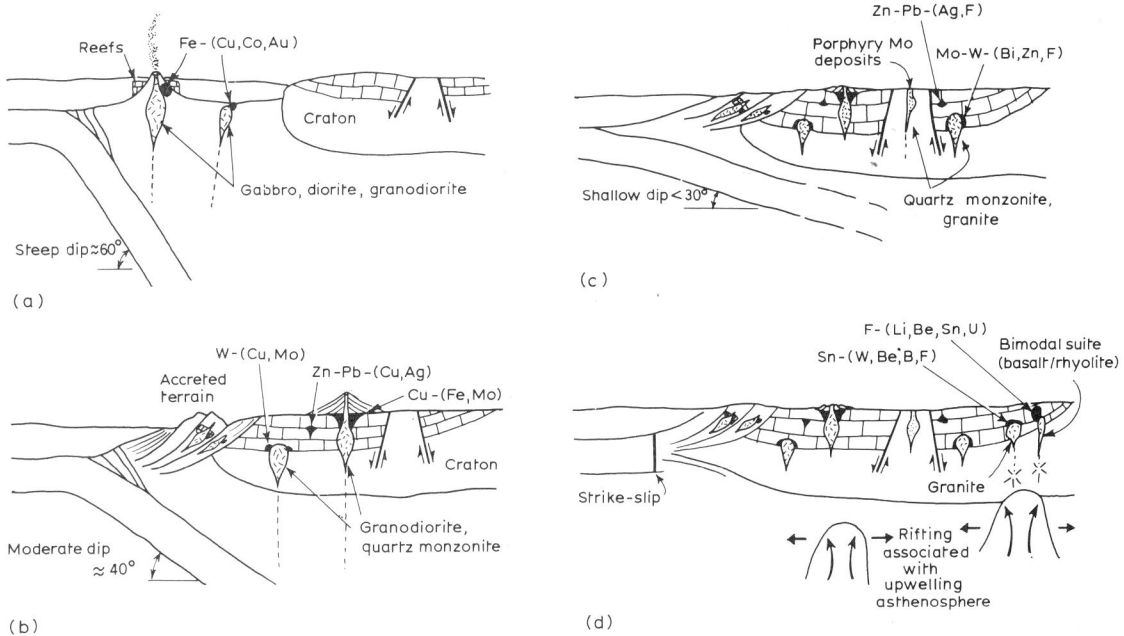

Fig. 10.5 Idealized tectonic models for skarn formation (a) An oceanic subduction environment; (b) a continental subduction environment; (c) a transitional low-angle subduction environment; (d) a post-subduction or incipient rifting environment. (From Meinert 1983.)

growth a period of sulphide deposition and retrograde alteration may, and often does, occur. The superimposition of all these events – metamorphic, metasomatic, sulphide deposition and retrograde alteration – leads to confusing paragenesis in the skarn and complicated mineralogy and zoning in it and the host lithology. For example, the Carr Fork skarn at Bingham, Utah, shows a copper skarn, and an early wollastonite, pyroxene and garnet zone developed in a cherty limestone, which were overprinted by actinolite–chalcopyrite and later by clay – calcite –sulphides (Atkinson and Einaudi, 1978). Furthermore the Carr Fork deposit illustrates the preferential formation of skarn in calcareous horizons – the Yampa and Highland Boy Limestones – and the fact that the highest-grade copper is reached at some distance from the contact with the causative pluton (the Bingham stock). This distribution in grade is related to the host rock lithologies, their stratigraphic and structural positions and their relative impermeability compared with the enclosing metasediments.

(e) Tectonic setting

Some attempt at unifying skarn occurrence may be attempted by placing them within a plate tectonic framework (Table 10.1). Although this may be an overgeneralization, and exceptions must occur, it could form a basis for regional exploration for specific skarn deposit types (Fig. 10.5).

Almost the only skarn deposits associated with the more mafic igneous rock types of oceanic island-arc terrains are calcic magnetite skarns; often having significant Cu, Co and Au (Fig.

10.5a). These may also have anomalously high nickel and zinc with an absence of tin, tungsten and lead. Overall they are iron-rich and deficient in magnesium, reflecting the primitive ocean crust, the wall rocks and the plutons associated with these skarns. On the other hand, continental margin orogenic belts contain magnesian magnetite skarn deposits.

Tungsten and base-metal (Cu, Zn–Pb, Mo) sulphide skarns are also typical of magmatic arcs related to moderate angle subduction beneath continental crust (Fig. 10.5b). Quartz monzonites and granodiorites are the most common plutons, the intrusions representing relatively great depths, say between 5 and 15 km. Evidence suggests that the tungsten-rich skarns formed at greater depths, and higher temperatures, than Cu and Zn–Pb skarns.

Most copper skarn deposits are in continental crust, few being associated with island-arc settings. Important calcic Cu skarns are associated with altered and mineralized porphyry stocks and plugs of granodiorite to quartz monzonite intruded at relatively high levels in the continental crust, 1–6 km (Fig. 10.5c). These skarns are sulphide-rich and are mined principally for copper, copper–iron and zinc–lead. Locally important by-products of molybdenum, gold and silver are extracted. The best-known examples of these are the porphyry-copper-related skarns.

Calcic zinc–lead skarn deposits are thought to form in mid- to late-orogenic stages of continental margin belts, associated with granodioritic or granitic magmatism. Skarns proximal to the pluton are generally less Mn-rich with lower Pb to Cu ratios than distal skarns.

Tin skarns are associated with ilmenite-series granites of both I and S types, particularly the latter, emplaced late in, or after, the orogenic cycle of continental magmatic arcs or in relatively stable or incipiently rifted cratonic environments (Fig. 10.5d). In both magnesian and calcic tin skarns, the extremely rare tin borates, and the much more common malayaite, are formed early, and cassiterite is not formed until more acidic, relatively low-temperature, retrograde alteration conditions are reached. The amount of cassiterite, and therefore the amount of recoverable tin, is directly related to the degree of retrograde alteration. It has been suggested that high tin values in massive sulphide replacement bodies in dolomite (for example Renison Bell, Tasmania) may represent the low-temperature distal analogues of tin skarns.

During the transition to post-subduction tectonics the magmatic arc may widen or migrate further inland resulting in intrusives which show more interaction with continental crust than did the earlier plutons. The magmas so formed are quartz monzonite to granite and have associated skarns mined for a variety of metals. Tungsten and molybdenum are generally dominant but with major amounts of copper and zinc and with minor amounts of bismuth, lead, silver and gold.

10.4.2 Tungsten skarn deposits

In 1983 the mine production of tungsten totalled 37 350 tonnes (*Mining Annual Review*, 1984). China was the largest producer, totalling some 10 000 tonnes. The USSR produced 9000 tonnes, Australia 2000 tonnes and the USA 1100 tonnes. These figures are lower than in many previous years because the period covered was one of world-wide recession; the figure for the USA is particularly reduced (it was 2700 tonnes in 1977 and over 3000 tonnes in 1980) as the major mines were particularly affected by the recession. In Europe, Austria and Portugal are the biggest producers. The two main ore minerals for tungsten are wolframite and scheelite.

Tungsten occurs in the following classes of deposit (Hosking, 1982): (1) tungsten-bearing banded granitoids; (2) tungsteniferous brines and evaporites; (3) placers; (4) stratabound and allied deposits; (5) pegmatite deposits; (6) hydrothermal deposits; (7) pyrometasomatic deposits. Of these only the last three are of any major economic significance and the last, which includes skarns, yields the largest tonnages. In skarns the main tungsten ore mineral is scheelite and this accounts for about 58% of the tungsten used in the Western World.

The bulk of the World's tungsten production, and reserves, in skarns comes from a very few, relatively large deposits – King Island, Tasmania; Sangdong, Korea; Pine Creek, California (Fig. 10.1) and two areas in Canada (Canada Tungsten, Northwest Territories and MacMillan Pass, Yukon). These deposits each contain more than 10 million tonnes of ore. The grade of WO_3 worked ranges from as low as 0.4% in open pits in the Western USA (Bateman, 1982) to 1% or more in Canada. In addition to the normal mining constraints on economic working of these deposits – location, mining method, by-products availability, etc. – the scheelite grain size is very important, as is the molybdenum content. In general early anhydrous skarn has moderate and relatively consistent grades of fine-grained, high-molybdenum scheelite; retrograde skarn has variable grades of medium- to coarse-grained low-molybdenum scheelite. In these latter skarns the occurrence of the erratic high-grade veins and alteration zones may favour small-scale selective mining. In Japan molybdenum-rich scheelite is associated with I-type granites, whereas molybdenum-poor scheelite is associated with S-type granites.

The presence of dolomite in the sedimentary succession has an adverse effect on tungsten grade because of the importance of Ca activity in the precipitation of scheelite (Newberry, 1980). The only important tungsten skarn in dolomite is Costabonne, France (Guy, 1980). Tungsten skarns are generally stratiform and less than 15 m thick. They may be continuous for hundreds of metres along lithological contacts; vein skarns are rare.

A division of tungsten skarns was suggested by Newberry (1979): (i) 'reduced' skarns, such as Canada Tungsten, which were formed in carbonaceous host rocks or at great depth and which have abundant ferrous iron assemblages including hedenbergitic pyroxene, almandine-rich garnet, biotite and hornblende and, (ii) 'oxidized' skarns, such as King Island, which were formed in non-carbonaceous or hematitic rocks or at lesser depth and which have an abundance of ferric iron assemblages, such as andraditic garnet and epidote.

Since the mineral compositions from tungsten skarns vary continuously from dominantly ferrous to moderately ferric this must reflect, at least in some cases, subtle differences in environment of formation. The early anhydrous assemblages (600–500°C) often suffer retrograde alteration and the hydrous assemblages are formed at temperatures between 450°C and 300°C. Very high grades of tungsten are inevitably associated with retrograde assemblages, and Newberry (1980) suggests that the abundant calcium released by the breakdown of pyroxene and garnet minerals during retrograde alteration contributes to the precipitation of tungsten from solution.

King Island scheelite deposits

The King Island scheelite deposits contribute approximately 65% of Australia's tungsten production. These deposits have been described by Edwards et al. (1956), Knight and Nye (1953, 1965), Danielson (1975) and Kwak et al. (1982). King Island is situated about 240 km south of Melbourne and the scheelite ore bodies lie in the south-east of the island. The ore bodies are startiform and the ore horizons vary from 5 to 40 m in thickness. Three ore bodies have been recognized, the No. 1 ore body and the Dolphin ore body lie near the contact with a granodiorite, and the Bold Head ore body is 3 km from these and is near the contact of an adamellite. Total production from 1911 until 1972 was approximately 5.7 million tonnes of ore averaging 0.53% WO_3; reserves quoted in 1975 were 7 million tonnes averaging 0.75% WO_3 (Danielson, 1975). In 1982 the Dolphin deposit was producing 60% of the yearly output (400 000 tonnes of ore at 0.69% WO_3) according to Kwak et al. (1982).

The scheelite deposits are in a 150–200 m thick sequence of contact metamorphosed and metasomatized sediments known as the 'Mine Series' which are probably early Cambrian or late Proterozoic in age. The Mine Series is overlain by a metavolcanic unit with agglomerates, basalts,

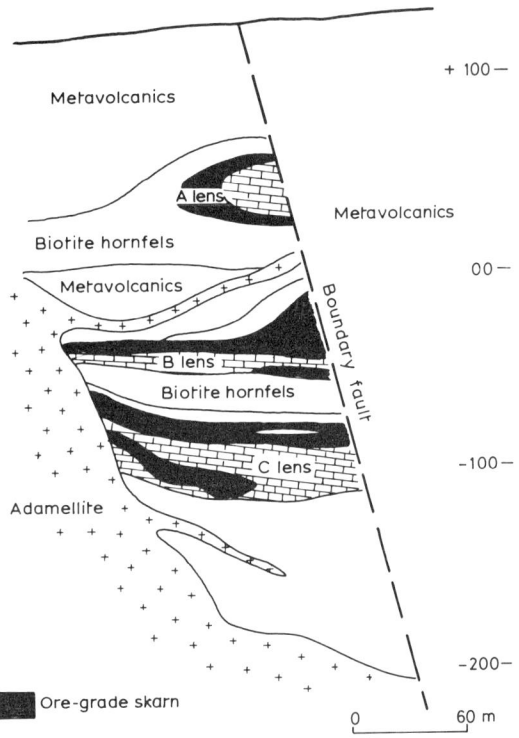

Fig. 10.6 Section through the Bold Head contact aureole looking north, King Island deposit (after Danielson, 1975).

tuffs and spilites. Underlying the Mine Series is a quartzite of undetermined thickness. The granodiorite and adamellite intrusive stocks are dated as late Devonian or early Carboniferous. The skarn rocks containing the ore bodies are thought to have formed by selective metasomatic replacement of limestone beds (Fig. 10.6). The scheelite is present as disseminated grains (average 0.05–0.2 mm) in and along the margins of andradite garnets, in quartz and, rarely, calcite. Coarse grains of scheelite, up to 50 mm, occur in joints, tension gashes and in quartz–calcite pods.

Three lenses of mineralization, A, B and C, occur with the lowest, 'C lens' containing the main mineable ore. This lens is divided into lower and upper components. The mineralization in the 'upper C lens' is reasonably uniformly disseminated while in the 'lower C lens' and the underlying banded footwall beds the scheelite is concentrated along bedding and is confined to the marble horizons. Often only the top and the bottom of a marble bed are mineralized, suggesting that mineralization has progressed laterally and along dip with little movement across bedding. In addition to scheelite, molybdenite is present but neither wolframite nor cassiterite has been recognized. Other sulphides present are pyrite, arsenopyrite, pyrrhotite and chalcopyrite (Danielson, 1975), although sphalerite, galena, bournonite and bismuthinite have also been recorded (Edwards et al., 1956).

The ore skarn in all three deposits has a primary assemblage of: andradite–grossularite garnet + hedenbergite–diopside–pyroxene + powellite–scheelite + quartz. This stage 1 assemblage is overprinted by varying amounts of ferrohastingsite amphibole + epidote + pyrrhotite + pyrite + calcite + quartz + low-Mo–scheelite + molybdenite.

Kwak et al. (1982) have examined the variation in garnet composition and have shown that generally the mole percent andradite in the core and outer core zones decreases consistently to the outer edge of the skarn. The outer edge of the skarn is taken at the Northern Boundary Fault some 500 m from the granodiorite contact. In the midsection and edge zones of the same garnets the andradite values decrease to about 400 m and then increase again.

When the scheelite and powellite ratios of scheelite occurring as inclusions in garnets and as

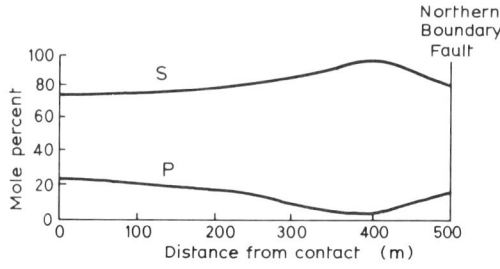

Fig. 10.7 Composition of interstitial scheelite, Dolphin deposit, King Island. S, scheelite mole fraction; P, powellite mole fraction. (After Kwak et al. 1982.)

Ores formed by metamorphism

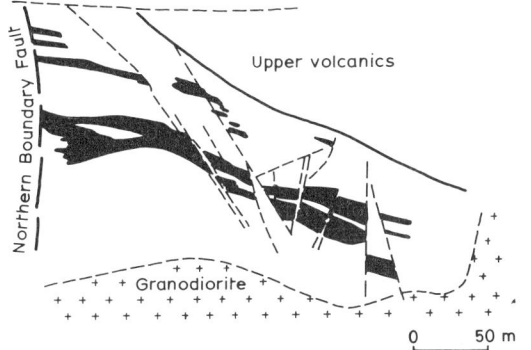

Fig. 10.8 Section 220–200 through the Dolphin deposit at King Island (after Kwak *et al.* 1982).

interstitial grains are also studied these show distribution patterns comparable to the midsection and edge zones of the garnets – reaching a maximum scheelite/powellite ratio at about 400 m from the contact (Fig. 10.7). Kwak *et al.* conclude that these systematic chemical variations within minerals suggest relatively high permeability during skarn genesis with a fairly simple hydrological system.

The tungsten values in the ore are generally in a sinusoidal distribution (Brown, quoted in Kwak *et al.*, 1982). The highest values are furthest from the contact and seem to suggest fluid flow out from the Northern Boundary Fault but the sinusoidal nature indicates contributions via the granodiorite–skarn contact and some of the faults shown in Fig. 10.8. Total mineable tungsten tends to increase towards the faults, near which a bismuth-rich zone occurs.

The skarns in the Dolphin deposit lie above a flat portion of the granodiorite contact (Fig. 10.8) and the ore lenses in the Bold Head deposit are also associated with shelving of the contact, this time with the adamellite at Bold Head (Fig. 10.6).

Genesis of the King Island deposits

Examination of the fluid inclusion filling

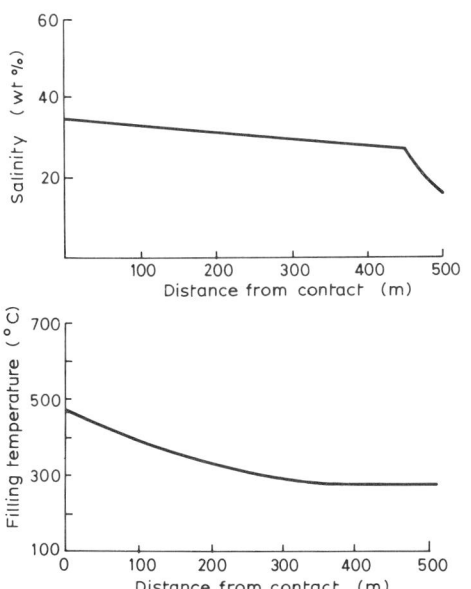

Fig. 10.9 Salinites and filling temperatures for primary fluid inclusions, interstitial scheelite, Dolphin deposit, King Island (after Kwak *et al.* 1982).

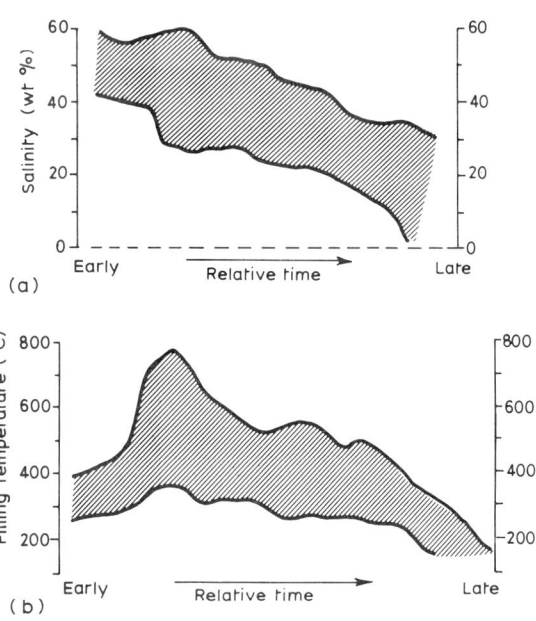

Fig. 10.10 Compilation of (a) salinity of fluid inclusions and (b) temperature, with time. From Dolphin deposit section 220–200 (Fig. 10.8). Minerals used in compilation – garnets, pyroxene and interstitial scheelite. (Modified from Kwak *et al.* 1982.)

temperatures and salinities shows systematic variations in space and time (Figs. 10.9 and 10.10). Laterally many inflections occur in the 300–400 m interval from the contact. Paragenesis shows that there is a systematic decrease of total salinity of the solutions with time in the different minerals studied (Fig. 10.10). Filling temperatures on the other hand, as shown by the inclusions from different portions of the garnets, were low initially (300–400°C), rising later to magmatic temperatures, when boiling occurred, and then decreased.

These data show that quite small quantities of hot saline magmatically derived solutions entered the marble in which contact skarns had already formed and cooled to about 380°C. Solution of $CaCO_3$ occurred in response to Fe-mineral deposition creating enhanced permeability until the skarn must have appeared like a calc-silicate 'sponge' (Kwak et al., 1982). Boiling occurred (600–800°C) nearest the contact, and the Northern Boundary Fault appears to have been active. Consequently primary fluid entered the evolving skarn from the fault and the granodiorite contact, meeting in the 300–400 m interval. Other faults in the sequence shown in Fig. 10.8 were either absent or too tightly closed at this time to contribute significantly to the plumbing system. Mo-rich scheelite began to precipitate during the boiling event.

Subsequently any of this Mo-rich scheelite not enclosed in garnet was dissolved and redeposited as Mo-poor scheelite and molybdenite at fairly low temperatures (±300°C). At this late stage some of the other faults may have provided channelways and the skarns adjacent to these faults in the upper (B lens) were formed. The fluid inclusion data indicate that the deposit formed at a depth of 2–3 km. This is shallower than many tungsten skarn deposits are thought to have formed (Einaudi et al., 1981).

10.4.3 Tin skarn deposits

In general the tin skarn deposits are of minor economic importance as they are of low grade and are small. The grade of tin in these deposits ranges from 0.1 to 0.7 wt% but much of the tin is in silicates and is not economically recoverable. Tonnages involved vary from a few million tonnes of ore to about 30 million tonnes as at Moina, Tasmania, where the skarn is over 1 km long and up to 100 m thick (Kwak and Askins, 1981).

A considerable number of tin-bearing silicates and borates are known from skarns but, with the possible exception of malayaite, they are extremely rare and of no economic significance except they contribute to the development of soil geochemical anomalies and the chemical grade of the deposits. Probably stanniferous garnets and other species containing tin as a non-essential component are generally more important in these respects. In skarns the only economic ore mineral for tin is the tin oxide cassiterite.

According to Burt (1978) cassiterite appears to form relatively late during the genesis of tin skarns and seems to owe its appearance to falling temperatures and the increase in acidity of the conditions (which show an enrichment in B_2O_3, CO_2 and HF). These changes in the environment of the skarn, which are related to the retrograde stage of skarn formation, result in the liberation of tin from the silicates and/or the borates and deposition of the tin as cassiterite (Burt, 1978):

$$CaSnSiO_5 + CO_2 > CaCO_3 + SiO_2 + SnO_2$$
Malayaite $\quad\quad > $ Calcite + Quartz + Cassiterite

$$CaSnB_2O_6 + 2SiO_2 > SnO_2 + CaB_2Si_2O_8$$
Nordens- + Quartz > Cassit- + Danburite
kioldine $\quad\quad\quad\quad$ erite

Although this reaction may occur, much of the field evidence suggests that the cassiterite was formed directly from the invasion of a stanniferous agent. In many of the skarns in South East Asia malayaite shows no signs of degradation although cassiterite may occur in the same ore body (K. F. G. Hosking, personal communication).

According to Burt (1978) cassiterite is the only stable tin mineral in the presence of quartz, fluorite, calcite and the other gangue minerals that are typical of skarn deposits but Hosking's evidence of co-existing non-degraded malayaite with cassiterite and other skarn species throws some doubt on this assertion. The late stage of tin skarn formation is often accompanied by the introduction of low-sulphur sulphides such as pyrrhotite and arsenopyrite and, rarely, loellingite, sphalerite and bornite. However high-sulphide values are not necessarily coincident with high-tin values as may be demonstrated at Mt Lindsay, Tasmania (Kwak, 1983).

Compared with skarns generally, those at Mt Lindsay are particularly anomalous in having high Ti and relatively high Sn. In the primary skarns the tin may be in cassiterite with limited amounts in scheelite; later the tin is almost exclusively in garnet (up to 0.51 wt%) and sulphides. According to Kwak (1983) subsequent skarn development may lead to the dissolving of the tin which may then be reprecipitated as chlorite–cassiterite–pyrrhotite–carbonate skarn as seen in the ore mined at Cleveland Mine, Tasmania. We are not totally convinced that the tin here is skarn derived. Kwak's line of reasoning suggests that this skarn type may have occurred at Mt Lindsay but has subsequently been eroded.

Kwak (1983) emphasizes the importance of the morphology of the pluton in the development of skarns. The skarns are preferentially developed above troughs and shelves of the upper contact of granitoids. Where the contact is steeply dipping, or vertical, only narrow skarn zones occur. The relationship of the undulations in the pluton contact to skarn location has already been mentioned in the case of the King Island tungsten deposits (Section 10.4.2). This morphological relationship between the pluton and the skarn deposits can be very important at the target appraisal stage of exploration for skarns.

10.4.4 Zinc–lead skarn deposits

The typical grades of zinc–lead skarn deposits are 6–12% Zn, lesser Pb, negligible Cu, also many contain silver (28–255 g/t Ag). Examples include Washington Camp, Arizona (Simons, 1974) and Santa Eulalia, Mexico (Hewitt, 1968). Zinc–lead skarns are distinguished from other skarn types by a manganese- and iron-rich mineralogy and because they commonly occur distal to intrusive contacts. Other distinctive features for this skarn type may be listed as follows:

(1) they usually occur along structural or lithological contacts;
(2) there is an absence of significant metamorphic aureoles centred on the skarn;
(3) they show pyroxene as the dominant calc-silicate mineral;
(4) there is an association of significant amounts of sulphide mineralization with pyroxene rather than with garnet or other silicate minerals;
(5) the retrograde mineralogy of these skarns is manganese-rich ilvaite, pyroxenoids, subcalcic amphiboles and chlorite.

The intrusives with which zinc skarn deposits appear to be associated range from granodiorite to leucogranite and from deep-seated holocrystalline batholiths to porphyritic hypabyssal stocks and dykes and even aphanitic rhyolite dykes and breccia pipes (Table 10.1). Zinc skarn deposits formed near contacts with equigranular batholithic intrusives tend to be smaller and less rich in manganese than other types of zinc skarns. Zinc skarn deposits formed near contacts with smaller intrusives include some of the largest of this type. If these skarns are formed near dykes it seems that the dyke acted largely as a structural pathway and the ultimate source of the metasomatic solutions was a deeper, possibly cogenetic, magmatic body.

In this type there is a continuous transition from skarn ore to massive sulphide replacement of carbonate rock. This latter ore type is important in most skarns of this class and may constitute the largest and highest grade part of the ore. The skarn is dominated by johannsenitic pyroxene, sometimes to the exclusion of garnet. Bustamite is common and garnet, where present, is andraditic. Metal zoning may be pronounced with increasing depth from lead to lead–zinc and, deeper still, to copper–iron.

The occurrence of zinc skarn distal to a probable, suspected, or even unknown, igneous source with no obvious spatial or temporal link to the igneous source emphasizes the importance of solution pathways in the genesis of zinc–lead skarn deposits. The role of such solution pathways may be illustrated by reference to the Ban Ban zinc deposit in south east Queensland, Australia (Ashley, 1980).

Here there is the peculiarity of a high-grade zinc zone associated with nearby low-grade copper and this may be a function of the initial fluid composition (high Zn:Cu ratio). The separation of the zinc zone from the low-grade copper (and iron sulphide) zone may be due to zinc remaining in solution longer than copper (and some iron) before being precipitated by hydrolysis reactions as sphalerite.

10.4.5 Iron skarn deposits

Although skarns were major sources of iron at the turn of the century their importance for this metal has decreased with the development of the very large tonnage iron-formations (Chapter 8). Economic concentrations of magnetite in skarn deposits are found in the complete range of skarn-forming environments, but two specialized environments appear to produce skarn with magnetite as the only ore mineral. These environments are island arcs (calcic skarn) and Cordilleran-type magnesian skarn (Table 10.1). Other skarns from which major amounts of magnetite have been, or could be, mined but from which iron is not the principal metal worked include the massive magnetite bodies replacing calc-shale and tin-bearing skarn at granite contacts in west Malaysia (Hosking, 1973), the inner garnet zones of zinc-bearing calcic skarn near Hannover, New Mexico (Hernon and Jones, 1968) and the magnesian skarn of several large porphyry copper skarns such as Ok Tedi, Papua New Guinea (Bamford, 1972).

(a) Island-arc calcic magnetite skarn deposits

These are distinguished from other skarn types by their association with gabbros and diorites in volcanic–sedimentary sequences and also the relatively large amount of skarn that is formed in the igneous rocks. Island-arc calcic magnetite skarns also show widespread calcium metasomatism and anomalous concentrations of cobalt and occasionally nickel.

In this setting the magnetite ore bodies characteristically show a close spatial association with garnet zones or occur in limestones beyond the skarn zone. They may be small irregular contact deposits, capable of yielding about 5–20 million tonnes, up to stratiform bodies hundreds of metres in thickness and several kilometres in strike length (the Cornwall deposit, Pennsylvania, contains in excess of 100 million tonnes at 41% Fe (Lapham, 1968)). Magnetite is the dominant primary iron oxide mineral; any haematite present normally forms through surface oxidation. Pyrite and chalcopyrite are present in minor amounts and normally the copper content is less than 0.2 wt% in the iron ores.

(b) Cordilleran magnesian magnetite skarn deposits

Magnetite may form readily in magnesian skarns since magnesian calc-silicates (for example forsterite, talc and serpentine) do not take up much iron in solid solution at the oxidation states normally considered to prevail in skarn-forming environments. Any iron in solution therefore tends to form magnetite whereas the same solution entering a limestone at high temperatures would tend to form iron-rich garnet or pyroxene.

Craigmont, British Columbia
Lithological control for skarns is also demonstrated by the Cu–Fe skarn ore at Craigmont, British Columbia (Morrison, 1980). Here the occurrence of copper- and iron-rich units in the host rocks controlled the distribution of much of the ore. The ore was therefore localized in favourable host rock facies adjacent to unfavourable facies that acted as physical and chemical barriers to ore fluid migration. In the case of Craigmont the favourable facies was characterized by interbedded limy and quartzo-feldspathic sedimentary rocks with concentrations of iron in pyrite and biotite; the copper occurred as

disseminated chalcopyrite. Bedding planes and fractures apparently acted as channelways for the fluids which leached copper and iron from the host rocks and concentrated them in the facies-bounded trap.

If such stratigraphic control exists in an area, such as that proposed by Morrison for the ore distribution at Craigmont, then this would be of value in mine-based exploration. The recognition of the favourable facies would be an important exploration aid. Assuming the necessary metal content existed prior to skarn formation then the exploration geologist must look for structural, stratigraphic and/or chemical traps in the contact aureole which could lead to ore-bearing skarn.

10.4.6 Molybdenum skarn deposits

The intrusions associated with molybdenum-bearing skarns are generally more evolved than those associated with iron, copper or most tungsten skarns. There is a strong similarity in the composition of igneous rocks associated with porphyry molybdenum and skarn molybdenum deposits (Table 10.3). This suggests similarities in igneous petrogenesis (Meinert, 1983). Granites are commonly associated with this skarn type (Einaudi et al., 1981).

In addition to molybdenum, these skarns may be mined for various metals including copper, tungsten and bismuth. Some molybdenum skarns may have important quantities of lead, zinc, tin or even uranium. An example of a molybdenum-copper skarn which contains pitchblende is Azegour in Morocco (Ridge, 1976). The ore minerals in these skarns are molybdenite, scheelite, chalcopyrite and bismuthinite. The presence of scheelite in this deposit was not recognized initially (Hosking, personal communication). This emphasizes the importance of using a short-wave ultraviolet light when prospecting skarns.

Minor amounts of sphalerite and galena may occur, although some molybdenum occurrences may have significant sphalerite in distal skarns, such as Lone Mountain, Nevada (Bonham and Garside, 1979). These polymetallic skarns seem to be characterized by a mineralogy of hedenbergitic pyroxene, grandite garnet and wollastonite. Even deposits that contain molybdenite as the only important sulphide, such as Little Boulder Creek, Idaho (Balla and Smith, 1980), have large amounts of hedenbergitic pyroxene although only minor grandite garnet. The skarn at Little Boulder Creek occurs in calcareous siltstone and such deposits as this one are considerably larger than the polymetallic skarns but have not yet been brought into production. The presence of dolomitic or silty rocks appears to be important in molybdenum skarn ore formation, as other very large deposits, as well as Little Boulder Creek, are associated with these rock types, such as Cannivan Gulch, Montana (Schmidt and Worthington, 1977).

The tonnage and grade of these molybdenum-bearing deposits are a reflection of the mining method used or proposed. Azegour has been selectively mined to produce grades in excess of 1% MoS_2, whereas the bulk mining method proposed for Little Boulder Creek would be based upon average grades between 0.1 and 0.2% MoS_2. This latter deposit would contain reserves close to 100 million tonnes while the higher-grade deposits would have reserves nearer 10 million tonnes.

10.4.7 Copper skarn deposits

Most skarn deposits mined for copper are associated with calc-alkaline granodiorite to quartz monzonite stocks emplaced in continental

Table 10.3 Average composition of igneous rocks associated with molybdenum deposits (from Meinert, 1983)

Porphyry	Mo deposits	Mo skarn deposits
SiO_2	75.2	74.8
Al_2O_3	12.3	14.3
CaO	0.80	1.2
Na_2O	3.31	3.1
K_2O	5.21	5.0
K_2O/Na_2O	1.61	1.61

margin orogenic belts. These are the stocks associated with the porphyry copper deposits (Section 3.2.2). A few copper skarns are found in island-arc settings associated with quartz diorite to granodiorite plutons.

As a group, copper skarns are characterized by: (i) an association with felsic, porphyry-textured stocks of hypabyssal character, (ii) proximity to stock contacts, (iii) high garnet/pyroxene ratios, (iv) relatively oxidized assemblages (andraditic garnet, diopsidic pyroxene, magnetite and haematite) and (v) moderately high sulphide contents. This group includes some of the World's largest skarn deposits – the porphyry copper-related skarns of south-west USA. These may contain 50–500 million tonnes of open pit copper ore in skarn and calc-silicate hornfels.

(a) Porphyry-related copper skarns

The copper skarns associated with porphyry copper range in age from Devonian to Pleistocene with the largest accumulation in the Laramide (70–50 million years) in the south-west USA. They reflect the environment of formation of the porphyry coppers: high level of emplacement and relatively oxidized fluids leading to the formation of ferric-rich garnet and ferrous-poor clinopyroxene. Early skarn formation occurred at 500–350°C which was probably too low for large-scale scheelite formation and too high for sphalerite to form.

Prograde skarn is correlated with the potassium silicate alteration in the pluton, whereas skarn destruction or silica–pyrite replacement of limestone is correlated with intense hydrolytic alteration in the plutons. With potassium silicate alteration the mineral association in carbonate rocks may be distinguished as follows: limestone develops andraditic garnet and diopsidic clinopyroxene with pyrite, chalcopyrite and magnetite; on the other hand, dolomite develops forsterite–serpentine–magnetite–chalcopyrite skarn.

Sulphides and iron oxides occur as disseminations and in veins in skarn and as massive replacements of marble at the skarn front. Overall, porphyry-related calcic skarns contain up to 10% iron oxides and 2–15% sulphides. Magnesian skarn is of relatively minor importance in porphyry-related deposits. Compared to calcic skarn, magnesian skarn has a high-magnetite content, low-sulphide content, and low pyrite to chalcopyrite ratios.

The variability of porphyry-related skarn deposits largely involves the degree of retrograde alteration, reflecting in turn the degree of sericitic alteration of the porphyry copper deposit. The dominant trend in retrograde alteration involves the formation of hydrous silicates that are progressively depleted in calcium as the intensity of alteration increases.

In addition to the retrograde alteration, a common alteration feature accompanying sulphide deposition in carbonate rocks associated with porphyry-related skarn may be various forms of silica, accompanied by pyrite or iron oxide. Therefore silica–pyrite, silica–haematite, or rarely, magnetite–chlorite may occur. Silica–pyrite may replace skarn, but more commonly it replaces limestone as massive irregular bodies, mantos or steep, structurally controlled, breccia pipes.

Average hypogene grades for mineralized sedimentary rocks as worked in open pits with a cut-off of 0.3–0.4% copper ranges from 0.6 to 0.9% Cu. With these grades the tonnage of ore contained in the sedimentary rocks associated with porphyry-related skarns ranges from 50 to 500 million tonnes. The copper/molybdenum ratio in the skarn is generally higher than in the associated pluton.

(b) Contrasts between porphyry-related and other copper skarn deposits

Porphyry-related skarns tend to consist of fine-grained to massive aggregates of calc-silicates due to the relatively high fluid flow in the porphyry systems which are normally emplaced at shallow depths in highly fractured rocks. In contrast non-porphyry-related skarns appear to form in less dynamic magmatic hydrothermal environments and perhaps at greater depths or greater distances from intrusive margins. In these

Ores formed by metamorphism

latter circumstances fluid movement is more restricted, and there is less likelihood of supersaturation and large crystals can grow slowly.

Porphyry-related skarns form during active structural deformation accompanying multiple intrusive events. The repetitive fracturing of brittle sedimentary rocks, hornfels, and already-formed skarn, is characteristic of this environment and this leads to vein densities greater than in non-porphyry skarns. One of the most characteristic types of veinlets in porphyry-related skarns is quartz-sulphide with actinolite alteration envelopes in diopside skarn or hornfels. These veinlets are mostly contemporaneous with biotite-orthoclase alteration of the associated plutons.

The quasi-monomineralic nature of skarn zones formed in the early stages, and usually preserved in simpler non-porphyry skarns may be largely destroyed in porphyry systems. In the non-porphyry skarns late-stage assemblages such as calcite-quartz-chlorite-pyrite or specularite are largely confined to vug fillings in garnetite, minor veins or immediate skarn intrusive contacts. The low water/rock ratios and low permeabilities in these skarns, compared to those that are porphyry-related, presumably block retrograde process. In contrast large volumes of porphyry-related skarns are commonly altered to late carbonate-hydrous silicate assemblages, and limestone may be replaced on a large scale by silica-pyrite.

Shelton (1983) describes skarn deposits from Mines Gaspé, Quebec. Here the skarn-type ore reserves are 11 million tonnes of average 1.25% Cu and 0.03% Mo with minor Pb, Zn, Ag, Au and

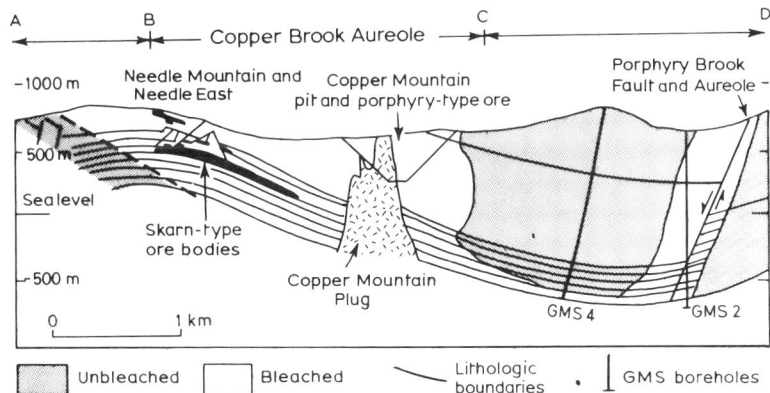

Fig. 10.11 Section across Copper Mountain intrusion and skarn-type ore bodies, Mines Gaspé, Quebec (Shelton 1983).

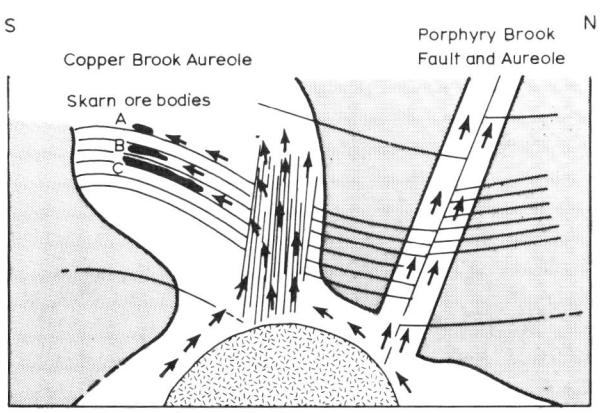

Fig. 10.12 Skarn ore formation. Schematic cross-section showing hydrothermal fluid flow patterns based on fluid inclusion and stable isotope data, Mines Gaspé, Quebec (Shelton 1983).

Ore deposit geology and its influence on mineral exploration

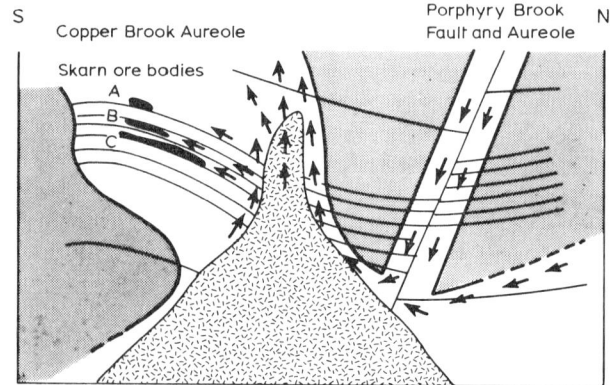

Fig. 10.13 Porphyry copper ore formation. Schematic cross-section showing hydrothermal fluid flow patterns based on fluid inclusion and stable isotope data, Mines Gaspé, Quebec Shelton, 1983).

Bi. Hydrothermal fluids were driven, by heat, upwards through fracture zones, the metasomatism leading to skarn-type rocks by reaction with the most calcareous sedimentary rocks. Subsequently, high-salinity magmatic fluids used the metamorphic equivalents of the most calcareous units as ore fluid aquifers. High-salinity fluids reacted with calcite and deposited early stage ore less than 1 km up dip from the fracture zones. Then less saline meteoric water mixed with the high-salinity fluid and formed the early skarn-type replacement ore bodies up to 2 km up dip from the fracture zone. This was subsequently followed by the late-stage porphyry-type ore formation associated with the quartz monzonitic Copper Mountain plug (Figs. 10.11, 10.12 and 10.13).

(c) Carr Fork, Bingham, Utah

The Bingham mining district, which contains the first recognized porphyry copper mine, has produced 13 million tonnes of copper, lead and zinc since 1896. Non-porphyry ore accounts for 23% of this total metal production (Atkinson and Einaudi, 1978). Large-scale underground mining began in the Highland Boy limestone copper-bearing skarn (Fig. 10.14) and this was followed by lead–zinc–silver ore being exploited from this limestone and the Yampa limestone. In the Carr Fork area the copper ore in skarn in the Yampa limestone was followed down dip to 1000 m below surface.

Geological interest in the Bingham area followed the trends in exploration and mining – the early work dealt with the mineralized sedimentary rocks on the western contact, then the emphasis changed to the disseminated copper ore in the intrusive rocks. Since the early 1970s more work has been published on the replacement deposits of lead–zinc ores and the skarns (Atkinson and Einaudi, 1978).

Contact effects at the Bingham stock fall into three stages:

(1) An early stage of contact metasomatism resulting in the formation of diopside-, wollastonite- and quartz-bearing assemblages close to the stock, and tremolite, talc, dolomite (a) calcite and quartz further away. This stage was accompanied by little or no sulphide deposition.

(2) The main stage of copper ore deposition accompanied by actinolite- and andradite-bearing assemblages.

(3) A late stage of pyritic mineralization accompanied by smectites, talc, chlorite and carbonates.

Overall the disseminated and stockwork copper mineralization may be regarded as an inverted cup-shaped structure draped over a low-grade core zone. This characteristic shape has been shown for porphyry molybdenum ore bodies in the western USA (Climax, Henderson – Section 3.4.2). This mineralization cuts across igneous and sedimentary rocks alike. The north-west side of

Ores formed by metamorphism

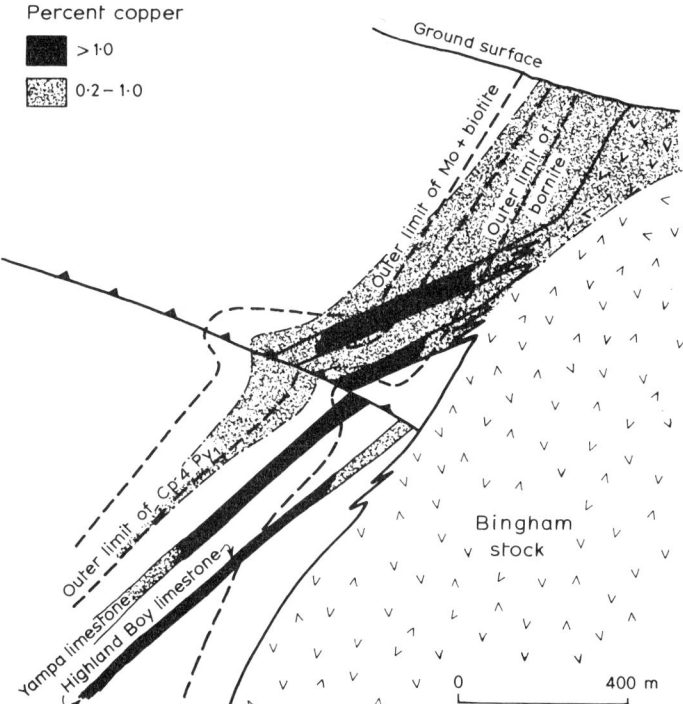

Fig. 10.14 North-west cross-section (looking north-east) of part of the Bingham stock and associated skarn deposits (after Atkinson and Einaudi, 1978).

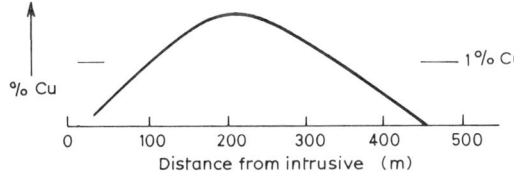

(a)

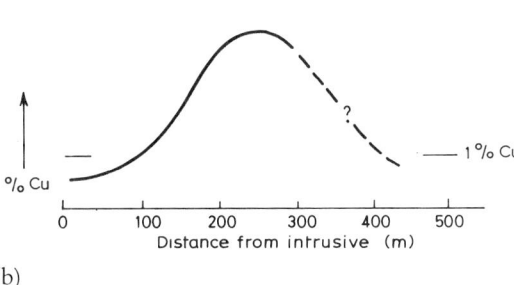

(b)

Fig. 10.15 Copper grade distribution in the skarn deposits in (a) the Yampa and (b) Highland Boy limestones (both 900–1300m elevation) relative to the Bingham quartz monzonite porphyry (after Atkinson and Einaudi, 1978).

the cup-shaped body is shown in Fig. 10.14 where it may be seen that the 0.2% copper tapers downwards and passes from being within the intrusion, quartzite and limestone, to being completely in the quartzite at 700 m below present pit bottom. On the basis of drill hole data this grade appears to bottom out at about 1200 m below surface and about 400 m away from the stock (which itself contains <0.1% copper)

At any given elevation the copper grades rise to a maximum and then decrease outward from the porphyry stock (Fig. 10.15). A similar distribution of grades in a tungsten skarn has been discussed earlier in the case of the King Island deposit (Section 10.4.2). The grades in the limestones, the Yampa and Highland Boy (Fig. 10.14), are much higher than in other lithologies because of their more favourable original composition. The position of the maximum grade in limestone is about 100 m closer to the intrusion than in the quartzite. This again may be explained by differing original lithologies and particularly the lack of fracture permeability in the limestones compared with the adjacent rocks (Meinert, 1983).

Although the copper distribution in the major skarn beds is erratic in detail, generalizations may be made (Fig. 10.15). This Figure shows curves for the elevations 900–1300 m, but lower elevations yield similar curves progressively shifted away from the contact and with copper grade maxima progressively lower with increasing depth. Atkinson and Einaudi (1978) explained the higher grades in the Highland Boy limestone as possibly being due to its lower stratigraphic position and therefore greater accessibility to mineralizing fluids which were thought to be moving upwards and outwards. The grade distribution is independent of the thickness of limestone units – the controlling factors appear to be distance from stock and the high permeability in the enclosing quartzite which allowed uniform entry of hydrothermal fluids. Furthermore, uniformity of original lithology reduced the possibility of fluid–wallrock reactions which would have affected ore deposition.

10.5 GENESIS OF SKARN DEPOSITS

From the evidence so far available it seems that there are three stages in the production of skarn deposits (Fig. 10.16): (1) emplacement of the causative pluton, giving rise to contact metamorphism; (2) prograde skarn growth, usually a metasomatic event; (3) retrograde alteration. These stages superimpose to form the complex mineralogy and zoning patterns of most large skarn deposits. The extent of the development of each stage will depend upon the local geological conditions. For instance stage 1 metamorphism will be much more extensive around skarns formed at greater depths, whereas stage 3 alteration will be intense in those skarns formed at relatively shallow depths and greatly affected by interaction with meteoric water.

(a) Stage 1 (Fig. 10.16b)

This may be considered the stage of ground preparation for skarn genesis and is normally barren of ore minerals. The contact metamorphism will tend to make the country rocks more brittle and more susceptible to the hydrofracturing process of stage 2. If intruded at a relatively great depth the carbonate rocks may behave in a more ductile manner rather than fracturing as they would at shallower depths.

(b) Stage 2 (Fig. 10.16c)

The crystallization of the magma releases magmatic hydrothermal fluids and causes hydrofracturing of the pluton and, often, the surrounding hornfels. Two varieties of metasomatic skarns have been distinguished:

(a) 'Reaction skarns' or those related to local metasomatism. These are generally confined to moderate- or high-grade metamorphic terrains and are often formed in sequences of interlayered shales or cherts and limestones. The chemical compounds for reaction are usually considered to be derived from the country rocks with no introduction of exotic material and the skarns are not often associated with ore formation.

(b) 'Igneous metasomatic' skarns which are thought to result from infiltration metasomatism related to magmatic hydrothermal systems. These are generally associated with ore deposition of some variety.

In the prograde skarn stage it is usually recognized that the inner zones (relative to the pluton) replace outer zones as these, in turn, move out. Some skarns have been identified where the inner zones formed first and later solutions flowed through them without equilibrating and reacted with the rock type beyond (Bartholome and Evrard, 1970).

(c) Stage 3 (Fig. 10.16d)

The retrograde stage is the time of sulphide deposition which is predated by the introduction of some oxide minerals such as cassiterite. The alteration of skarns is typically structurally controlled. This alteration cuts across earlier skarn patterns and the sulphide deposition may extend beyond skarn into marble and hornfels. The alteration products reflect

Ores formed by metamorphism

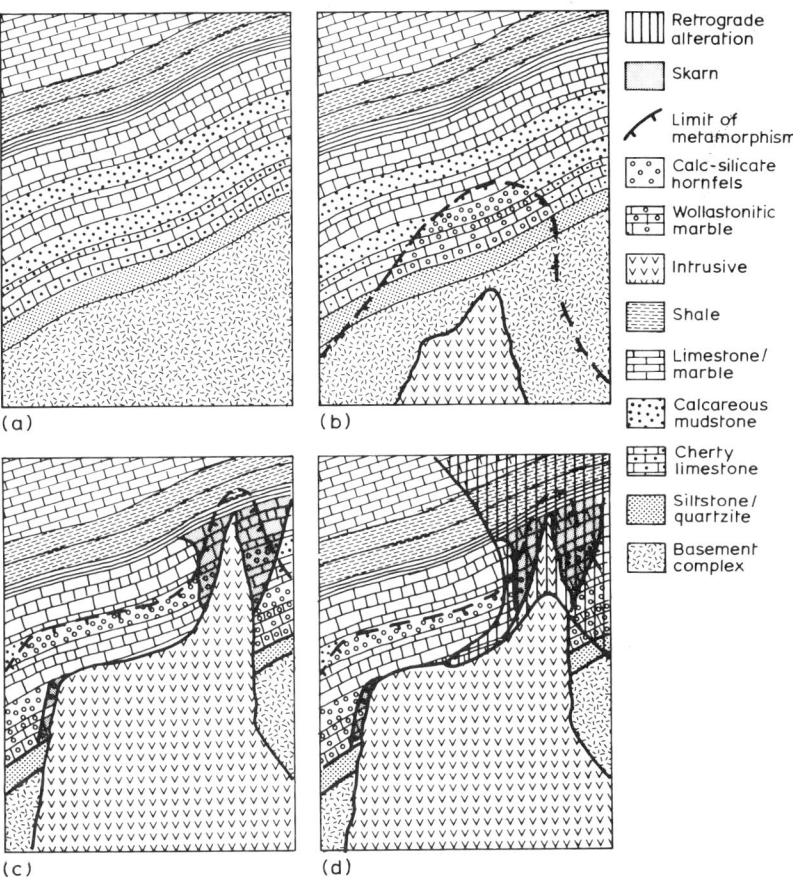

Fig. 10.16 Schematic illustration of sequential stages of skarn formation for relatively deep (left side of pluton) and relatively shallow (right side of pluton) skarn deposits. (a) Typical sedimentary section consisting of interbedded siltstone, shale, limestone and calcareous mudstone prior to intrusion. (b) Initial intrusive activity causing metamorphism (inside dashed line) of calcareous mudstone to calc-silicate hornfels, cherty limestone to wollastonitic marble, siltstone to quartzite and limestone to marble. (c) Release of metasomatic solutions from intrusive causing skarn overprint (shaded area) of earlier metamorphic minerals (note pockets of wollastonitic marble and hornfels in massive skarn) and skarn formation in purer carbonate beds. (d) Retrograde alteration (vertical ruled pattern) of high-temperature calc-silicate minerals caused by cooling and possible entry of meteoric water. In the deeper skarn the metamorphic aureole is more widespread than prograde skarn and retrograde alteration is minor. In the shallower skarn system, prograde skarn partly overruns the metamorphic aureole and both are overprinted by retrograde alteration. (From Meinert 1983.)

the original skarn silicate composition modified by the leaching of calcium and the introduction of volatiles. This usually means the replacement of calcium-rich calc-silicates by an assemblage of: (i) calcium-depleted silicates, (ii) iron oxides or sulphides and (iii) carbonates or albitic plagioclase.

The calcium released into solution may result in late precipitation of scheelite (Newberry, 1980).

The sulphides will precipitate due to declining temperature, local oxidation–reduction reactions in specific zones of the earlier skarn or at the marble contact due to neutralization of the hydrothermal solutions. This latter may give rise to high-grade sulphide bodies in skarn deposits.

Skarns apparently formed at greater depths, for example, tungsten skarns show less late alteration than those formed at shallower depths such as those associated with porphyry copper deposits. Skarns close to igneous bodies show more alteration than distal skarns, as for example zinc-bearing systems. A period of extensive skarn destruction can occur if significant hydrothermal circulation continues at low temperature. In some deposits this late period is limited to sparse vugh fillings of quartz–carbonate–sulphides while in others large portions of the skarn are converted to mixtures of quartz, chalcedony, clays, carbonates, sulphides and iron oxides. In some areas massive silica–sulphide replacement bodies form in limestone beyond the limits of the skarn as at Ely, Nevada (James, 1976).

Summarizing therefore, skarn formation spans the hydrothermal circulation of fluids from barely submagmatic to virtually hot spring. The trend goes from dominantly metamorphic through magmatic to meteoric.

10.6 EXPLORATION FOR SKARNS

Since most skarn deposits, with the possible exception of those containing lead and zinc, occur close to plutonic intrusions, exploration for them will be concerned initially with the recognition of suitable contact metamorphic terrain combined with suitable original lithologies for the location of ore minerals. The most favourable tectonic settings for the development of the different types of skarns are summarized in Table 10.1 and this table also indicates the type of igneous masses most likely to be associated with the skarn. Consequently any regional exploration programme for skarns must be based on these factors and should be guided by the current state of knowledge of the genesis of skarns. We consider that the empirical model will dominate exploration for a specific class of skarn whereas local factors affecting genesis will become increasingly important as the target area reduces to the level of an individual deposit within that class.

Careful geological mapping and sampling would be preceded by the use of regional geophysical surveys, photogeological interpretation and, possibly, use of satellite imagery. Using aerial photography, marked vegetation and topographic contrasts are often visible at the carbonate–igneous body contact. Potentially economic skarn environments have been recognized from these features in peninsular Thailand (K. F. G. Hosking, personal communication) Fracture analysis is another technique worth examining during the regional survey since skarn development is favoured by enhanced permeability and fractured lithologies. An example of skarn development in severely fractured and brecciated rocks is the 47 million tonne, 2.4% Cu deposit at Ertsberg East, Irian Jaya, Indonesia (Ward, 1984); the original Ertsberg deposit (which was discovered accidentally) formed a very prominent topographic feature.

Gravity surveys, to determine the distribution of batholiths or larger plutons, would also be useful in indicating the pluton morphology for smaller stocks and cupolas if these are sufficiently near-surface. As has been shown earlier, certain skarn types tend to occur where the intrusion contact flattens or shelves; these undulations may be recognizable from gravity surveys. Since magnetite may form an appreciable component of skarns (Table 10.1), regional magnetic and aeromagnetic surveys may be used to outline the skarn area. Kwak and Askins (1981) show the distribution of wrigglite skarn at Moina, Tasmania, using fluxgate magnetometer data (Fig. 10.17). Here the skarn was delineated beneath a covering of Tertiary basalt and, east of the Bismuth Creek fault, below a reverse fault, Hugo's fault.

Geochemistry may be used, as in the East Greenland area (Hallenstein et al., 1981), where heavy mineral concentrates from rivers and moraines were examined for scheelite content. If scheelite is the target, then it is important that heavy mineral concentrates are examined under ultraviolet light. In fact in the case of the East Greenland work prospecting was undertaken at night as local conditions dictated. When prospecting using ultraviolet light, specimens that fluoresce yellow (or yellowish) may be scheelite

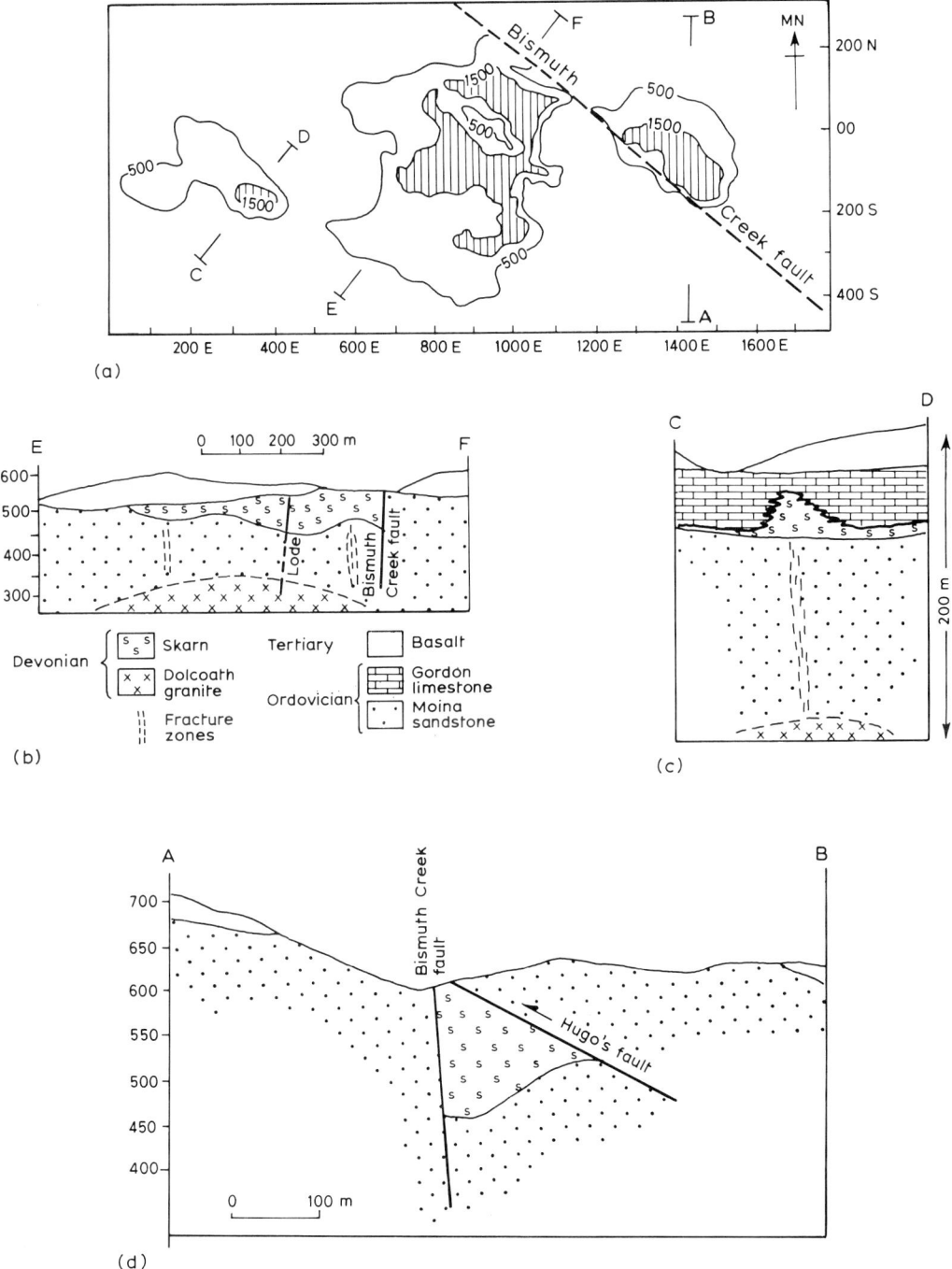

Fig. 10.17 (a) Magnetic anomaly map of the Moina Laminar skarn area, Tasmania. Values are in nanoteslas. (b), (c), (d), Geological sections taken along lines E–F, C–D, A–B respectively. C–D is schematic. (Modified from Kwak and Askins 1981.)

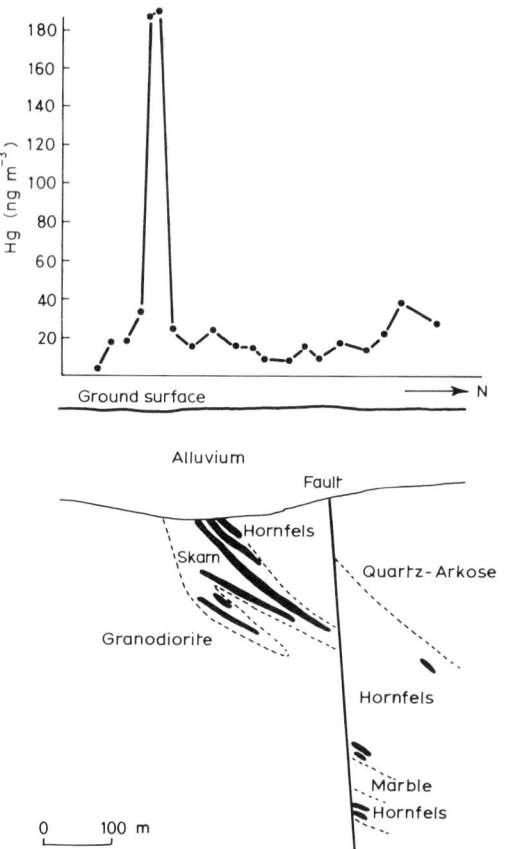

Fig. 10.18 Mercury content in soil gas over copper skarn deposit near Shanghai (modified from Zonghua and Yangfen 1981).

with a few percent molybdenum or powellite or malayaite (Fig. 10.2) A review of the theory and practice in tungsten prospecting is given by Yan (1982).

When prospecting in areas of residual overburden, normal soil geochemistry may be used where, for instance, malayaite or tin-bearing garnets are good indicator species for tin skarns. Another geochemical approach, using soil gas mercury was described by Zonghua and Yangfen (1981). They showed that a distinct recognizable mercury anomaly was discovered over the deposit (Fig. 10.18). In this case the method was successful despite the deposit being overlain by a thick transported overburden. To define the primary dispersion of mercury in the ore bodies and host rocks, drill samples were analysed for mercury and copper (Fig. 10.19). Copper formed a distinct halo around the body whereas the centre of the mercury halo is near the upper margin of the ore body. This upward zoning for mercury has a very important influence on the accuracy with which the method could locate the suboutcrop of the skarn below the alluvium (Fig. 10.18).

The occurrence of the highest grades at intermediate distances from the contact, within the metamorphic aureole, and concentrated within specific lithologies has been noted, for example in the Highland Boy limestone at Carr Fork. Therefore geological models need to be established with this in mind and exploration targets may be defined at some distance from the causative pluton. The targets will be based on the correct stratigraphic and structural level within the host rocks. Recognition of the approach to this zone, based solely on drilling, may be difficult but a careful study of mineralogical changes and distinct metamorphic alteration may prove useful guides.

10.7 CONCLUDING STATEMENT

The variations that we have noted in skarns include those in mineralogy, size and texture. These variations arise from skarn formation taking place at different depths, and are caused by differences in igneous intrusions, protoliths and tectonic environments. In view of the possible combinations of these factors, it is not surprising that skarns, as a deposit type, are more diverse than any other we have described in this textbook. Fortunately there are some systematic variations within these controlling factors which permit division of the skarns and the erection of models for their exploration.

Differences in the host rock major element composition give rise to the twofold division into calcic skarn and magnesian skarn. Further subdivision of these, based upon the main economic metal present, is now generally

Ores formed by metamorphism

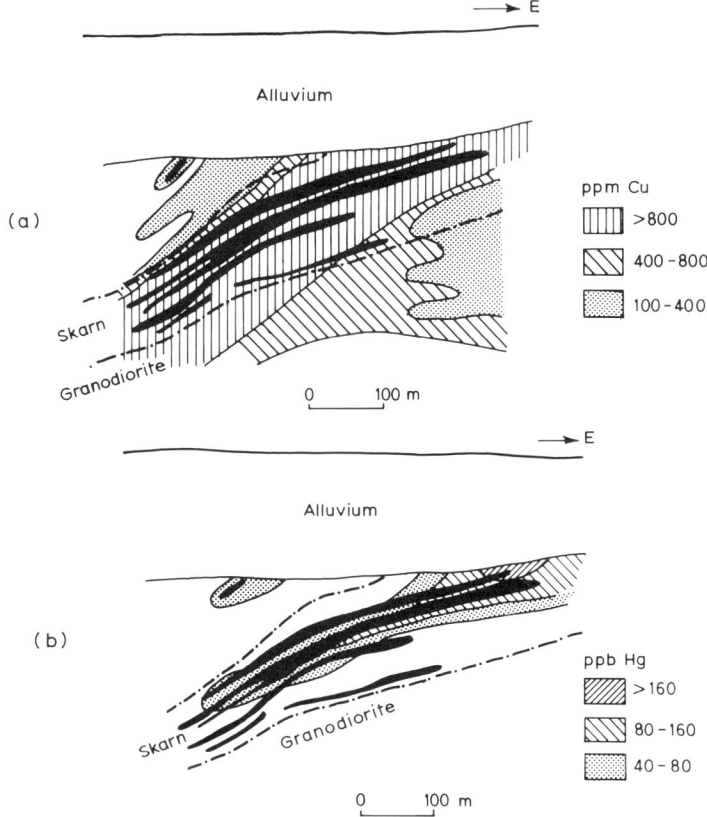

Fig. 10.19 Skarn copper deposit near Shanghai. Primary dispersion of (a) copper and (b) mercury, from borehole samples. (After Zonghua and Yangfen 1981.)

accepted. Although differences exist in the composition of the intrusions associated with the main metal skarn classes, these differences are reasonably systematic (Meinert, 1983). Tin skarns are associated with plutons with high silica values (about 76.6%) whereas, at the other end of the scale, iron skarns are found with plutons with lower silica values (about 61.5%).

There are sufficient similarities between the deposits within each metal class to assign that class to a plate tectonic setting (Table 10.1). As we have discussed in the previous section, this tectonic setting would form the starting point for any exploration programme. In the case of skarns, this exploration must be metal-specific. Empirically it may be said that skarns formed at greater depths will be small and vertical with extensive metamorphism whereas those formed at shallower depths will be larger, extensively altered, and with limited metamorphism. An empirical model, based on intrusions, lithologies and fracture patterns, can now be used to define prospective targets.

We believe that once the exploration programme has become site-specific then the genesis of that particular deposit must be examined in great detail to establish which factors, structural, stratigraphic or chemical, have controlled the occurrence of skarn and the distribution of payable grades within it.

REFERENCES

Ashley, P. M. (1980) Geology of the Ban Ban Zinc Deposit, a sulphide-bearing skarn, Southeast Queensland, Australia. *Econ. Geol.*, **75**, 15–29.

Atkinson, W. W. and Einaudi, M. T. (1978) Skarn formation and mineralisation in the Contact Aureole at Carr Fork, Bingham, Utah. *Econ. Geol.*, **73**, 1326–1365.

Balla, J. C. and Smith, R. G. (1980) Geology of the Little Boulder Creek deposit Custer County, Idaho (abs). *Geol. Soc. Am.* (Abstracts with programmes), **12**, 95.

Bamford, R. W. (1972) The Mount Fubilan (Ok Tedi) porphyry copper deposit, Territory of Papua New Guinea. *Econ. Geol.*, **67**, 1019–1033.

Bartholome, P. and Evrard, P. (1970) On the genesis of the zoned skarn complex at Temperino, Tuscany. *Int. Union Geol. Sci.* (pub.) Ser. A, **2**, 53–57.

Bateman, P. C. (1982) Scheelite-bearing skarns of the East-Central Sierra Nevada, California, USA. *Tungsten Geol. Symp., Jiangxi China* (United Nations) Economic and Social Commission for Asia and the Pacific Regional Mineral Resources Development Centre (ESCAP RMRDC), Bandung, Indonesia, pp. 23–31.

Bender, F. (1979) The tungsten situation: supply and demand, present and future. *Proc. 1st Int. Tungsten Symp., Stockholm*, pp. 2–17.

Bonham, H. F. and Garside, L. J. (1979) Geology of the Tonopah, Lone Mountain, Klondike and northern Mud Lake quadrangle, Nevada. *Nev. Bur. Mines Geol. Bull.*, **92**, 142 pp.

Burt, D. M. (1978) Tin silicate–borate–oxide equilibria in skarns and greisens – The system CaO–SnO_2–SiO_2–H_2O–B_2O_3–CO_2–F_2O_{-1}. *Econ. Geol.*, **73**, 269–282.

Danielson, M. J. (1975) King Island scheelite deposits. In *Economic Geology of Australia and Papua New Guinea* (ed. C. L. Knight), *Australas. Inst. Min. Metall. Monogr. No. 5*, **1**, 592–598.

Edwards, A. B., Baker, G. and Callow, K. J. (1956) Metamorphism and metasomatism at King Island scheelite mine. *J. Geol. Soc. Austr.*, **3**, 55–98.

Einaudi, M. T. (1977) Petrogenesis of copper-bearing skarn at the Mason Valley mine, Yerington district, Nevada. *Econ. Geol.*, **72**, 769–795.

Einaudi, M. T. and Burt, D. M. (1982) Terminology, classification and composition of skarn deposits. *Econ. Geol.*, **77**, 745–754.

Einaudi, M. T., Meinert, L. D. and Newberry, R. J. (1981) Skarn deposits. *Econ. Geol. 75th Anniv. Vol.*, 317–391.

Evans, A. M. (1980) *An Introduction to Ore Geology*, Blackwell, Oxford, 231 pp.

Guy, B. (1980) Etude géologique et pétrologique du gisement de Costabonne. Bureau de Recherches Géologiques et Minières, Paris, **99**, 237–250.

Hallenstein, C. P., Pedersen, J. L. and Stendal, H. (1981) Exploration for scheelite in East Greenland – a case study. *J. Geochem. Explor.*, **15**, 381–392.

Hernon, R. M. and Jones, W. R. (1968) Ore deposits of the Central mining district, New Mexico. In *Ore Deposits of the United States 1933–1967 Graton–Sales Volume* (ed. J. D. Ridge), American Institute of Mining, Metallurgy and Petroleum Engineers, New York, pp. 1211–1238.

Hess, F. L. (1919) Tactite, the product of contact metamorphism. *Am. J. Sci.*, **48**, 377–378.

Hewitt, W. P. (1968) Geology and mineralisation of the main mineral zone of the Santa Eulalia district Chihuahua, Mexico. *Soc. Min. Eng. AIME Trans.*, **241**, 228–260.

Holmes, I., Chambers, A. D., Ixer, R. A., Turner, P. and Vaughan, D. J. (1983) Diagenetic processes and the mineralization in the Triassic of central England. *Miner. Deposita*, **18**, 365–377.

Hosking, K. F. G. (1973) The search for tungsten deposits. *Bull. Geol. Soc. Malaysia*, **5**.

Hosking, K. F. G. (1982) A general review of the occurrence of tungsten in the World. *Tungst. Geol. Symp., Jiangxi, China*, (ESCAP RMRDC), Bandung, Indonesia, pp. 59–86.

Hutchinson, R. W. (1979) Evidence of exhalative origin for Tasmanian tin deposits. *CIM Bull. Metall.*, **72(808)**, 90–104.

Hutchinson, R. W. (1980) Author's reply 'Evidence of exhalative origin for Tasmanian tin deposits, *CIM Bull.*, **73(815)**, 167–168.

James, L. P. (1976) Zoned alteration in limestone at porphyry copper deposits, Ely, Nevada. *Econ. Geol.*, **71**, 488–512.

Jensen, M. L. and Bateman, A. M. (1979) *Economic Mineral Deposits*, 3rd edn, Wiley, New York, 593 pp.

Jonasson, I. R., Jackson, L. E. and Sangster, D. F. (1983) A holocene zinc orebody formed by supergene replacement of mosses. *J. Geochem. Explor.*, **18**, 189–194.

Knight, C. L. and Nye, P. B. (1953) The King Island scheelite mine. In *Geology of Australian Ore Deposits* (ed. A. Edwards), 5th Empire Mining & Metallurgy

Congress, Melbourne, pp. 1222–1232.

Knight, C. L. and Nye, P. B. (1965) Revised by staff of King Island Scheelite (1947) Ltd. The scheelite deposit of King Island. In *Geology of Australian Ore Deposits* (ed. J. McAndrew) 8th Commonwealth Mining & Metallurgy Congress, Melbourne, pp. 515–517.

Kwak, T. A. P. (1983) The geology and geochemistry of the zoned Sn–W–F–Be skarns at Mt. Lindsay Tasmania, Australia. *Econ. Geol.*, **78**, 1440–1465.

Kwak, T. A. P. and Askins, P. W. (1981) Geology and genesis of the F–Sn–W (–Be–Zn) skarn (wrigglite) at Moina, Tasmania. *Econ. Geol.*, **76**, 439–467.

Kwak, T. A. P., Plimer, I. R. and Taylor, R. G. (1982) Australian tungsten deposits. In *Symposium on Tungsten Geology* Jiangxi, China, Oct. 1981. (ESCAP RMRDC), Bandung, Indonesia, pp. 127–153.

Lapham, D. M. (1968) Triassic magnetite and diabase of Cornwall, Pennsylvania. In *Ore Deposits of the United States 1933–1967, Graton-Sales Volume* (ed. J. D. Ridge), American Institute of Mining, Metallurgy and Petroleum Engineers, New York, pp. 72–94.

Lindgren, W. (1905) The copper deposits of the Clifton–Morenci district, Arizona. *US Geol. Surv. Prof. Pap.* **43**, 375 pp.

Lindgren, W. (1922) A suggestion for the terminology of certain mineral deposits. *Econ. Geol.*, **17**, 292–294.

Meinert, L. D. (1983) Variability of skarn deposits: guides to exploration. In *Revolution in the Earth Sciences* (ed. S. J. Boardman), Kendall/Hunt, Dubuque, IA, pp. 301–317.

Morrison, G. W. (1980) Stratigraphic control of Cu–Fe skarn ore distribution and genesis at Craigmont British Columbia. *CIM Bull.*, **73**, 109–123.

Newberry, R. J. (1979) The importance of stratigraphy and structure of Sierran metamorphic rocks in the development of economic tungsten deposits. *Geol. Soc. Am.* (Abstracts with programmes), **11**, 120.

Newberry, R. J. (1980) The geochemistry of tungsten deposition in skarn deposits – a field and theoretical approach. *Geol. Soc. Am.* (Abstracts with programmes), **12**, 492.

Park, C. F. and MacDiarmid, R. A. (1975) *Ore Deposits*, 3rd edn, Freeman, San Francisco.

Pintz, W. S. (1984) *Ok Tedi: Evolution of a Third World Mining Project*. Mining Journal Books Ltd., London, 206 pp.

Ridge, J. D. (1976) *Annotated Bibliographies of Mineral Deposits in Africa, Asia (exclusive of the USSR) and Australasia*, Pergamon Press, Oxford, 545 pp.

Schmidt, E. A. and Worthington, J. (1977) Geology and mineralisation of the Cannivan Gulch molybdenum deposit, Beaverhead County, Montana (abs): *Geol. Assoc. Canada, Progr. with Abs.*, **2**, 46.

Shelton, K. L. (1983) Composition and origin of ore-forming fluids in a carbonate-hosted porphyry copper and skarn deposit: A fluid inclusion and stable isotope study of Mines Gaspe, Quebec. *Econ. Geol.*, **78**, 387–421.

Simons, F. S. (1974) Geologic map and sections of the Nogales and Lochiel quadrangles, Santa Cruz County, Arizona. *U.S. Geol. Surv. Misc. Geol. Inv. Map* 1–762.

Solomon, M. (1980) Evidence of exhalative origin for Tasmanian tin deposits (R. W. Hutchinson, CIM Bull., August 1979), *CIM Bull*, **73(815)**, 166–167.

Stumpfl, P. (1977) Mineralogical aspects of ores. Sediments ore and metamorphism: new aspects. *Philos. Trans R. Soc. London*, **A286**, 507–525.

Ward, M. H. (1984) Freeport Indonesia goes underground and expands production. *Min. Mag.*, Nov., 481–487.

Yan, M. (1982) Basic theory and chief methods applied in tungsten prospecting in Jiangxi Province. *Tungst. Geol. Symp. Jiangxi, China*, (ESCAP RMRDC), Bandung, Indonesia, pp. 489–502.

Zonghua, W. and Yangfen, J. (1981) A mercury vapour survey in an area of thick transported overburden near Shanghai. China. *J. Geochem. Explor.*, **15**, 77–92.

11

The design and implementation of exploration programmes

11.1 INTRODUCTION

Every mining operation begins with an exploration phase. The exploration procedure adopted varies with the finance available, the type of mineral deposit sought and the geological and geographical environment. Well-organized exploration programmes follow an established pattern beginning with a desk study and ending with the target selection. The target is then evaluated and feasibility studies on the deposit are undertaken. Some companies precede the full feasibility study with a pre-feasibility study when the geological and mining factors are drawn together to outline the likely extent of the project and the form the full feasibility study should take. If the feasibility studies are favourable and all factors are positive the project proceeds to development and eventually to production. The generalized sequence is illustrated in Fig. 11.1 and the component parts are elaborated upon in Table 11.1.

In this text we have presented the general characteristics of the major ore deposit types and current ideas on ore genesis. These are some of the factors that exploration managers and geologists will consider when deciding upon exploration strategy. The market requirements, mineral

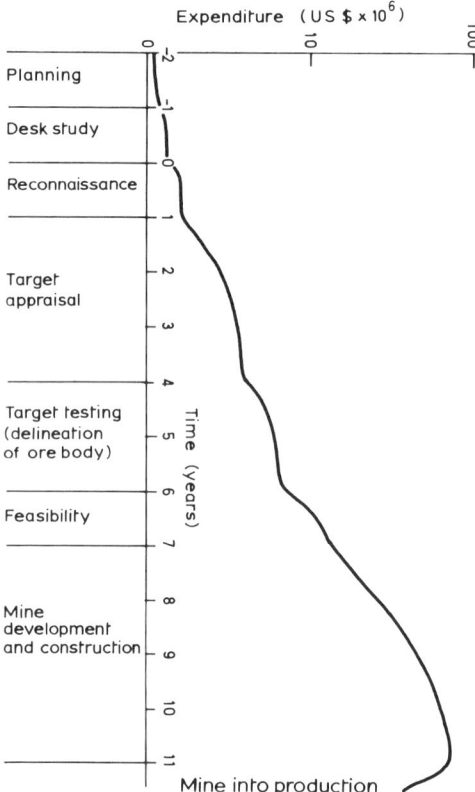

Fig. 11.1 Stages and expenditure in a mineral exploration programme (timing and expenditure are for a medium-sized mine). (After Eimon 1980.)

Table 11.1 Typical stages in an exploration programme, techniques employed and personnel involvement

Time (years)	Stage	Techniques	Personnel	Concession size (max. in 1000 km^2)	Expenditure (US$ million)
−2	Planning	Data base search Economic modelling Economic forecasting Exploration model review	Senior management Economic analysts Mineral economists Market specialists Senior geologists	1000	
−1	Desk study	Data base search (computer) Selection of data Literature search Remote sensing data Compilation of data & reports Land acquisition	Experienced explorationists Geological assistants Information specialists Lawyers	1000	Up to 0.3
0	Reconnaissance	Photogeology Remote sensing Regional geological mapping Aeromagnetic surveys Electromagnetic surveys Radioactivity surveys Gravity surveys Stream sediment geochemistry Prospecting	Field geologists Geochemists Geophysicists Photogeologists Prospectors	100	0.5–1.5
1	Target selection (target appraisal)	Geological mapping Photogeology Soil geochemistry Lithogeochemistry Ground magnetic surveys Electrical and electromagnetic surveys Surface radiometric surveys Geological model analysis	Field geologists Geochemists Geophysicists Photogeologists	<10	2.5–50
4	Target testing	Drilling Trenching Pitting Test mining Mine sampling Geophysical logging Geostatistics Geological model analysis	Geologists Drilling team Geophysical loggers Laboratory chemists Field and laboratory assistants	<1	2.5–50

processing techniques and possible mining methods must also be recognized. However, it must be borne in mind that only 15% of exploration expenditure is devoted to Third World countries (Willcox, quoted in *Mining Journal*, June 1980). The economic constraints on the future development and mining of the deposit must constantly be borne in mind. The only object of exploration is to find a mineral deposit that can be worked economically.

11.2 WHO UNDERTAKES EXPLORATION?

Exploration is undertaken at all levels, from the individual who stakes a claim and pans for gold in Canada, or searches for opals in Australia, to the international mining companies with extensive exploration programmes in each continent. Most geologists will be part of a team involved in an exploration programme and consequently we will concentrate on organizations which undertake exploration rather than individuals involved in exploration.

11.2.1 Government agencies

Typically governments or State agencies carry out geological research and mapping. This work provides basic surveys and method evaluation which are useful to companies in helping them conduct more effective exploration. Some governments are more actively involved in exploration activities *sensu strictu*. In Finland state-funded exploration constitutes 20% of the total exploration for the country whereas in Australia government exploration expenditure averages only 5% of the total (*Australian Bureau of Statistics*, 1983). Japanese government agencies have played a much greater role, financing 75% of Japanese domestic exploration in 1979 (*Report of Metal Mining Agency of Japan*, 1981).

Government agencies are also involved with exploration beyond their national boundaries, in joint ventures with the governments of developing nations. The Overseas Development Agency in the UK, the Bureau de Recherches Géologiques et Minières in France and the United Nations Development Programme are in this category. About half the UN funds for exploration are currently spent in Africa, another third in Asia and the remainder in Latin America. The World Bank is also a major source for funding for exploration programmes.

11.2.2 Consultancy groups and contractors

Consultancy facilities exist as groups within larger corporate structures or as separate external consultancy companies. Consultancy companies may provide exploration expertise, either as an entire exploration package or, more commonly, as specialist services in some aspect of exploration such as remote sensing, geochemical or geophysical surveys. The economic geologist may work for one of these companies or may be required to liaise with the consultants on an employer's behalf. In the latter situation the geologist must have sufficient appreciation of the methods being used to evaluate the results, reports and interpretations supplied by the consulting group. During the exploration programme, specialized drilling contractors are likely to be used, even by the larger mining companies, and the exploration geologist must be prepared not only to supervise the drilling and to log core but also to understand and draft, if required, drilling contracts.

Under this heading we may include small mining or exploration companies which undertake individual exploration programmes at grassroots level. These may be focused on specific deposit types. The companies normally greatly dilute their interests or sell out when they find viable deposits. At this stage larger more established mining companies move in to carry out detailed exploration and feasibility study as a prelude to mining.

11.2.3 Mining companies

The major mining companies have divisions solely

The design and implementation of exploration programmes

responsible for exploration. Teams from the exploration department may move around the World as corporate strategy dictates. International companies may have resident exploration teams in every continent and these will be used to supplement each other as required or may be advised and augmented from company head office. The degree of interchange for geological staff between exploration and exploitation varies from company to company but we believe that such interchange is very important in the development of the geologist and will be of long-term benefit to the company concerned. If possible, all exploration geologists should spend some of their professional career as mine geologists and vice versa.

Major mining companies are able to finance large exploration programmes internally without recourse to external finance. Furthermore the exploration team from a major mining house has advantages over its competitors from smaller companies in the corporate back-up available to it – large data banks, computer software, and access to in-house staff for specialist surveys. Despite these advantages larger companies do not necessarily find more ore deposits than their smaller counterparts (Lindemann, 1982). What every successful exploration company needs are inspired geologists backed by dynamic exploration managers.

11.3 FACTORS AFFECTING EXPLORATION PROGRAMMES

At the outset the objectives of the programme must be defined. What is sought, why is it being sought, where is it to be sought and in what time period? In designing the exploration programme with these questions in mind, certain factors must be considered which will affect the planning stage and subsequent operation of the programme.

11.3.1 The political aspects

Governments may be very heavily involved in their minerals industry either through nationalization of some or all of their mining industry or by allowing foreign companies to operate only under very stringent restrictive regulations and tax constraints. The wave of nationalizations in the late 1960s and 1970s forced many foreign-owned mining companies to withdraw from less-developed countries. Such nationalization did not leave a complete exploration vacuum, however, as the new state-owned companies partly stepped in. The copper mining industry in Chile may be quoted as one example of nationalization. As an example of tax disincentives to outside companies we may cite the renegotiation of the Bougainville operation, and its effects on the Ok Tedi project in Papua New Guinea (Davies, 1978).

Land may be withdrawn from exploration for internal political reasons, or internal tax laws may change which in turn reduces exploration. Derry and Booth (1978) show that from 1950 until 1970

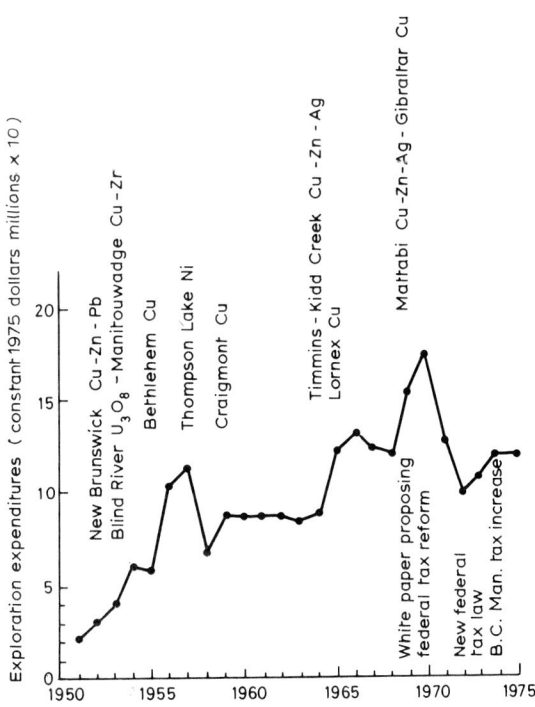

Fig. 11.2 Effect of taxation on exploration expenditure (from Derry and Booth 1978).

there was a gradual increase in the exploration expenditure in Canada (Fig. 11.2) but three tax changes between 1970 and 1975 show a marked effect on this expenditure.

In the developed nations, for example USA, Canada and UK, environmental legislation may prevent mineral exploration in designated areas such as National Parks or areas of outstanding natural beauty. Walthier (quoted in Hutchinson, 1980) concluded that 85% of the public lands in the United States had been withdrawn from mineral exploration. Despite the restrictions that may be imposed by environmental considerations on the working of any deposit found, there has been an increasing concentration of effort in exploration within the developed nations (USA, Canada, Australia and South Africa) in recent years. Hutchinson (1980) quotes this increase from 60% of total exploration expenditure in 1966 to 85% in 1977. Much has been written in the 1970s about the run down of exploration in less-developed countries, apparently in response to adverse changes in fiscal regimes and increased political risks. We believe that this run down is partly explained by the world-wide weakening of exploration, outside the fuel minerals, as a result of market conditions. As the major ore deposits of the less-developed countries are metallic minerals they have suffered most noticeably from this cut-back.

11.3.2 Selection of minerals

In the past, certain mining companies have almost restricted their operations to particular sectors of the market, for example molybdenum, gold or lead and zinc. Exploration in these circumstances is confined to programmes aimed at maintaining, or increasing, reserves of the metal or metals produced by the mining group. This strategy has the advantage of developing exploration teams skilled in defining, recognizing and evaluating new mineral discoveries of the chosen metal type. The policy has the disadvantage of little flexibility in response to movements on the World metal markets and little opportunity to diversify rapidly operations in response to different demands for mineral commodities.

Although some companies still lead the field in exploration and production of particular metals, the tendency for 'one-metal' companies (e.g. Amax – molybdenum) has declined with companies diversifying. Some of the major metalliferous mining companies have developed very successful coal divisions (e.g. Anglo American). Furthermore there has been an increased interest on the part of oil companies in the minerals sector. Shell, for example, took over Billiton in 1970, and the latter company's employed capital increased from about £60 million in 1973 to around £885 million in 1983 (Baxendell, 1983). Some of the major oil companies with metal interests are shown in Table 11.2. Exploration philosophy, lead times and overall expenditure are markedly different between the two industries and these take-overs will lead to changes within the organization and implementation of exploration programmes.

Market analysts are employed to observe, and forecast, trends in metal markets and possible changes in metal-usage patterns. Companies might expect exploration strategists to have similar understanding of developments. The exploration geologist is usually presented with a *fait accompli* – search for a particular metal or mineral deposit type. This is the main role of the exploration geologist. Geologists must, however, have an input into the construction and evaluation of 'economic' mineral deposit models upon which target types may be based. The usual process would be to consider grade–tonnage profiles of deposit types (such as those we have presented in this text) with respect to metal prices, mining, processing and marketing costs. These factors will govern the attractive nature of the target.

11.3.3 Choice of region for exploration

The choice will be either (a) which region offers the maximum possibility of discovering an economic mineral deposit of the type sought or (b) given a geographical constraint, what deposits of

The design and implementation of exploration programmes

Table 11.2 Major oil companies and metal interests (from Baxendell, 1983)

Top nine oil companies	Principal metals connection	Major metals
Exxon	Exxon Minerals	Cu (Zn)
Royal Dutch/Shell	Billiton	Al, Pb, Zn, Ni, Sn, W
Mobil	—	—
British Petroleum	BP Minerals International (Selection Trust)	Ni, Zn, Cu, Au, Sn
Texaco	Getty	
Socal	Amax (20% holding)	Mo, Cu, Pb, Zn, Ni, W
Gulf Oil	Gulf Mineral Resources	Exploration only
Standard Oil of Indiana	Amoco Minerals	Cu, Au, Mo
Atlantic Richfield	Anaconda	Al, Cu, Mo
Some other key companies		
Sohio	Kennecott Minerals	Cu, Pb, Zn
Pennzoil	Duval	Cu/Mo

the commodity sought are likely to exist, and which types would have the highest probability of being economically exploitable.

Exploration must be directed towards areas where the geology is likely to be favourable to the occurrence of economic deposits of the mineral type required. To ensure this is the case the exploration geologist will have a good knowledge of ore deposit geology and will base the decision on the premise that certain types of mineral deposit occur in certain, often quite clearly defined, geological environments. Why the deposit occurs there may not be absolutely determined and may not necessarily concern the explorationist at this stage. The selection of the area will be based upon an assessment of the chances of success, where success is measured as discovery of ore deposits likely to offer the maximum financial return for the money to be invested.

11.3.4 Exploration philosophy

Exploration generally falls into two broad divisions – grassroots, to locate new mineral deposits, and mine-based exploration to prove additional ore reserves and to ensure all the available ore is worked before the mine is abandoned. Most companies will have programmes underway simultaneously in both categories but the relative weighting will vary with company strategy, finance, and time. A third category of exploration strategy exists. This is the purchase or acquisition approach where the company 'buys into' an existing project.

(a) Grassroots exploration

Geologists who are required to explore for previously unknown economic mineral deposits must understand the features by which the deposits are recognized and the methods by which such deposits are detected. In general we agree with Ridge (1983) that it is not essential that the ore-finding geologist should know how and why the ore deposit formed provided he or she is well acquainted with exploration techniques and the physical/chemical characteristics of ore deposits. We believe the genesis is important in that certain geological environments or geological periods are preferentially mineralized with specific mineral deposit types. Successful grassroots exploration rests on detailed examination of geological environments, mainly the stratigraphy and structure. The flow-chart for a grassroots exploration programme will follow that described in Section 11.4, from reconnaissance survey through to target evaluation with a diverse array of techniques employed en *route*.

(b) Mine-based exploration

Mine-based exploration utilizes not only detailed stratigraphic and structural knowledge and an understanding of the ore body type, but also a working hypothesis of the genesis of the ore body. These factors will all be considered and interpreted to indicate where continuations of the ore body are likely to lie. Detailed geochemical or geophysical surveys may be undertaken but most of the ore discoveries will rely upon very careful geological control of the exploration drilling programmes. There is a very hazy dividing line here between exploration and development. There is no hard and fast rule but when drilling is within the mine itself it may be classified as development, but step-out drilling to outline adjacent ore bodies should be classed as exploration. Facts on this type of exploration are difficult to collect but we would contend that there is little doubt that expansion to existing mines, allied with increased processing facilities, is often a greater contributor to increase in total output than completely new mines.

(c) Purchase/acquisition approach

This third alternative, where the company buys into a project which has already reached one of the later stages of exploration, may offer the best business objectives for the organization. This approach requires that the geologist is skilled at evaluation techniques – both geological and financial/economic. The quality of the data available as well as its quantity must be very carefully assessed.

11.3.5 Exploration budget

On a corporate basis the exploration budget is the anticipated expenditure on exploration necessary to ensure the company achieves its goals. The budget is calculated by discounting the expected income from the achievement of the goals. When individual projects are considered, a risk factor is added into the calculation, but the budget must be realistic.

Once established the budget must cover costs of geological, geochemical and geophysical surveys as necessary, plus the cost of drilling, pitting, acquisition of mineral rights, mineral processing test work, administrative and other overheads. Only when feasibility studies are completed is the decision to mine usually made. Many feasibility studies have been made without the prospect being mined. For this reason a mining company may include feasibility study costs in exploration but they will be covered by a defined budget head within the overall exploration budget. Exploration and evaluation of mineral deposits is a time-consuming, expensive process (Fig. 11.1) and it is important for the geologist to realize that throughout this period, which may extend to a decade, the cash flow is negative for the company (Fig. 11.3). During development and into production the cash flow situation improves and provided continued mine-based exploration is undertaken the eventual lifetime of the mine may be many decades beyond the original quoted reserves.

An example of the derivation of a specific exploration budget is given by Binon (1983). He quotes the case where the budget is required for an exploration programme to discover an ore body of 100 million tonnes. The profit per tonne is taken as $0.60 and the minimum rate of return is to be 20%. The exploration and development are forecast to be 8 years and the life of the mining operation is considered to be 35 years. The budget allowance is calculated as follows,

Discount factor for 8 years at 20% = 0.233
Discount factor for 35 years at 20% = 0.143
Budget = $100 \times 10^6 \times \$0.60 \times 0.233 \times 0.143 \times$ risk factor (Assume risk factor = 0.05)

Budget = $99 957

Studies by Mackenzie (1981) report data on the amount of funds per year and persistence required for given levels of probability of success and this is another way of establishing necessary budgets.

The design and implementation of exploration programmes

Once the budget is determined subtraction of the acquisition costs will give the exploration allowance. The exploration manager then apportions the allowance to maximize the chances of finding an ore deposit. In most cases the manager estimates the amount of geological, geochemical, geophysical and drilling work on the basis of general knowledge of the mineral occurrence, and costings are based on previous work. Table 11.3 provides some typical programme costings for Western Australia in 1984. Budgets are reviewed regularly, at least annually, for each project.

11.4 THE EXPLORATION PROGRAMME

11.4.1 Stage 1 – planning

Careful planning is as important to an exploration geologist as it is to a military strategist. The ground must be carefully studied and the target accurately selected. Each exploration programme will have individual variations but the general scheme will be that shown in Fig. 11.4. The techniques and personnel requirements at each stage are shown in Table 11.1. The factors we have

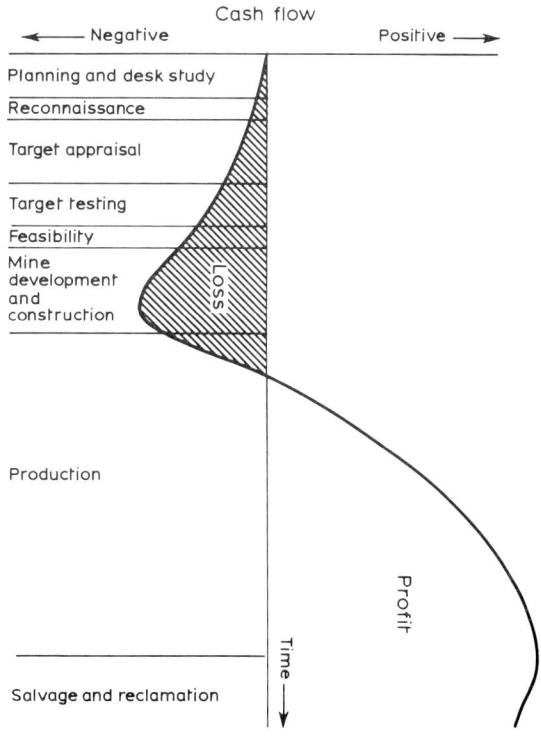

Fig. 11.3 Cash flow through exploration to production (after Lindemann 1982).

Table 11.3 Examples of exploration costs.

	Typical programme costings (1984) in A$ × 10^3		
	A	B	C
Personnel	276	301	55
Field costs	197	118	22
Leasing	20	10	15
Drilling	60	500	50
Analytical facilities	90	124	17
Remote sensing	13	—	—
Geological laboratory	8	6	5
Contracts (exc. drill.)	53	18	1
Corporate overheads would have to be added to these costings.			

Programme types:
A = Remote exploration site with several grassroots programmes
B = One project mainly utilizing drilling (around an old gold mine)
C = One project using reconnaissance geophysics followed by percussion drilling
Source of data: Western Mining Corporation Ltd., personal communication.

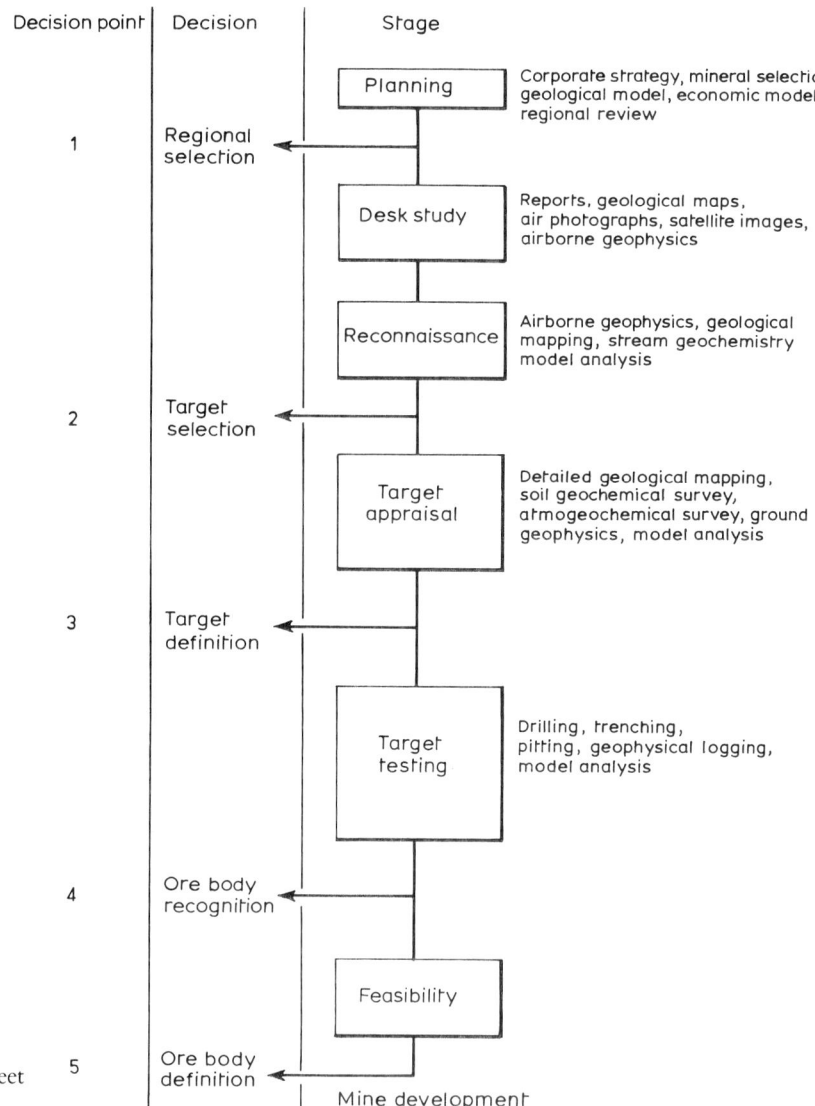

Fig. 11.4 Exploration flow-sheet with decision points.

already described will be considered during this stage and economic and exploration models will be reviewed.

The exploration programme that follows the planning stage will be based on a concept or hypothesis and it is important that at the end of each stage – reconnaissance, target appraisal, etc. – the results from that stage are analysed with respect to the concept. The concept itself must not become too rigid, it will usually change and may be regarded as a 'working hypothesis'. It is essential to realize that very important decisions are taken upon the results of each stage; these decisions have far-reaching consequences in financial and human terms. The most crucial decision will be to abandon the project at that stage or continue. A decision to continue will carry with it the commitment to further expenditure and this must be viewed in the context that each successive stage is more expensive than the one completed (Fig. 11.1). If it is considered that insufficient data have been collected the decision to return to an earlier part of the programme may be taken.

Summarizing, at each stage the decision is 'abandon and risk the possibility of missing an ore deposit or continue and commit increased expenditure'. An exploration geologist must be an optimist but optimism must be tempered by realism. The objective of careful planning, as of every other stage in the programme, is to locate a mineral deposit that can be worked economically.

11.4.2 Stage 2 – literature search, data accumulation – the desk study

All exploration programmes should start with a careful examination of available information. This stage is relatively cheap and is very cost effective. Correct decisions at this stage, based upon systematically collected and analysed data, can save considerable time and money in later, more expensive, stages. Evidence available at the desk study stage will be diverse in nature and probably variable in quality. The veracity and background of each source must be examined. Amongst the information available may be old mine records, geological maps, sampling records, existing geophysical and geochemical data, air photographs and satellite images. At one extreme all this, and perhaps more, may be available while at the other extreme there may be only one small-scale geological map available. With the increasing availability of satellite imagery, however, this latter state of affairs is becoming rarer.

Either during the desk study or arising from it an empirical geological model for the mineralization will evolve. Today most exploration programmes are based on such geological models of varying degrees of sophistication and complexity. The model may not be so heavily dependent upon the ore genesis of the deposit type as some geologists would wish, or like to believe, but will rely on the exploration team having a firm grasp of geological environments in which the deposit types occur.

Among the more important and widely accepted models, developed and used in the last thirty-five years the basis of which we have described in this text, are porphyry copper – major use commenced early 1950s modified by the Lowell and Guilbert 'model' 1970; roll front uranium – major use from mid-1950s; volcanogenic sulphides – major use from early 1960s.

When exploration of an area is envisaged, the target is subject to corporate goals and may be ore-mineral deposit type. Some authorities believe it is much more desirable, however, to look for any deposit type which is present – the concept of 'Total Exploration' (Ridge, 1983). Most explorationists believe that to concentrate on one particular deposit type increases the odds of success. However, during the desk study, evidence of other mineral deposit types, or environments suitable for their occurrence, may be discovered even if not actively sought.

In the synthesis of available data the geologist must be able to interpret geological maps but may have to rely on other specialists to present interpretations of the airborne geophysical data, remote sensing data, etc. The more familiarity the geologist can establish with the methods of interpretation of these data the more proficient an economic geologist he or she will become. We will briefly review some of the information sources that may be available and the types of information the desk study should yield from these data sources.

(a) Geological maps

Maps may be available at many scales depending upon the location (in the UK the British Geological Survey produce maps at 1:50 000 or even 1:10 000 while in other parts of the World the only geological maps available may be at a scale of 1:1 000 000). The detail contained in and reliability of the maps are also very variable and therefore no general statement is possible on this data source. The regional structure, stratigraphy and lithological units must be ascertained at this stage.

(b) Air photographs

For decades the air photograph has been a fundamental source of information before, and during, exploration programmes. Individual air photographs may be used to complement, supplement, or even replace base maps. Mosaics of air photographs may be used to define regional structure, extent of outcrop, river systems and other watercourses as well as the most convenient route of access. Air photographs are also vitally important in planning the infrastructure for any future mining operation based on the mineral deposit discovered. The interpretation of air photographs is included in most undergraduate geology degrees but readers requiring more detail on the subject are referred to the work by Paine (1981).

(c) Remote sensing

Satellite images provide another very useful component of the exploration programme. A review of remote sensing for exploration is given by Goetz et al. (1983). Landsat imagery has been readily available to exploration geologists for more than a decade. The main sensor is the multi-spectral scanner (MSS) with a resolution of approximately 8 m in four different bands of the spectrum. Two of the bands, those sensing green and red, may be useful in distinguishing between ferrous and ferric oxides in well-exposed rocks and soils (Press, 1983). The spectral bands were chosen for agriculture-related applications (vegetation mapping) rather than strictly geological uses. At the time this text is being written Landsat 5 is the latest satellite launched and this has on board a thematic mapper (TM) with spectral bands optimized for geological observations. It is considered possible that this sensor may be able to distinguish hydrothermal alteration effects. This would be extremely useful for mineral deposits that have large alteration haloes such as porphyry copper deposits. An attempt to demonstrate the application of satellite imagery to the exploration for porphyry copper deposits in Southern Arizona is given by Abrams et al. (1983). The extreme alteration around the large porphyry copper deposits in the south-west USA may be readily recognized from satellite imagery (Elkin, personal communication).

With radar-imaging satellites becoming available the definition in humid areas will be improved since in these areas infra-red and visible spectra do not provide the high-quality images necessary. At present, however, satellite images are used to produce good-quality maps usually at the scale of 1:50 000 and to plan further ground work (in the way air photographs are used). For large areas satellite images have advantages over the mosaics of air photographs in both these uses but they have the disadvantage that they do not, normally, provide stereo images which are useful in, for example, accurate measurements of dip. However, the metric camera on Spacelab 1 yielded some stereo pairs with spatial resolution of 20 m (Williams and Southworth, 1984).

Structural trends and lineaments are often obvious on satellite images where they may be obscured during surface mapping by their large scale, by surface materials or general lack of exposure. Much further work needs to be done on the recognition of these structures and their association with the discovery of mineral deposits before we are convinced that satellite images can be used in more than a 'basic tool' capacity. Readers requiring more information on remote sensing are referred to the two volumes edited by Reeves (1983).

(d) Airborne geophysical data

Multisensor airborne surveys incorporating electromagnetic, magnetic and radiometric systems have been undertaken for many years. These, in conjunction with geological maps, have been the basic framework for the exploration programme. As the flight line spacing will probably be in excess of 1 km the data are likely to be insufficiently detailed to allow the distinction of individual mineral deposit targets but will be adequate to dismiss apparently barren ground. Furthermore the type of data and depth of search will be different – radiometrics effectively map the variations in the distribution of radioactive elements in the top 1 m, or less, while the magnetics may receive responses from depths in excess of 1 km. The last few years have seen the use of multi-frequency, multi-coil airborne electromagnetic systems to give information on general resistivity distribution in the ground and, in suitable circumstances, overburden thickness variations as well as recognition of responses from good conductors (Cornwell, 1983).

11.4.3 Stage 3 – reconnaissance

The desk study may have defined areas that are potentially mineral-bearing, or the desk study will have rejected areas as barren within the limits of the data and the geophysical methods used. If no previous airborne geophysical data are available at the desk study stage, and the mineral deposit type is amenable to discovery using geophysics, then a regional survey must be flown to allow definition of the areas most likely to contain mineralization. It is likely at this stage that large numbers of anomalies will appear but only some of these will be potentially mineralized targets. These must be further investigated, evaluated and ranked in order of likely economic significance. This will be achieved by ground surveys following the pattern described below.

If, however, a target area has been reasonably accurately defined at this stage in the programme then it is conceivable that the decision might be taken to test the target with a drill hole or holes instead of devoting more time and expense to further relatively expensive indirect methods such as those described below. Drilling may be the most effective next step in any case, for instance in the Canadian Shield, airborne electromagnetic (AEM) anomalies are often test drilled prior to geological mapping as there is often less than 5% outcrop.

As previously mentioned, each successive stage in the exploration operation is more expensive than the preceding one. Consequently the ground surveys must be concentrated in those areas defined as most favourable or 'anomalous' from the desk study, airborne survey and satellite imagery. The exact procedure in the ground survey, the methods used and their order of application, will depend on the mineral deposits being sought and the geological environment. Therefore we will illustrate the general arrangement and mention specific applications of methods to certain ore deposit types as they occur. The activities in this stage may be sequential or some of the parts of the survey may take place simultaneously and a critical path type of analysis may be applied to the activities to illustrate the timing of each stage within the programme. It must be emphasized that the indirect methods considered here, geochemical and geophysical, are used to *locate* target areas and not to test them.

(a) Stream geochemical sampling

Assuming a drainage system exists in the area then either stream water – or stream sediment – sampling, or both, may be undertaken. In practice sediment sampling is much more widely used because:

(1) Water is not always present.
(2) Element concentration in water tends to vary with seasons.
(3) The transport and storage of water samples are more difficult.
(4) Sediments can be treated by selecting sizes or partial analysis to enhance metal values.
(5) The success of water sampling depends upon the solubility of the indicator metals or complexes being sought.

Drainage surveys are the most widely used geochemical reconnaissance technique. The value of the survey is that the sampled material reflects a large catchment area. The elements may be grouped as mobile (S,Mo,U) and semi-mobile (Zn,Cu,Ni,Co) and if these are considered as well as those metals that may occur in the heavy mineral fraction of stream sediments (Au,Sn,W) then every type of ore deposit should have some form of drainage anomaly signature. The parameters affecting the occurrence of mineral species in stream sediments have been discussed in Chapter 5. Of interest here is the decay pattern of the anomalous mineral species or metal content from the source mineral deposit. Some attempts have been made to express, in simple mathematical terms, the relationship between the exposed size and grade of the source, the anomaly strength and the measured area of drainage as reported by Rose *et al.* (1979). We feel that such a mathematical treatment is liable to give field workers a false sense of security as nature is not that predictable.

Although, as we have stated above, sediment sampling is usually preferred to the sampling of stream water, one of the important characteristics of stream water is its homogeneity, usually achieved through mixing by turbulence. An increase in the metal concentration of the dissolved salts is therefore readily detectable given sufficient concentration or sensitivity in the analytical technique. It should be noted that analyses quoting ppb may not be routine requirements on the volume basis usual in some exploration programmes.

In stream sediment surveys, the recognition of such an anomalous increase in metal content may rely upon analysis of material from the optimum sieve size. Most anomalous metal in stream sediments is concentrated in the finer-size fractions, indeed an inverse relationship between metal content and grain size is observed in a variety of climates (Rose *et al.*, 1979). Most of the resistant primary ore minerals are relatively heavy and tend to be enriched in the heavy mineral fraction of the sediment. A panned concentrate may therefore provide the most rapidly collected useful sample. Contrast between the anomalous value and the background will be enhanced if the ore mineral does not occur as an accessory in the host rock.

Care must be taken to distinguish between material derived from stream banks adjacent to the sampling station (colluvial sediment) and material of truly alluvial origin (the active sediment). In arid or semi-arid areas stream sediment analysis may not be applicable because of severe contamination from bank collapse and wind-blown material. Organic matter tends to scavenge metals from solution, therefore this may provide the best sampling medium in terrains with poorly developed drainage systems.

For reconnaissance sampling the sample interval may be determined empirically through orientation surveys or by reference to previous work. In virgin terrain the frequency of sample stations may be controlled by ease of access. DeGeoffroy *et al.* (1967, 1968) show that one or two samples per 10 km^2 were adequate at the reconnaissance stage for the recognition of zinc anomalies in south-west Wisconsin (Fig. 11.5). In the case of large deposits such as porphyry coppers (Section 3.3), copper, molybdenum and even tungsten anomalies may be detectable tens of kilometres downstream from the deposits.

During this reconnaissance work, as in any other procedure, the explorationist must constantly question the assumptions upon which the survey is based and check the validity of the data. For instance is the field sample collected representative of the material and for the metal sought? Is it likely that the metal or mineral sought will produce a pronounced 'nugget effect', for example gold or diamonds? Is the analytical method appropriate and of sufficient quality for the exploration problem at hand? It is important

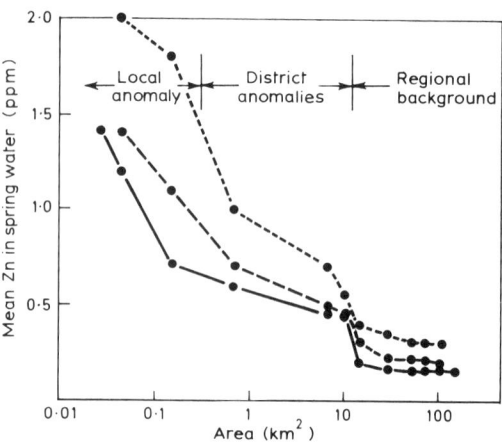

Fig. 11.5 Relation of mean zinc content of spring waters to area, averaged for three areas centered on known anomalies in south-west Wisconsin (from DeGeoffroy *et al.* (1968).

that the exploration geologist is aware of the accuracy and limitations of the data upon which subsequent interpretations are based.

When the sampling programme has been completed, and all analyses carried out, maps are drawn to show the concentration of metals in the stream water or sediments at each of the sample stations. Cut-offs in the streams are recognized where the metal enters the stream system either from tributaries not yet sampled or from the groundwater system as springs. A sharp cut-off will be recognized where the metal values gradually increase to a peak and then fall rapidly to background values; more commonly the metal enters the stream at numerous points and a more diffuse cut-off 'area' is recognized.

If exploration is taking place in previously mined areas the possibility of contamination due to mining, smelting and tailings must be considered. The exploration geologist will be required to examine the geochemical maps and decide which of the anomalous areas are likely to be 'significant' (possibly related to mineralization), and which are likely to be 'non-significant' (originating from causes other than mineralization). Non-significant anomalies may arise from other sources beside mining contamination and these include high background source rocks, sampling error, analytical error, pH variations and varying Fe–Mn oxides in the sediment.

Geochemistry, or any indirect approach, must be explicable in terms of geology – surficial or bedrock. Therefore the anomalous geochemical values will be reviewed in the light of local topography, likely subsurface drainage and regional geological knowledge to define the area of greatest interest in which detailed follow-up work should be concentrated. Readers requiring further information on geochemical exploration are referred to the texts by Levinson (1980), Reedman (1979) and Rose *et al*. (1979).

(b) Geological mapping

Fundamental to any exploration programme is the production of adequate, accurate geological maps. While these may be based on the airborne geophysical data, remote-sensing data and aerial photographs, the detail is derived from the ground survey. To undertake a stream geochemical survey a drainage system must exist; if it does not then primary geological mapping, possibly along widely spaced traverse lines, may be undertaken. The aim will be the same, to define the limits of evidence of mineralization. In the ideal case this geological mapping will be based upon exposures, but may of necessity have to rely on 'float mapping'.

Float mapping is the recognition of lithologies and lithological boundaries from isolated rock fragments in the overburden. If this is to be successful the nature of the overburden – residual or transported – must first be determined. Float mapping is most successful in country overlain by residual overburden but modifications to this technique, including such methods as boulder tracing, may be possible in areas with transported overburden. Techniques applicable for use in areas of glaciated terrain are described in a series of publications by the Institution of Mining and Metallurgy (1984).

If geological mapping is undertaken on defined survey lines the geological data are interpolated between the traverse lines. As with all exploration undertaken on traverse lines, the accuracy of the map produced will depend on the spacing of the traverses and the sample points. Normally the entire geological structure and stratigraphy are mapped and extra information is collected using indirect methods such as geophysics and geochemistry.

11.4.4 Stage 4 – target appraisal

The preceding work should have indicated the localized areas that may contain mineralization. These 'targets' must now be defined as accurately as possible using geological mapping as the prime approach and going to geophysics or geo-chemistry as needed to establish the targets for testing. If the targets have already been sufficiently well outlined prior to this stage then the decision

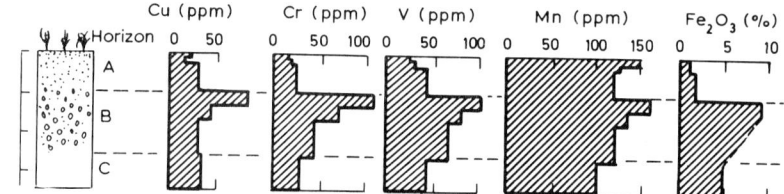

Fig. 11.6 Variation in metal content with soil horizon, latosol profile, Zambia (from Rose *et al.* 1979).

may be taken to proceed directly to drilling to prove the deposit.

(a) Geological mapping

The target appraisal stage is very site specific and the geological mapping is very detailed. All evidence of mineralization, alteration phenomena and lithological or structural features is accurately recorded. Rock specimens are collected for analysis. The geological maps produced will form the basis for all data presentation for subsequent stages of the programme including drill site location and trenching and pitting activities.

The production of accurate geological maps presupposes reasonably adequate outcrop frequency. If this is not the case then greater emphasis is placed upon the indirect methods of geochemistry and geophysics. The geologist will then be required to produce geological maps based upon data from these methods and an understanding of the regional structure.

(b) Soil geochemical survey

To outline the target more closely systematic geochemical sampling may be undertaken. Most commonly this will be soil sampling. Again the distinction between residual and transported soils is critical. Residual soil surveying is very widely used as the soil anomalies detected are usually a reliable guide to mineralization occurrence. The method is very dependable and cost effective when compared with other techniques available at this stage. Most of the common ore metals may form recognizable anomalous patterns in residual soil but the strongest contrast will be developed for relatively immobile elements such as Pb. In areas of transported overburden more specialized sampling is necessary and these techniques of deeper sampling and partial extraction are better developed for areas of glacially transported overburden than for other types of transported overburden. We have shown the efficacy of this type of survey in base-metal exploration in Ireland (Section 6.3.7).

The spacing of traverses, frequency of sampling station and sample depth have to be determined on local conditions, time available and financial budgets. An orientation survey must be undertaken to determine the type of overburden, optimum sample spacing and depth. Figures 11.6 and 11.7 illustrate the variation in detail of the information obtained from varying these parameters. On the basis of the orientation survey a soil-sampling programme can be phased, closer spacing and/or profile sampling being undertaken over areas of greater potential.

Data from soil geochemical surveys are usually treated statistically to determine the background, threshold and anomaly values for the area under survey. Readers can find full descriptions of these

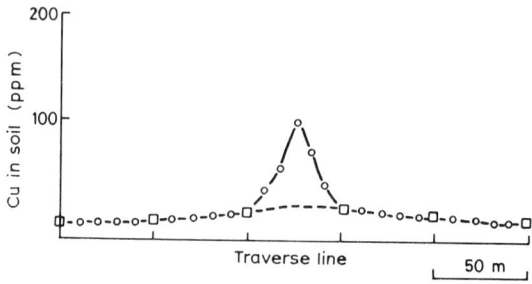

Fig. 11.7 Effect of sample interval on recognition of narrow vein ○, 10 m sample interval; □ 50 m sample interval.

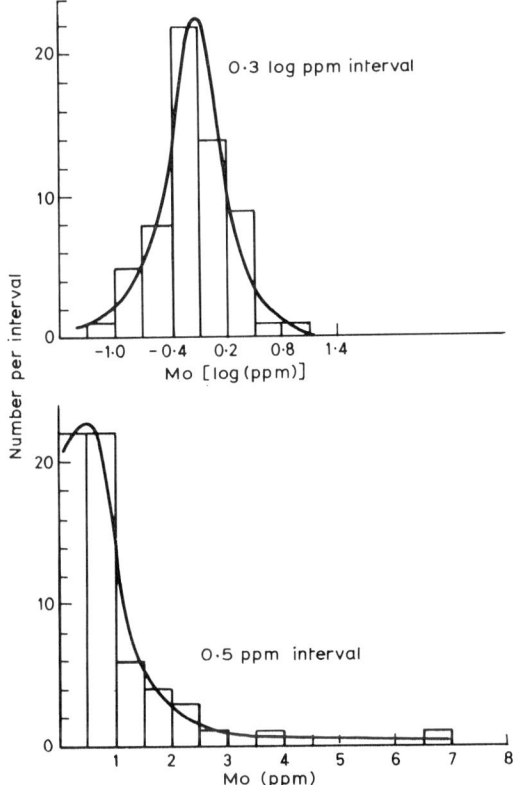

Fig. 11.8 Plot illustrating approximately log-normal frequency distribution of molybdenum in granite (top) contrasted with marked skewness of raw data (bottom). (After Ahrens 1954.)

terms in geochemistry texts, such as Rose *et al.* (1979) but we would make the following observations on the concepts:

(a) *Background* is a range of values characterizing unmineralized material.
(b) *Threshold* is the upper limit of this range.
(c) *Anomalies* are abnormal values $(+/-)$ from the background range.
(d) *Threshold* is a dynamic concept given the mixture of geological entities encountered in a typical survey.

Natural populations of soil samples may show a normal distribution although most exploration data suggest that the distribution is more commonly log-normal (Fig. 11.8). Although purely statistical methods of deriving threshold values exist (background $+2$ or 3 standard deviations), we would always recommend the construction of a histogram to recognize the possible occurrence of multiple populations. A bimodal population such as that shown in Fig. 11.9, which arises from soil samples taken overlying two lithologies, necessitates the recognition of two backgrounds and two threshold values.

Geochemical data are contoured, as this is invaluable in defining areas for further sampling, geophysics or drilling. Usually only samples containing concentrations of elements above threshold values are contoured and the appearance of the anomalous areas on the map will illustrate if a suitable threshold value has been used. If the anomalous values appear at isolated stations then it is likely that too high a threshold has been chosen.

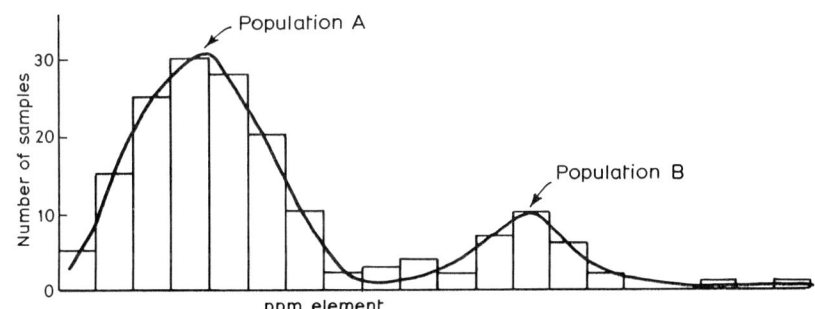

Fig. 11.9 Bimodal distribution from sampling over two lithologies A and B.

Ideally the threshold chosen should have resulted in restricted, but continuous, zones of anomalous values.

(c) Atmochemical methods

Geochemical soil surveys are sometimes referred to as 'pedochemical methods'; if surveys are based upon the analysis of gases emanating from mineral deposits then these are sometimes called 'atmochemical methods' (Fig. 10.18). Most of the work dealing with the application of vapour geochemistry to the location of sulphide deposits is still at an experimental stage but the use of these techniques in the search for deeply buried radioactive deposits is widely reported (Boyle, 1982b). The most widely used gas is radon (section 9.3.4) but helium has also been used. An example of a survey using helium in soil gas in the vicinity of a uranium deposit in Colorado is shown in Fig. 11.10; the deposit lies under 10 m of overburden.

Radon sampling is carried out by two different procedures. The radon may be directly extracted and measured from the sampling medium (either soil or water); alternatively measuring apparatus or films are implanted in the stream sediment or soil and the alpha-active radon is detected and determined by instrumental or alpha-track etching procedures.

The major advantage of soil gas radon and helium surveys over standard soil sampling and over other types of radioactive surveys is the recognition of very deeply buried deposits. One potential disadvantage is the route by which the soil gas reaches the surface; this route may be largely controlled by structural features such as faults which may not be related to the mineral deposit sought and may give rise to the surface expression being sited away from the deposit.

(d) Geophysical surveys

A wide variety of geophysical techniques is available based on the physical properties of rocks and minerals. Details of the methods will not be included here, they are available in specialist textbooks by Parasnis (1983), Griffiths and King (1981) and Telford *et al.* (1976). Readers who wish

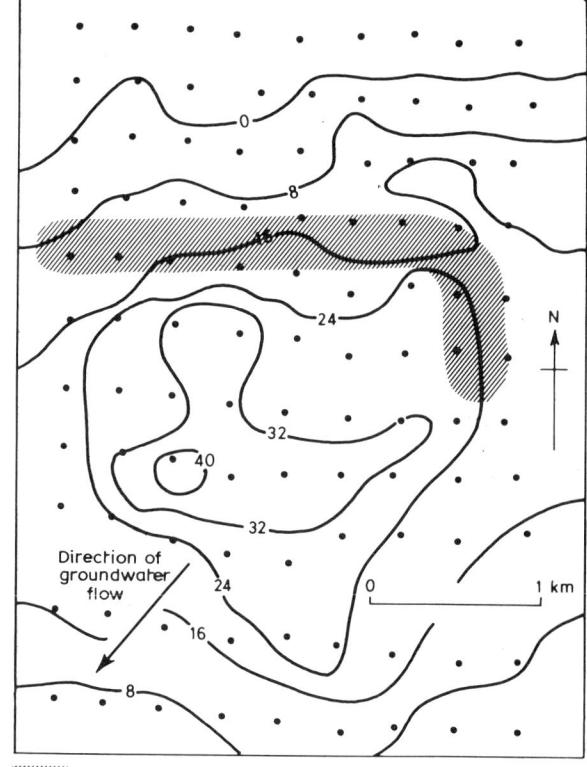

Fig. 11.10 Contour map of helium in soil gas in the vicinity of the Weld County, Colorado, roll-type uranium deposit. Base value for the atmosphere is 5.239 ppm (from Boyle 1982b).

to follow the changes that occur in instrumentation are advised to read the annual reviews of exploration methods in the *Canadian Mining Journal* and the *Mining Annual Review*.

The definition of geophysical anomalies relies upon there being a measurable difference in the physical properties of the mineral deposit, or its wall-rock alteration, and the surrounding barren host rock. Consequently the location of a mineral deposit by geophysical methods is dependent upon the selection of the most appropriate geophysical technique. In this, geophysics is no different from geochemistry or remote sensing;

the explorationist must always use the most applicable tool for the circumstances. If many line-kilometres of geophysical work are not to be wasted the planning and mapping must be thoroughly carried out and the ore-finding problems must be adequately defined to the geophysicist.

There is a marked difference between the exploration programmes for metal deposits and for oil or gas. In the latter, seismic reflection dominates, with gravity and to a lesser extent electromagnetic methods also used. In mineral exploration, with the possible exception of coal, and recently in South Africa, gold, seismic methods are rarely used. They are used in exploration for marine placers as described in Section 5.5.4.

In target appraisal ground magnetic surveys are commonly used to assist in structural and geological mapping. Although gravity surveys may be undertaken, they are expensive and require very accurate survey control and extensive data reduction processing. In suitable circumstances gravity surveys can define targets very accurately (e.g. the Pyramid deposit, Pine Point – Fig. 6.31). Resistivity and induced polarization surveys are extensively used and from the latter pseudo-sections are produced which may be used in guiding subsequent drilling.

(e) Model analysis

The geological mapping, geochemistry and ground geophysics should have defined the restricted areas within which mineralization is considered to exist. These areas will be relatively small by comparison to the original area selected for reconnaissance or examined in the desk study.

When all the target appraisal data are collated the exploration team must again review the geological model for the proposed mineralization. In the light of this reassessment the decision will be taken as to which target or targets should proceed to the target-testing stage. If the programme moves to this next stage it will become increasingly obvious whether the team has discovered mineralization, significant mineralization or a potential ore body.

11.4.5 Stage 5 – target testing

The next step in exploration will be deeper sampling, usually through drilling, although some deposits such as bauxites and kaolin may be examined by pits and trenches. The object of deeper drilling is to establish if a mineral deposit exists. If it does, further in-fill drilling will be carried out to evaluate the ore body. In some cases bulk samples will be collected from declines, adits and raises driven into the ore body. As well as establishing preliminary ore reserves, the extensive sampling programme will provide material for mineral processing test work.

Conventionally surface exploration diamond drilling is carried out in the search for metallic mineral deposits but in the case of non-metallic deposits other drilling techniques are often employed (Table 11.4) (Morris, 1983). One of the first decisions taken by the geologist at this stage, having decided upon the type of drilling to be carried out, is the angle at which the hole or holes should be drilled. When testing a geophysical anomaly a vertical hole is preferred since an anomaly may not be an ore and it must be tested as cheaply as possible. The problem with an angled hole in this situation is that it may pass over the cause of the anomaly if the overburden is thicker than expected or it may pass under the anomaly if the pitch of the causative body is flat. Inclined drilling should be against the dip of the structure. Drilling under a dipping vein from the footwall is only justified where overburden or rough terrain make this imperative. Deviation of the hole is frequently increased by entering the structure at the flat angles imposed by this approach. Base lines from which inclined drilling is undertaken should be parallel to the strike of the structure and the holes are best drilled perpendicular to this base line.

The orientation and spacing of the drill holes will be determined by the supposed shape of the deposit as interpreted from the geological model, regional information and the detailed surveys already undertaken. As with most aspects of exploration, the exploratory drilling will be phased:

Table 11.4 Exploration drilling techniques (Morris, 1983)

Mineral	Chip dust sampling						Coring				Flush				
	Open hole	CSR	Auger cont flight	Auger hollow stem	DTH	Shell and auger	Conventional	Wireline	Counterflush	Long hole	Water	Mud	Brine	Air	Foam
1. Salt	O				X	X	O/L	O/L			X	X	O		
Salt U/G	O						O	O/L							
2. Ball clay	O	X	O		X	X									
Ball clay U/G	O		O	O		O	O	O/L				O		O	O
3. Fullers earth	O	O	O	O			O	O				O		O	
4. Potash	O	O	O				O	O/L				O			
Potash U/G	O						O	O/L	O	O		O			
5. Gypsum	O	O	O				O	O/L			O	O		O	
Gypsum U/G	O		O				O	O			O	O		O	
6. Brick clay	O	X	O	O	X	O	O				O	O		O	O
7. China clay	O		O			O	O	O			O	O		O	
8. Fire clay	O						O				O	O		O	
9. Barytes	O	O	O		O		O	O			O	O		O	
Barytes U/G	O				O		O				O	O		O	
10. Fluorspar	O	O	O		O		O	O			O	O		O	
Fluorspar U/G	O				O		O				O	O		O	
11. Iron ore	O	O	O		O		O	O			O	O		O	
Iron ore U/G	O		O		O		O			O	O	O		O	
12. Coal	O	O	O		O		O	O/L		O	O	O		O	O
Coal U/G	O				X		O	O			O	O		X	
13. Limestone	O	O			O		O/M	O/M			O	O		O	
Limestone U/G	O				O		O	O			O	O		O	
14. Igneous rocks	O				O		O/S	O/S			O	O		O	
Igneous rocks U/G	O				O		O				O	O		O	
15. Industrial sands	O	O	O	O	X	O	O				O	O		O	O
16. Sand and gravel			O			O					O	O		O	

O, Commonly used; X, not recommended; L, large diameter; M, medium diameter; S, small diameter; CSR, centre sample recovery; DTH, down the hole; U/G, underground.

The design and implementation of exploration programmes

(1) *Preliminary drilling* is used to determine the attitude, ore-body width, density and nature of the hanging- and foot-walls. This stage will attempt to prove the persistence of the mineralized body to a moderate depth and provide information as to mineral content.

(2) Closer spacing or *'infill' drilling* will be used to establish the length and depth extent of the ore deposit. The presence of ore shoots within the mineralized structure may also be recognized at this stage. From this drilling the grade and tonnage to the lowest proposed mining depth will be computed.

The main responsibility of the geologist is to supervise the drilling and to log the core. Adequate core-logging is a vital part of any exploration programme. The geologist must visually log the lithologies and mineralization in the core using conventional abbreviations; the core lengths are measured and core losses are recorded. The core is split and lengths are assayed to determine the grade of the deposit intersected.

We believe that thorough geotechnical logging should also be carried out. From these measurements estimates of rock quality designation (RQD) are made. It is becoming increasingly common to photograph core. This is useful for subsequent re-examination of lithological variations but is only two-dimensional and therefore must be used with caution for any geotechnical measurements.

When the core is extracted from the core barrel it is placed in specially constructed core boxes. There are three ways in which core is placed in the box (Fig. 11.11): (i) 'book fashion'; (ii) 'snake fashion'; (iii) 'reverse book fashion'. Fortunately, reverse book fashion is very uncommon and book fashion is the most commonly used arrangement. The top of the core must be marked on the box and regular markers showing the depth of the hole are inserted between core fragments; some geologists also mark the sides of the core troughs. Readers wishing to learn more of the techniques of diamond drilling are advised to read the handbook by Cumming and Wicklund (1980).

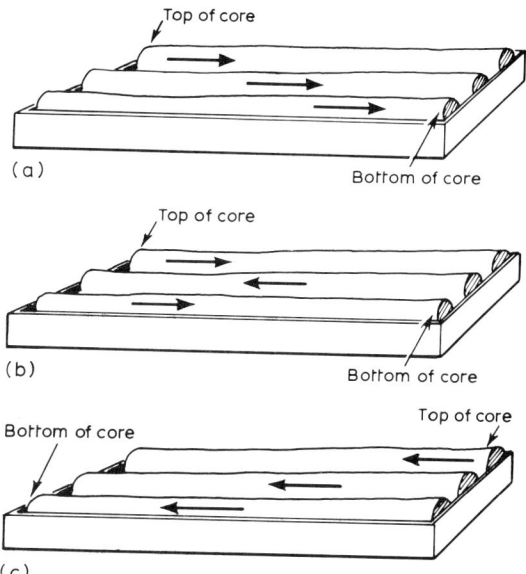

Fig. 11.11 Core-logging system for reading stored cores. Cores should be photographed upon extraction from barrel (if possible) and placed in boxes. Markers must be inserted to record depths. (a) Book fashion – core is read from left to right. (b) Snake fashion – core is read forwards and backwards as shown. (c) Reverse book fashion – very rarely used.

The primary reason for diamond drilling is to outline the ore deposit in the case of grassroots exploration, and to define future mining development in the case of mine-based exploration. In both cases ore reserve definition is the main aim. The accuracy of the estimates arising from exploration drilling will be governed by the number of holes drilled relative to the size of the property, the core recovery, the ore body type and the statistical or geostatistical technique used to produce the ore reserve from the drill hole data. Examples of overvaluing and undervaluing deposits occur in the literature. In the latter category Rogers (1982) describes comparisons between grades calculated from about 1200 km of drill core and production grades in the Dome Mine ore bodies, Timmins, Ontario (Table 11.5).

(a) Geophysical logging

Drilling a cored diamond drill hole is very

Table 11.5 Comparison of diamond-drill-indicated grades with (a) actual stope muck grades (multiple-vein structures) and (b) lateral development (single-vein structures). Data from Dome gold mine, Ontario. (Rogers, 1982)

(a)

Stope (year started)	Ore type	No. of DD holes	Length of DHH through zone		% DDH indicated ore	Grade (dwt)				Final Production	
			Total	>1.00 dwt		Total DDH in stope		DDH >1.00 dwt		Tons	Grade (dwt)
1013 No. 3 (1960)	II (b)	12	419.0	155.5	37	4.88 (uncut) 1.80 (cut)		13.10 (uncut) 4.58 (cut)		72 745	3.70
1272 No. 2 (1957)	II (b)	9	250.0	174.5	60	5.43 2.61		8.26 3.57		37 088	4.12
1372 Nos. 1,3 (No. 4 Panel) (1975)	II (b)	6	394.0	162.0	42	2.28 1.54		5.00 3.21		9 833	3.82
1388 No. 15 Panel (1978)	II (b)	3	162.0	129.0	80	7.33 3.90		9.10 4.75		7 065	3.95
1548 No. 3 (1967)	II (a)	5	144.0	51.5	36	1.85 1.24		5.05 3.33		45 621	4.66
1903 No. 2 (1952)	II (a)	12	420.0	199.0	47	2.53 1.70		5.08 3.29		73 773	5.35
2415 No. A Panel (1977)	II (b & c)	8	423.0	129.5	31	2.47 1.32		7.78 4.02		24 918	2.20
2415 No. 8 Panel (1975)	II (a & b)	7	325.0	134.0	41	2.05 1.93		4.14 3.88		22 326	4.20
2527 No. 3 No. 9 Panel (1975)	II (c)	4	261.0	181.0	70	2.55 2.38		3.24 2.96		7 105	3.39

Ore types: II (a) veins in massive lavas; II (b) stockworks in sediments; II (c) stockworks in porphyries and altered wall rocks. Note how drill cores fail to indicate true size or grade of orebody eventually mined.

(b)

Location (zone width, ft)	Length of zone (ft)	Ore type	Holes through zone		% DDH indicated ore	DDH grade (dwt)	Development sample		Indicated reserve grade
			Total	>1.00 dwt			Chips (dwt)	Muck (dwt)	
1281 No. 16 Dr. (10)	150	I (c)	10	2	20	2.60 (uncut) 2.20 (cut)	27.43 (uncut) 4.47 (cut)	6.96	5.72
1308 No. 6 Sill Dr. (6)	200	I (c)	5	1	20	2.66 1.80	31.96 5.46	5.44	5.45
1308 No. 6 Stope (9)	200	I (c)	4	2	50	7.75	(Current stope 7604 tons at 5.38 dwt)		
2006 No. 8 Dr. (10)	250	I (a)	5	2	40	0.90 0.90	2.75 2.66	3.35	3.00
2010 No. 10 Dr. (10)	420	I (a)	9	3	33	0.80 0.80	3.73	1.95	2.84
2351 Sill Dr. (5)	1000	I (a)	11	3	27	1.75 1.75	4.93 3.53	3.20	3.37
2615 No. 6 Dr. (10)	550	I (a)	6	2	33	1.38 1.38	4.32	4.32	4.32

DHH, Diamond drill hole; dwt, penny weight. Ore types: I (a) ankerite veins; I (c) quartz-fuchsite veins.
The planar nature of veins means they are easier to locate than the multiple-vein structures in (a) but economic gold values are rarely recognized from diamond drilling (particularly in ore type I (a)). Compare DDH grade with development sampling and indicated reserve.

expensive; the total drilling stage is likely to be up to 25–30% of the typical exploration programme (Table 11.3). The average exploration drilling costs in the US for uranium were about $16/m in 1982. In Australia it was US$81/m for core drilling and for other forms of drilling the average was US$21/m. Because of this the maximum amount of information must be derived from both the core and the hole itself. As we have briefly mentioned above, core logging must be thorough and should include geotechnical as well as geological/mineralogical/assay information. In the minerals industry the hole itself may be overlooked as a source of information. The hole will be accurately surveyed, the water level will be 'dipped' and recorded but very little else may be done before the hole is plugged. With increasing drilling costs and the advent of micro-electronics we believe that increasing use will be made of down-the-hole instrumentation in mineral exploration.

Obviously radiometric techniques may be used in uranium exploration (Chapter 9) and down-hole EM is used in sulphide exploration (Chapter 3). Down-hole IP surveys are also underway (Ogilvy, 1984). The advantage of these techniques is that they may indicate a target missed by a diamond drill hole – so-called 'off hole mineralization'.

Down-the-hole geophysical information is used by the coal industry in the UK to determine fracture density anomalies and RQD using a technique known as 'RocTec' (Kusznir and Whitworth, 1983). A major use for a variety of geophysical logs is within the evaporites industry. The geologist entering the evaporite mining industry may have to become familiar with reading neutron porosity index, bulk density (gamma–gamma log or density log), natural gamma, sonic and caliper logs to recognize not only if evaporites occur but whether potentially economic potash minerals are present. At the present time no technique is available for down-the-hole assaying for a wide range of metals and until such techniques are available there will be no substitute for the completely cored hole in determining grades for a potential mineral deposit.

(b) Geological model analysis

The drilling should have provided the mineral inventory for the ore deposit, the rock mechanics information and the indicated reserves. These data will allow the geologist to modify the model and to decide if further exploration, such as more drilling, bulk sampling or trial mining is required. It is at this stage that exploration passes into feasibility studies and ore body development. The geologist must ensure that sufficient data are available for other professionals – mining engineers, geostatisticians and mine planners – to make their decisions regarding the future of the project. Of course the geological involvement does not cease here; the geologist should be involved throughout production to final reclamation of the mined area.

11.5 CONCLUDING STATEMENT

We have reviewed the process through which an exploration programme goes, indicating the likely costs at each stage. Wherever possible we have suggested appropriate techniques that may be used but we have avoided detailed descriptions of these techniques as these may be found in the relevant specialized textbooks. In our discussions we have omitted any reference to the success rate for exploration programmes. If success is measured as the number of commercially exploitable deposits discovered as a percentage of the total number of grassroots exploration programmes conceived then the success rate must be very low. Figures are not readily available but we would estimate that of all exploration programmes at the planning stage, considerably less than 0.1% reach a mining venture within the timespan shown in Fig. 11.1. For this reason mining companies have many diverse exploration programmes operating simultaneously.

This success rate is not the complete picture, however, as deposits will be discovered that cannot be brought into production in the time frame of Fig. 11.1. Here the exploration team will have been successful in defining the deposit, which was the original objective. Financial conditions will have changed between the planning stage and discovery such that the feasibility studies indicate the deposit is no longer economically extractable. The files of most major mining companies contain many such examples. Possibly a large number of the deposits discovered in the early 1970s will never be mined because of greatly escalating costs. Exploration around operating mines has a much higher success ratio if development drilling to upgrade reserve categories is included. The exploration work here is usually in a much shorter time frame than for grassroots exploration.

In the first chapter of this textbook we reviewed the way in which exploration had evolved over this century. In view of the low success rate quoted above, in what ways has geological knowledge advanced in the last decades to help the geologist find more ore bodies? Firstly there has been the development of the geological models, whether they are genetic or empirical. These have been useful in assisting the selection of regions for search and for defining sites within those regions most suitable for mineralization. Genetic models are conceived to explain the observed facts; often those facts themselves are sufficient guides for ore search. Collating those facts may be a better guide for the explorationist than an elegantly conceived genetic model that is difficult to apply in the field.

Secondly, the advent of plate tectonic theory must have an effect on the explorationist. It is only a little over a decade since geologists began to establish links between plate tectonic processes and metallogenesis. This is still very much in its infancy and although it may be applicable on a grand global scale, we feel the narrower regional applications require much more definition.

Thirdly there have been advances in remote-sensing data and indirect methods. With modern electronics these must develop at an ever-increasing pace. What may be lacking is sufficient time for geologists to interpret accurately the vast quantities of data so rapidly collected. Desk-top computers which will handle large volumes of data will assist here. Exploration and mining geologists must become conversant with the varied applications of computers within their industry. One of the many ways that computers are being applied is for observed data on particular deposit types to be fed into the computer. The computer then collates the data and produces what we would call an empirical model – the geological factors required for the occurrence of that deposit type. Desk-study data are then fed into the computer and the computer performs the match between the search area data and its 'empirical model'. The computer then highlights those areas that best match the combination of features characteristic to the occurrence of the deposit type. This will rapidly increase the coverage of search areas but the geologist must ensure that the computer model is also a sensible geological model.

There have been major advances in laboratory techniques – fluid inclusions, isotope studies, etc. These promise improvements in defining target areas for mineralization but the techniques must be reduced in complexity and cost before they will become routine parts of an exploration programme. The type of chemical identification of potentially productive intrusions such as we described in Chapter 3, if further developed, will provide a practical approach that can assist at the planning desk-study stage.

Lastly there is the development of improved drilling technology and the increase that has taken place in down-the-hole geophysical techniques. The mining industry has used diamond drilling for many decades but we wonder if the time is approaching when the dependence upon all cored holes will be reduced. Sampling from drill chips is much cheaper and could be equally informative. Coring would then be restricted to the required length of the hole in subsequent holes. The use of fewer cored holes will have to be accompanied by much improved in-hole techniques for assaying, detection of off-hole mineralization and fracture analysis.

With this immensely improved armoury, the exploration geologist should increase the success rate. But even with all this instrumentation, electronics and theory there can be no substitute for the inspired explorationist with geology 'under the skin', a thorough understanding of ore deposits, a knowledge of exploration techniques and, last but by no means least, luck!

REFERENCES

Abrams, M. J., Brown, D., Lepley, L. and Sadowski, R. (1983) Remote sensing for porphyry copper deposits in Southern Arizona. *Econ. Geol.*, **78**, 591–604.

Ahrens, L. H. (1954) Log normal distribution of elements. *Geochim. et Cosmochim. Acta*, **5**, 49–73; **6**, 121–131.

Baxendell, P. (1983) The diversification of oil companies into mining. *Trans. Instn. Min. Metall.*, **92**, A80–A84.

Binon, L. C. (1983) Determining the exploration budget. *Min. Eng.*, March, 257–260.

Boyle, R. W. (1982a) Geochemical methods for the discovery of blind mineral deposits: part 1. *CIM Bull.*, **75(844)**, 123–142.

Boyle, R. W. (1982b) Geochemical methods for the discovery of blind mineral deposits: part 2. *CIM Bull.*, **75 (845)**, 113–132.

Cornwell, J. D. (1983) Geophysical methods in mineral exploration. *Trans. Instn. Min. Metall.*, **92**, B80–B82.

Cumming, J. D. and Wicklund, J. D. (1980) *Diamond Drill Handbook*, J. K. Smit & Sons Diamond Products Limited, Toronto, 547 pp.

Davies, H. L. (1978) History of Ok Tedi porphyry copper project, Papua New Guinea, Pt. 1. *Econ. Geol.*, **73**, 796–809.

DeGeoffroy, J., Wu, S. M. and Heins, R. W. (1967) Geochemical coverage by spring sampling method in the southwest Wisconsin zinc area. *Econ. Geol.*, **62**, 679–697.

DeGeoffroy, J., Wu, S. M. and Heins, R. W. (1968) Selection of drilling targets from geochemical data in the southwest Wisconsin zinc area. *Econ. Geol.*, **63**, 787–795.

Derry, D. R. and Booth, J. K. B. (1978) Mineral discoveries and exploration expenditure – a revised review 1966–1976. *Min. Mag.*, **138**, May, 430–433.

Eimon, P. I. (1980) Planning and establishing exploration programs. *1st Int. Min. Plan. Dev. Symp.*, China.

Goetz, A. F. H., Rock, B. N. and Rowan, L. C. (1983) Remote sensing for exploration: an overview. *Econ. Geol.*, **78**, 573–590.

Griffiths, D. H. and King, R. F. (1981) *Applied Geophysics for Geologists and Engineers*, Pergamon Press, Oxford, 230 pp.

Hutchinson, R. D. (1980) A mineral exploration strategy in the 1980s. *CIM Bull.*, **73(820)**, 26–29.

Institution of Mining & Metallurgy (1984) Prospecting in areas of glaciated terrain. Also Volumes published in 1977, 1979, 1982, London.

Kusznir, N. J. and Whitworth, K. R. (1983) RocTec: A new technique for determining bulk rock strength from borehole geophysical logs. In *Prospecting and Evaluation of Non-Metallic Rocks and Minerals* (eds K. Atkinson and R. Brassington), Institution of Geologists, London, pp. 69–96.

Levinson, H. (1980) *Introduction to Exploration Geochemistry*, Applied Publishers, Wilmette, Illinois, 924 pp.

Lindemann, J. (1982) The minerals exploration process – an overview. *The Mines Mag.*, October, 2–5.

Mackenzie, B. W. (1981) Looking for the improbable needle in a haystack: the economics of base metal exploration in Canada. *CIM Bull.*, **74(829)**, 115–125. 125.

Morris, R. O. (1983) Drilling techniques in mineral prospecting and evaluation. In *Prospecting and Evaluation of Non-Metallic Rocks and Minerals* (eds K. Atkinson and R. Brassington), Institution of Geologists, London, pp. 97–106.

Ogilvy, R. D. (1984) Down-hole IP surveys applied to off-hole mineral exploration – some design considerations. *Geoexploration*, **22**, 59–73.

Paine, D. P. (1981) *Aerial Photography and Image Interpretation for Resource Management*, Wiley, New York, 571 pp.

Parasnis, D. S. (1983) *Mining Geophysics*, Elsevier, London, 395 pp.

Press, N. (1983) The use of Landsat imagery in prospecting for industrial minerals. In *Prospecting and Evaluation of Non-Metallic Rocks and Minerals* (eds K. Atkinson and R. Brassington), Institution of Geologists, London, pp. 15–20.

Reedman, J. H. (1979) *Techniques in Mineral Exploration*, Applied Science Publishers, London, 533 pp.

Reeves, R. G. (ed.) (1983) *Manual of Remote Sensing*. Two Volumes, American Society of Photogrammetry, Virginia.

Ridge, J. D. (1983) Genetic concepts versus observational data in governing ore exploration. *CIM Bull.*, **76(852)**, 47–54.

Rogers, D. S. (1982). Diamond drilling as an aid in ore definition at the Dome Mine. *CIM Bull.*, **75(842)**,

98–104.
Rose, A. W., Hawkes, H. E. and Webb, J. S. (1979) *Geochemistry in Mineral Exploration*, Academic Press, London, 657 pp.
Telford, W. M., Geldart, L. P., Sheriff, R. E. and Keys, D. A. (1976) *Applied Geophysics*, Cambridge University Press, Cambridge, 860 pp.
Williams, R. S. and Southworth, C. S. (1984) Remote sensing makes important gains. *Geotimes*, August, 13–15.

Appendix

MINERAL LIST

Name	Composition	Uses	Remarks
Actinolite	$Ca_2(Mg, Fe)_5Si_8O_{22}(OH)_2$	–	Amphibole
Albite	$NaAlSi_3O_8$	–	Plagioclase
Almandine	$Fe_3Al_2(SiO_4)_3$	–	Garnet; commonly found in gneisses and schists
Alunite	$KAl_3(SO_4)_2(OH)_6$	Has been used in alum manufacture and as a source of potash.	Forms by hydrothermal alteration of volcanic rocks, for example in western USA.
Anatase	TiO_2	–	–
Andalusite	Al_2SiO_5	May be mined for use in manufacture of refractories, especially sparking plugs	Mined from primary deposits (France) or placer deposits (S. Africa).
Andradite	$Ca_3Fe_2(SiO_4)_3$		Garnet
Anhydrite	$CaSO_4$	Was mined in UK as a source of sulphuric acid.	An important rock-forming mineral.
Ankerite	$Ca(Fe, Mg, Mn)(CO_3)_2$	–	A common gangue mineral in hydrothermal veins.
Anthophyllite	$(Mg, Fe)_7Si_8O_{22}(OH)_2$		Amphibole
Apatite	$Ca_5(PO_4)_3(F,Cl)$	A source of phosphate.	May occur in igneous, sedimentary, or metamorphic rocks.
Argentopyrite	$AgFe_2S_3$	Minor source of silver.	
Arsenopyrite	$FeAsS$	Not mined for arsenic content as most commercial arsenic is recovered from smelter fumes.	An abundant mineral, most commonly found in high-temperature hydrothermal veins.
Augite	$(Ca, Na)(Mg, Fe, Al, Ti)(Si, Al)_2O_6$		Pyroxene; important rock-forming silicate.
Autunite	$Ca(UO_2)_2(PO_4)_2 \cdot 10\text{–}12H_2O$	Source of uranium.	Typically found in the zone of weathering over primary uranium deposits.

Ore deposit geology and its influence on mineral exploration

Name	Composition	Uses	Remarks
Azurite	$Cu_3(CO_3)_2(OH)_2$	Minor copper ore mineral	Occurs in zone of oxidation above primary copper sulphides.
Baddeleyite	ZrO_2	Main use is in ceramics.	Mined from alkaline igneous intrusives and placer deposits derived from them.
Barite	$BaSO_4$	Principal use is as drilling mud.	Distributed in a range of geological environments.
Bastnaesite	$(Ce, La)(CO_3)F$	Important source of rare earths	Mined from alkaline igneous intrusions and placer deposits derived from them.
Beryl	$Be_3Al_2Si_6O_{18}$	Major source of beryl.	Mined principally from pegmatites and placer deposits derived from them.
Biotite	$K(Mg, Fe)_3(Al, Fe)Si_3O_{10}(OH, F)_2$	–	Important rock-forming silicate.
Bismuthinite	Bi_2S_3	Not a commercial source of bismuth	Occurs chiefly in low to high temperature hydrothermal vein deposits.
Bixbyite	$(Mn, Fe)_2O_3$	–	–
Boehmite	$AlO(OH)$	Important ore mineral of aluminium.	Major constituent of bauxite; results from weathering of aluminium silicate rocks.
Bornite	Cu_5FeS_4	Major ore mineral of copper	Widely distributed copper ore mineral.
Bournonite	$PbCuSbS_3$		Sulphosalt: occurs in hydrothermal veins
Brannerite	$(U, Ca, Ce)(Ti, Fe)_2O_6$	Important uranium ore mineral	Particularly important in Blind River/Elliot Lake district, Ontario.
Braunite	$3Mn_2O_3 \cdot MnSiO_3$	Ore mineral for manganese.	Primarily results from metamorphism of manganese oxides and silicates.
Bravoite	$(Ni, Fe)S_2$	–	Pyrite in which nickel substitutes for >50% of Fe content.
Bronzite	$(Mg, Fe)_2Si_2O_6$	–	Ferroan enstatite, pyroxene group. Gangue mineral in some magmatic ore deposits.

Appendix

Name	Composition	Uses	Remarks
Bustamite	$(Mn, Ca)_3Si_3O_9$	–	–
Calcite	$CaCO_3$	Calcite as the chief constituent of limestone has a wide range of industrial uses.	Commonly found as a gangue mineral in carbonate-hosted Pb/Zn deposits.
Carnotite	$K_2(UO_2)_2(VO_4)_2 \cdot 3H_2O$	Uranium and vanadium ore mineral.	Principally mined from Colorado Plateau, USA
Carrollite	$Cu(Co,Ni)_2S_4$	Main ore mineral of cobalt.	Principally mined in Zaire and to a lesser extent in Zambia.
Cassiterite	SnO_2	Main ore mineral of tin.	The bulk of world production is from placer deposits in S.E. Asia.
Celestite	$SrSO_4$	Principal source of strontium.	Occurs chiefly in sedimentary deposits.
Chalcocite	Cu_2S	Significant copper ore.	Occurs principally in zone of supergene enrichment.
Chalcopyrite	$CuFeS_2$	Main ore mineral of copper.	Distributed in a wide range of geological environments.
Chlorite	$(Mg, Fe)_5Al(AlSi_3)O_{10}(OH)_8$	–	Commonly found as a gangue mineral in hydrothermal deposits.
Chromespinel	$Mg(AlCr)_2O_4$	Members of the spinel group may be of gem quality.	Cr spinel occurs mainly as an accessory mineral in basic igneous intrusions.
Chromite	$FeCr_2O_4$	Main ore mineral of chromium.	Mined mainly from layered basic intrusions and ophiolitic complexes.
Chrysocolla	$(Cu, Al)_2H_2Si_2O_5(OH)_4 \cdot nH_2O$	Minor copper ore mineral	Forms in the oxidized zone of copper deposits.
Cliachite		–	Amorphous form of hydrated aluminium oxide.
Clinopyroxene		–	Sub-division of pyroxene family which crystallizes in the monoclinic system.
Clinozoisite	$Ca_2Al_3(SiO_4)_3(OH)$	–	= Epidote. Major mineral in reaction skarn deposits.

Ore deposit geology and its influence on mineral exploration

Name	Composition	Uses	Remarks
Coffinite	$USi(O, OH)_4$	Important uranium ore mineral.	Particularly important in some sandstone-hosted deposits.
Columbite-tantalite	$(Fe, Mn)(Nb, Ta)_2O_6$	Important source of niobium and tantalum.	Mainly obtained from placer deposits, particularly in Nigeria.
Cordierite	$Mg_2Al_4Si_5O_{18}$	Clear blue varieties may be cut as gemstones.	Formed by metamorphism of aluminium-rich rocks.
Corundum	Al_2O_3	Valuable as a gem and as an abrasive.	Gemstone quality occurs in re-crystallized limestone. Near-gem quality from placers in Sri Lanka. May result from metamorphism of bauxite (emery deposits).
Covellite	CuS	Copper ore mineral	Mainly occurs in the zone of supergene enrichment above primary sulphide ore.
Cryptomelane	KMn_8O_{16}	–	–
Cuprite	Cu_2O	Copper ore mineral.	Mainly occurs in the zone of supergene enrichment above primary sulphide ore.
Danburite	$CaB_2(SiO_4)_2$	–	–
Diamond	C	Gemstone, near-gem and abrasives	Production from kimberlites and placer deposits.
Diaspore	$AlO(OH)$	Not generally regarded as a commercial source of alumina but some plants process ore with more diaspore than boehmite or gibbsite.	Occurs as massive fine-granular material with boehmite and gibbsite in bauxites.
Digenite	Cu_2S	–	Abundant at Butte, Montana and Kennecott, Alaska otherwise in minor quantities in copper ore deposits.
Diopside	$Ca(Mg, Fe)Si_2O_6$	Cr diopside is an indicator mineral used in exploration for kimberlites.	Occurs chiefly in medium- and high-grade metamorphic rocks, especially those rich in calcium.

Appendix

Name	Composition	Uses	Remarks
Djurleite	$Cu_{31}S_{16}$	–	Principal occurrence is in supergene blankets of sulphide deposits.
Dolomite	$CaMg(CO_3)_2$	Main uses are as a flux in blast furnaces and a source of magnesia for refractory bricks.	Of widespread occurrence in sedimentary strata. An important alteration product in Mississippi Valley-type deposits.
Enargite	Cu_3AsS_4	Minor ore of copper	Occurs in vein and replacement deposits.
Enstatite	$(Mg, Fe)SiO_3$	–	Orthopyroxene
Epidote	$(Ca)_2(Al, Fe)_3(SiO_4)_3(OH)$	–	Commonly occurs in limestones which have been affected by contact metamorphism.
Feldspathoids	–	–	Restricted to alkaline-rich igneous rocks deficient in silica.
Ferrochromite	–	–	A variety of chromite with very low Al and Mg and low, but significant, Ti, Mn, and Zn contents. Occurs in association with nickel sulphides in volcanic–peridotite ores in Western Australia.
Ferrohastingsite	$NaCa_2(Fe, Mg, Al)_5(Al_2Si_6)O_{22}(OH)_2$		Amphibole in the hornblende series.
Fluorite	CaF_2	Main uses as a flux in steel manufacture and a basis for the manufacture of hydrofluoric acid.	Occurs in a wide range of geological settings.
Freibergite	$(Ag, Cu, Fe)_{12}(Sb, As)_4S_{13}$	Important source of silver.	A silver-rich variety of tetrahedrite-tennantite.
Fuchsite			Chromian muscovite
Galena	PbS	Major lead ore.	Occurs in a range of geological environments but particularly those of sedimentary affiliation.

Name	Composition	Uses	Remarks
Garnierite	$(Ni, Mg)_3Si_2O_5(OH)_4$	Ore of nickel.	General term applied to a number of hydrous nickel silicates.
Geikielite	$MgTiO_3$	Indicator mineral for kimberlites	Mg-rich variety of ilmenite
Gersdorffite	$NiAsS$	Minor nickel ore mineral.	
Gibbsite	$Al(OH)_3$	Ore mineral for aluminium.	Important constituent of bauxite.
Goethite	$FeO.OH$	Important source of iron.	Normally found in supergene zone of iron ore deposits.
Gold	Au	–	See 'native gold'.
Grandite	the series Grossular-Andradite, Garnet group	–	–
Graphite	C	Used in foundry facings, crucibles, lubricants, paints and electrodes.	Occurs in rocks resulting from regional or contact metamorphism. Can result in strong geophysical anomalies.
Grossularite	$Ca_3Al_2(SiO_4)_3$	–	Garnet, formed by contact or regional metamorphism of impure limestones and dolomites.
Gummite	general term for secondary uranium oxides.	Minor source of uranium.	Generic name for gum-like uranium minerals.
Gypsum	$CaSO_4.2H_2O$	Used mainly in construction industry especially in plaster wallboard.	Occurs as extensive sedimentary deposits interbedded with limestones and rock salt.
Halloysite	$Al_2Si_2O_5(OH)_4$	Clays with a substantial halloysite content may be used in ceramics.	Made up of an irregular sequence of kaolin layers plus interlayer water.
Hausmannite	Mn_3O_4	Source of manganese?	Important constituent of manganese deposits formed by circulating meteoric waters.
Haematite	Fe_2O_3	Major source of iron.	Main source is enrichment ores derived from banded iron-formation.

Appendix

Name	Composition	Uses	Remarks
Hedenbergite	$Ca(Fe, Mg)Si_2O_6$	–	Commonly associated with skarn ore deposits formed at high temperatures.
Hemimorphite	$Zn_4Si_2O_7(OH)_2 \cdot H_2O$	Minor ore of zinc.	Occurs in the oxidized zone of zinc deposits.
Hornblende	Amphibole with complex composition.	–	Common rock-forming silicate.
Huebnerite	$MnWO_4$	Source of tungsten.	Occurs in contact metamorphic deposits adjacent to granitic intrusives.
Hydrogoethite	–	Iron ore mineral.	Component of ore body at Krivoi Rog deposit, USSR.
Hydrohaematite	–	Iron ore mineral.	Component of ore body at Krivoi Rog deposit, USSR.
Illite	$(K, H_3O)(Al, Mg, Fe)_2(Si, Al)_4O_{10}(OH)_2 \cdot H_2O$	–	An important constituent of argillaceous sediments and pelites.
Ilmenite	$FeTiO_3$	Important source of titanium.	Mined principally from beach sand deposits.
Ilvaite	$CaFe_3(SiO_4)_2(OH)$	–	
Jacobsite	$Mg(Fe, Mn)_3O_4$	–	Weakly magnetic member of the magnetite series.
Jadeite	$Na(Al, Fe)Si_2O_6$	Decorative stone.	Found in associations with serpentine and low-grade metamorphosed greywackes. Jadeitic diopside occurs in cognate xenoliths in some kimberlites.
Jarosite	$KFe_3(SO_4)_2(OH)_6$	–	Secondary mineral occurring as coating on iron-bearing material especially where pyrite is oxidizing.
Johannsenite	$CaMnSi_2O_6$	–	Pyroxene
Kaolinite	$Al_2Si_2O_5(OH)_4$	Diverse uses but mainly sold to the paper industry.	The bulk of kaolin deposits are formed by chemical weathering of granitoids.
Kenomagnetite	$Fe_{8/3}O_4$	Minor iron ore	Found in the zone of supergene enrichment of banded iron-formation in W. Australia.

Name	Composition	Uses	Remarks
Kyanite	Al_2SiO_5	Used in the manufacture of mullite refractories.	Occurs in aluminium-rich sediments which have been altered by medium-grade regional metamorphism. Mined mainly from beach sands.
Leucoxene	general term for alteration products of ilmenite.	Source of titanium.	A mixture of titanium minerals such as rutile, anatase and sphene.
Limonite	general term for hydrous iron oxides.	–	A term commonly, but incorrectly, used for yellow-brown variety of goethite.
Loellingite	$FeAs_2$	–	Occurs chiefly in medium-temperature hydrothermal veins.
Magnesite	$MgCO_3$	Calcined for manufacture of refractory bricks. Used for making magnesium metal.	Alteration product of peridodtite, dolomite or limestone.
Magnetite	Fe_3O_4	Important source of iron.	Mainly exploited from iron-formation.
Malachite	$Cu_2CO_3(OH)_2$	Copper ore mineral.	Occurs in the oxidized zone above primary sulphide deposits.
Malayaite	$CaSnSiO_5$	Not a commercial source of tin.	Found in skarn tin deposits, may fluoresce yellow.
Marcasite	FeS_2	–	Low-temperature polymorph of pyrite.
Martite	Fe_2O_3	Ore mineral of iron	Pseudomorph of haematite after magnetite.
Metatorbernite	$Cu(UO_2)_2(PO_4)_2 \cdot 8H_2O$	Ore mineral of uranium.	Occurs in weathered zone over primary uranium deposits.
Millerite	NiS	Nickel ore mineral	Normally of minor significance but attains importance in some Canadian nickel deposits of volcanic-peridotite association.

Appendix

Name	Composition	Uses	Remarks
Minnesotaite	$(Fe, Mg)_3 Si_4 O_{10} (OH)_2$	Minor source of iron in the Lake Superior district USA.	
Molybdenite	MoS_2	Principal source of molybdenum	Mined principally from porphyry deposits. Also occurs in veins and contact-metamorphic deposits.
Monazite	$(La, Ce, Nd, Th) PO_4$	Important source of thorium and cerium	Exploited mainly from placer deposits.
Montmorillonite	$(Na,Ca)_{0.33}(Al,Mg)_2 Si_4 O_{10} (OH)_2 \cdot nH_2O$	Diverse industrial end uses including drilling mud, binder for pelletizing iron ore, plasticizer in moulding sands for foundries.	Formed by the alteration of volcanic ash.
Montroseite	$(V, Fe) O (OH)$	Important source of vanadium	Occurs in the Uravan belt in western Colorado, USA.
Muscovite	$KAl_2 Si_3 O_{10} (OH)_2$	–	Rock-forming silicate.
Native copper	Cu	Minor source of copper.	An important source of copper in prehistoric times.
Native gold	Au	Bulk of gold is purchased by investors and Central Banks. Also used in jewellery, electronics and dentistry.	Bulk of Western World production from Witwatersrand Basin. Placer deposits and veins also important.
Native silver	Ag	Diverse uses which include photography, electronics components and jewellery.	Occurs in small amounts in the oxidized zones of some sulphide deposits. Also found in veins.
Niccolite	NiAs	Minor nickel ore mineral	Occurs mainly in association with pentlandite/pyrrhotite in ores of magmatic affiliation.
Nontronite	$Na_{0.33} Fe_2 (Si, Al)_4 O_{10} \cdot nH_2O$	–	Iron-rich montmorillonite species.

Ore deposit geology and its influence on mineral exploration

Name	Composition	Uses	Remarks
Nordenskioldine	$CaSnB_2O_6$	–	Tin-borate species found in skarn deposits.
Olivine (group)	$(Mg, Fe)_2SiO_4$	Mg-rich olivine is used in the manufacture of refractory bricks.	Important rock-forming silicate.
Opal	$SiO_2 \cdot nH_2O$	Precious opal is a gemstone.	Fine opals come from sandstones in Australia.
Orthoclase	$KAlSi_3O_8$	Sometimes exploited for use in ceramics	Common rock-forming silicate.
Orthopyroxene	Generic name for pyroxenes with orthorhombic structures.	–	Common group of rock-forming silicates.
Pentlandite	$(Fe, Ni)_9S_8$	Main ore of nickel	Occurs mainly in association with mafic and ultramafic rocks.
Petalite	$LiAlSi_4O_{10}$	Minor source of lithium.	Occurs in granite pegmatites and as a minor constituent of some placers.
Phlogopite	$KMg_3Si_3AlO_{10}(F, OH)_2$	–	Essential mineral in kimberlite.
Pitchblende	Massive uraninite UO_2	Major uranium ore mineral	A variety of uraninite which commonly occurs as colloform crusts.
Plagioclase (subgroup of feldspars)	$(Na, Ca) Al (Al, Si) Si_2O_8$	–	Important rock-forming silicate.
Platinum	Pt commonly with Pd, Ir, Fe, Ni.	Used mainly in the chemical industry	Exploited mainly from layered basic intrusions but significant quantities also mined from alluvial placers in the Urals, USSR.
Potash feldspar	Microcline or orthoclase	Used as a flux in the ceramics industry	Important rock-forming silicate.
Powellite	$CaMoO_4$ can contain W	–	Mo-bearing variety of a partial series from scheelite.
Prehnite	$Ca_2Al_2Si_3O_{10}(OH)_2$	–	Occurs chiefly in cavities in basic igneous rocks.

Appendix

Name	Composition	Uses	Remarks
Pseudomalachite	$Cu_5(PO_4)_2(OH)_4 \cdot H_2O$	Minor copper ore mineral	Occurs in zone of oxidation
Pyragyrite	Ag_3SbS_3	Important ore of silver.	Occurs in stratiform sediment-hosted Pb/Zn deposits and in veins. May also form as a result of supergene enrichment.
Pyrite	FeS_2	Occasionally mined as a source of sulphur.	A common associate of ore minerals in a wide range of geological environments.
Pyriboles	–	–	Group term for pyroxenes and amphiboles.
Pyrochlore	$(Ca, Na, Ce)(Cb, Ti, Ta)(O, OH, F)_7$ can contain Th.	Important source of niobium	Occurs in alkaline igneous complexes and placers derived from them.
Pyrolusite	MnO_2	–	Occurs as a secondary mineral resulting from the alteration of manganese minerals.
Pyrope	$Mg_3Al_2(SiO_4)_3$	Minor production for gemstones in Czechoslovakia.	Important indicator mineral for kimberlites.
Pyrrhotite	$Fe_{1-x}S$	Mined as by-product at Sudbury, Ontario and converted to high-grade iron ore.	Closely associated with pentlandite. The magnetic susceptibility of pyrrhotite makes magnetometer surveys relevant to nickel exploration.
Quartz	SiO_2	Diverse industrial uses: mainly in construction. Also as a flux in metallurgy and in manufacturing of glass.	Important rock-forming silicate.
Rammelsbergite	$NiAs_2$	Minor nickel ore	Occurs chiefly in veins with other nickel and cobalt minerals.
Realgar	AsS	–	A minor constituent in some hydrothermal veins.

Ore deposit geology and its influence on mineral exploration

Name	Composition	Uses	Remarks
Rhodochrosite	$MnCO_3$	–	Forms a gangue mineral in hydrothermal veins and as a secondary mineral in residual manganese deposits.
Rhodonite	$(Mn, Fe, Mg, Ca)SiO_3$	–	May form protore for supergene manganese ores, for example Amapa area, Brazil.
Rutile	TiO_2	An important source of titanium	Mined mainly from beach sands.
Saleeite	$Mg(UO_2)_2(PO_4)_2 \cdot 10H_2O$	Uranium ore	Occurs in the zone of weathering above primary uranium deposits. Notably in Northern Territory, Australia.
Scheelite	$CaWO_4$ (can contain Mo)	Important tungsten ore.	Mined mainly from contact metamorphic deposits.
Sericite	Mica similar to muscovite	–	Fine-grained variety of muscovite.
Serpentine	$Mg_3Si_2O_5(OH)_4$	Chrysotile is main source of commercial asbestos.	Group which includes chrysotile, antigorite, and lizardite.
Siderite	$FeCO_3$, (can contain Mn, Mg, Ca)	Ore mineral of iron	Mined from iron-formation. Also occurs as a gangue mineral in veins.
Sillimanite	Al_2SiO_5	–	Formed in aluminium-rich rocks under conditions of high-grade metamorphism.
Sklodowskite	$Mg(UO_2)_2Si_2O_6(OH)_2 \cdot 5H_2O$	Minor uranium ore mineral	Occurs in the Granite District, USA and Shinkolobwe, Zaire.
Smectite	$(Na, Ca)(Al, Mg)_4Si_8O_{10}nH_2O$	Very diverse industrial uses.	Group term which includes montmorillonite, saponite and hectorite.
Specularite	Fe_2O_3	–	Variety of haematite.
Spessartine	$Mn_3Al_2(SiO_4)_3$	–	Occurs in granite pegmatites and in metamorphosed manganese-bearing rocks.

Appendix

Name	Composition	Uses	Remarks
Spessartite	= Spessartine	–	
Sphalerite	(Zn, Fe,) S	Major source of zinc	Mined predominantly from sediment-hosted deposits. Volcanogenic setting also significant.
Sphene	$CaTiSiO_5$	–	Occurs as accessory mineral in acid and intermediate igneous rocks. Minor component of some placers.
Spodumene	$LiAlSi_2O_6$	Source of lithium and in ceramics	Mined principally from granite pegmatites.
Stannite	Cu_2SnFeS_4	–	Minor tin mineral.
Staurolite	$(Al_2SiO_5)_2(Fe, Mg)(OH)_2$	–	Occurs in aluminium-rich gneisses and schists and as a component of some placer deposits.
Stephanite	Ag_5SbS_4	Minor source of silver	Occurs in sediment-hosted base metal deposits (Queensland) and silver-bearing vein deposits (Colorado).
Stibnite	Sb_2S	Main antimony ore mineral	Mainly occurs in low-temperature veins.
Stilpnomelane	$K(Fe,Al)_{10}Si_{12}O_{30}(OH)_{12}$	A minor source of iron in the Lake Superior district, USA.	Occurs dominantly in regional metamorphic rocks derived from greywackes.
Talc	$Mg_3Si_4O_{10}(OH)_2$	Diverse industrial uses: as a lubricant, in ceramics, as a filler in paint, paper and rubber.	Occurs mainly in association with ultrabasic rocks which have been altered by low-grade regional metamorphism.
Tantalite	$(Fe, Mn)(Ta, Nb)_2O_6$	–	Occurs in granite pegmatites.
Tennantite	$(Cu, Fe)_{12}As_4S_{13}$	Minor copper ore mineral.	May contain appreciable content of silver – see freibergite.

Ore deposit geology and its influence on mineral exploration

Name	Composition	Uses	Remarks
Tetrahedrite	$(Cu, Fe)_{12}Sb_4S_{13}$	Minor copper and antimony ore mineral.	Occurs in a range of geological environments.
Thucholite	A mixture of uranium and thorium with carbonaceous matter	Uranium ore	Associated with quartz pebble conglomerates in Witwatersrand Basin and Blind River district, Ontario.
Titanomagnetite	–	Source of titanium and iron	Mined as a by-product from the Palabora carbonatite, S. Africa. Also obtained from marine placers off the coast of New Zealand.
Topaz	$Al_2SiO_4(OH, F)_4$	Transparent forms may be used as a gemstone.	Mainly occurs in pegmatites and high-temperature granite veins. May also be found in some placers.
Tourmaline	$(Na,Ca)(Mg, Fe, Li)_3B_3Al_3((Al, Si)_3O_9)_3(OH, F)_4$	Some gemstones exploited in Brazil. Some varieties have valuable piezoelectric properties.	Widely distributed in granitoids and in high-temperature hydrothermal veins.
Tremolite	$Ca_2Mg_5Si_8O_{22}(OH)_2$	–	Amphibole.
Tyuyamunite	$Ca(UO_2)_2(VO_4)_2(H_2O)_{10}$	Uranium ore mineral	Secondary uranium mineral – widely distributed in Colorado Plateau area, USA.
Uraninite	UO_2	Major uranium ore mineral.	Occurs in a wide range of geological environments.
Uranophane	$Ca(UO_2)_2Si_2O_7 \cdot 6H_2O$	Uranium ore mineral	An important ore mineral at the Rossing deposit, Namibia.
Wolframite	$(Fe, Mn)WO_4$	Important tungsten ore.	Mined dominantly from high-temperature vein deposits.
Wollastonite	$CaSiO_3$	–	Formed by high-grade metamorphism of siliceous limestones.

Appendix

Name	Composition	Uses	Remarks
Xenotime	YPO_4	A minor source of yttrium	Widespread occurrence in acidic and alkalic igneous rocks. Forms a component of some beach sand deposits.
Zircon	$ZrSiO_4$	Principal source of zirconium and hafnium.	Mined mainly from beach sands.

Index

Abitibi Belt, Canada 52, 116–18, 13, 133, 153, 156, 167
Accreting plate margin 116
Acid leaching
   lateritic nickel 291, 289
   uranium 205
Acoustic reflection profilers 200
Actinolite 282, 426
Adiabatic expansion 162
Adit xiv, 87, 415
Adsorption
   uranyl ions 340, 353, 362, 363
Aerial photography 50, 54, 61, 62–3, 89, 186, 194, 227, 244, 256, 297, 394, 409, 410
Aerobic bacteria 363
Africa 55, 293, 332, 333, 402
Age
   bauxite deposits 281–3
   cratons 60
   Hamersley Basin deposits 326
   hydrothermal gold deposits 146–7
   iron-formation 317
   kaolin deposits 301
   kimberlites 54–5
   lateritic nickel deposits 292–3
   Mississippi Valley-type lead-zinc deposits 259–60
   Nickel sulphide deposits 40
   Palaeo-placers 201
   Placer deposits 176–7
   Podiform chromite deposits 32–3
   Porphyry copper deposits 70, 72–3
   Porphyry molybdenum deposits 94
   Sandstone-hosted uranium deposits 359–60
   Sediment-hosted copper deposits 216–17
   Sedimentary-exhalative deposits 230, 231
   Skarn deposits 374–5
   Stratiform chromite deposits 19
   Volcanogenic sulphide deposits 109–11
Aggneys area, South Africa 230
Agnew deposit, Western Australia 40, 45, 47–9
Aiken, South Carolina 303
Airborne electromagnetic anomalies 411
Airborne electromagnetic methods 50, 62, 104, 106, 128, 130, 133, 164, 165, 194–5, 209, 332, 354–5, 394, 411
Airborne gamma-ray spectrometry 288, 354
Airborne geophysical surveys 12, 410–11
Airborne multispectral scanning 61
AK 1 pipe, Western Australia 54, 57, 62, 64
Alaska 73
Alaskite 302
Albania 19, 32, 33
Albite 79, 393, 429
Algeria 260
Algoma iron-formation 316–17
Alkali-calcic
   porphyry molybdenum deposits 93
Alkaline associated
   porphyry copper deposits 74
Alligator River deposit, Northern Territory 347, 349, 353
Alluvial fans 358
   alluvium 184, 186, 200
   alluvial placer deposits 178, 181–8, 199–200, 211
Almandine 381, 429
Alpine mining district 260, 262
Alpine orogeny 32, 33, 177
Alteration
   bauxite deposits 284–5
   bleaching associated with 364
   hydrothermal gold deposits 156, 167
   Mississippi Valley-type lead-zinc deposits 261–2
   Pine Point deposits 264
   porphyry copper deposits 78–82, 92
   porphyry molybdenum deposits 96–9
   retrograde 392–3

Alteration – *cont.*
  sandstone-hosted uranium deposits 364
  skarn deposits 377–9, 388, 389, 393–4
  Teutonic Bore 121
  unconformity-type uranium deposits 346–7
  vertical extent in porphyry copper deposits 80
  volcanogenic sulphide deposits 113, 115, 131, 134, 138
  wall-rock 150–1, 155, 160–1, 254
Alumina 276, 283, 287, 288, 325, 332
Alumina grade bauxite 283
Aluminium 276, 281, 285, 306, 371, 378
Aluminium hydroxides 276
Aluminosilicates 283, 285
Alunite 429
Amapa Territory, Brazil 307
Amax 102, 402
Amazon Basin, Brazil 297
Ambrosia Lake district, New Mexico 363
American Bayer process 276
Ammonia complexing with gold 159
Amorphous silica 80
Amphiboles 161, 377
Amulet deposit, Noranda 117, 132
Anaerobic bacteria 362
Analysis
  bauxites 288–9
  iron ores 323–5
  nickel laterites 297
Anatase 283, 429
Andalusite 429
Andean porphyry copper province 82
Andean subduction zone 220–1
Andesite 81, 82, 84
Andradite garnets 372, 376, 381, 382, 388, 390, 429
Angola 55, 221
Anhydrite 79, 80, 220, 226, 233, 262, 264
Anhydrous skarn 381
Ankerite 80, 327, 429
Anomaly hunting 12
Anomaly values
  soil geochemistry 4, 5, 130, 167, 267, 384, 395, 396, 414–15
Anorthosite layers
  Critical Zone, Bushveld Complex 26, 27, 28
Anthophyllite 429
Antimony 246
Apatite 79, 80, 90, 429
Appalachians 72
Aquifers 10, 228, 229
Archaean greenstone belt 39, 40, 42, 49, 64, 119, 146–56, 167, 169
Archaean iron-formation 323
Archaean Shield, Western Australia 146–52, 160–1, 164–9, 169–70, 302, 323, 326, 333
Archaean volcanogenic sulphide deposits 109–10, 115, 137–8
Area of influence
  grade estimation xii, 87, 297
Argentina 73
Argentopyrite 235, 429
Argillic zone
  porphyry copper deposits 80, 84
  porphyry molybdenum deposits 98, 100
Arizona 94, 259, 364, 408
Arkansas 284
Arsenic 159, 166, 167, 246
Arsenopyrite 112, 154, 235, 382, 385, 429
Asbestos 370
Ashanti mine, Ghana 201
Asia 400
Assay cut-off 102, 234
Astrobleme xii, 39
Athabasca Formation, Canadian Shield 341, 342, 347, 348, 349, 350, 351, 352, 354, 355
Atmochemical methods 416
  see also Soil gas sampling
Au subtype
  hydrothermal vein deposits 145
Auger drilling 290–1, 355
Augusta, Georgia 303
Auriferous conglomerates 4, 176, 179, 202–3
Australia 170, 332
  bauxite deposits 279, 281, 283, 284
  beach sand deposits 176, 189
  exploration 402, 404, 422
  gold mining 146, 147, 151
  iron ore production 314, 315, 325, 333
  marine placers 197
  nickel production 38
  shale-hosted lead-zinc deposits 216, 232–47
  tungsten production 380, 381
  uranium mining 7, 338–9, 354, 365
  see also Northern Territory, Queensland, Western Australia
Australian Bauxite Province 281
Autolith
  defined xii, 35
Automated X-ray spectrometry 288
Autunite 340, 429
Avoca copper mine 4
Azegour, Morocco 387
Azurite 223, 430

Back-arc region 107, 115, 356
  western USA 99
Background values
  soil geochemistry 414–5
Bacteria 253, 362–3
Baddeleyite 179, 430
Bahia, Brazil 201
Bakwanga deposit, Zaire 176, 181
Ball clay 300
Ballinalack deposit, Ireland 256
Baluba, Zambia 224
Ban Ban deposit, Queensland 386
Banded granitoids 380
Banded haematite quartzite 316
Banded iron-formation 150, 321–2, 324, 325, 326–31, 333–4
Bangweulu massif 222
Barberton district, South Africa 52, 147
Barite 152, 238, 247, 252, 262, 265, 430
Barium 246
Barney Creek Formation, McArthur River 238
Barrier Complex, Pine Point district 263–4, 366
Basal Reef, Witwatersrand Basin 205, 206, 209
Basalt-rhyolite magmatism 99
Basement Complex, Zambian Copperbelt 224, 226
Basinal compaction model 241
Bastard Unit, Merensky Reef 29
Bastnaesite 179, 430
Batholiths 93, 96, 385, 394
Bathurst mining district, Canada 112, 131, 136
Batten Trough, McArthur River 232, 238, 242
Baux de Provence, France 281
Bauxite 275, 276–9, 286, 301, 302, 311, 417

Index

analysis 288–9
Bauxite deposits 276–85
  exploration 288–91
  genesis 285–8
Bauxite grades 286
Bauxitic clay 283
Bauxitization 285, 287, 324
Bayer process xii, 276, 289
Beach sand deposits 18, 176, 180–1, 188–97
Beaverlodge area, Saskatchewan 354
Beisa uranium mine 8
Belingwe, Zimbabwe 52
Bell Channel 117
Beneficiation
  iron ore 314, 315, 319, 320
Berg, British Columbia 92
Bernabe Montano, New Mexico 363
Beryl 2, 182, 430
Besshi Mine, Japan 107, 108, 115
'Billiard ball' model of nickel sulphide genesis 47
Billiton 420
Bingham Mine, Utah 69–70, 71, 74, 75, 81, 82, 89, 377, 390
Biogeochemical sampling 90
Biological activity 324
Biotite 78, 79, 81, 82, 84, 85, 97, 98, 102, 161, 230, 371, 372, 381, 386, 389, 430
Birds Reef, Witwatersrand Basin 205
Bismuth 90, 387, 390
Bismuthinite 382, 387, 430
Bitumen 238, 262
Bixbyite 307, 430
Black Angel Mine, Greenland 259–60
Black sands
  titanium-bearing 188
Black shale 317
Blast furnaces 315
Blasthole
  defined xii
  sampling at Panguna, Bougainville 87
Blind River-Elliot Lake district, Canada 201, 202
Block caving
  defined xii, 74
Blows
  kimberlite 57
Bluebush sequence
  Kambalda mining district 43

Boart
  diamonds 54
Boehmite 276, 283, 284, 288, 430
Boké Mine, Guinea 283
Bold Head deposit, King Island 382, 383
Bolivia 107
Bonnett Plume district, Yukon 267
Borates 384
Borneo 199
Bornite 74, 76, 79, 80, 84, 106, 112, 118, 218, 219, 222, 225, 236, 385, 430
Bornite-molybdenum mineralization 81
Boss Mountain, Canada 98
Botswana 38, 39, 54, 55
Bougainville, Papua New Guinea 84, 137, 401
Boulder Beds, Kinta Valley 184
Bournonite 382, 430
Boxwork
  defined xii, 295
Brakspruit-type potholes, Merensky Reef 32, 33
Barney Creek Formation, Australia 233
Brannerite 202, 339, 347, 430
Braunite 307, 430
Bravoite 347, 430
Brazil
  Amapa Territory 307
  Amazon Basin 297, 331
  bauxite deposits 281
  chromite deposits 33
  enriched haematite deposits 323, 329–31
  gold mining 146
  Minas Gerais 293, 307, 317, 325
  nickel laterites 291–2
  podiform chromite deposits 33
  Quadrilatro Ferriferro 181, 310, 325
  tin palaeo-placers 175
Brazilian Shield 110
Breccias 65–7, 78, 83, 84, 85, 94, 102, 103, 104, 122, 236
Brecciation 249, 262, 347
  explosive 107, 127
Brine expulsion model 233
Brines 226, 261, 262, 349, 380
British Columbia 96
British Geological Survey 409
Brockman iron-formation, Hamersley Basin 326, 327, 331

Broken Hill, Australia 231
Bronzite 26, 430
Bucket-wheel excavators 193
Budgets
  exploration 406–7
Bulk mining methods
  bauxite 289
  skarns 387
Bureau de Recherches Géologiques et Minières 402
Bushveld Complex, South Africa 19, 22, 23–4, 24–32, 39
Bushveld Granite 26
Bustamite 385, 431
Butte, Montana 308
By-product metals xiii, 10, 39, 74, 87, 106, 107, 112, 185, 205, 218, 260, 261, 293, 359, 370

Cadmium 90, 260, 261
Calc-alkaline molybdenum deposits 93, 101
Calc-alkaline porphyry copper deposits 74
Calc-alkaline rocks 51, 74, 77, 93, 99, 101, 107, 130, 154, 387
Calc-silicate rocks 371, 377–8, 393
Calcic skarns 372, 377, 379–80, 396
Calcite 80, 99, 349, 379, 384, 389, 431
Calcium 371, 381, 393
Calcium-aluminium silicates 377
Calcium-magnesium silicates 377
Calcium metasomatism 386
Calcrete 191
  see also Caliche layer
Caledonian orogeny 33
Caliche layer xiii, 90
  see also Calcrete
California 37, 50, 181, 293
Calp 248
Camborne School of Mines Geothermal Project 241
Cameroon Bauxite Province 281
Campbell Mine, Ontario 152, 153
Campo Formoso, Brazil 22
Canada 7, 55, 337
  Abitibi Belt 52, 116–18, 131, 153, 156, 167
  Archaean greenstone belt 39, 72
  Athabasca Formation 341, 342, 347, 348, 349, 350, 351, 352, 354, 355
  Dome Mine, Timmins, Ontario 153, 154, 155, 419

Canada – *cont.*
　exploration 54, 403–4
　hydrothermal gold deposits 146, 147, 148, 157, 170
　iron ore production 315
　Key Lake Deposit, Saskatchewan 343, 344, 346–7, 350, 352, 354
　Kidd Creek deposit 12, 109, 117–18, 199, 133, 136
　nickel mining 38, 41, 42
　Noranda mining district 109, 112, 113, 116–17, 118, 128, 136
　Pine Point mining district 260, 261, 262–6
　Porcupine mining district 146, 148, 149, 153–6, 163
　porphyry copper deposits 72, 73, 74, 75, 79
　porphyry molybdenum deposits 93
　Rabbit Lake deposit 343, 346, 347, 349, 350, 352, 354
　skarn deposits 375, 381
　Superior Province 147
　Timmins mining district 117–18, 156, 164, 166, 167
　uranium 175, 202, 207, 337, 339, 355, 365
　volcanogenic sulphide deposits 112
Canada Tungsten, Northern Territories 376, 381
Canadian Cordillera 81, 82, 98, 100
Canadian Shield 12, 40, 51, 109, 147, 164, 166, 167, 168, 323, 341, 411
Canga 293
Cannivan Gulch, Montana 387
Capel beach sand deposits, Western Australia 188, 194
Captains Flat deposit, New South Wales 130
Carbon 206, 220, 235
Carbon dioxide 363
Carbon Leader, Witwatersrand Basin 205, 206
Carbonaceous material 359, 361–2, 364
Carbonate facies
　iron-formation 320, 321
　silica-dolomite rock 235
Carbonate-hosted lead-zinc deposits
　Ireland 216, 247–59
　Mississippi Valley-type 259–67
Carbonates 99, 112, 154, 161, 220, 247–8, 259, 261, 262–3, 264, 279, 345, 349, 352, 353, 381, 385, 392, 393, 394
Carbonatization xiii, 151, 155, 161, 166
Caribbean Bauxite Province 281
Carnotite 340, 359, 431
Carr Fork, Bingham, Utah 379, 390–2
Carrollite 218, 224, 226, 431
Carswell structure, Saskatchewan 354
Cascade, Washington 81
Cascades, Canada 73
Cash flow
　exploration period 406, 407
Cassiterite 118, 178, 181, 182, 195, 197–9, 377, 380, 384–5, 431
Catoca, Angola 58
CDM 190–3
Celestite 262, 431
Cell-by-cell replacement 370, 371
Cementation
　bauxites 290–1
Central Rand group, Witwatersrand Basin 205, 206, 209
Central Tisdale anticline, Porcupine mining district 154, 155
Ceresco ore body, Climax mine 101
Cerro Matoso deposit, Colombia 293
Chalcedony 394
Chalcocite 77, 84, 112, 218, 219, 223–4, 431
Chalcocite blanket 84
Chalcophile metallogeny xiii, 99
Chalcopyrite 18, 24, 42, 74, 76, 79, 80, 84, 94, 106, 112, 113, 117, 118, 120, 121, 127, 137, 155, 218, 219, 222, 235, 236, 261, 308, 377, 379, 382, 386, 387, 388, 431
Chambishi, Zambia 225
Channel lag deposits 356
Chemical analysis
　bauxites 289
　iron ores 331
Chemical composition
　fluid inclusions 159
Chemical fingerprints 105, 138
Chemical weathering 275, 287, 300–1, 310, 348
Chibuluma, Zambia 224
Chile 73, 82, 401
Chililabombwe, Zambia 224
China 276, 281, 283, 285, 315, 375, 380
China clay xiii, 298, 300, 306
Chingola, Zambia 223, 229
Chinle Formation, Colorado Plateau 358, 361
Chlorides 159
Chlorite 80, 81, 99, 112, 293, 346, 350, 353, 359, 364, 389, 431
Chloritic alteration 346, 347, 350, 351
Chloritization 113, 131, 347, 351
Chrome diopside 62
Chrome spinel 62, 293, 431
Chromite 18, 36, 42, 208, 431
　Bushveld Complex 26–9
　deposits 18 et seq.
　strategic mineral 36
Chromitite layers 22–3, 26–9
Chromium 18, 26–29
Chrysocolla 223, 431
Chuquicamata, Chile 70, 74, 75, 309
Churchill province, Canadian Shield 40
Chute bars 183, 358
Classification
　bauxite deposits 276
　chromite deposits 19
　hydrothermal vein deposits 144–6
　kaolin deposits 301
　nickel sulphide deposits 38
　porphyry molybdenum deposits 92–3
　sediment associated copper 219–20
　skarn deposits 373–4
　volcanogenic sulphides 107
Clays 62, 80, 98, 275, 279, 283, 340, 347, 351, 352, 353, 358, 359, 379, 394
Cliachite 276, 431
Climate
　effects on ore deposition 182–3, 281, 287, 291, 293, 331
Climax, Colorado 74, 75, 94, 95, 98, 99, 100, 101, 390
Climax-type molybdenum deposits 92, 94–5, 95–6, 97, 104, 137
Clinopyroxene 23, 388, 431
Clinopyroxenite 35
Clinozoisite 378, 431
Close boring
　exploration 194, 195
Cluff Lake deposit, Saskatchewan 350, 354
Coal 404, 417, 422

Index

Coalified wood 362, 364
Cobalt 18, 112, 218, 224, 225, 226, 229, 293, 295, 297, 379, 386, 411
Cobalt-gold-nickel subtype
  hydrothermal vein deposits 145
Cobaltiferous pyrite 224
Coed-y-Brenin copper deposit 7
Coffinite 339, 340, 346, 347, 359, 361, 432
Collins Bay, Saskatchewan 350
Colloform texture xiii, 249, 262
Colluvial placer deposits 178, 181–8
Colombia 181
Colorado 94, 95, 359, 364, 416
Colorado Plateau 355–6, 358, 359, 361, 362
Colour aerial photography 194
Columbite-tantalite 185, 432
Combined grade 112
Comminution
  ore processing 11
Commodities 4
  ore types and 6
Commoner Mine, Zimbabwe 151
Complex resistivity 37
Complexes
  gold transport 159
  uranyl ions 339, 340, 352
Concentrates 11, 87
Conglomerates 201, 202, 203, 205, 208, 209, 210, 329, 335
Conglomerate-hosted uranium deposits 357
Congo craton 55
Constructive plate margins 35
Consultancies
  exploration 400
Contact metamorphism 102, 377, 390, 391, 392
Contact metasomatism 390
Continental margins
  orogenic belts 387–8
  placer deposits 178, 181
  skarn formation 380
Continental platforms 285
Convective system 233
Convergent plate margin 82
Conzinc Riotinto 87
Cooley dolomite, McArthur River 236, 237
Copper 4, 7, 9, 10, 18, 34, 53, 94, 107, 113, 121, 130, 131, 137, 152, 225, 226, 235–6, 243, 245, 261, 365, 370, 374, 377, 379, 380, 386, 387, 389, 390, 392, 411

Copper group
  volcanogenic sulphides 107, 108
Copper-cobalt ores 221
Copper-iron skarn ores 386–7
Copper/nickel ratios
  sulphide mineralization 51
Copper skarn deposits 371, 374–5, 380, 387–92
Copper/zinc ratios
  sulphides 127
Copper-zinc-silver deposits 118
Coral debris 259
Cordierite 372, 432
Cordilleran magnesian magnetite
  skarn deposits 386
Core drilling 201, 290–1, 365
Core-logging 402, 419
Cornwall, UK 106, 178, 197–8, 303
Cornwall deposit, Pennsylvania 386
Cornwallis district, Canada 267
Corocoro deposit, Bolivia 220
Corundum 80, 276, 283, 284, 432
Costabonne, France 372
Costs
  drilling 419, 422
Covellite 77, 432
Coxco deposit, McArthur River 260
Crackle breccias 83
Crackled zones 84
Craigmont, British Columbia 386–7
Cratons 55, 59, 60, 285
Creta deposit, USA 226
Cristina deposit, Guatemala 298
Critical Zone, Bushveld Complex 22, 23, 26–7, 32
Crushing
  iron ore 315
Cryptomelane 307, 432
Crystallization
  intrusions 82–3, 84
Cu-Pb-Au-Ag subtype
  hydrothermal vein deposits 145
Cu:Pb:Zn ratio
  sediment-hosted copper deposits 281
Cuba 33, 35, 293, 294
Cumulates
  ultramafic 35
Cumulus layers
  chromitite 29
Cupreous Pyrite type
  volcanogenic sulphides 108, 109, 112, 113, 115, 116, 124, 137
Cupriferous mica 223
Cuprite 77, 432

Cut-off grades 9–10, 102, 180, 234, 289, 293, 301, 319, 388, 413
Cyprus 35, 111, 112, 116
Cyprus type
  volcanogenic sulphides 107, 108
Czechoslovakia 301

Daldyn-Alakit district, Yakutia 57
Dales Gorge, Hamersley Basin 326, 327, 331
Damba deposit, Zimbabwe 42, 48
Dambos 227
Danburite 384, 432
Darling Range, Australia 283
Data accumulation 409
Data analysis
  exploration programmes 409–11
Decline xiii, 415
Deep overburden sampling 257
Deformed volcanogenic sulphide deposits 138
Deilmann ore body, Saskatchewan 344, 347
Deloro Group, Porcupine mining district 153, 154, 155
Desert varnish
  sampling 90
Desk studies
  exploration programmes 409–11
Destor-Porcupine Fault, Porcupine mining district 154, 155
Destructive plate margins 82, 137
Devon, UK 303
Dewatering
  aquifers 229
  sediments 358
Diagenesis 226, 325, 370
Diamond drilling 87, 92, 133–4, 169, 209–10, 245, 257, 332, 333, 355, 417, 419, 420–21
Diamonds 175, 181, 193, 432
  boart 54
  gems 54
  industrial 54
  near-gem 54
Diamondiferous deposits 190–3
Diaspore 80, 176, 283, 284, 432
Diatremes 57, 84, 361
Dickenson Mine, Ontario 152, 153
Digenite 77, 432
Diopside 62, 387, 382, 390, 432
Diopsidic pyroxene 388
Direct reduction method
  iron manufacture
Dispersion trains 89, 167

449

Dissemination
   sulphides 42, 45, 48–9, 74, 76
Distal accumulations
   volcanogenic sulphide deposits 128, 134
Djurleite 77, 84, 433
Doline
   defined xiii, 350
Dolomite 220, 262, 264, 266, 279, 307, 317, 372, 380, 381, 387, 388, 433
Dolomitization 259, 261, 262, 264
Dome Mine, Timmins, Ontario 153, 154, 155, 419
Dome Mine, diamond drilling at 418–19
Dominican Republic 293
Dominion Group, Witwatersrand Basin 205
Down-the-hole geophysics 419, 423
Dredging
   tin mining 186
Drilling 12, 37, 87, 88, 92, 104, 133–4, 166, 167–9, 188, 194, 195, 201, 22, 243, 257, 267, 290–1, 297, 305–6, 309, 311, 332, 333, 354, 355, 363, 364–5, 396, 417–19, 423
Drilling contractors 402
Drills
   portable 167, 168–9
Drowned beaches 197
Ducktown, Tennessee 110
Dugald River area, Australia 245
Dune deposits 190
Dunite 22, 35, 36, 48, 64
Dunite-peridotite class
   nickel sulphide deposits 38, 40–9, 64
Dykes 35, 36, 57, 60, 385

Early Proterozoic volcanogenic sulphide deposits 110
East Alligator district, Northern Territory 347, 350, 354, 355
East Greenland 394, 396
East Pacific Rise 116
East Shasta, USA 111
Eastern Goldfields province, Western Australia 42, 152, 160, 161, 164, 168
Economic factors
   in mineral exploitation 7–11
Ecuador 73

E_h–pH variations 127, 225, 308–9, 413
El Salvador, Chile 78, 82
El Salvador ore body 78
Electrical generation
   nuclear power 141
Electrical resistivity 188
   see also resistivity
Electrochemical reactions
   sulphide ores 310–11
Electromagnetic methods 54, 61, 62, 134, 169, 245, 354–5, 410, 422
Eli, Nevada 70, 394
Elsburg Reef, Witwatersrand Basin 205
Eluvial placer deposits 177, 178, 181–8, 184, 199
EM see Electromagnetic methods
Emerugga Dolomite, McArthur River 236, 237
Empire ore body, Marquette Range 332
Emu Fault, McArthur River 237
Enargite 80, 430
Endako, British Columbia 94, 98, 99
Endoskarn 369, 371–2
Eneabba beach sand deposit, Western Australia 189–90, 195, 196, 197
England
   kaolin deposits 301, 303–5
English China Clays 301, 306
Enriched haematite ore deposits 325–6
Enstatite 433
Environmental Impact Statements xiii, 104
Environmental legislation
   effects on mineral exploration 404
Environmental stability
   uranium 339–41
Epeiric sea xiii
Epidote 80, 371, 381, 433
Epigenesis 13–14, 154, 216, 359
Epithermal deposits 144
Epoch deposit, Zimbabwe 48
Erosion 91, 93, 275
Erosional surfaces, association with Mississippi-Valley-type deposits 261
Erstberg East, Irian Jaya, Indonesia 394
Escape River deposit, Queensland 290

Estimated additional resources
   uranium 337
Estuarine deposits 190
Euboea, Greece 293
Europe 333
European Bayer Process 276
European Province 281
Evaluation
   bauxites 283, 288–290
   beach sand deposits 193–195
   iron ores 317–8
   kaolin deposits 305–6
   kimberlites 59
   lateritic nickel deposits 297–299
   Panguna porphyry copper deposit 85
   weathered sulphide deposits 309
   see also Exploration
Evander area, Witwatersrand Basin 205, 210
Evaporites 4, 220, 265, 380
Exmibal deposit, Guatemala 293, 299
Exogenous-epigenetic model 163–4
Exogenous genetic model 13–14
Exoskarns 371–2, 377
Exploration 1, 12
   bauxite deposits 288–91
   beach sand deposits 193–7
   budgets 406–7
   factors affecting 403–7
   genetic models 13–15
   hydrothermal gold deposits 164–9
   Irish lead–zinc deposits 256–9
   iron ore deposits 331–2
   kaolin deposits 305–6
   kimberlites 60–3
   lateritic nickel deposits 297–8
   marine placers 200–1
   methods 12
   Mississippi Valley-type lead-zinc deposits 266–7
   nickel sulphide deposits 49–54, 64
   placer deposits 186–8, 211–12
   podiform chromite deposits 36–7
   porphyry copper deposits 85, 87–92
   porphyry molybdenum deposits 104–6
   programmes 398–20
   sandstone-hosted uranium deposits 363–5
   skarn deposits 394–6
   unconformity-type uranium deposits 353–5

Index

volcanogenic sulphide deposits 128–37
weathered sulphide deposits 309–311
Witwatersrand Basin 209–11
Zambian copperbelt 227

Falcondo, Dominican Republic 293
False bottoms 184
False colour photography 50, 164
Farallon plates 82
Faulting 152, 154, 161, 210–11, 238, 249, 251–2, 64, 267, 343, 347, 359
  gain of ground 210–11
  loss of ground 210–11
Feasibility studies 7–8, 398, 399, 406
Feldspar 74, 81, 97, 98, 302, 378
Feldspathoids 182, 433
Felsic volcanics 115, 116–17, 120, 130, 148–9, 156, 361
Ferric hydroxides 293
Ferric iron assemblages 381
Ferric oxides 408
Ferrochrome 18, 19
Ferrochromite 42, 430
Ferrohastingsite 382, 433
Ferromagnesian minerals 306
Ferromanganese 332
Ferrous iron assemblages 381
Ferrous oxides 410
Filling temperatures of fluid inclusions 383–4
Finland 38, 402
Fitula ore body, Zambia 227
Flameless atomic absorption analysis 166
Flin Flon, Canada 110
Float mapping 413
Flora
  copper exploration 227
  nickel exploration 50
Flotation xiii, 315, 70
Fluid inclusions 69, 83, 158, 154–5, 165–6, 254, 265, 349, 353, 361, 383–4, 423
Fluorine 103
Fluorite 99, 102, 103, 185, 261, 262, 265, 433
Fluorite-dominant subtype
  Mississippi Valley-type lead–zinc deposits 261
Fluvial placer deposits 177, 178, 181–8, 190
Fluviatile sandstones 358, 361

Fluxgate magnetometry 394, 395
Folding 39, 43, 238, 249
Fond du Lac deposit, Saskatchewan 354
Footwall xiii, 45, 113, 120, 127, 238, 249, 279, 280
Footwall alteration 113, 134
Footwall in bauxites, effect on mining 279, 280
Fore-arc trough 115
Forecasting
  metal prices 8–9
Foreign companies
  restrictions 403
Forrestania district
  Western Australia 48
Forsterite 372, 386, 388
Fractional crystallization 29, 52
Fractional segregation 52
Fractures
  analysis 88, 394
  porphyry copper mineralization 74, 76, 78, 82, 83, 88, 89
Framboidal texture xv, 235
France 4, 281, 283, 314, 315
Freibergite 235, 248, 433
Froth flotation xv, 70, 315
Fuchsite 433
Fugacity xiv, 29

Gabbin, Western Australia 302, 303, 306
Gabbroic intrusions 51
Gabbroid class
  nickel sulphide deposits 38
Gaertner ore body, Saskatchewan 347
Gain of ground, faults 210–11
Galena 84, 112, 154, 218, 230, 235, 236, 238–9, 248, 260, 261, 262, 264, 265, 308, 377, 382, 387, 433
Gallium 260, 261
Gamma-ray spectrometry 165, 195, 209, 354, 365
Gamsberg, South Africa 231
Garnet 62, 208, 371, 376, 377, 379, 381, 382–3, 385, 386, 388, 396
Garnet zone
  porphyry molybdenum deposits 99
Garnetite xiv, 389
Garnierite 291, 295, 434
Garon Lake 117
Gas Hills district, Wyoming 358
Gases

mineral deposits 361, 416
Gaspé Copper Mines Ltd 9, 389–90
  see also Mines Gaspé
Gaspé Copper Mines effect of cut-off grade on reserves 9
Geiger-Muller counters 355
Geikielite 62, 434
Gemstone quality
  diamonds 54
Genesis 1, 403
  bauxite deposits 285–8
  beach sand deposits 188–9
  hydrothermal gold deposits 156–64
  Irish lead-zinc deposits 252–5
  iron-formation 324–5, 331, 334
  kaolin deposits 300–1
  kimberlites 57–60, 64
  lateritic nickel deposits 294–5
  marine placers 197–9
  Merensky Reef 30–2
  nickel sulphide deposits 45–9, 64
  northern Australian lead-zinc deposits 238–42
  Pine Point deposits 264–6
  placer deposits 181–4
  podiform chromite deposits 35–6, 63
  Porcupine mining district 154
  porphyry copper deposits 82–4, 85, 87
  porphyry molybdenum deposits 99–101
  sandstone-hosted uranium deposits 360–3
  skarn deposits 383–4, 386, 391, 392–4
  unconformity-type uranium deposits 348–53
  volcanogenic massive sulphide deposits 125–7
  Witwatersrand Basin 207–9
  Zambian copperbelt 224–7
Genetic models 12–15, 127–8, 163–4, 255, 266, 351–3, 364, 365, 421
Geobotanical sampling 90, 128, 227
Geochemical data
  exploration programmes 407
Geochemical-cell mineralization 357, 361, 362–3
Geochemistry 12, 50–3, 61, 87, 88, 89–92, 130, 131, 138, 166–7, 211–12, 227, 242, 243, 245, 267, 268, 309–10, 339–41, 360–1, 384, 394, 396, 413, 414

Geographical distribution
  bauxite deposits 281–5
  hydrothermal gold deposits 146–7, 148
  iron-formation 317, 318–19
  kaolin deposits 301
  kimberlites 54–5
  lateritic nickel deposits 392–3
  Mississippi Valley-type lead-zinc deposits 260
  nickel sulphide deposits 40
  palaeo-placers 201
  placer deposits 176–7
  podiform chromite deposits 32–3
  porphyry copper deposits 70, 72–3
  porphyry molybdenum deposits 72–3, 93–4
  sedimentary-exhalative deposits 231
  sediment-hosted copper deposits 217
  skarn deposits 374–5
  stratiform chromite deposits 19, 22
  unconformity-type uranium deposits 337, 338–9
  volcanogeneic sulphide deposits 109–11
Geological mapping 37, 87, 92, 131, 164, 165, 187, 211, 242, 244, 246, 256, 332, 333, 355, 402, 409, 413, 414
Geological research
  government 402
Geologists
  mine 1–2, 227–30
Geophysics 12–13, 37, 53–4, 88, 89, 130, 131, 133, 138, 165–6, 187–8, 194–5, 200, 109, 211, 227, 244–5, 256, 258–9, 267, 268, 305, 310–11, 394, 409, 411, 413, 414, 416–17, 419, 423
Georgia, USA 301, 302–3
Geostatistics
  exploration 288, 333
Geosynclinal sediments 343, 345, 347
Geotechnical logging of cores 419, 423
Geotechnics 227, 229
German-Polish Kupferschiefer 215, 217
Germanium 261
Germany 106, 314, 315

Gersdorffite 347, 431
Gibbsite 276, 283, 284, 287, 298, 350, 434
Glacial drift
  Canada 54, 167
Glacial sediments
  Canadian Shield 164, 165, 167
Glacial till
  sampling 167
Glaciated terrain
  exploration in 12
  mapping 413
Globe-Miami, Arizona 70
Goethite 276, 283, 288, 293, 295, 330, 431
Gold 18, 74, 76, 84, 87, 89, 90, 105, 107, 112, 113, 116, 131, 143, 156, 159–60, 175, 180, 188, 182, 183, 184, 185, 188, 197, 201, 211, 212, 379, 389, 404, 411, 412, 417, 434
Gold deposition
  thermal control 161–2
Gold deposits
  Witwatersrand Basin 203–11
Gold mineralization 106, 146, 152, 203
  timing 155–6
Gold-silver tellurides 154
Gold tellurides 147, 151
Golden Grove Prospect, Western Australia 112, 120
Golden Mile, Western Australia 147
Gondite xvi, 307
Gortdrum deposit, Ireland 247, 256
Gossans xvi, 53, 85, 89, 92, 242, 243, 244, 245, 257, 309–10, 311
Gove deposit, Australia 290
Governments
  role in exploration 402
  role in mining industry 4, 7
Grab sampling
  exploration 200
Grade
  bauxite 283, 286, 289
  control 195, 333
  copper deposits 69–70
  cut-off 9–10, 180, 234, 289, 301, 319, 413
  Hamersley Basin 327
  hydrothermal gold deposits 147
  Irish lead-zinc deposits 248
  iron-formation 314, 318–20
  kimberlites 57
  lateritic nickel deposits 193, 197

Mississippi Valley-type lead-zinc deposits 260
nickel sulphide deposits 40
palaeo-placers 201
Pine Point deposits 262
placer deposits 177–70
podiform chromite deposits 33
porphyry copper deposits 74, 75, 87
porphyry molybdenum deposits 94
porphyry tin deposits 107
prediction 135–7
sandstone-hosted uranium deposits 359
sediment-hosted copper deposits 217
shale-hosted lead-zinc deposits 233–4
skarn deposits 375–7, 381, 384, 385, 387, 388, 390, 391
stratiform chromite 23, 24
supergene manganese deposits 307
–tonnage model 135
unconformity-type uranium deposits 346, 347
volcanogenic sulphide deposits 111–12, 114
Witwatersrand Basin 205
Zambian copperbelt 224–5
Grandite garnet 387, 434
Granite 81, 82, 95, 188, 241, 302, 304–5, 349, 361, 364, 380, 381, 386, 387, 415
Granite Mountains, Wyoming 361
Granitic melts
  potassium-rich 99
Granitoids 145, 184, 349, 380
Granodiorite 77, 84, 96, 380, 382, 383, 385
Grant district, Colorado Plateau 355–6, 357, 358, 359
Graphite 370, 434
Graphitic metasediments 342, 350
Grassroots exploration 405–6
Gravel pumping
  tin mining 186, 187
Gravitational settling 29, 31, 35–6, 47
Gravity surveys 12, 61, 130, 209, 259, 267, 268, 305, 332, 394, 417
Great 'Dyke', Zimbabwe 19, 22
Great Slave Lake, Canada 262, 263, 265

Index

Greece 32, 33, 276, 283
Green biotite 99
Green strength xiv, 298
Green Tuff Belt 121
Greenland 93
Greenschist 158, 161
Greenstone belts 12, 39, 40, 42, 45, 49, 50, 64, 115, 116, 119, 146–56, 151, 161, 167, 169, 302
Greenvale deposit, Queensland 293, 299
Greisen zone xvi, 99, 185, 199
Greywackes 154, 222
Grinding
  iron ore 315
Griqualand West Basin, South Africa 322
Grossularite 378, 382, 434
Ground magnetic surveys, 62, 106, 209
Groundwater 144, 145, 352
  oxidizing 353, 361
  sampling 92, 209
  uranyl complexes 339, 340
Guatemala 293, 297
Guiana Shield Province, South America 283, 281
Guinea 281, 283, 294
Guinea Shield Province, Africa 281
Gummite 347, 431
Guyana 55, 279, 281, 283, 284, 301
Gypsum 158, 220, 223, 262, 431

Haematite 77, 80, 202, 208, 283, 288, 317, 320, 325, 327, 386, 388, 434
Haematite-goethite ores 330, 331
Haematite ore deposits 325–6
Haematite-rich ores 330, 331, 334
Haib, Namibia 72
Halimba Basin, Hungary 283
Halite 158
Halloysite 283, 302, 435
Hamersley Basin, Western Australia 317, 318, 319, 320, 325, 326–31
Hanging wall xvi, 45, 120
  alteration zones 113, 115, 127
  bad conditions 32
  stability 227
Hannover, New Mexico 386
Harberton Bridge area, Ireland 257, 258
Hartebeestfontein gold mine, Witwatersrand Basin 210
Hartley Complex, Great 'Dyke', Zimbabwe 22

Harzburgite 22, 35–6, 63
Haulageways
  mines 227
Hausmannite 307, 435
Hawaii 288
Heavy minerals 187, 189, 190
Hedenbergite 372, 381, 382, 387, 435
Hemimorphite 242, 245, 246, 435
Hemlo deposit, Canada 169
Henderson mine, Colorado 12, 74, 80, 94, 95, 96, 97, 98, 99, 100, 101, 106, 137, 390
Hercynian orogeny 33
High-level bauxites 276–9
High-temperature Bayer Process
Highland Boy deposit, Utah 391, 396
Highland Boy Limestones 379
Highland Valley area 12
Hilton Mine, Australia 232, 233, 235, 238, 242, 243, 244, 245
Hokuroko district, Japan 111, 122
Hollinger Fault, Porcupine mining district 153, 154, 155
Holocene zinc orebody 371
Homogenization temperatures 349
Hornblende 78, 381, 382, 433
Hornfels 82, 102, 103, 372, 389, 392
Host rocks
  bauxite deposits 283, 287
  carbonate-hosted lead-zinc deposits 247–8
  hydrothermal gold deposits 148–50
  kaolin deposits 302
  lateritic nickel deposits 293
  Mississippi Valley-type lead-zinc deposits 261
  palaeo-placers 202–3
  Pine Point deposits 262–4
  podiform chromite deposits 34–5
  porphyry copper deposits 77–8
  sandstone-hosted uranium deposits 358, 364
  sediment-hosted copper deposits 219–20
  skarn deposits 377–9, 386–7
  unconformity-type uranium deposits 341–5
  volcanogenic sulphide deposits 115
Howard's Pass, Yukon 230, 231, 371
Hoyle Township, Canada 166
Huebnerite 93, 185, 211, 380, 382, 433

Humic materials 340, 362
Hunt Mine, Western Australia 160, 161, 163
HYC deposit, McArthur River 232, 233, 234, 237, 238, 239, 241, 242, 371
Hydraulic jets
  kaolin mining 302
Hydrofracturing xvi, 83, 107, 162
Hydrogen sulphide 357, 362
Hydrogeology 229
Hydrogoethite 325, 435
Hydrohaematite 325, 435
Hydrology 2, 287–8
Hydrostatic pressure 83
Hydrothermal convective systems 127–8, 129, 241–2
Hydrothermal fluids 100, 121, 215, 240, 254, 255, 349, 392
Hydrothermal gold deposits 146–56, 169
  exploration 164–9
  genesis 156–64
Hydrothermal kaolin deposits 301, 303–5
Hydrothermal metasomatism 390
Hydrothermal tungsten deposits 380
Hydrothermal vein deposits 4, 143–6, 197
Hypogene mineralization 74, 76–7, 80, 81, 84, 363
Hypogene-type nickel deposits 291, 292
Hypothermal deposits 144

Iapetus Ocean xiv, 251
Iberian Pyrite Belt 111
Igneous metasomatic skarns 378, 392
Illinois-Kentucky district, USA 260, 261
Illite 80, 350, 435
Ilmenite 18, 62, 185, 188, 190, 197, 202, 380, 435
Ilvaite 385, 378, 435
Inco Metal 293
India 332
  bauxite deposits 281, 285
  hydrothermal gold deposits 146, 148
  podiform chromite deposits 19, 33
  supergene manganese deposits 308
Indicator minerals 61, 62

Indigenous
  genetic model 13–14
Indigenous/epigenetic model 164
Indigenous/syngenetic model 14, 164
Indium 260, 261
Indonesia 180, 181, 197, 199–201, 285, 293, 297
Induced polarization (IP) 245, 258, 267, 268, 417, 423
Inductive coupling xiv
Industrial diamonds 54, 133, 166, 243, 245, 258–9, 267, 268, 415, 420
Industrial minerals 370
Infill drilling 306, 417
Intracratonic setting
  chromite deposits 19
  kimberlites 57–9
  Mississippi Valley-type lead-zinc deposits 262
  palaeo-placers 206–9
  sedimentary-exhalative deposits 230
Intrusions
  Climax-type ore bodies 100
  cyclical nature 96
  large layered 38, 39
  pipe-shaped 54
  porphyry copper deposits 77–8, 82, 83
  porphyry molybdenum deposits 95–6
  quartz diorite 80–1
  skarns 379–80, 385
  ultrabasic magma 35
Intrusive dunite association
  nickel sulphide deposits 38, 40, 41, 42, 45, 48–9
Intrusive kimberlite breccia 56–7
Intrusive mafic/ultramafic complexes 50
Investment
  return on 164
Inyan Kara Group, Wyoming 358
Iodides 159
IP (see Induced polarization)
Iran 33
Ireland 4, 14, 216, 230, 412
Iridium 53
Irish Central Plain 14, 247, 249, 256
Irish Geological Survey 256
Iron 79, 81, 162, 248, 266, 284, 291, 293, 295, 297, 306, 324–5, 326, 370, 371, 374, 375, 377, 380, 385, 386
Iron-formation 149, 164, 181, 202, 251, 314, 315, 316–26
Iron hydroxides 317, 340
Iron ores 314–17, 371
Iron oxides 81, 388, 393, 394
Iron Quadrilateral, Brazil 181, 318, 325
Iron-rich metamorphic calc-silicates 378
Iron-rich sediments 316
Iron silicates 320
Iron skarn deposits 376, 386–7, 397
Iron sulphides 393
Ironstone 316, 321
Island-arc calcic magnetite skarn deposits 386
Island-arc setting 77, 107, 137, 294, 295, 379–80
Isopachs 290, 292
Isoquality maps 290, 292
Isotope studies 15, 131–3, 144, 240, 421
Itabira district, Brazil 317
Itabirite 316–17, 325

J-M Reef, Stillwater Complex 24
Jabiluka deposit, Northern Territory 350, 352, 353, 355
Jacobina district, Brazil 201
Jacobsite 307, 435
Jadeite 56, 435
Jagersfontein, South Africa 57
Jamaica 281, 283
Japan 111, 112, 375, 400
  polymetallic type of volcanogenic sulphides in, 121
Jarosite 77, 433
Jasper 317
Jensen Cation Plot 51
Johannsenite 377, 378, 385, 435
  see also Johannsenitic pyroxene
Johannsenitic pyroxene 38
Joint formation 242

Kafue anticline, Zambia 222
Kalahari craton 55
Kalgoorlie area, Western Australia 147, 148, 160, 167
Kambalda mining district, Western Australia 42–5, 52, 53, 164
Kandite group 298
Kaolin 275, 301, 415
Kaolin deposits 298, 300–6
Kaolinite 80, 99, 276, 283, 298, 302, 340, 347, 435
Kaolinitic mylonite 347
Kaolinization 300–1, 303
Kapvaal craton, South Africa 24–5
Karst bauxites 276–8, 279–81, 283, 284–5
Karst topography 352
Karstification 261, 262, 266, 347
Katanga Supergroup, Zambian copperbelt 220, 221, 222
Kaverong Quartz Diorite 84, 85
Kazakhstan 72
Keg River Formation, Pine Point district 263
Kemi, Finland 22
Kempirsai mining area 33
Kenomagnetite 330, 435
Key Anacon deposit, New Brunswick 131
Key Lake deposit, Saskatchewan 343, 344, 346–7, 350, 352, 354
Key Tuffite 117, 118
Keystone Mine 104
Kibaran Massif, Zambia 222
Kidd Creek deposit, Canada 12, 109, 117–18, 119, 133, 136
Kieslager type
  volcanogenic sulphides 108, 109, 113, 115–16
Killingdal Mine, Norway 111
Kimberley, Western Australia 60, 62
Kimberley district, South Africa 55, 56, 209
Kimberlites 4, 18, 54–63, 64, 193
  intrusive-breccia 56
  massive 56
Kimberlitic tuff 57
King Island scheelite deposit, Tasmania, 374, 376, 381–4, 385, 392
Kinta valley tin field, Malaysia 184–6, 188
Kitwe, Zambia 229
Klerksdorp gold field, Witwatersrand Basin 205, 207, 210
Klipfontein Hills, South Africa 307
Klippen
  defined xiv, 35
Klondike area, Yukon 183, 188
Kniest facies xvi, 231
Koffiefontein, South Africa 58
Koidu pipe, Sierra Leone 57
Kolar gold field, India 162

Index

Komatiite 40–1, 48, 50, 51, 64, 152, 156
Komatiitic magmas
   computer modelling 52
Kombolgie Formation, Northern Territory 345, 349, 350, 351, 352, 353
Konkola, Zambia 224, 229
Koodaideri deposit, Hamersley basin 327, 330
Koolpin Formation, Northern Territory 349
Koongarra deposit, Northern Territory 352, 355
Krivoi Rog, USSR 317, 318, 325
Kunene River, Namibia 190
Kupferschiefer, East Germany 220, 225
Kuroko deposit, Japan 111, 113, 115, 121–4, 136
Kuroko type
   volcanogenic sulphides 107, 108, 109, 121–4
Kyanite 182, 436
Kyushu Island, Japan 197

La Caridad deposit, Mexico 89–90
La Gloria deposit, Indonesia 298
Labile xiv, 340–1
Labrador trough, Canada 317
Lacustrine deposits 190
Lady Annie deposit, Australia 243, 245
Lady Loretta deposit, Australia 230, 232, 239, 243, 245, 246
Lady Loretta deposit, Australia, reserves 243
Lake Superior district USA 317, 319–20, 332
Lakehurst Area, New Jersey 188
Lamaque Mine, Canada 167
Laminated Marginal Zone, Bushveld Complex 26
Land's End, UK 372
Landsat imagery 50, 89, 410
Laptev Sea, USSR 197
Laramide age 82, 359, 388
Larymna, Greece 293
Lateral facies change
   Mississippi Valley-type lead-zinc deposits 261
Laterite 283
Lateritic crust bauxites 276–9, 284
Lateritic nickel deposits 37, 275, 291–8, 311

Lateritic weathering 199, 342
Latin America 400
Layered intrusions
   mafic/ultramafic 19, 22
   nickel sulphide deposits 38, 39
Layered Series, Bushveld Complex 26, 27
Leached cappings
   porphyry copper deposits 92, 106
   see also Gossans
Leaching
   carbonate beds 261
   gold 163, 164
   gossans 309–10
   lateritic nickel deposits 291, 295
   ore genesis 240, 349, 353, 354
   skarn silicates 393
   volcanic ash 361
Lead 90, 107, 113, 130, 245, 248, 254, 257, 259, 260, 262, 265, 267, 370, 374, 377, 380, 387, 389, 390, 394, 414
Lead isotope studies 246
Lead times 404
Lead-zinc skarn deposits 371, 374–5
Leichardt River Fault Trough, Australia 232, 238, 240, 242
Lena River, Siberia 181
Lens-shaped bodies
   dunites 41
   kaolin 303
Lesotho 55, 56, 60
Less-developed countries 401
Leucocratic quartz diorite 84, 85
Leucoxene 99, 206, 276, 433
Liberia 332
Ligands xiv, 159
Lignite 188, 340
Limassol Forest Complex, Cyprus 35
Limestone 184, 185, 259, 261, 279, 283, 379, 382, 388, 392
Limonite 77, 293, 302, 340, 436
Lindgren classification
   ore deposits 2–4, 144–5
Lisbon Valley district, Colorado Plateau 355, 358, 359, 364
Literature search
   exploration programmes 409–11
Lithocap rocks 80
Lithogeochemistry 92, 130, 138, 167, 257, 258, 267
Lithophile metallogeny xv, 99
Lithostatic pressure 83
Lithostructural domains

Canadian Shield 341–2
Little Boulder Creek, Idaho 387
Little Long Lac district, Ontario 149
Load-haul-dump units 227
Loellingite 385, 436
London Metal Exchange prices 9
Lone Mountain, Nevada 387
Loss of ground, faults 210–11
Low-level bauxites 276–9
Lowell and Guilbert model
   porphyry copper deposits 79–82, 87, 409
   variations from 80
Lower Dominion Creek, Klondike 184
Lower Redwell deposit 101
Lower Roan, Zambian copperbelt 221–2, 224, 229
Lower Zone, Bushveld Complex 26, 27
Luano ore body, Zambia 227
Lubin-Sieroszowice, Poland 220
Lufilian arc 221
Lunnon ore shoot, Kambalda 45
Luzon Island, Philippines 197

McArthur River deposit, Queensland 230, 232–47
McIntyre Mine, Canada 155
Mackenzie Basin, Canada 265, 266
McLean deposit, Saskatchewan 354, 355
McMillan Pass, Yukon 381
Macon, Georgia 302
Macrobands 322, 327
Madsden gold deposit, Canada 167
Mafic rocks 18, 19, 24–5, 45, 64, 115, 116–17, 119–20, 145, 151, 153, 156, 162, 168, 361, 379
Magcobar, Ireland 248, 249
Magma chamber
   podiform chromite origin 35–6
Magmas
   calc-alkaline 101
   silicate 51
   ultrabasic 35, 47, 48
Magmatic arcs 380
Magmatic deposits 18–64
Magmatic events
   Bushveld Complex 25–6
   kimberlite genesis 58
   Merensky Reef 30–2
   podiform chromite genesis 63
   porphyry copper genesis 82–3

Magmatic events – *cont.*
  porphyry molybdenum genesis 99, 100, 101
  volcanogenic sulphide genesis 127–8, 129
Magmatic fluids 82, 101–102, 144, 145, 392
Magmatic hydrothermal deposits 69
  see also Porphyry copper deposits; Porphyry molybdenum deposits; Volcanogenic sulphide deposits
Magmatic plumes 82
Magnesia 295, 297
Magnesian calc-silicates 386
Magnesian ilmenite 62
Magnesian skarns 372, 374, 380, 388, 396
Magnesite 34, 436
Magnesium 291, 293, 352, 371, 378
Magnetic induced polarization 133
Magnetic separation
  iron ore 315
Magnetic surveys 37, 50, 54, 61, 62, 104, 105, 106, 169, 187, 244–5, 332, 394, 395, 410, 417
Magnetite 42, 77, 79, 84, 97, 102, 104, 106, 112, 125, 197, 202, 291, 317, 320, 325, 327, 332, 377, 386, 388, 436
Magnetite zone
  porphyry molybdenum deposits 99
Magnetometry *see* Magnetic surveys
Main Reef, Witwatersrand Basin 205
Main Zone, Bushveld Complex 26, 27
Malachite 223, 436
Malartic gold field 167
Malayaite 376, 384–5, 396, 436
Malaysia 175, 181, 281, 285
Malmbjerg, Greenland 94
Manganese 90, 249, 266, 306, 332, 371, 385
Manganese carbonate 307
Manganese deposits, supergene 306–8
Mansfeld, East Germany 220
Manitoba nickel belt 41
Mantle-derived gases 361
Mantle-derived sulphur 47
Marcasite 124, 436
Marine placer deposits 178, 197–201, 417
Marine transgressions 220

Market requirements 300, 306, 400
Marquette Range, Lake Superior 332
Marra Mamba deposit, Hamersley Basin 326, 333
Martite 325, 436
Martite-kenomagnetite-goethite ore 330
Mason Valley Mine, USA 372
Massive kimberlite 56
Massive sulphide deposits 47, 48, 107
  see also Volcanogenic sulphide deposits
Masson Formation 344, 354, 355
Matagami mining district 117, 118, 136
Matchless-Otjihase, Namibia 110
Matsumine Mine, Japan 124
Mavovoumi deposit, Cyprus 124
Mediterranean-type karst bauxites 281
Meggen, Germany 231
Merensky Reef, Bushveld Complex 12, 24, 29–32, 39
Mesobands 322, 324, 327
Mesothermal deposits 144
Metabasalts 45
Metagabbros 45
Metal contour maps 250, 255
Metal markets 137
Metal zoning in skarns 385
Metaline, Washington, USA 262
Metallurgical factors affecting mining 11
Metallurgical grade bauxite 283
Metamorphic water 144, 145
Metamorphism 222, 223, 370–1, 377
Metasomatism 371, 373, 378, 382, 386, 390, 391
Metatorbernite 347, 436
Meteoric hydrothermal fluids 82, 99–100, 126
Meteoric water 162, 264, 330, 348, 349, 354, 361
Methane 353, 363
Mexico 93
Mica 223, 302, 350, 359
Michipicoten district, Canada 315
Microbands 322, 327
Microcline 302
Migmatite 342
Millenbach deposit, Canada 117, 127, 134
Millerite 42, 48, 291, 347, 436
Minas Gerais State, Brazil 293, 307,

317, 325
Mine-based exploration 134–5, 406
Mine development
  design 229
Mine geologists 1–2, 227–30
Mine plans 210–11
Mine records 407
Mineral, Virginia 128
Mineral Butte 90
Mineral exploration
  see Exploration
Mineralization
  bauxite 276, 285, 286, 288
  beach sand deposits 190
  hydrothermal gold deposits 151–2
  Kambalda district 44, 45
  kaolin deposits 301, 302, 303
  kimberlites 58
  King Island scheelite deposits 382
  lateritic nickel deposits 294, 295
  lead-zinc deposits 230, 249
  McArthur River 236–7
  modelling 409, 417, 422
  off-hole 134, 422
  palaeo-placers 202
  Pine Point district 264
  placer deposits 182–3, 197
  podiform chromite deposits 34, 63
  porphyry copper deposits 74, 76–7
  porphyry molybdenum deposits 96
  sandstone-hosted uranium deposits 359–60
  skarn deposits 377, 381, 391–2
  supergene manganese deposits 308
  temperature 158
  timing 215–16
  unconformity-type uranium deposits 346–7, 351, 352
  volcanogenic sulphide deposits 113, 115, 120–1, 123–4
  Western States-type 256
  Witwatersrand Basin 206
  Zambian copperbelt 220
Mineralogical analysis
  bauxites 289
Mineralogy
  bauxite deposits 283–4
  Eneabba deposit 190
  Hamersley Basin 328, 330–1
  hydrothermal gold deposits 147
  Irish lead-zinc deposits 248–9
  kaolin deposits 302

Index

lateritic nickel deposits 293
Mississippi Valley-type lead-zinc deposits 260–1
nickel sulphide deposits 41–2
palaeo-placer deposits 201–2
Panguna deposit 84–5
placer deposits 180
podiform chromite deposits 34
Porcupine mining district 154
porphyry molybdenum deposits 94–5
sandstone-hosted uranium deposits 359
sediment-hosted copper deposits 218–19
sedimentary-exhalative deposits 230
shale-hosted lead-zinc deposits 234–5
skarn deposits 377, 385
stratiform chromite deposits 22–4
unconformity-type uranium deposits 346–7
volcanogenic sulphide deposits 112–13
Witwatersrand Basin 205–6
Zambian copperbelt 222–4
Mines Gaspé, Quebec 9, 389–90
Mining 10–11
  bauxite 289, 290
  beach sands 195–7
  chromite deposits 19, 30, 32
  diamonds 192, 193
  iron ore deposits 331
  kaolin deposits 302, 306
  Kinta Valley, Malaysia 186
  Panguna, Papua New Guinea 85, 87
  Zambian copperbelt 227–30
Mining companies 402–3
Mining method, choice of, 10–11
Mining, pilot 11, 289, 290, 298
Minnesotaite 320, 437
Miogeosynclines 358
Mir, USSR 58
Mississippi Valley-type lead-zinc deposits 12, 216, 252, 259–67
Missouri 260, 261
Moa Bay, Cuba 293
Moho 35
Mokta deposit, Ivory Coast 307
Molasse 356
Molybdenite 74, 76, 94, 97, 102, 382, 384, 387, 437
Molybdenum 74, 84, 89, 90, 106, 137, 364, 370, 380, 381, 388, 389, 396, 404
Monazite 185, 195, 197, 437
Montmorillonite 62, 80, 98, 340, 364, 437
  montmorillite transition 241, 242
Montroseite 359, 437
Monument Valley-White Canyon district, Colorado Plateau 355, 364
Morenci, Arizona 81, 85, 88
Morrison Formation, Colorado Plateau 358, 361
Moroccan Anti-Atlas 220
Motovasu ore body, Japan 125
Mount Bischoff deposit, Tasmania 373
Mount Bruce Supergroup, Hamersley Basin 326–7
Mount Charlotte Mine, Western Australia 160
Mount Cleveland, Tasmania 373
Mount Emmons, Colorado 94, 95, 96, 97–8, 99, 101–4, 105, 106, 137
Mount Gordon Arch, Queensland 238
Mount Isa mine, Queensland 2, 12, 230, 232, 233, 235–6, 238, 241, 242, 245
Mount Keith, Western Australia 40
Mount Lindsay deposit, Tasmania 385
Mount Lyell, Australia 112
Mount Saddleback, Western Australia 288
Mount Tolman, Washington 94
Mount Tom Price, Western Australia 322
Mount Whaleback, Hamersley Basin 330
Mufulira, Zambia 222, 225
Multi-element analysis 267
Multi-spectral scanner 410
Murphy tectonic ridge, Australia 232, 238
Muscovite 99, 302, 437
Muskeg Formation, Pine Point district 264
Mwadui, Tanzania 57, 58

Nabarlek deposit, Northern Territory 347, 353, 355
Nacimiento deposit, USA 226
Nagyharsany deposit, Hungary 281
Namibia 56, 188, 197, 337, 365
National Parks, mining in, 7
Nationalization
  mines 4, 403
Native copper 224, 437
Native gold 147, 149, 150, 437
Native silver 147, 235, 437
Navan deposit, Ireland 247, 248, 249, 250, 254, 255, 256, 257, 258, 259
Nchanga area, Zambia 221–2, 224, 227
Near gem diamonds 54
Net texture
  sulphide ores 47, 48
Neutron activation analysis 156, 166
New Brunswick 111
New Caledonia 33, 36, 37, 292, 293, 294, 295–7
New Mexico 357
New South Wales 72
New Zealand 32, 176, 177
Newfoundland 111
Newman deposit, Western Australia 326
Niccolite 347, 437
Nickel 18, 37–8, 112, 125, 261, 293, 295, 346–7, 380, 386, 411
Nickel depletion in magmas 51
Nickel/forsterite ratio in olivines 51
Nickel laterite deposits 275, 291–8
Nickel silicates 37
Nickel sulphide deposits 9, 18, 37–54, 64, 291
Niger 365
Nigeria 175, 183, 184
Niobium 103, 175, 181, 211
Nkana, Zambia 225
Nodular texture 34
Nome, Alaska 197
Nontronite 293, 437
Noranda mining district, Canada 109, 112, 113, 116–17, 118, 128, 136
Nordenskioldine 384, 437
Norita deposit, Matagami, Canada 134, 135, 138
Norite 26, 27
Norseman mine, Western Australia 167
Norseman-Wiluna Belt, Western Australia 41, 42, 45, 119, 121
North American plates 82
North Island, New Zealand 197
North Pennine ore field, UK 252

Northern Boundary Fault, King Island 382, 384
Northern Brazilian Shield Province 281
Northern Territory, Australia
  Jabiluka deposit 350, 352, 353, 355
  Kombolgie Formation 345, 349, 350, 351, 352, 353
  Koolpin Formation 345
  Koongarra deposit 352, 355
  Olympic Dam deposit 7, 338, 365
  Pine Creek geosyncline 343–5, 347, 349, 350, 352, 355
  Rum Jungle deposit 347, 349, 350, 355
Norwegian Caledonian Belt 110
Nsuta deposit, Ghana 307
Nuclear power 143, 337, 365
Nugget effect 412

^{18}O 132–3
Oceanic crust
  chromite genesis 63
  origin of ophiolites 34–5
Ochre horizon
  sulphide deposits 124
'Off-hole' mineralization 422
Oil companies with metal interests 404, 405
Ok Tedi copper deposit, Papua New Guinea 10, 73, 74, 137, 275, 377, 386, 403
Old Alluvium, Kinta Valley 184
Old Leadbelt, USA 266
Oligoclase 302
Oligomict conglomerates 202–3
Oliphants River, South Africa 190
Olivine 23, 34, 35, 41, 42, 45, 49, 51, 52, 124, 291, 293, 438
Olympic Dam uranium deposit, Australia 7, 338, 365
Oman 111
Oolitic iron ores 314, 315, 321
Opal 402, 438
Open-hole rotary drilling 195
Open pit mining 10, 11, 74
Ophiolites 32–3, 34–5, 37, 63, 293
Orange Free State 209
Orange River, Namibia 186, 188, 190, 193
Oranjemund, Namibia 190–3
Orapa pipe, Botswana 54, 58
Ore fluids 126, 144–5, 156–61, 161–2, 251, 252, 254–5, 265–6, 349, 353, 361, 388, 390

Ore reserves
  see Reserves
Oregon, USA 293
Organic adsorbents
  uranyl ions 340
Organic mud 184
Orogenic belts 181, 317
Orthoclase 79, 81, 103, 389, 438
Orthopyroxene 23, 438
Orthoquartzite 203
Otago, New Zealand 176, 179, 181
Overburden 10, 166, 193, 257
Overseas Development Agency 400
Oxide iron-formation 327
Oxidization 10, 308, 317, 320, 325, 364
Oxidized skarns 381
Oxygen fugacity
  chromitite layer formation 29
Oxygen isotope ratios 162

Pacific Basin 73
Pacific plates 82
Pakistan 33
Palaeo-highs 220, 226, 248
Palaeo-placer deposits 175, 176, 201–11, 212
Palaeozoic age 72, 284, 374
Palladium 29, 53
Palynology 188
Pamour Mine, Canada 153
Panguna Andesite 84, 85
Panguna deposit, Papua New Guinea 74, 75, 81, 84–7
Papua New Guinea 33, 73, 90
Paraburdoo deposit, Hamersley Basin 326, 327, 332
Parent rock, rôle in bauxitization 287
Parnassos-Giona deposits, Greece 288
Passive continental margin 180
Pathfinder elements 166, 246
Paystreaks 183
Peat 184, 340
Pebble dykes 78, 83, 84, 85
Pedochemical methods 416
Pegmatite 348, 380
Pegmatitic texture, Merensky Reef 31
Penalty clause (smelter) 11, 248
Peneconcordant uranium deposits 357, 358, 361–2
Peneplain bauxites 276–9
Peneplains 288, 295
Pennine ore field, UK 261

Pentlandite 18, 24, 42, 291, 438
Percussion drills 257
Peridotite 22, 35, 45, 52, 56, 59, 63, 295
Perm mining area 33
Permeability 78, 255, 261, 262, 264, 347, 350, 361, 378, 384, 392
Perseverance ultramafic intrusion, Agnew deposit 45
Persian Gulf 322
Peru 73, 82, 93
Pervasive silica zone
  porphyry molybdenum deposits 99
Petalite 182, 438
PGE *see* Platinum group elements
pH *see* E_h–pH
Phase spectral induced polarization 259
Phenocrystal origin of diamonds 58
Philippines 9, 33, 73, 82
Phlogopite 438, 56
Phosphates 340, 347
Phosphorus 297, 325, 332
Photochemical oxidation 325
Photogeology 63
Photography 50, 63, 164
Photosynthesis xvii, 317, 325
Phyllic zone
  porphyry copper deposits 79–80, 81, 84
  porphyry molybdenum deposits 98, 101, 102
Physical weathering 275
Pickle-Crow district, Ontario 149
Picrites 45
Piedmont placers 199
Pilbara Block, Western Australia 147, 151
Pillow lavas 35, 124
Pilot mining 11
  bauxites 289, 290
Pilot plants
  ore processing 11
Pine Creek, California 381
Pine Creek geosyncline, Northern Territory 343–4, 345, 347, 349, 350, 352, 355
Pine Point district, Canada 260, 261, 262–6
Pipe-dredge sampling 200–1
Pipes
  footwall alteration 113
  kimberlites 54, 55, 57, 58
Pisolitic bauxite 279

Index

Pitchblende 346, 387, 438
Pitting 188, 297, 417
Placer deposits 54, 175–88, 211–12, 380
Plagioclase 23, 26, 80, 97, 98, 302, 303, 371, 393, 438
Planning
  exploration programmes 407–9
Plant debris 357
Plant material
  fossilized 358
Plants
  distribution 90
Plate boundaries
  porphyry copper deposits 82
Plate tectonics 4, 5, 6, 35, 42, 82, 87, 115, 221, 137, 423
Platinum 29, 175, 211, 438
Platinum group elements 18, 23–4, 29–32, 34
Platreef, Bushveld Complex 24, 29, 30
Plutonic environment
  tin deposits 107
Plutonic intrusions 394
Plutonic stockworks 102
Plutonic-volcanic
  hydrothermal vein deposits 146
Plutons 78, 184, 392, 397
Podiform chromite deposits 4, 19, 32–7, 63–4
Point bars 183, 358
Political factors
  in exploration 403–4
  in mining development 4, 7
Polygons method xv, 87
Polymetallic type
  volcanogenic sulphides 108, 109, 113, 115, 121–4, 137
Pomalaa, Indonesia 293
Porcupine Group, Porcupine mining district 153, 154
Porcupine mining district, Canada 146, 148, 149, 153–6, 163
Porcupine syncline 154, 155
Porphyritic intrusions
  porphyry molybdenum deposits 95–6
Porphyry 94, 101, 102, 103
Porphyry copper deposits 9, 10–11, 69–84, 95, 99, 100, 106, 137, 275, 309, 311, 375, 377, 386, 394
  exploration 87–92, 409, 410
  Panguna, Papua New Guinea 84–7

Porphyry gold deposits 106
Porphyry molybdenum deposits 72–3, 74, 80, 92–106, 137, 387, 391
Porphyry-related copper skarns 388–90
Porphyry tin deposits 106–7
Portugal 111
Potash feldspar 74, 97, 98, 102, 438
Potassic zone
  porphyry copper deposits 79, 80, 81, 84
  porphyry molybdenum deposits 97–8, 102
Potassium 297
Potassium feldspathization 104
Potassium-rich granitic melts 101
Potassium silicate 388
Potholes
  Merensky Reef 29–30, 32, 33
Powder River district, Wyoming 358
Powellite 382, 383, 396, 438
Precambrian Shield 54, 55, 317
Predictive grade/tonnage model 135
Prehnite 438
Preliminary drilling 417
Premier, South Africa 58
Presqu'ile dolomite 262, 264
Presqu'ile Reef, Pine Point district 266
Pressure changes
  chromitite layer formation 29
Prices
  metals 8–9, 164, 169, 234, 243, 247, 316, 355
Primitive type
  volcanogenic sulphides 108, 109, 113, 115, 116–21, 137
Primos intrusive
  Urad ore body 102
Prince Lyell, Australia 112
Processing
  ores 11, 235
Propylitic zone
  porphyry copper deposits 80, 81, 82, 84
  porphyry molybdenum deposits 98
  porphyry tin deposits 107
Prospectors 12, 64, 164, 167, 229
Proterozoic iron-formation 323–4
Proterozoic volcanogenic sulphide deposits 110, 138
Protolith xvii, 373, 378

Protore 307
Proximal accumulations
  volcanogenic sulphide deposits 131, 121, 134, 135
Pseudolodes 295
Pseudomalachite 439
Puerto Rico 90
Pulau Tujuh area, Indonesia 200
Pulse-type induced polarization 258–9
Pyragyrite 235, 248, 439
Pyramid deposit, Pine Point district 267, 417
Pyriboles 182, 439
Pyrite 18, 42, 74, 76–7, 79–80, 81, 89, 94, 98, 99, 101, 102, 112, 117, 118, 120, 121, 124, 147, 154, 155, 159, 160, 161, 202, 206, 218, 219, 225, 226, 230, 235, 236, 238–9, 248, 252–3, 262, 264, 302, 303, 308, 341, 382, 386, 388, 389, 439
Pyritic quartzite 203
Pyritization xvii, 151, 155, 262
Pyrochlore 179, 436
Pyrolusite 307, 439
Pyrometasomatism 373, 380
Pyrope 439
Pyroxene 31–2, 78, 293, 371, 377, 379, 381, 382, 385, 387, 388
Pyroxenite 26, 27
Pyrrhotite 18, 24, 42, 48, 77, 81, 102, 112, 117, 128, 147, 154, 159, 160, 166, 230, 235, 245, 262, 308–9, 382, 385, 439

Quadrilatero Ferrifero, Brazil 181, 318, 325
Quartz 74, 79, 80, 97, 98, 99, 102, 106, 202, 276, 283, 293, 302, 304, 349, 371, 384, 389, 391, 394, 439
Quartz-banded ore 316
Quartz diorite 80–1, 84, 85, 96
Quartz Hill, Alaska 94
Quartz–molybdenite stockwork 94
Quartz-monzonite molybdenum deposits 92, 94–6
Quartz monzonite 77, 380
Quartz-pebble conglomerates 201
Quartzite 205, 222, 307, 317, 342, 382, 392
Queensland 282
  Ban Ban deposit 386
  Barney Creek Formation 238

Queensland – *cont.*
  Batten Trough 232, 238, 242
  Cooley dolomite 236
  Emu Fault 237
  Escape River deposit 290
  Greenvale deposit 293, 299
  Hilton Mine 232, 233, 235, 238, 242, 243, 244, 245
  Leichhardt River Fault Trough 232, 238, 240, 242
  McArthur River deposit 230, 232–47
  Mount Isa Mine 2, 12, 230, 232, 233, 235–6, 238, 241, 242, 245
  Urquhart Shale 233, 234, 235, 236, 241, 242, 245
Questa 99

Rabbit Lake deposit, Saskatchewan 343, 346, 347, 349, 350, 352, 354
Radar 186
Radar imaging satellites 408
Radiometric surveys 326, 355, 360, 363, 365, 410–11, 422
Radon surveys 354, 355, 416
Raise xv, 87
Rammelsberg, Germany 230, 231
Rammelsbergite 347, 437
Ramu River
  chromite deposits 33
Ranger deposit, Northern Territory 347, 348, 352, 355
Reaction skarn 378, 392
Realgar 150, 439
Reconnaissance
  exploration 16, 128, 130, 186, 194, 264, 408, 411–13
Recoveries
  minerals 10, 11
Red-bed copper deposits 265
Red Lady Basin 101, 103, 105
Red Lake district, Canada 148
Red Mountain, Arizona 80, 101
Red Sea 116
Redox reactions
  ore genesis 346, 357, 362–3
Reduced skarns 381
Reduction
  uranyl ions 340–1
Redwell Basin, Colorado 101, 102, 103
Reef carbonates 259
Reflection seismic
  exploration 188
Refractory cupriferous mica 223

Refractory grade bauxite 281, 283
Regional exploration 60, 130, 404–5
Regoliths 354
Remote sensing
  exploration 50, 89, 106, 128, 164–5, 187, 188, 244, 256, 297, 410, 423
Renison Bell, Tasmania 373, 380
Rennell Island deposit, Australia 288, 290–1
Replacement deposits 390
Reserves 1, 7–8
  bauxite 283, 287
  calculation 135–7, 188, 195, 289, 290
  chromite 19
  copper 70, 216, 217
  gold 147
  Hamersley Basin 327
  Irish lead-zinc deposits 248
  iron-formation 318–20
  kaolin 301, 306
  lateritic nickel 292–3
  Mississippi Valley-type lead-zinc deposits 259, 260
  Mount Emmons 102, 103
  natural diamonds 54
  palaeo-placer deposits 201
  Panguna deposit 87
  Pine Point district 262
  porphyry copper 74, 75
  porphyry molybdenum deposits 94
  sediment-hosted copper deposits 217
  shale-hosted lead-zinc deposits 233–4
  skarn deposits 375–7, 386, 387, 389
  supergene manganese deposits 307
  uranium 143, 337, 338
  volcanogenic sulphide deposits 111–12, 114, 124–5
Residual deposits 275, 301
Residual kaolin deposits 302, 303
Residual manganese deposits 306–8
Residual soil surveys 414
Resistate minerals 181, 187, 202
Resistivity
  exploration 61, 166, 187, 188, 258–9, 305, 365, 417
Retrograde alteration
  skarn deposits 379, 380, 381, 388–9, 392–3

Retrograde boiling 82
Return on investment
  gold mining 164
Reverse-circulation drilling 169, 257
Rhenium 74
Rhodochrosite 307, 440
Rhodonite 307, 440
Rhyolite 102, 117, 128, 326
Rift-phase
  greenstone belts 152
Rift related
  porphyry molybdenum deposits 93, 95, 99–100, 104
Rift-ridge environment 116
Rifting
  ore bodies 220–1, 262
Ring structures 186
Rio Algom mine, Colorado Plateau 358
Riofinex 7
Ripping xv, 307
River gravels
  placer deposits 183
Roan Antelope, Zimbabwe 12, 227
Robb Lake district, Canada 266
Roberts Victor, South Africa 57
Roc-Tec 422
Rock geochemistry 245–6
  see also Lithogeochemistry
Rock quality designation (RQD) 419, 422
Rock sampling 60, 90
Rokana/Mindola, Zambia 224
Roll-type uranium deposits 357, 360, 361, 362–3, 409, 416
Rondonia district, Brazil 181, 186
Rubidium 90, 103
Rum Jungle deposit, Northern Territory 347, 349, 350, 355
Rutile 90, 99, 190, 197, 283, 440

St Austell, Cornwall, UK 304, 305
St Helena, Orange Free State 209
St Ives bay, Cornwall, UK 197
Saleeite 347, 440
Salinity
  fluid inclusions 157, 241, 264–5, 349, 353, 383–4, 390
Sampling 200, 201, 288, 290, 297, 332–3, 394
  deep overburden 257
San Manuel mine, Arizona 74, 75, 76
San Manuel-Kalamazoo, Arizona 81, 91
Sanbagawa Belt, Japan 125

Index

Sandstone-hosted lead deposits 265
Sandstone-hosted uranium deposits 337, 338–9, 355–65
Sangdong, Korea 381
Santa Eulalia, Mexico 385
Santa Rita, New Mexico 70
Saprolite xv, 295
Saprolite zone
  lateritic nickel deposits 293
Satellite imagery
  exploration 61, 89, 128, 297, 394, 409, 410
Saudi Arabia 116
Scandinavia 39, 115
Scheelite 94, 154, 187, 211, 376, 377, 380, 381–4, 385, 387, 393, 394, 440
Schumacher mine, Canada 153
Scintillometry 61, 195
Scout boring
  exploration 194, 195
Scrapers
  mining 196, 197
Scree
  see Talus
Scully mine, Labrador 333
Sea
  effects on beach sand deposits 189
Seawater 126–7, 133, 253, 255
Secondary enrichment see
  Supergene enrichment
Sediment-hosted copper deposits 216–21
Sediment-hosted lead-zinc deposits 216, 230–267
Sediment sampling
  exploration 51, 62, 130, 166, 200, 411–13
Sedimentary bauxites 281, 285
Sedimentary-exhalative lead-zinc deposits 230–1, 248
Sedimentary kaolin deposits 301, 302–3
Sedimentation
  Hamersley Basin 326
  Irish lead-zinc deposits 248
  sedimentary-exhalative deposits 232–3
  Western States-type mineralization 358
  Witwatersrand Basin 207–8
Sediments 275
  cassiterite distribution 197–9
  chemical 156
  geosynclinal 254

heavy metal collecting 188
metal sources 254
Molasse type 358
reworking 182
uranium contents 340
Seepage areas 227
Segregation
  sulphides 45, 47, 48
Seismic methods 37, 61, 166, 187–8, 199, 200, 332, 417
Selective alteration 78–9
Selenium 90
Self-potential methods, 310–11
  see also Spontaneous potential
Semiconformable alteration zones
  volcanogenic sulphide deposits 113
Separation
  ore processing 11
Sericite 79, 98, 99, 112, 160, 440
Sericitization xvii, 79–80, 113, 131, 151
Serpentine 293, 372, 386, 388, 440
Serpentinite 48
Serpentinization 48
Serra dos Carajas district, Brazil 315, 325
Seward Peninsula, Alaska
Shaba, Zaire 222
Shakanai Mine, Japan 124
Shale-hosted lead-zinc deposits 216, 232–47
Shale-hosted sulphide deposits 107
Shangani deposit, Zimbabwe 41
Sheeted dykes 35
Shell 404
Shirley Basin district, Wyoming 358, 361, 362
Side-scanning radar 186
Siderite 160–1, 302, 320, 327, 440
Sierra Leone 55, 57
Sierrita-Esperanza, Arizona 82, 89
Silcrete 302
Silica 80, 288, 295, 297, 298, 324–5, 326, 332
Silica–dolomite rock 235–6, 245
Silica removal
  from bauxites 287
Silica zones
  porphyry molybdenum deposits 98
Silicate facies
  iron-formation 320
Silicate magmas
  composition 51

Silicates 26, 47, 161, 230, 291, 295, 384
Siliceous ore
  Kuroko deposit 123
Silicification 103, 131, 155, 236, 262
Siliclastics 248
Sillimanite 182, 440
Sills
  dolerite 60
  kimberlite 57
Silver 74, 76, 87, 90, 107, 112, 113, 116, 121, 130, 131, 143, 147, 154, 230, 235, 245, 248, 261, 262, 377, 389
Silvermines deposit, Ireland 230, 247, 248, 249, 251, 253, 254, 255
Single element geochemical survey 245
Single refractive seismic profiling 188
SIROTEM xv, 134, 166
Size characterization a – predictive model 135
Skarn deposits 372–80
  exploration 394–6
  genesis 392–4
Skarns 371–2
Sklodowskite 347, 440
Skouriotissa deposit, Cyprus 124
Smectite 293, 440
Smelters 11
  see also Penalty clause
Soil gas sampling 92, 396, 397
Soil geochemistry 51, 53, 89, 106, 130, 166–7, 245, 246, 267, 384, 396, 414–16
Soil sampling 53, 62, 87, 90, 92, 130, 166–7, 227, 242, 256–7
Soils
  dispersion patterns 51
  trace metals 128
  uranium adsorption 340
Sokoman Formation, Labrador 322
Solomon Islands 84
Sonia Mark 1 acoustic reflection profiler 200
Soroako deposit, Indonesia 293, 299
Source rocks
  placer deposits 181–2
South Africa
  Basal Reef, Witwatersrand Basin 205, 206, 209
  Bushveld Complex 19, 22, 23–4, 24–32, 39
  Central Rand Group,

461

South Africa – *cont.*
  Witwatersrand Basin 205, 206, 209
  chromite reserves 36
  Critical Zones, Bushveld Complex 22, 23, 26–7
  exploration 404, 417
  gold deposits 146, 147, 157, 175
  Kimberley District 55, 56, 209
  Lower Zone, Bushveld Complex 26, 27
  Main Reef, Witwatersrand Basin 205
  Main Zone, Bushveld Complex 26, 27
  massive sulphide deposits 109
  Merensky Reef 12, 24, 29–32, 39
  nickel production 38
  palaeo-placer deposits 175
  Swaziland Supergroup 205, 318
  Transvaal Basin 317, 322
  Upper Zone, Bushveld Complex 317, 322
  uranium deposits 175, 337, 365
  Witwatersrand Basin 12, 201, 202, 203–10, 212
South Alligator Trough, Saskatchewan 344, 345
South America 73, 74, 75, 79, 82, 137, 279, 293, 332
South Carolina 301, 302–3
South China Sea 199
South Dakota 355, 356, 364
South Primor'ye, USSR 197
South Tisdale anticline, Porcupine mining district 154, 155
Southern Cross Province, Western Australia 149
Spacelab 1 410
Spain 111
Specularite 325, 389, 440
Spessartine 99, 378, 440
Spessartite 307, *see* Spessartine
Sphalerite 84, 112, 117, 118, 121, 154, 218, 230, 235, 236, 238–9, 248, 253, 260, 262, 264, 265, 266, 267, 308, 377, 382, 385, 386, 387, 441
Sphene, 182, 441
Spinifex texture
  komatiites 40–1, 50
  peridotites 45, 52
Spodumene 182, 441
Spontaneous potential (SP) 310–11
Spur sampling 90

Stability
  uranium 339–41
Stable isotopes 15, 239
Stack uranium deposits 357
Stacked lenses in volcanic sulphide deposits 135
Stainless steel 18, 38
Stannite 118, 441
Starratt-Olsen deposit, Canada 167
Statistical methods
  soil geochemistry data 414–15
Staurolite 189, 441
Steel
  prices 316
Steel industry 314–15
Steelpoort seam, Bushveld Complex 27, 28
Stephanite 235, 438
Stepwise discriminant analysis 246
Steyn Reef, Witwatersrand Basin 205
Stibnite 150, 441
Stillwater Complex, Montana 19, 22, 23–4
Stilpnomelane 320, 441
Stockwork xv
  hydrothermal vein deposits 145
  plutonic 101
  porphyry molybdenum deposits 93, 100, 102
  quartz–molybdenite 94
  tin deposits 106, 178, 185
  volcanogenic sulphide deposits 113, 124, 132
Stratabound ore deposits 249, 344, 380
Stratabound type
  sediment-hosted metal deposits 215–16, 249
Strategic minerals xvi, 18
Stratiform chromite deposits 19–29, 32
Stratiform ore deposits 249
Stratiform type
  sediment-hosted metal deposits 215–16
Stratigraphy
  exploration 266
  iron enrichment control 331
  iron skarn deposits 386–7
  Pine Point deposits 262–3
  Porcupine mining district 153–4
  sandstone-hosted uranium deposits 358
  unconformity-type uranium deposits 341–5

  volcanogenic sulphide deposits 127
Stratovolcanoes 85, 107
Stream sediment sampling 62, 87, 89–90, 130, 166, 245, 256–7, 267, 411–13
Stringer ores 124
Stringer zones 117–18
Structural trends
  kapvaal craton 25, 26
Structural control
  Irish base metal deposits 249–52
  McArthur River–HYC deposits 238
  volcanogenic sulphide deposits 127
Structure
  Agnew deposit 45, 46
  Bushveld Complex 26, 27
  Climax ore body 97
  Eneabba beach sand deposit 189–90
  Irish lead–zinc deposits 249–52
  iron enrichment control 331
  Kambalda mining district 43–5
  Kinta Valley tin field 184, 185
  McArthur River deposit 233, 237–8
  Mt Emmons deposit 98, 102
  Namibia beach sand deposit 191, 192
  New Caledonia deposit
  Panguna deposit 84, 86
  Pine Point district 264
  Porcupine mining district 154, 155
  sandstone-hosted uranium deposits 359
  sediment-hosted lead–zinc deposits 237–8
  unconformity-type uranium deposits 347
  Witwatersrand Basin 204–5, 206
  Zambian copperbelt 222, 223
Stylolite xvi, 320
Subduction 82, 221
Subduction related
  porphyry molybdenum deposits 93, 95, 99, 104
  skarn deposits 379
Subtropical climate
  bauxite formation 287
Sudbury Basin, Canada 110
Sudbury Complex, Ontario 38, 39
Sulawesi, Indonesia 298
Sullivan, Canada 230, 231

Index

Sulphate
  seawater 253
Sulphate reduction 127
Sulphide facies
  iron-formation 320
Sulphide mineralization 81–2
Sulphides 52, 159, 222, 225, 235, 248–9, 363, 384, 385, 388, 394, 416
  disseminated 42, 45, 48–9, 74, 76
  immiscible 51
  supergene enrichment 275
Sulphur 47, 79, 127, 238–40, 252–3, 262, 264, 340–1
Sulphur dioxide 253
Sulphur isotope studies 127, 225, 236, 237, 238–40, 253, 264, 266
Sumatra 199
Sunda Shelf 199
Supergene enrichment 80, 81, 84, 85, 275, 308–11, 320, 330, 331, 334, 354
Supergene iron ore 327
Supergene manganese deposits 275, 306–8
Supergene origin
  kaolin deposits 303–4
Supergene sulphide deposits 308–11
Supergene zone
  porphyry copper deposits 77, 309, 311
  see also Chalcocite blanket
Superior iron-formation 317
Superior Province, Canadian Shield 40, 147, 149, 152, 168
Surinam 281
Swaziland 55
Swaziland Supergroup, Witwatersrand Basin 209, 318
Sweden 371
Sybella granite, Mount Isa 241
Sylvite 158
Syndiagenesis
  Irish zinc–lead deposits 255
Syngenetic,
  copper ores 235–6
  hydrothermal gold deposits 157
  pyrite 227
  sediment-hosted metal deposits 216
  skarn deposits 373
  Zambian copperbelt 224–5
Synsedimentary ore minerals 225
Synthetic aperture radar 297
Synvolcanic faults 121

Synvolcanic fracture zones 117

Taconite 317, 319–20
Tailings 11, 248
Talc 230, 293, 370, 386, 439
Talus xvi, 371
Tambao deposit, Upper Volta 307
Tantalite 179, 441
Tantalum 175
Tanzania 54, 55
Target appraisal
  exploration 130–1, 400, 407, 408, 413–22
Tarkwa area, Ghana 201, 202–3
Tasmania 373, 375
Taxation 7, 12, 247, 403–4
Techniques employed in exploration 401
Tectonic setting 423
  bauxite deposits 284–5
  hydrothermal gold deposits 151–3
  iron-formation 322–4
  lateritic nickel deposits 293–4
  Mississippi Valley-type lead–zinc deposits 262
  palaeo-placer deposits 203
  placer deposits 180–1
  Porcupine mining district 154
  porphyry copper deposits 82
  porphyry molybdenum deposits 99
  sandstone-hosted uranium deposits 358
  sediment-hosted copper deposits 220–1
  sedimentary-exhalative deposits 230, 232
  skarn deposits 379–80, 394, 397
  volcanogenic sulphide deposits 115–16
Tellurium 90
Temperature
  homogenization 349
  mineral formation 158, 384
Temperature gradient
  mines 10, 161
Tennantite 80, 112, 218, 441
Tennessee mining district, USA 260, 262
Tethys Ocean 262
Tetrahedrite 112, 236, 248, 442
Teutonic Bore, Western Australia 118–21, 130, 133
Texas 355, 356, 357, 362
Textures

podiform chromite 34
Thailand 181, 197, 394
Thematic mappers
  exploration 89, 408
Thermal gradient in gold mines 161
Thermal metamorphism
  intrusions 82
Tholeiite 50, 51, 107, 115, 121, 148, 162, 226
Threshold values
  soil geochemistry 414–15
Thucholite 347, 442
Timing
  gold mineralization 155–6
Timmins mining district 117–18, 156, 164, 166, 167
Tin 74, 118, 130, 143, 175, 180, 181, 184, 197, 211, 370, 374, 387, 411
Tin-bearing garnets 396
Tin-bearing silicates 384
Tin deposits 184–6, 197, 373
Tin mining 178, 180
Tin skarn deposits 372, 376, 380, 384–5, 397
Tisdale Group, Porcupine mining district 153–4, 155
Titania 347
Titanium 18, 175, 180, 188, 211, 385
Titanium-bearing beach sands 176
Titanomagnetite 197, 441
Titanomagnetite sand 197
Tom Price deposit, Hamersley basin 326, 327, 329, 332
Tonnage, *see* Reserves
Topaz 89, 442
Topaz zone
  porphyry molybdenum deposits 98–99
Topira deposit, Guyana 306
Topography
  bauxite formation 287–8
  effects on deposition 294
  in exploration 394
  placer deposit formation 181
Total Exploration concept 409
Tourmaline 225, 302, 304, 442
Tournaisian age 249
Trace elements
  plants 90
  release from minerals 370–1
  soils 128
Transient electromagnetic methods
  exploration 166
Transitional environment
  placer deposits 178, 180–1

463

Transport
  ore genesis 12, 159–60, 266
Transported bauxites 281
Transvaal Basin, South Africa 317, 322
Transvaal Griquatown region, South Africa 317
Tremolite 230, 442
Trenching
  exploration 166, 167–8, 188, 227, 355, 417
Trento Valley, Italy 260
Tri-State district, USA 260, 261
Trincheras deposit, Guatemala 298
Trombetas deposit, Brazil 281, 283
Troodos massif, Cyprus 124
Tropical climate
  bauxite formation 287
True colour photography 50
Tuffs 14, 57, 117, 118, 122, 240, 363
Tungsten 74, 90, 103, 143, 370, 373, 374, 375, 376, 377, 380, 383, 387, 396, 411
Tungsten-bearing banded granitoids 380
Tungsten skarn deposits 371, 380–4, 393
Tungsten world production 380
Tungsteniferous brines 380
Tunisia 260
Turbidites 149
Turee Creek Group, Hamersley Basin 326
Turkey 33, 111
Twin Buttes 377
Tynagh deposit, Ireland 14, 247, 248, 249, 251, 252–3, 255, 256, 257, 259
Tyuyamunite 340, 359, 442

U subtype
  hydrothermal vein deposits 145
UG2 reef 24
Ultrabasic magmas 35, 47, 48
Ultrabasic rocks 35, 58, 107
Ultramafic rocks 18, 19, 40–1, 45, 63, 64, 145, 148, 153, 156, 168, 291, 295
Ultramafic zone
  ophiolite complexes 37
Ultraviolet light
  exploration 387, 394, 396
Uncompahgre Uplift, Colorado Plateau 359, 359
Unconformity-type uranium
  deposits 337, 338–9, 341–53, 357
Underground mapping 229
Underground mining 10, 74, 76, 228, 229, 281
United Kingdom 4, 301, 314, 315, 404, 409, 422
United Nations Development Programme 60, 402
Upland bauxites 276–9
Upper Mississippi Valley district, USA 261
Upper Proterozoic sediment-hosted copper deposits 217, 218
Upper Redwell deposit 102, 103
Upper Roan Group, Zambian copperbelt 222, 229
Upper Zone, Bushveld Complex 26, 27
Urad ore body 96, 100, 101
Urals 33
Uraniferous carbon 206, 359
Uraniferous limonite 240
Uraniferous pegmatite 348
Uraninite 206, 339, 340, 345, 347, 353, 359, 361, 442
Uranium 7, 143, 175, 201, 202, 203–11, 212, 337–8, 361, 363, 387, 411, 422
Uranium–lead ages
  in sandstones 359
Uranium mining
  government policy 354
Uranophane 340, 443
Uranous ions 339–40
Uranyl complexes 339, 340, 352
Uranyl ions 339, 353, 361, 363
Uravan Belt, Colorado Plateau 355, 357, 359, 364
Urquhart Shale, Mount Isa 233, 234, 235, 236, 241, 242, 245
US producer prices
  metals 8–9
USA
  Arizona 94, 259, 364, 408
  bauxite deposits 281
  Climax, Colorado 74, 75, 94, 95, 98, 99, 100, 101, 391
  Climax-type molybdenum deposits 92, 94–5, 95–6, 97, 104, 137
  Colorado 94, 95, 359, 364, 416
  Colorado Plateau 355–6, 358, 359, 361, 362
  exploration 404, 422
  Georgia 301, 302–3
  Grant district, Colorado Plateau 355–6, 357, 358, 359
  Henderson mine, Colorado 12, 74, 80, 94, 95, 96, 97, 98, 99, 100, 101, 108, 137, 391
  Illinois–Kentucky district 261
  Illinois–Wisconsin district 260
  kaolin deposits 301
  kimberlite deposits 55
  lateritic nickel deposits 293
  Mississippi Valley-type lead–zinc deposits 12, 216, 252, 259–67
  Missouri 260, 261
  Mt Emmons, Colorado 94, 95, 96, 97–8, 99, 101–4, 105, 106, 137
  New Mexico 357
  porphyry copper deposits 72, 73, 74, 75, 79
  porphyry molybdenum deposits 92–106
  Redwell Basin, Colorado 101, 102, 103
  San Manuel mine, Arizona 260, 262
  San Manuel-Kalamazoo, Arizona 81, 91
  skarn deposits 375
  Tennessee mining district 260, 262
  Texas 355, 356, 357, 362
  Tri-State district 260, 261
  tungsten production 380
  uranium production 337, 355
  Uravan Belt, Colorado Plateau 355, 357, 359, 364
  Utah 359, 364
  Viburnum Trend, Missouri 259, 260, 261, 262
  Wyoming 355, 356, 357, 358, 359, 362, 364
USSR
  bauxite deposits 281, 285
  chromite production 19, 33
  Daldyn-Alakit district 57
  iron ore production 315, 325
  Kazakhstan 72
  kimberlite deposits 54, 55, 57, 64
  Krivoi Rog 317, 318, 325
  Laptev Sea 197
  Lena River 181
  Mir 58
  nickel production 38, 39
  Perm 33
  porphyry copper deposits 72

Index

porphyry molybdenum deposits 93
skarn deposits 375
South Primor'y 197
tungsten production 380
Uzbek region 72
Yakutia 57, 60
Utah 359, 364
UTEM xvi, 133
Uzbek region, USSR 72

Vaal Reef, Witwatersrand Basin 205, 206
Val d'Or district, Canada 148
Valley-Copper, Canada 12
Vanadates 340, 359
Vanadiferous magnetite 18
Vanadium 18, 357, 359, 361, 363
Variable-frequency induced polarization 258–9
Vegetation mapping 408
Vegetation sampling
  exploration 50, 62–3, 87, 89, 90, 297
Vein silica zone
  porphyry molybdenum deposits 99
Veining
  porphyry copper deposits 84–5
Veins
  classification of 154
  mining from 4
Venezuela 55
Ventersdorp group, Witwatersrand Basin 205
Viburnum Trend, Missouri 259, 260, 261, 266
Visean age 248, 249
VLF–EM
  geophysical exploration 37
Volcanic–peridotite association
  nickel sulphide deposits 38, 40–1, 41–2, 42–5, 45–8
Volcanic rocks 115, 116–17, 119–20, 126–8, 130, 145, 156, 238, 279, 326
Volcanism
  ore genesis 82, 323, 324
Volcano-sedimentary origin
  hydrothermal gold deposits 146
Volcanogenic sulphide deposits 14, 107–16, 137, 152, 409
  exploration 128–36
  genesis 125–8
  types 116–25

Volcanogenic sulphides
  classification 107–9
Vredefort Dome, Witwatersrand Basin 206

W–Sn subtype
  hydrothermal vein deposits 145
Wagina Island, Papua New Guinea 290–1
Wall-rock
  alteration 150–1, 155, 160–1, 346–7, 416
  composition 81
  reactions with ore fluid 162–3
Wasatch formation, Wyoming 358
Washington Camp, Arizona 385
Waste disposal
  kaolin deposits 306
Water
  incursion into mines 229
Water sampling
  exploration 166
Water table 331
Water table lowering 229
Water Tank Hill, Western Australia 166, 168
Watson Lake 117
Watts Mountain Formation, Pine Point district 264
Waulsortian reefs 248, 249
Waves
  heavy mineral concentration 193
Wawa district, Canada 320, 321
Weathering 37, 274–5, 307, 324
  kimberlites 62
  lateritic nickel deposits 294
  sulphides 308–9
Wehrlite xvi, 35
Weipa, Queensland 283
Welkom gold field, Witwatersrand Basin 210
West Africa 284, 307
West African craton 55
West Rand Group, Witwatersrand Basin 206, 209
West Shasta, USA 111
Western Australia 72, 158
  Agnew deposit 40, 45, 47–9
  AK1 pipe 54, 57, 62, 64
  Archaean Shield 146–52, 160–1, 164–9, 169–70, 302, 323, 326, 333
  Eastern Goldfields Province 42, 147, 152, 160, 161, 164, 168
  Gabbin district 302, 303, 306

  gold 158, 163
  Golden Grove Prospect 112, 120
  Hamersley Basin 317, 318, 319, 320, 325, 326–31
  Hunt Mine 160, 161, 163
  Kalgoorlie Area 147, 148, 160, 167
  Kambalda mining district 39, 42–5, 52, 53, 164
  Kimberley 60, 62
  kimberlite deposits 54, 57
  Marra Mamba deposit 326, 333
  Mount Bruce Supergroup 326–7
  Norseman-Wiluna Belt 41, 42, 45, 119, 121
  Paraburdoo deposit 326, 327, 332
  Pilbara Block 147, 151
  Tom Price deposit 322, 326, 327, 329, 332
  Yilgarn Block 40, 42, 43, 46, 49, 109, 147, 149, 151, 166, 168, 302, 323
Western Fault, McArthur River 237
Western States-type mineralization 356
Westland, New Zealand 179, 181
Wheal Jane Mine, Cornwall 15
White River Formation, Wyoming 361
Wind
  effects on beach sand deposits 109
Wind River Formation, Wyoming 358
Wire-line drilling xvi, 306
Witwatersrand Basin, South Africa 12, 201, 202, 203–11, 212
Witwatersrand Supergroup, South Africa 212
Wolframite 93, 185, 211, 380, 382, 442
Wollaston Domain, Canadian Shield 341–2
Wollastonite 370, 377, 379, 387, 391, 442
Woodlawn deposit, New South Wales 130, 131, 133
World Bank 402
Wrens, Georgia, USA 302
Wrigglite 394
Wyloo group, Hamersley Basin 326
Wyoming 355, 356, 357, 358, 359, 362, 364

X-ray diffraction analysis 288, 350
X-ray spectrometry 288
Xenocrystal origin of diamonds 58

Xenothermal deposits 144
Xenotime 185, 443

Yakabindie, Western Australia 52
Yakutia, USSR 57, 60
Yampa deposit' Utah 379, 390
Yandera, Papua, New Guinea 81, 89, 90
Yellow ore
  Kuroko deposit 123
Yerington, Nevada 73
Yilgarn block, Western Australia 40, 42, 43, 46, 49, 109, 147, 149, 151, 166, 168, 302, 323
Young Alluvium, Kinta Valley 184
Yugoslavia 32, 33, 93, 283, 293

Zaire 55, 178, 218, 221, 227
Zambia 220, 222, 224, 226, 227–30
Zambian copperbelt 216, 217, 221–30
Zarnitsa, USSR 58
Zimbabwe 161
  Archaean Shield 110, 146, 147
Chromite production 19, 36
Damba deposit 42
gold 146, 147, 148, 149, 163, 164, 170
Great 'Dyke' 19, 22
nickel sulphide deposits 38, 39, 40, 41, 42, 48, 50
Shangani deposit 41
Zinc 90, 107, 113, 124, 125, 130, 131, 137, 245, 248, 254, 257, 259, 260, 262, 265, 266, 267, 370, 371, 374, 377, 380, 387, 389, 390, 394, 404, 411, 412
Zinc-bearing calcic skarn 386
Zinc–copper deposits 116
Zinc–copper group
  volcanogenic sulphides 107, 108
Zinc-dominant subtype
  Mississippi Valley-type lead–zinc deposits 261
Zinc–lead–copper group
  volcanogenic sulphides 107, 108
Zinc–lead skarn deposits 376–7, 380, 385–6
Zircon 62, 175, 180, 185, 190, 195, 197, 208, 211, 225, 443
Zone refining 58
Zoning
  beach sand deposits 189
  hydrothermal gold deposits 151, 155, 160–1
  Irish lead–zinc deposits
  Kuroko deposit 123–4
  McArthur River 237, 240
  porphyry copper deposits 74, 76–7, 79–80
  porphyry molybdenum deposits 96–9
  sandstone-hosted uranium deposits 364
  sediment-hosted copper deposits 218–19
  sedimentary-exhalative deposits 230
  skarn deposits 371, 379, 385, 386
  Teutonic Bore 120–1
  vertical 80
  volcanogenic sulphide deposits 112–13, 134
  Zambian copperbelt 225